FORENSIC INVESTIGATION OF STOLEN-RECOVERED AND OTHER CRIME-RELATED VEHICLES

FORENSIC INVESTIGATION OF STOLEN-RECOVERED AND OTHER CRIME-RELATED VEHICLES

Eric Stauffer, MS
Monica S. Bonfanti, PhD

ELSEVIER

AMSTERDAM • BOSTON • HEIDELBERG • LONDON
NEW YORK • OXFORD • PARIS • SAN DIEGO
SAN FRANCISCO • SINGAPORE • SYDNEY • TOKYO
Academic Press is an imprint of Elsevier

AP

Academic Press is an imprint of Elsevier
30 Corporate Drive, Suite 400, Burlington, MA 01803, USA
Linacre House, Jordan Hill, Oxford OX2 8DP, UK

Library of Congress Cataloging-in-Publication Data
Stauffer, Eric, 1975–
Forensic investigation of stolen-recovered and other crime-related vehicles / Eric Stauffer, Monica S. Bonfanti.
p. cm.
Includes bibliographical references and index.
ISBN 0-12-088486-0 (alk. paper)
1. Automobile theft investigation. 2. Forensic sciences. I. Bonfanti, Monica S. II. Title.
HV8079.A97S735 2006
363.25′962—dc22

2006003752

British Library Cataloguing-in-Publication Data
A catalogue record for this book is available from the British Library.

ISBN 13: 978-0-12-088486-5
ISBN 10: 0-12-088486-0

For information on all Academic Press publications visit our Web site at www.books.elsevier.com

Printed in the United States of America
Transferred to Digital Printing, 2011

La justice sans la force est impuissante, la force sans la justice est tyrannique. [. . .] Il faut donc mettre ensemble la justice et la force, et pour cela faire que ce qui est juste soit fort ou que ce qui est fort soit juste.

Blaise Pascal, La justice et la raison des effets, Pensées (1670).

Justice without force is powerless and force without justice is tyrannical. [. . .] Therefore, one must put justice and force together, so that what is just is strong or what is strong is just.

CONTENTS

CONTRIBUTORS

ERIC STAUFFER, MS
Atlanta, Georgia, USA

Eric Stauffer is a criminalist presently residing in Atlanta, Georgia. In 1998 he obtained his Bachelor of Science degree in forensic sciences from the *Institut de Police Scientifique et de Criminologie* of the University of Lausanne in Switzerland. In 1999 he moved to the United States and, two years later, obtained a Master's Degree in forensic science from Florida International University in Miami, Florida. Mr. Stauffer is also a Fellow of the American Board of Criminalistics and a Certified Fire and Explosion Investigator.

After graduating with his Bachelor degree, Mr. Stauffer worked temporarily as a Crime Scene Officer for the Fribourg State Police in Switzerland. Simultaneously, he worked as a firearms and toolmarks examiner at the University. In 2001 he moved to Atlanta, Georgia and joined the private sector as a forensic scientist. As such, his duties included the examination of fire scenes (both residences and vehicles), stolen-recovered vehicles, the laboratory examination of physical evidence, and the review of crime scene investigations and forensic laboratory examinations. During the last several years, Mr. Stauffer has been studying police forensic and investigative procedures from different countries.

Mr. Stauffer is a recognized speaker and instructor in the field of forensic sciences and has presented his work at several conferences in both national and international forums. He has also published several articles in peer-reviewed journals, as well as book chapters. Since 2003, he has been a member of the International Association of Auto Theft Investigators.

MONICA S. BONFANTI, PhD
Police Cantonale Genevoise, Geneva, Switzerland

Since 2000, criminalist Dr. Monica S. Bonfanti has been employed as the Technical Chief of the forensic laboratory and crime scene unit *(Brigade de Police Technique et Scientifique)* at the Geneva State Police *(Police Cantonale Genevoise)* in Switzerland. She is in charge of all technical matters, investigates all types of cases, and is frequently called to major crime scenes in order to lead the crime scene officers. She also educates crime scene officers about the investigation of stolen-recovered vehicles.

In 1993, Dr. Bonfanti obtained her Bachelor of Science in forensic sciences at the *Institut de Police Scientifique et de Criminologie* of the University of Lausanne in Switzerland. While earning her PhD, she worked part-time for five years at the firearms and gunshot residue laboratory of the Zurich Police Department (Switzerland). She investigated numerous cases involving firearms, gunshot residue, and toolmarks in Switzerland and the Netherlands, where she worked at the Ministry of Justice in Rijswijk in 1993.

She has presented her work at several conferences in both national and international forums, as a member of the European Network of Forensic Science Institutes (ENFSI) and a Fellow of the working group on firearms and gunshot residue analysis. She published numerous articles about firearms, gunshot residue, and toolmarks, as well as a book chapter and a book on the same topics.

ALAIN G. BARBIER, MS
Interpol, Lyon, France

Mr. Alain Barbier is a Commissioner of the Belgian Federal Police seconded at the International Criminal Police Organisation (Interpol). As the Assistant Director, he is responsible for Database Management and Forensic Support. He first joined Interpol in the spring of 1999 as a Vehicle Crime Program Manager, Strategic Project Officer for the Police I-24/7 Telecommunication Network Deployment, and Advisor of the Executive Director for Police Services.

Mr. Barbier joined the police in 1988 and became a Senior Police Officer with the Belgian Federal Police. He also worked in Belgium as Platoon Commander and Program's Chief for National and International Vehicle Crime.

Mr. Barbier holds a Masters in Law and Criminology from the State University of Liège, in Belgium, a Degree in Police Management and Administration from the Royal Gendarmerie Academy for Officers in Belgium, and a Masters in Military Science from the Royal Military School, in Belgium. He is also an expert in police management of crime phenomenon at national and international levels.

DIDIER BROSSIER
Institut de Recherche Criminelle de la Gendarmerie Nationale, Rosny-Sous-Bois, France

Adjudant-Chef Didier Brossier is the Chief of the Mechanical Identification Unit of the Vehicle Section *(Unité d'Expertise Identification Mécanique du Département Véhicules)* at the *Institut de Recherche Criminelle de la Gendarmerie Nationale (IRCGN)*, located in Rosny-Sous-Bois near Paris, France.

He took his first position at the Gendarmerie with the mobile squad in Versailles Satory more than 30 years ago. In 1981, he was moved to the department station located in Montaigu and then in Monfort L'Amaury. After five years, he specialized in air transportation policing at the airport of Paris-Orly.

In 1991, he was attached to the Vehicle Section of the IRCGN. His first task was to create a database to identify the make/model of vehicles involved in hit-and-run road traffic accidents. His present duties include the identification of vehicles, particularly the ones that have been re-VINed. He also performs the restoration of VINs. Adjudant-Chef Brossier also specializes in the examination of vehicle license plates in order to determine their authenticity and, when applicable, the manner in which they have been counterfeited.

Adjudant-Chef Brossier is also a member of the European Working Group that designed the European Vehicle Identification Database (EuVID). EuVID is an electronic database that collates identification information on different types and models of vehicles.

JEAN-FRANÇOIS CHEVALLEY
Police Cantonale Vaudoise, Lausanne, Switzerland

Mr. Jean-François Chevalley is an inspector with the criminal police *(police de sûreté)* of the Vaud State Police *(Police Cantonale Vaudoise)* based in Lausanne, Switzerland. He has extensive experience investigating all types of crime and conducting interviews with suspects, victims, and witnesses.

He started his law enforcement career approximately 20 years ago as a uniformed officer for the city police of Lausanne. Later, he was able to join the investigation division. He obtained extensive experience through the juvenile squad, the narcotic squad, and the violent crime squad. In 1991, he traveled to New Scotland Yard in London to study the police methods used in England.

In 2001, he attained the position of inspector at the Vaud State Police, and his present duties involve the investigation of all types of violent crimes. He also acts as an investigation supervisor for other types of crime.

Besides his regular duty, Mr. Chevalley is also a member of the bomb squad *(Groupe des spécialistes en dépiégeage)* as a specialist in ammunition. He is also a member of the Disaster Victim Identification team, and as such, participated in extensive identification operations conducted in Thailand following the 2004 tsunami.

MARC DEMIERRE, BS
Police Cantonale Genevoise, Geneva, Switzerland

Mr. Marc Demierre is an investigator with the criminal investigation division *(police judiciaire)* of the Geneva State Police *(Police Cantonale Genevoise)* in Geneva, Switzerland. He began attending the police academy in January 2001. After completion, gained experience in the domestic crime and drug crime units before joining the crime scene (forensic) unit in January 2003. Mr. Demierre consults on many different types of crime scenes, including burglaries, stolen-recovered vehicles, and homicides. More recently, he has been promoted to deputy leader of the questioned documents unit.

Prior to his engagement with the police, he completed a Bachelor of Science in forensic science at the *Institut de Police Scientifique et de Criminologie* at the University of Lausanne in

Switzerland. During his studies, he held internships with the traffic unit of the Lausanne police department, the forensic unit of the Vaud State Police *(Police Cantonale Vaudoise)*, and the forensic laboratory of the French Military Police, in Rosny-sous-Bois, near Paris, France. More recently, Mr. Demierre participated in an international police exchange and spent one month studying American police methods with the New York City Police Department Crime Scene Unit.

EMMANUEL FIVAZ, BS
Police Cantonale Neuchâteloise, Neuchâtel, Switzerland

Mr. Emmanuel Fivaz is a scientific inspector with the crime scene unit *(service d'identification judiciaire)* of the Neuchâtel State Police *(Police Cantonale de Neuchâtel)* in Switzerland. As a criminalist, he is often requested to examine stolen-recovered vehicles and specializes in lock examination. Similarly, he has extensive experience in the examination of vehicles used by criminals to carry out illegal activities such as homicides, robberies, or kidnappings.

In 1998, Mr. Fivaz obtained his Bachelor of Science in forensic sciences at the *Institut de Police Scientifique et de Criminologie* of the University of Lausanne in Switzerland. As a student, he conducted extensive research into the duplication process of automotive keys. His research concentrated on the persistent traces left by the mechanical process of key duplication.

After graduating, Mr. Fivaz worked for three years with the Zoug State Police in Switzerland as a criminalist with the criminal investigation division. He then joined the Neuchâtel State Police. He is currently in charge of the examination of questioned documents, and more particularly counterfeit identity documents.

MOIRA JOHNSON, BS
Australian Federal Police, Canberra, Australia

Moira Johnson is currently the Discipline Team Leader of Crime Scenes, Forensic and Technical Services with the Australian Federal Police (AFP) in Canberra, Australia. Prior to joining the AFP as a Senior Scientific Officer in 2002, Ms. Johnson was a police officer with the New South Wales Police Service for 14 years, with 11 years as a Crime Scene Investigator with the Forensic Services Group. In 2002, Ms. Johnson graduated with a Bachelor of Applied Science (Forensic Investigation) from the Canberra Institute of Technology.

As a crime scene investigator, Ms. Johnson has examined a large number of crime scenes ranging from burglaries to multiple murder scenes. She was a member of the Disaster Victim Identification Team that attended the Thredbo Landslide Disaster in 1997. Between November 2002 and April 2003, she was the Crime Scene Team Leader of the forensic team involved in the investigations of the Bali Bombings and the JW Marriott Hotel Bombing in Jakarta in August 2003.

In her current role, Ms. Johnson is responsible for the coordination of discipline training including general crime scene investigation, shoeprint, tire track, and toolmark examination and comparison, bloodstain interpretation, vehicle examination and identification, fire investigation, and post-blast scene examination. She is also responsible for writing and updating training and procedure manuals and maintaining quality assurance procedures within the section.

HORST KATTERWE, PhD
Bundeskriminalamt, Wiesbaden, Germany

Dr. Horst Katterwe is a forensic scientist with the Forensic Science Institute of the federal criminal investigation service (*Bundeskriminalamt* [BKA]) in Wiesbaden, Germany. He studied physics at the Technical University (*Technische Universität* [TU]) Berlin in Germany. He worked as a physical scientist first at the Institute of Physics at the TU Berlin and then at the University of Kaiserslautern, where he received his doctorate degree. In 1976, he joined the BKA as a forensic scientist and head of the Materials Technology division.

Dr. Katterwe is head of the cooperation of marks examiners between the German states and the federal government. He is a member and chairman of the steering committee of the ENFSI Working Group Marks. His research interests include marks examination and identification, testing of new casting materials, image processing, fracture matching, evidence interpretation, probability theory model calculations, and recovery of erased numbers in metallic and polymeric materials.

Dr. Katterwe is a member of the German Physics Society *(Deutsche Physikalische Gesellschaft).* He has been awarded "Best Presentation" by the Association of Firearms and Toolmarks Examiners in 1992, 2001, and 2004. In 1993, he received the prize of the academy of police commanders *(Preis der Polizeiführungsakademie)* for his forensic science research on "Entropy-Elasticity and Mechanical Memory" (restoration of erased numbers in polymers).

STEPHANE KUMMER
Police Cantonale Genevoise, Geneva, Switzerland

Mr. Stéphane Kummer is a crime scene investigator with the criminal investigation division *(police judiciaire)* of the Geneva State Police *(Police Cantonale Genevoise)* in Geneva, Switzerland. He has worked at the crime scene (forensic) unit *(Brigade de Police Technique et Scientifique)* since January 1994 and investigated hundreds of crime scenes including burglaries, homicides, stolen-recovered vehicles, rapes, and fires. He also specializes in forensic ballistics, examination of explosive devices, examination of mechanical devices, and identification of victims of mass casualties.

Prior to his engagement with the police, he obtained degrees in electronics and mechanical engineering. He also worked as an engineer in the space industry, studying and developing micro-mechanisms for communication and observation satellites.

Mr. Kummer held several internships at national and international levels. He worked with crime scene (forensic) units in the states of Fribourg, Neuchâtel, Tessin, Valais, and Vaud in Switzerland. He also completed a training internship in forensic ballistics with the Royal Military School *(Ecole Royale Militaire)* in Brussels, Belgium. He attended classes on selected topics at the Institut de Police Scientifique of the University of Lausanne. He holds the Swiss federal explosive handling permits B and P *(permis d'emploi d'explosifs)*.

MIKEL LONGMAN, BS
Arizona Department of Public Safety, Phoenix, Arizona, USA

Mikel Longman is chief of the Criminal Investigations Division of the Arizona Department of Public Safety in Phoenix, Arizona. As a career law enforcement officer with extensive background in both patrol and criminal investigations, he served as the executive director of the Arizona Automobile Theft Authority. During his 30 years with the State of Arizona, Chief Longman has served in a variety of assignments including resident patrol officer on the Navajo Indian Reservation, supervisor of a motorcycle squad on the Phoenix Metropolitan Freeway System, Highway Patrol Officer in rural Arizona, and supervisor of an undercover narcotics squad. He served as a patrol district commander, motorcycle district commander, organized crime unit commander, and commander of the Arizona Vehicle Theft Task Force.

Chief Longman is a graduate of the FBI National Academy, has a Bachelor of Science in Public Safety Administration, and is an Arizona Peace Officer Standards and Training (POST) certified instructor. He is an active member in numerous fraternal and professional associations, including the Arizona and International Associations of Chiefs of Police, the North American Export Committee, where he serves as a board member, and the International Association of Auto Theft Investigators (IAATI), where he serves as 3rd vice-president. He is also a member of Police International Sonora/Arizona (PISA), the 100 Club, Associated Highway Patrolmen of Arizona (AHPA), Arizona Auto Theft Investigator Association, and the National Motor Vehicle Title Information System (NMVTIS) law enforcement response subcommittee.

ROBERT F. MANGINE
NorthAmerican Technical and Forensic Services, Las Vegas, Nevada, USA

Robert Mangine is a member of the American College of Forensic Examiners, a Certified Forensic Consultant, a Certified Fire and Explosion Investigator, and a Certified Automotive Locksmith. He has been involved in various criminal and homicide investigations with law enforcement agencies nationwide and was accepted in court as an automotive forensic expert on 42 occasions in seven US states and the District of Columbia. Mr. Mangine has extensive and ongoing training in forensic locksmithing, vehicle fire investigation, and steering column, ignition lock, and anti-theft system examination. He also received training in automotive mechanics from the US Department of Defense and in explosives/demolitions training while in the US Marine Corps. During his career, Mr. Mangine examined over 12,000 vehicles and conducted over 375 auto theft training seminars for insurance companies, law enforcement, military and federal agencies, and professional organizations.

Mr. Mangine studied criminal justice and security administration in college. Between 1975 and 1990, he was employed by several corporations to design and maintain security alarm systems and locking systems, and acted as a regional manager for investigations. He has also served as Assistant Director of Public Safety for Seton Hall University in South Orange, New Jersey.

In 1991, Mr. Mangine founded NorthAmerican Technical and Forensic Services in Frederick, Maryland. Twelve years later, the company relocated its headquarters to Las Vegas, Nevada, and currently has six automotive forensic examiners located around the United States. Mr. Mangine has also been a member of IAATI for more than ten years.

DIANA OMBELLI, BS
Sdu Identification, Haarlem, The Netherlands

Mrs. Ombelli joined Sdu Identification in Haarlem, the Netherlands, in 2001 as Head of the laboratory. Her duties involved the supervision of testing activities incoming raw materials, semi-manufactured products, and end products. Since 2002, she manages projects on the development and implementation of new identity documents and related IT systems. More recently she has been charged to handle issues concerning the ISO formatting and quality assessment of digital photographs to be saved on chip in electronic passports and identity documents.

Mrs. Ombelli graduated with a forensic science degree from the *Institut de Police Scientifique et de Criminologie* of the University of Lausanne, Switzerland. After graduation, she worked for three years as a forensic scientist at the Police Laboratory in Bern (Switzerland). Then, she worked at the Swiss Federal Aliens Office where she headed a feasibility study on national information desks for travel documents. Later, she coordinated the design and manufacture of a Swiss visa sticker issued electronically. In 2000, she had advisory roles in the Swiss Passport project and the Dutch project of new Travel Documents.

She is a member of the International Association for Identification and was the Swiss representative member of the New Technology Working Group within the International Civil Aviation Organization. In 2000 she was Chairwoman of the European Interpol Conference on Fraudulent Travel Documents.

MANUEL POZA
Police Cantonale Vaudoise, Lausanne, Switzerland

Mr. Manuel Poza is an inspector with the criminal police *(police de sûreté)* of the Vaud State Police *(Police Cantonale Vaudoise)* based in Lausanne, Switzerland. Since 2002, he has headed the stolen vehicles research team *(Groupe Recherches Véhicules Volés [GRVV])*.

Prior to his engagement with the police, Mr. Poza held an apprenticeship under a business employee in a bank. In 1986, he entered the police academy and joined the criminal police department after graduation.

After 10 years of experience in several different investigative divisions, Mr. Poza became affiliated with the GRVV, which is part of the white-collar crime investigation division. In 2002, he was promoted to head of the group. The GRVV investigates traffic of stolen vehicles, extortion, and blackmailing involving stolen vehicles. Since 2001 Mr. Poza has lead the interstate criminal police working group against crimes involving vehicles *(Groupe de travail intercantonal des polices judiciaires suisses contre les délits véhicules).*

JERRY RATCLIFFE, PhD
Temple University, Philadelphia, Pennsylvania, USA

Dr. Jerry Ratcliffe is an associate professor in the Department of Criminal Justice, Temple University, Philadelphia. Previously he served for 11 years as a police officer with the Metropolitan Police in London (UK) where he worked on patrol, in an intelligence and information unit, and with the Diplomatic Protection Group. He completed a BSc (Hons) in Geography at the University of Nottingham, but due to an ice-climbing accident left the police and remained in academia.

As a lecturer in policing (intelligence) at the New South Wales Police College in Australia, he ran graduate programs in criminal intelligence, and for a number of years coordinated Australia's National Strategic Intelligence Course.

A Fellow of the Royal Geographical Society, he has a PhD in spatial and temporal crime analysis techniques (Nottingham). Dr Ratcliffe is the creator of HotSpot Detective, an add-on crime mapping and analysis program for MapInfo. He has published over 20 articles and three books: "Strategic Thinking in Criminal Intelligence" (Federation Press, 2004), "GIS and Crime Mapping" (Wiley, 2005), and "Policing Illegal Drug Markets" (Criminal Justice Press, 2005). He publishes and lectures on environmental criminology, intelligence-led policing and crime reduction.

SIMONE REYNOLDS, BS
Canberra Institute of Technology, Canberra, Australia

Simone Reynolds has been employed since September 2005 as a crime scene investigation teacher with the Canberra Institute of Technology in Canberra, Australia.

Ms. Reynolds graduated from the University of Canberra in 1995 with a Bachelor of Science with Honors in Medical Laboratory Science. Soon after graduating, she commenced employment as a microbiologist with the Australian Capital Territory (ACT) Government Analytical Laboratory and then the Therapeutic Goods Administration.

In 1999, Ms. Reynolds decided to take a different career path and commenced employment as a Scientific Officer (Crime Scene Investigator) with the Australian Federal Police Forensic Services. As a Scientific Officer, Simone has assisted with and managed a number of serious crime investigations within the ACT and overseas. In 2002, she was a member of the disaster victim identification and investigation teams for the Bali bombings in Indonesia. Also the same year, she commenced work as an Assistant Quality Manager, which included the role of Occupational Health and Safety Manager within the Forensic Services.

FRANCESCO SAVERIO ROMOLO, PhD
Università degli Studi di Roma "La Sapienza", Rome, Italy

Dr. Francesco Saverio Romolo is a Professor at the Università degli Studi di Roma "La Sapienza" in Rome, Italy. His present duties include cases and research programs in several areas of forensic chemistry, including analysis of explosives, gunshot residue detection, and analysis of drugs of abuse.

In 1990, Dr. Romolo obtained his Master of Science in Chemistry from the University "La Sapienza," and three years later, he completed another Master of Science in Pharmacy at the same university. In 2004, he completed his doctoral degree at the Institut de Police Scientifique of the University of Lausanne, Switzerland. His thesis involved the examination of organic gunshot residue from lead-free ammunition and was received with the highest honors.

In 1993, he entered the Academy of the Italian National Police and attended the Course for Technical Directors. In 1997, he became Deputy Head of the Explosives Analysis Laboratory within the Criminal Police HQ of the National Police Department in Rome. In 2000, he became Head of the Gunshot Residue Analysis Laboratory. Dr. Romolo is also a guest lecturer and consultant for the University of Lausanne. He presented the results of his research in several international meetings and has authored many articles in international journals.

WILLIAM T. SMYLIE
Davie, Florida, USA

William Smylie was a highly decorated member of the Miami Police Department (MPD) in Miami, Florida for 25 years. He spent several years as a patrol officer before being promoted first to criminal investigator in 1974, and then to Sergeant/supervisor in vehicle theft. As one of the original founding members, he was assigned to the Miami-Dade Auto Theft Task Force from its inception as an investigative supervisor for the final 10 years of his Miami PD career.

Following a full service retirement in 1995, Mr. Smylie worked for the State Attorney's Office in Miami for two years as a criminal investigator, assigned to the Miami-Dade Auto Theft Task Force. This was followed by employment as a Special Agent with the National Insurance Crime Bureau (NICB) for six years, assigned full-time to the Auto Theft Task Force until retirement again in late 2003.

A lifelong observer of the automobile industry and devoted student of automotive history, Mr. Smylie has instructed many classes over the years in vehicle theft investigation and vehicle identification techniques. He has been qualified as an expert witness in state and Federal courts numerous times to testify regarding stolen vehicle detection and identification of professionally altered stolen vehicles during an investigative career spanning 33 years.

MARC STAUFFER
Phenix Assurances, Lausanne, Switzerland

Mr. Marc Stauffer is responsible for the special investigation unit at Phenix Assurances, an insurance company member of the Allianz Group, located in Lausanne, Switzerland. His duties include the investigation of suspicious claims for the entire Swiss territory. He has a daily involvement in the investigation of auto thefts and damages resulting from fire, natural elements, animals, and accidents.

Marc Stauffer studied at the *Institut de Police Scientifique et de Criminologie* of the University of Lausanne in Switzerland before joining the Vaud State Police as a Crime Scene Officer. During that time, he was able to investigate various crime scenes including stolen-recovered vehicles.

In 1972, he joined the insurance industry as a claim specialist and obtained the Swiss Federal Insurance Diploma *(diplôme fédéral d'assurances)*. Following that achievement, he was able to obtain other positions within the insurance industry both as claim specialist and agent. In 1986, he joined Phenix Assurances first as the claim service direction inspector and then as the head of the information technology system. Finally, he was chosen to head the special investigation unit for the Swiss territory.

Mr. Stauffer regularly participates in seminars such as the ones organized by the IAATI and IASIU. He is also presently involved with the development of direct liaisons between the SIUs of the Allianz Group's French-speaking countries.

GREG TERP, MPA
Miami-Dade Police Department, Miami, Florida, USA

Lieutenant Greg Terp has been the Commander of the Miami-Dade Multi-Agency Auto Theft Task Force since January 1996. He joined the Miami-Dade Police Department in 1976 and has worked in Tactical Operations, Canine, Explosive Detection, Bomb Disposal, and Investigations.

In 1993, Lt. Terp obtained his Bachelor of Arts in Criminal Justice from St. Thomas University in Miami, Florida. Two years later, he earned his Master's Degree in Public Administration from the University of Miami.

Lt. Terp is actively involved in the fight against the theft of automotive vehicles. While leading his Task Force, he initiated the container imaging system Stolen Auto Recovery System (STARS) at the Port of Miami, which became fully operational by 2000. He is the current Chairperson of the North American Export Committee (NAEC). This committee, comprised of international law enforcement and private industry professionals, works to find programs and technology to address the growing illegal exportation of stolen vehicles. He is also a member of other professional organizations such as the FBI National Academy Associates (FBINAA), the International Association of Auto Theft Investigators (IAATI), The National Motor Vehicle Theft Information System Law Enforcement Sub-committee (NMVTIS), the Florida Anti-Car Theft Committee (FACT), and the Florida Auto Theft Intelligence Unit (FATIU).

JEAN-FRANÇOIS VOILLOT, MS
Institut de Recherche Criminelle de la Gendarmerie Nationale, Rosny-Sous-Bois, France

Captain Jean-François Voillot is in charge of the serious crime unit of the national gendarmerie forensic laboratory *(Institut de Recherche Criminelle de la Gendarmerie Nationale [IRCGN])* located in Rosny-Sous-Bois near Paris, France.

After graduating from the French air force academy, Captain Voillot joined the national gendarmerie in 1992. During his career, he was first assigned to an anti-riot platoon in Marseille for four years. Then, he was promoted and became the head of the firearms section of the national forensic laboratory (IRCGN).

After this four year assignment, he studied at the *Institut de Police Scientifique et de Criminologie* of the University of Lausanne in Switzerland and obtained a Master of Science in forensic sciences. His thesis research involved the investigation of underwater crime scenes. Upon his return to the French Gendarmerie, he commanded a company in the south of France. Then, he was put in charge of the national serious crime unit at the IRCGN. Throughout these various occupations, he received extensive training in underwater police diving. Captain Voillot specializes in the underwater application of forensic principles.

GLENN WHEELER, BS
Bloomington, Illinois, USA

Mr. Glenn Wheeler earned his Bachelor of Science at the Lawrence Institute of Technology in Southfield, Michigan. His career with State Farm Insurance Companies began in 1960, and he retired after 43 years of service. During his tenure, he earned a Chartered Property & Casualty Underwriter (CPCU), Associates in Management (AIM), and Claims Law designation. He spent his last 25 years with State Farm managing SIU operations, capping his career as an SIU Corporate Consultant where he worked with SIU representatives in the United States, Canada, and Mexico.

He teaches insurance contracts and identification and investigation of insurance fraud at the corporate level for claims personnel, agency, and underwriting. He has instructed at the Illinois and Michigan State Police Academies, New South Wales, Australia law enforcement units, the National Insurance Crime Bureau (NICB), the International Association of Auto Theft Investigators (IAATI) and the North Central Regional Chapter (NCRC) seminars.

Mr. Wheeler has been credited with being a founding father of the NICB Special Investigations Academy, responsible for helping develop the curriculum. Additionally, he chaired the North American Export Committee (NAEC) for two years. Under his leadership, by-laws were established for the fledging organization and Mexico was added as a partner. He is the current President for NCRC and serves on the Boards of IAATI and NAEC.

FOREWORD

In these chapters, world-renowned experts have come together to provide a comprehensive guide to the forensic aspect of auto theft investigation. The authors' approach to the subject matter is easily understood and practical for immediate use in your investigations. The information presented within is right on target to handle auto theft investigations worldwide. For what I believe to be the first time, the many different facets surrounding the forensic investigation of stolen-recovered vehicles have been pulled together in one volume.

The "traditional" role of the auto theft investigator has undergone a considerable transformation over the years. As anti-theft devices have become more sophisticated, so have the thieves in defeating them. Auto theft investigators have become far more aware of the value of forensic techniques in the investigation of stolen vehicles, recovered vehicles, organized auto theft rings and the investigation of crimes resulting from the theft and use of stolen vehicles.

The auto theft investigator is experiencing a major shift in emphasis; a departure from routinely handling auto theft reports within individual jurisdictions, emphasis is shifting to multi-jurisdictional and worldwide investigations. Of even greater importance, investigators are constantly working with professionals from many different agencies to pool their resources and expertise. This teamwork is vital to gain an upper hand on the vehicle theft problems facing the world today.

In an effort to achieve the highest degree of effectiveness in handling their responsibilities, auto theft investigators are encouraged to expand their knowledge base in the use of forensic auto theft investigation. This comprehensive publication contains a wealth of important information and reminds us that knowledge and education are still our most powerful tools in performing our jobs. After reading this publication, I have become an avid fan of the works that Eric Stauffer and Monica Bonfanti have organized for you, the reader. Thanks to Eric and Monica for focusing our attention on the vital forensic aspect of auto theft investigation.

Karen L. Metz
President IAATI, 2004–2005
Retired Ft. Lauderdale Police Officer

PREFACE

Eric Stauffer and Monica S. Bonfanti

On March 25, 1912, a De Dion-Bouton was violently stolen from its chauffeur in Montgeron, France, by Jules Bonnot and his gang (known as *la bande à Bonnot*), a group of French anarchists reputed for stealing cars. The *Société Générale* (a national bank) was then robbed and the culprits used the car to escape. This is one of the numerous vehicles stolen or carjacked by the Bonnot Gang. As a matter of fact, carjacking, the act of violently stealing an occupied car, started with the crimes committed by these gangsters. However, the simple theft of vehicles did not originate with Jules Bonnot. In fact, it started as early as when the first vehicles were produced.

Today, auto theft is a hot topic as it concerns many citizens and affects the entire society. In the United States alone, one million vehicles are stolen every year. The city of Modesto in California holds the sad and impressive US record for the highest rate of stolen vehicles (more than 1,500 per 100,000 people) in 2004. The theft of automotive vehicles also dramatically increased in Europe in the early 1990s after the fall of the Berlin Wall and the opening of Eastern Europe. In France, Germany, and some other Western European countries, the rate of auto theft became unbearable and forced manufacturers and insurance companies to work together with law enforcement agencies to impose radical measures. In Europe, more than one million vehicles are stolen every year, the same figure as in the United States, while Australia "only" sees about 100,000 vehicles stolen every year. Nowadays, the business of auto theft is a colossal one and organized-crime groups have taken control of it.

The repercussions of this crime are very serious and influence everyone's life. Governments have responded to crime by writing laws and enforcing them through a justice system, charged with the prevention and repression of criminal activities. This is conducted in order to protect law-abiding citizens and to ensure a safe society. With the scientific and technical advances in forensic sciences and criminal investigation techniques since the beginning of the century, it would appear logical that all possible means be applied to the investigation of auto theft and more particularly, stolen-recovered vehicles. Unfortunately, it is not so. Many entities do not grant much importance to this crime, assuming it as a petty one. They are incorrect and oblivious to its modern violence. Additionally, it has major consequences for the financial platform and for the general population's feeling of safety or lack thereof. The repression of this crime greatly suffers from this unfortunate attitude. Not only is auto theft a very serious crime, but also it is a crime that can be thoroughly and scientifically investigated by collecting available forensic evidence.

The crime of auto theft has evolved over time. It is nowadays a commonly encountered crime, which presents strong ties with drugs of abuse, firearms, and human trafficking as well as terrorism. The latter is probably the most concerning topic for law enforcement agencies from around the world. There is strong evidence that terrorist groups are financed by the international trafficking of stolen vehicles. Anti-theft technology present in modern vehicles has also forced thieves to adapt and, as a result, auto theft becomes more violent on a daily basis.

There have been very few books dealing with the investigation of auto theft, and those are now mostly outdated. While there are many books on crime scene investigation and general forensic sciences, there are none dealing specifically with the forensic examination of a vehicle, stolen-recovered or more simply crime-related. The goal of this book is not only to fill this gap, but also to provide much more comprehensive information surrounding the investigation of auto theft. We hope that this volume will provide the most valuable information to conduct proper examination of stolen-recovered and crime-related vehicles.

This work is primarily intended for crime scene investigators, criminalists, police officers, and, of course, auto theft investigators (both public and private). The book is also ideal for personnel in training. It provides seamless transitions that will ensure the student a broad understanding of the topic. Experienced investigators will also be able to glean new information and obtain a different perspective on the overall issues, because the information presented in this publication is unique and internationally relevant. Other police investigators, private investigators, insurance adjusters, claim representatives, students in forensic sciences, and attorneys will find the information presented here useful and interesting, even if it does not directly cover the scope of their work. Some parts of this book can be directly used to describe how to examine a vehicle. Other chapters are more about the concepts behind the issues presented, and they provide the reader with the proper reference in order to perform the "how-to".

We also feel optimistic that this book will revive the motivation of law enforcement personnel, particularly higher management, in investigating auto theft. There is so much that can be done that is not presently done. The reasons are probably numerous, but mostly we think that it is due to a lack of means dedicated to the crime of auto theft and a lack of education regarding what can be obtained from the examination of stolen-recovered vehicles. We ask law enforcement personnel to encourage supervisors and politicians in combating the larceny of vehicles.

We tried to make this book as international as possible. The result: 22 authors from the United States, five European countries, and Australia. When studying police and forensic methods from foreign countries, one might discover better solutions and consequently want to change how things are done in his or her jurisdiction. This is the goal behind the international character of this book. While it is difficult to keep a perfect uniformity between these different approaches and writings, we hope that this will be largely overcome by the gain in the information provided. We have gathered some of the foremost experts in their specialties, many of whom have dedicated their lives to combating auto theft by improving police investigations and forensic sciences.

When planning the book, we understood that the topic of auto theft investigation could easily span several volumes with hundreds of pages. While we generally tried to confine the work to the forensic aspect of the investigation, we were also compelled to include some more traditional investigative portions, offering a broad view of possible problems and solutions. Additionally, because this book is available to the public at large, including the culprits stealing vehicles, we could not include the full breadth of information that would serve the investigators. However, whenever possible, we indicated to the reader where this secured information could be obtained. We are confident that readers will understand the need for these limitations and be patient with them.

Chapter 1 defines and presents the problem of auto theft. It is followed by general concepts of interviewing people and gathering circumstantial information (Chapter 2). Then, the reader is taken into the heart of the book: the examination of vehicles, treated in detail in Chapters 3 and 4. A basic review of the different physical evidences and their forensic value is presented in Chapter 5. Chapter 6 tackles a very specific issue—vehicle identification. This is followed by a chapter on recovery of erased serial numbers. Chapters 8, 9, and 10 deal with locks, keys, and anti-theft systems, along with issues specifically related to vehicular crime scenes. Then, the analysis of vehicle fluids, which can bring pertinent information to the investigation, is covered (Chapter 11). This is followed by two chapters dealing with the examination of vehicles discovered under particular circumstances: burned (Chapter 12) and underwater (Chapter 13). Chapters 14 and 15 present the criminalistics approach to the examination of vehicle tags and license plates, two topics often ignored. The proper methodology to perform a thorough search of a vehicle is presented in Chapter 16. Chapter 17 considers the specific involvement of auto theft in terrorism—a very current topic. Chapters 18 and 19 present the investigation of auto theft from both public and private perspectives. The manner in which vehicles are tracked is introduced in Chapter 20. Chapter 21 describes modern mapping techniques used to analyze auto theft and help in its reduction. Finally, Chapter 22 presents the work Interpol has put in place to fight against vehicle theft on an international level.

We hope you appreciate the attention to scientific detail and international perspectives presented here. We strongly believe that the investigation of auto theft—specifically stolen-recovered vehicles—can only improve in time. We have faith that this work will aid the fine police and private investigators from around the world in their dedicated fight against auto theft.

ACKNOWLEDGMENTS

The most difficult part of the acknowledgments is not to forget anyone. Unfortunately, *errare humanum est* (to err is human) and we apologize to the person(s) whom we forget to thank.

First of all, we would like to thank Jennifer Soucy, Acquisitions Editor at Elsevier Academic Press, for trusting us from the beginning of the project and guiding us in the early stage of the process. She was available every time we needed her, and she spent countless hours on the phone (to the point that she has committed our phone numbers to memory) improving the project and resolving every single issue we had. Without you, Jenn, this book would have never seen the light . . . and we would have spent several thousand hours enjoying life instead of being stuck in front of the computer monitor. Thank you so much! Then, we would like to thank the fantastic editorial crew of Elsevier Academic Press. Pam Chester, who brought great help and guidance at the beginning and at the end of the project. Pam, we temporarily lost you to mathematic books, but thank goodness you were rescued and came back to the forensic team. Your help was also invaluable and we thank you for that. Thank you to Kelly Weaver who responded very quickly to our "very few" requests and Mark Listewnik who has always been there and also placed his confidence in us. Jenn, Pam, Kelly, and Mark: you have done a remarkable job; you are some the easiest people to work with and we give you an outstanding ovation for that. But getting the material to put in the book is only half of the job. Once in their hands, the publisher must make this material look good. And that's where the production people at Elsevier Academic Press enter the game. Thanks to Heather Furrow who also had a chance to spend her share of hours on the phone, trying to resolve problem after problem. The book looks fantastic and we extend the ovation to you too. Cate Barr also contributed to the extraordinary cover (front and back) of this book. Thank you so much for providing such quality artwork.

If the publisher were the only entity involved in the creation of this book, the results would be a very nice looking hardback of approximately 600 blank pages. Fortunately, this book has much more to offer and the only people responsible for that are the contributors. Each contributor must be thoroughly credited. You have done such a terrific job that there are no words that can properly express the amount of gratitude we owe you. This book would be absolutely nothing without your contribution. You delivered the highest quality of material possible and we are very proud of you. Thank you so much! This is no small achievement. Now, you can return to your family and enjoy life again!

Many other people have contributed to this book by providing information, illustrations, expertise, guidance, and review. Most have been recognized within each chapter or for each contributed figure. We would also like to thank Stewart Mosher for his review and Karen Metz for her review and writing the foreword. Finally, we would like to particularly thank Sarah Brown who spent a considerable number of hours proofreading the text. She has done an outstanding job.

This sentence represents the last few minutes of a great journey that we started more than two years ago. We would never have believed how good it would feel to write it.

Eric Stauffer
Monica S. Bonfanti

CHAPTER 1

THE PROBLEM OF AUTO THEFT

Mikel Longman

1.1 INTRODUCTION

An old proverb states that "Money is the root of all evil," and some suggest that greed is an inherent human condition. Thieves have always been a scourge upon a civilized society, and the theft of means of transportation has been a problem throughout history. Passenger cars were stolen as soon as their production started. Starting in 1906, criminals such as Bonnot in France and Dillinger or Bonnie and Clyde in the United States stole vehicles to commit their misdeeds [1]. Vehicle theft has evolved from people simply stealing from another for their own personal use to a highly complex criminal endeavor. Generally, there is a clear distinction between property crime and violent crime. Vehicle theft is obviously a property crime, but it is more appropriate to recognize it as an economic crime and acknowledge that it becomes a hybrid crime when violence is used, such as in the case of carjacking.

Vehicle theft and its related criminal activities are epidemic throughout the world. They account for significant economic loss and affect the overall quality of life in communities. Vehicle theft is more than just a nuisance crime or about a piece of property. The real impact is the victimization that it causes to the modern and mobile society. A vehicle is no longer considered a luxury but a necessity for many people. Personal vehicles have become an integral component of everyday life and economic survival. The high cost of vehicles, insurance, and deductibles and the potential waiting periods for insurance settlements create a significant financial hardship for many victims. In some places, insurance is not mandatory or only liability coverage is required. These victims suffer the total loss if their vehicle is not recovered or is recovered but with severe damage. Thus, auto theft leaves countless victims without transportation, financially burdened, and feeling violated. In some cases, such as with carjacking, victims face direct confrontation with perpetrators, leaving them terrified, injured, or even dead.

This crime not only affects the quality of life of innocent citizens, but adversely impacts legitimate businesses, insurance companies, and governments. Legitimate businesses lose their clientele when organized crime groups sell similar but less expensive vehicles to unsuspecting clients. Car manufacturers must constantly increase security features and equip their vehicles with more expensive and more reliable anti-theft systems. Insurance companies have to handle a great amount of auto thefts, which decreases their productivity and increases their premiums. Police forces have to handle a great volume of reported stolen vehicles, which increases their already charged caseloads. Auto theft is a real burden to the modern society, and its fight requires serious preventive, investigative, and repressive measures.

1.2 OVERVIEW

1.2.1 Motives

Fundamentally, vehicles are stolen either for profit or for convenience. The high profit potential with minimal risks is particularly attractive for professional thieves. Organized criminal groups have diversified and consider vehicle theft, insurance fraud, and other similar activities very lucrative. Vehicles are sold either as a whole or in separate parts. Other criminals steal vehicles to commit other crimes, thus for convenience.

A/ Insurance Fraud

Historically, during economic downturns, crime rates, including vehicle thefts, increase. Insurance fraud has become a component of the monetary benefit of vehicle theft. As the cost of new vehicles increases, some owners overextend their finances or otherwise decide to dispose of their vehicles. Once disposed, a fraudulent theft report is filed with the police, and a fraudulent claim is filed with the insurance company. This scheme is often encountered in Europe, where it is easy for an owner to bring his or her vehicle to another country, sell it, and then declare the theft. The vehicle is almost never retrieved in such instances, and the owner obtains monetary gain from both the sale and the insurance settlement (see Chapter 19).

B/ Resale and Export

The theft of vehicles for domestic resale or for resale after export is a very lucrative activity, largely controlled by organized crime groups (see Chapters 17 and 18). Exportation of stolen vehicles is not readily resolved, because investigations are hindered by inadequate or nonexistent communications between law enforcement agencies in different countries. Border guards and police officers share a similar problem in encountering suspicious vehicles and having limited or no access to needed databases to determine whether or not a vehicle is stolen. Interpol, the premiere international police organization, states the following [2]: “Illicit trafficking of vehicles is a form of organized crime, which generates large profits for the perpetrators (estimated at 19 Billion USD which disappears into a parallel economy) and a feeling of insecurity that affects the public particularly due to the increased use of violence. A key aspect of this form of crime is the need to legalize stolen vehicles in order for the criminal to achieve a monetary gain”.

Thieves also attempt to legalize or conceal the identity of stolen vehicles by vehicle identification number (VIN) switching (also called re-VINing or ringing) with wrecked or salvaged vehicles in order to sell them to unsuspecting buyers. The VIN is unique to a single vehicle (see Chapter 6). A phenomenon referred to as "cloning" has become extremely problematic. This occurs when the VIN is copied from a donor vehicle and then replicated and applied to a similar make and model of a stolen vehicle. Utilizing counterfeit documents, the stolen vehicle assumes the identity of the original vehicle. Many times, multiple stolen vehicles use the VIN from the same donor. These vehicles are then distributed to different states, provinces, or countries and legitimized with new documents.

Figure 1-1 shows a common scenario involving the purchase at an auction of a severely damaged vehicle, generally considered a total loss by the insurance industry. This particular vehicle is a 2000 Lincoln Navigator that was damaged (burned) beyond repair. The legitimate reason behind this purchase is to salvage undamaged component parts and then ultimately dispose of the remainder of the vehicle at a recycling plant. Enterprising criminals often purchase these vehicles to simply obtain their VIN, other identifying serial numbers, and ownership documents for fraudulent purposes.

Figure 1-2 shows how the identity of a salvaged vehicle is reapplied to a stolen vehicle of similar make, model, and year, which results in the "rebirth" of the total-loss vehicle. This vehicle is a stolen 2000 Lincoln Navigator that has assumed the identity of the burned vehicle in Figure 1-1.

Criminals capitalize on the inadequate and ineffective communication systems between registration and titling agencies and jurisdictions. In one case, a VIN on a new vehicle at a dealership in Ontario, Canada was copied, replicated, and applied to at least four other stolen Canadian vehicles. Fraudulent documents were produced, and then the cloned stolen vehicles were taken into several different states in the United States.

Vehicle theft has become a high-tech crime. Criminals have also become more technologically literate. It is common to find sophisticated computers, metal presses, and other devices used to produce VIN plates and fraudulent documents within a thief's "toolkit" (see Chapter 15). At this time, luxury vehicles stolen in Europe are exported to the Eastern bloc, West Africa, and the Middle East. Less expensive vehicles are exported to North Africa, but this trend is changing. These vehicles mainly transit through the harbors of Marseille (France), Genoa (Italy), and Antwerp (Belgium).

C/ Resale and Export of Vehicle Parts

It is often more profitable for professional thieves to steal vehicles and to sell the parts separately. It is estimated that the net value of component parts, particularly on older vehicles, is often two to three times greater than the value of the whole vehicle. To sell parts, these professional thieves operate "chop shops" where vehicles are stripped and their component parts sold to unsuspecting buyers or unscrupulous auto repair shops (see Chapter 18). Nevertheless, modern vehicles are equipped with ever-increasing expensive electronic equipment such as navigation systems and entertainment systems. These devices, along with

a b

Figure 1-1

View of the 2000 Lincoln Navigator that was a total loss after fire damage.

Figure 1-2

View of a stolen 2000 Lincoln Navigator to which the VIN of the vehicle in Figure 1-1 was applied.

expensive parts such as airbags, are also highly attractive to professional thieves and result in the targeting of some specific vehicles.

D/ Commuter Theft or Joyriding

Vehicles are also stolen simply as a means of temporary transportation, often referred to as commuter theft, or joyriding. Suspects abandon the vehicle when they get to the intended destination and/or when they feel that they may get caught. They may steal another vehicle to get to the next destination, and this cycle continues as long as transportation is needed. Many teenagers, even without a driver's license, commit this type of theft. Also, in Europe there are many organized crime groups of burglars originating from the former Soviet bloc who are very active in this type of theft. These thieves come illegally into a country and steal one or more vehicles. Then they move very rapidly, most at nighttime, and regularly change vehicles.

E/ Commission of Another Crime

Criminals often steal vehicles to facilitate other more egregious crimes, such as burglaries, armed robberies, drive-by shootings, kidnapping, smuggling activities, and so forth. Also, some criminals use heavy and powerful vehicles, sometimes with a reinforced trunk, to ram into luxury shops, such as jewelry stores, in order to commit burglary (see Figures 4-29 and 4-30). Again, many of these organized crime groups originate from the former Soviet bloc. An added danger with vehicle theft is the propensity for high-risk behavior by suspects fleeing from the police. Attention to this problem is addressed in a Canadian program referred to as Project 6116 [3].

The government of Canada and a coalition of public and private sector organizations support Project 6116, the National Committee to Reduce Auto Theft. The tragic death of a Sudbury, Ontario police officer in a traffic collision caused by a juvenile driving a stolen vehicle was the catalyst for action in the formation of this committee. The initiative was named Project 6116 in honor of Sergeant Rick McDonald's badge number.

The unfortunate event mobilized the slain police officer's sister, Marlene Viau, and other Canadians to seek solutions to the problem of vehicle theft, especially in the area of prevention and deterrence of young people from getting involved in this criminal activity. Ms. Viau commented that "auto theft robs citizens of the right to feel safe and secure in their own communities," and that "innocent people like my brother lose their lives or are seriously injured each day in Canada as a result of this crime" [3].

Drug involvement and vehicle theft are also closely associated. Because drug users have difficulty maintaining employment, they find it necessary to steal to meet personal and addictive needs. In the United States, it is estimated that 50% of those arrested in possession of stolen vehicles are involved in drug activity. A disturbing trend has also developed where vehicles are stolen and used by terrorists to deliver weapons of mass destruction (see Chapter 17).

1.2.2 Modus Operandi

The modus operandi of car thieves has dramatically changed over the last several years. In the early 1990s, a thief would simply break into a parked car, hotwire it, and leave. The old method of hotwiring the vehicle is no longer applicable because of current ignition with anti-theft and computer-controlled systems. Therefore, criminals have become diversified (and violent) in their methodology.

More sophisticated criminals take advantage of inadequate internal controls at authorized automotive dealerships by obtaining keys simply by recording a VIN and purchasing a replacement key. In many cases, co-conspirators working at dealerships facilitate the theft of vehicles. It is anticipated that this modality will continue to be more common as anti-theft systems improve.

Carjacking is a violent method of choice used by criminals. Carjacking is stealing a vehicle by forcing it to stop and pulling the owners out of the vehicle by use of threat, weapons, knives, sprays, and possibly force or violence. In this manner, thieves can take possession of the vehicle directly with the original keys, without having to worry about anti-theft systems. Vehicles most often targeted are luxury powerful cars such as the Audi RS4 and the BMWs. Because these cars are usually equipped with the latest technology regarding electronic anti-theft systems, they are almost impossible to steal without the ignition key. This type of theft is relatively recent in Europe, with the first reported cases starting in 2000. Carjacking is also frequently encountered in the United States, particularly in Florida, where a number of tourists, unaware of the danger, have been attacked. Carjacking has been reported in increasing numbers in Belgium, Spain, and France in the past few years [4, 5]. In Europe, many victims were famous stars of show business or sports, making these thefts important media events. Along with carjacking, residential burglaries and, in some cases, home invasions are being perpetrated for the purpose of taking keys to steal luxury vehicles. This method is called homejacking and is also spreading rapidly throughout Europe. Carjacking and homejacking are emerging trends partially due to enhanced anti-theft applications present in vehicles.

In Italy, as in some other countries, a trend of "highwaymen" purposely crashing into vehicles emerged a few years ago [5]. Once the driver comes out of the vehicle to assess the damage, an accomplice jumps into it and drives away, leaving the owner on the street. In the United States in 2000, a study determined that about 35% of vehicles are stolen while parked at home, about 23% while in a parking lot or garage, and about 18% while on a road or highway [6].

1.2.3 Perpetrators

Juveniles are disproportionately responsible for auto thefts. Many jurisdictions report that juveniles (age varies between jurisdictions but generally considered under the age of 18) account for nearly 50% of all arrests in auto thefts. Vehicle theft is considered a "gateway" crime, which refers to the first serious crime engaged in by young offenders. Vehicles are stolen by juveniles for various reasons and then abandoned or often destroyed entirely for entertainment. In the United States in 2004, 26.5% of all arrests for motor vehicle theft were of juveniles (under age 18) and 59.9% of the arrestees were younger than 25 years old [7].

1.3 STATISTICAL DATA AND INTERNATIONAL PERSPECTIVE

1.3.1 Global Picture

To comprehend the magnitude of the global problem, consider that approximately four million vehicles are stolen annually worldwide, at an estimated economic loss in excess of USD 50 billion [8]. According to the 2004 Uniform Crime Report, one motor vehicle is stolen every 25.5 seconds in the United States [7]. This leads to a total of 1,237,114 vehicles stolen (and reported to the Federal Bureau of Investigation) for the year 2004, or 421.3 vehicles stolen per 100,000 inhabitants. Although this figure is slightly lower than that in the previous year, it remains quite impressive.

While all these figures initially appear astronomical, it is interesting to break down these numbers according to different parameters and to study the phenomenon from different perspectives. For example, the following questions can be answered with data and statistics: "How do auto theft rates vary from one country to another?" "Which vehicles are the most often stolen?" "Was auto theft more prominent 10 years ago?" The auto theft investigator will find some background figures regarding stolen vehicles pertinent to know.

1.3.2 Evolution in Number of Stolen Vehicles with Time

Figure 21-1 (see Chapter 21) shows the number of stolen vehicles per 100,000 inhabitants in the United States between 1983 and 2004. Interestingly, there is almost as much auto theft today as there was 20 years ago. The year 1991 was the most intense year for auto theft within this range. Because the theft rate continued to increase and because it became a very serious problem, Congress enacted the Anti Car Theft Act of 1992 based upon the recommendations of the Department of Transportation [9]. Several measures were taken, in addition to the measures already in place under the 1984 Act (which issued the Federal Motor Vehicle Theft Prevention Standard), and a report on the progress and efficiency of both standards was issued. The theft rates plateaued after 1991 and began dropping consistently until 2001, when a slight increase was noted. The implementation of marking parts and installing anti-theft devices played important roles in this decrease.

In Europe, a similar trend in the number of auto thefts started in 1990. Between 1990 and 1991, auto theft progressed 30.8% in Germany, 28.3% in Belgium, 20.0% in Italy, and 17.8% in Great Britain [1]. The most commonly cited explanation for this sudden rise is the opening of Eastern Europe and the subsequent freedom to move people and merchandise. The rise in auto theft was so important that insurance companies throughout Europe placed pressure on automobile manufacturers to develop and include anti-theft systems, known as transponders, in their vehicles (see Chapter 8). This, combined with other measures, successfully reduced the number of auto thefts throughout Europe.

It is important to remember that Figure 21-1, as in most auto theft statistics, is based upon the number of people rather than the number of vehicles. When the number of stolen vehicles is compared per 100,000 vehicles rather than 100,000 inhabitants, the resulting rate is higher. For example, in 2003 there were 1,261,226 motor vehicle thefts reported in the United States [7]. This corresponds to a rate of 433.7 thefts per 100,000 inhabitants. As there were 236,760,033 vehicles registered, the same number of motor vehicle thefts would correspond to a rate of 533 thefts per 100,000 vehicles [10].

Thus, to compare values from 1983 and 2002, it is assumed that the number of vehicles per inhabitant is constant. However, the number of vehicles per inhabitant in the United States increased slightly from about 740 to 780 vehicles per 1,000 inhabitants between the years 1983 and 2002. This means that for the same auto theft rate in those years, the total number of stolen vehicles is higher in 2002 than in 1983. Although the approximation is still feasible, one must use caution when comparing statistics over a longer period of time or across countries; the number of vehicles per inhabitant must be taken into account.

1.3.3 Comparison Between Countries

Figure 1-3 shows the number of stolen vehicles reported to Interpol in 2003 for the 20 countries with the greatest number of stolen vehicles [8]. This graph shows the United States ahead of all other countries, with well over 3 times more stolen vehicles than the next group of countries, the United Kingdom.

Although these statistics are impressive, they are also misleading. They do not take into account that some countries have more vehicles than others. The United States, with more than 215 million vehicles, is likely to have more vehicles stolen than Switzerland, where there are only 3.5 million vehicles. These data do not discern whether a vehicle in the United States is more likely to be stolen than a vehicle in Switzerland or Mexico. It is possible to attempt to calculate rates rather than absolute numbers. If the number of stolen vehicles is expressed per 100,000 vehicles, the classification from Figure 1-3 radically changes, as shown in Figure 1-4.

From this perspective, Israel leads with the greatest proportion of stolen vehicles with approximately 2,000 vehicles stolen per 100,000 vehicles in the population. Switzerland is just behind with about 1,800 vehicles stolen, and the United States falls to 17th position, with about 540 vehicles stolen per 100,000 vehicles. Japan, which exhibits approximately the same total number of stolen vehicles as Switzerland, appears to be much safer, because less than 100 vehicles per 100,000 vehicles are actually stolen. It is critical to take these values with a grain of salt. As a matter of fact, it is not certain if the data provided by Switzerland to Interpol is limited to the stolen motor vehicles; it is possible that this data includes the theft of bicycles. In such instances, the rate provided in Figure 1-4 would be skewed and highly exaggerated. Enormous differences between the classification in the number of stolen vehicles and in the rate such as with Israel, Switzerland, and Malaysia deserve a much more detailed investigation as to the exact cause, which is outside the scope of this chapter. Figure 1-4 is a perfect example of how difficult it is to perform cross country comparison of crime statistics. Each country records crime in a different manner, which is sometimes not compatible. While looking at rates rather than absolute numbers of stolen vehicles is better for understanding the intensity of auto theft between countries, such a cross comparison is not always feasible.

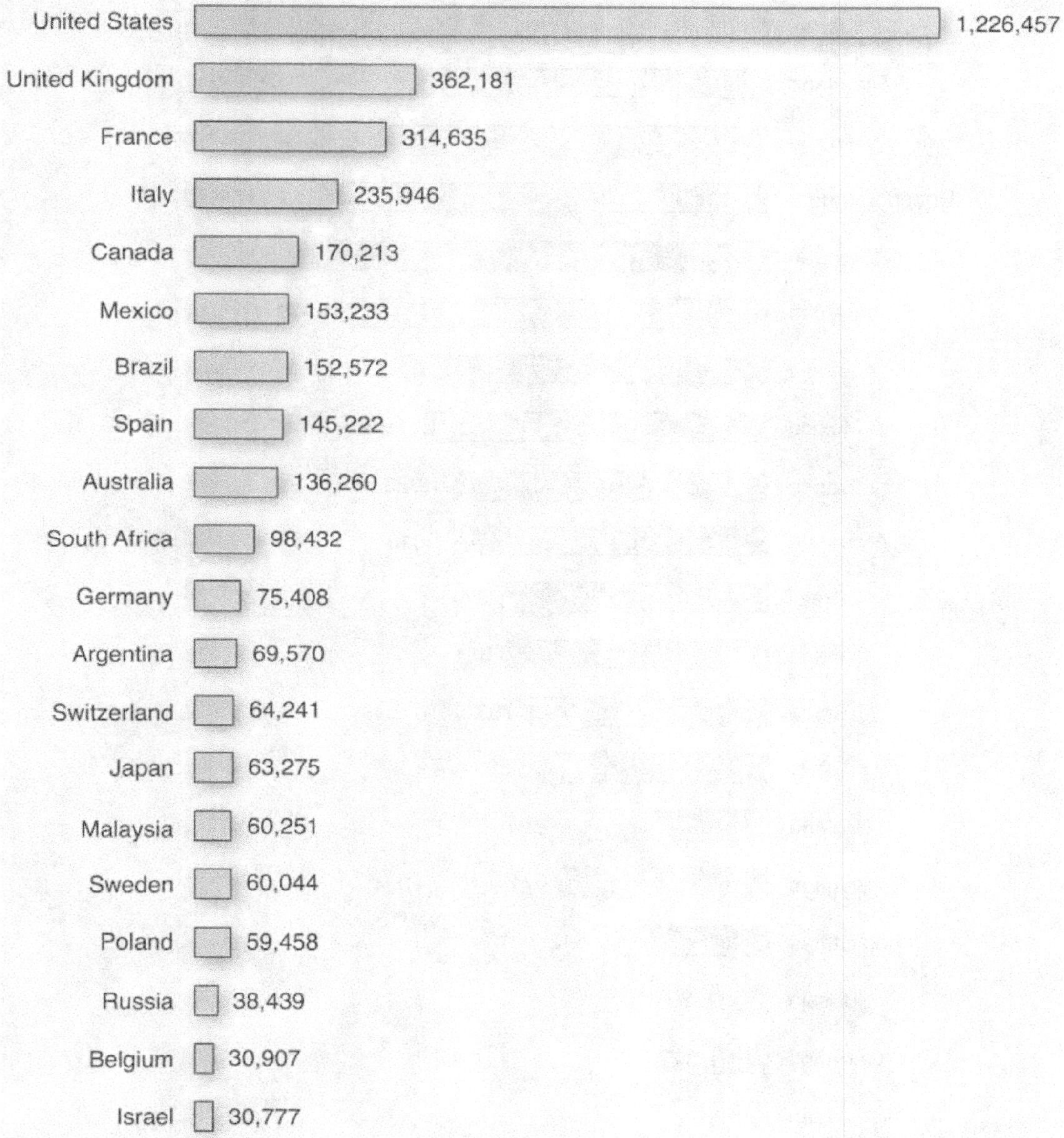

Figure 1-3

The 20 countries with the most stolen vehicles in 2003 according to Interpol [8].

1.3.4 US Geographical Statistics

A/ Ranking by States

Table 1-1 shows the disaggregation of the number of stolen vehicles per state (including the District of Columbia and Puerto Rico) in the United States for 2004 and the corresponding

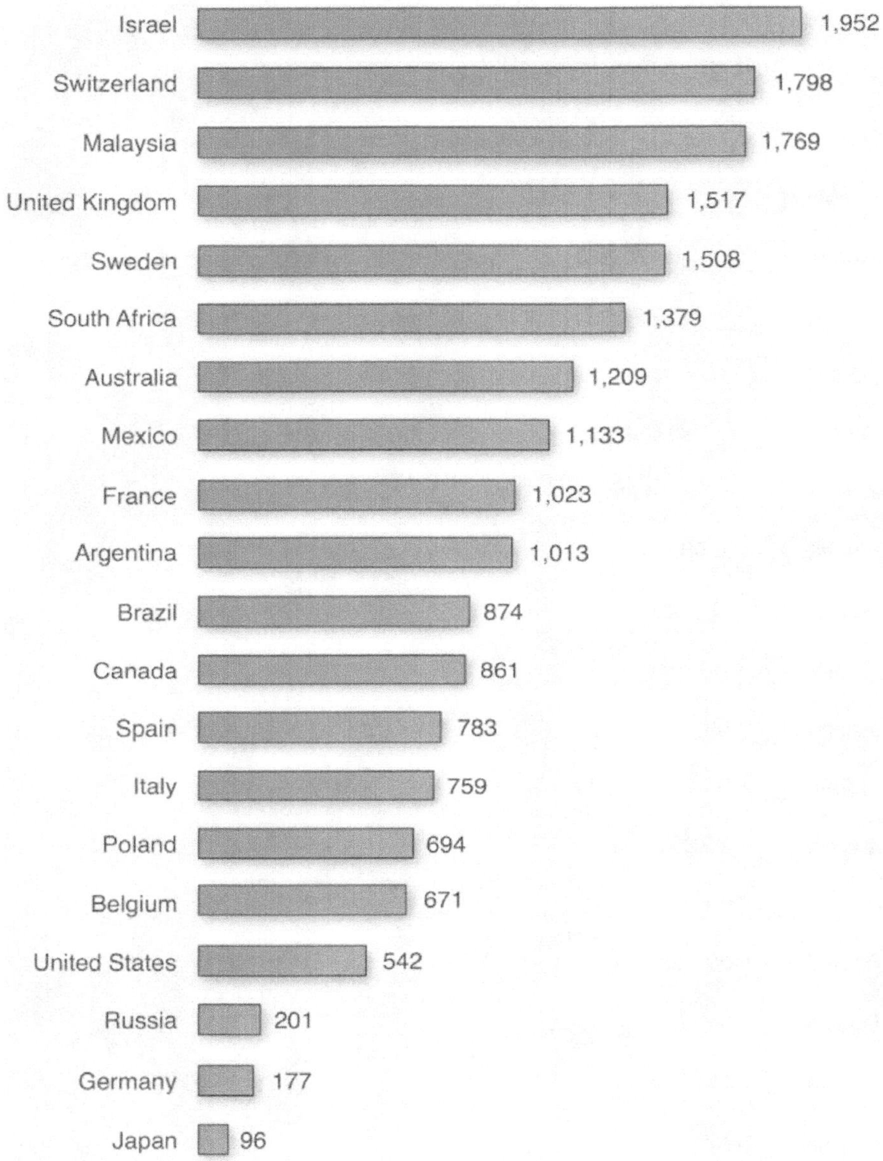

Figure 1-4

Rates of stolen vehicles per 100,000 vehicles for the same 20 countries as in Figure 1-3.

rate [7]. Unfortunately, these rates are based upon 100,000 inhabitants, since the rates based upon 100,000 vehicles are not published. California is the state with the greatest number of stolen vehicles (252,604), but ranks only fourth with regard to the rate. District of Columbia had only 8,408 vehicles stolen in 2004 but ranks first with 1,519 vehicles stolen per 100,000 inhabitants. Maine and Vermont close the ranking with rates just below 100.

Table 1-1

Number of stolen vehicles and rate (number of stolen vehicles per 100,000 inhabitants) for the 50 states of United States, plus the District of Columbia and Puerto Rico in 2004 [7].

Rank	State	Stolen vehicles	Rate
1	District of Columbia	8,408	1,519.0
2	Nevada	22,635	969.5
3	Arizona	55,306	962.9
4	California	252,604	703.8
5	Washington	43,233	696.9
6	Hawaii	8,620	682.6
7	Maryland	35,858	645.2
8	Colorado	24,063	522.9
9	Oregon	18,535	515.6
10	Georgia	44,238	501.0
11	Michigan	50,555	499.9
12	Florida	78,325	450.2
13	Missouri	25,893	450.0
14	Louisiana	19,714	436.6
15	Tennessee	24,749	419.4
16	Texas	94,077	418.3
17	New Mexico	7,902	415.2
18	Rhode Island	4,078	377.4
19	South Carolina	15,637	372.5
20	Oklahoma	12,957	367.7
21	Ohio	40,853	356.5
22	New Jersey	30,306	348.4
23	Massachusetts	22,053	343.7
24	Alaska	2,240	341.8
25	Indiana	21,091	338.1
26	Utah	7,651	320.3
27	Illinois	40,355	317.4
28	North Carolina	26,988	316.0
29	Connecticut	11,025	314.7
30	Alabama	14,024	309.6
31	Kansas	8,435	308.4
32	Nebraska	5,287	302.6
33	Mississippi	7,879	271.4
34	Minnesota	13,518	265.0
35	Puerto Rico	10,128	260.0
36	Delaware	2,147	258.6
37	Pennsylvania	30,969	249.6
38	Arkansas	6,491	235.8
39	Virginia	17,411	233.4
40	New York	41,002	213.3

Table 1-1

Continued.

Rank	State	Stolen vehicles	Rate
41	Kentucky	8,772	211.6
42	Wisconsin	11,374	206.5
43	West Virginia	3,739	206.0
44	Idaho	2,724	195.5
45	Iowa	5,404	182.9
46	Montana	1,618	174.6
47	Wyoming	799	157.7
48	New Hampshire	1,942	149.4
49	North Dakota	906	142.8
50	South Dakota	846	109.7
51	Maine	1,303	98.9
52	Vermont	575	92.5

Table 1-2

The 10 cities in the United States presenting the highest rate (number of stolen vehicles per 100,000 inhabitants) of stolen vehicles.

Rank	Metropolitan	Stolen vehicles	Rate
1	Modesto, California	7,024	1,571
2	Stockton-Lodi, California	8,163	1,448
3	Las Vegas, Nevada	19,794	1,266
4	Phoenix-Mesa, Arizona	40,371	1,241
5	Sacramento, California	18,747	1,151
6	Oakland, California	24,855	1,039
7	Visalia-Tulare-Porterville, California	3,800	1,033
8	San Diego, California	27,396	974
9	Fresno, California	8,770	951
10	Seattle-Bellevue-Everett, Washington	22,807	945

B/ Ranking by Cities

Table 1-2 shows the 10 cities in the United States that present the highest rates of auto theft. Note that 7 of 10 cities are in California and that 8 of 10 are located on the West Coast.

1.3.5 Most Commonly Stolen Vehicles in the United States

Statistics vary from year to year regarding the most often stolen vehicles in the United States. Also, there are different means by which the most often stolen vehicles are evaluated.

For 2004 in United States, CCC Information Services offer the classification shown in Table 1-3 [11].

This list is based upon the rate of theft as a percentage of the total number of registered vehicles of the same year and model. These data of stolen vehicles were obtained from more than 350 insurance companies. There are many different factors that influence the type of vehicles stolen. One factor cited is that some manufacturers keep the same parts on a given model for several years, which is the case with the Acura Integra. This makes it appealing for thieves to steal these cars for parts. In addition, it appears that powerful vehicles, such as the BMW M Roadster and the Audi S4, are becoming more and more targeted by thieves.

Another study from the National Insurance Crime Bureau (NICB) does not take into account the number of vehicles available on the road. Thus, Table 1-4 reports the 10 most often stolen vehicles in the United States based on the absolute numbers of vehicles stolen.

Table 1-3

The 25 most stolen vehicles in 2004 in the United States according to CCC Information Services [11].

Classification	Vehicle description
1	1999 Acura Integra
2	2002 BMW M Roadster
3	1998 Acura Integra
4	1991 GMC V2500
5	2002 Audi S4
6	1996 Acura Integra
7	1995 Acura Integra
8	2004 Mercury Marauder
9	1997 Acura Integra
10	1992 Mercedes-Benz 600
11	2001 Acura Integra
12	1989 Chevrolet R25
13	1993 Cadillac Fleetwood
14	1994 Acura Integra
15	1996 Lexus GS
16	2000 Acura Integra
17	1999 Mercedes-Benz CL
18	1996 Lexus SC
19	2004 Cadillac Escalade
20	1996 BMW 750
21	1996 Land Rover Range
22	1994 Audi Cabriolet
23	2001 BMW M Roadster
24	2003 Cadillac Escalade
25	2000 Honda Civic

Table 1-4

The 10 most stolen vehicles in 2004 in the United States according to the NICB.

Classification	Vehicle description
1	2000 Honda Civic
2	1989 Toyota Camry
3	1991 Honda Accord
4	1994 Chevrolet C/K 1500
5	1994 Dodge Caravan
6	1997 Ford F-150
7	1986 Toyota Pickup
8	1995 Acura Integra
9	1987 Nissan Sentra
10	1986 Oldsmobile Cutlass

1.3.6 Recovery Rates and Other Parameters

One significant indicator of the changing criminality of this offense is the decreasing recovery rate of stolen vehicles in many jurisdictions. For example, 20 years ago the typical recovery rate was in excess of 80% in the United States, yet now it is common to experience recovery rates in the range of 60% or less. This disturbing trend is common worldwide and signifies that up to 40% of stolen vehicles disappear and are never returned to their lawful owners.

Experience indicates that vehicles stolen by opportunist criminals, such as for temporary use to facilitate other crimes or by juveniles, are generally recovered within 24 to 48 hours. A discussion is warranted to explore what happens to stolen vehicles that are never recovered in whole or in parts. Although the crime of vehicle theft originates in one political jurisdiction, the suspects often come from another jurisdiction, and the proceeds of the crime may go to yet another jurisdiction.

According to the Uniform Crime Report of the Federal Bureau of Investigation, passenger cars account for 72.8% of all vehicles stolen in United States in 2004 [7]. Also, only 13% of these cases were cleared by law enforcement agencies, and 16% of them involved juveniles.

1.3.7 International Trafficking

Some vehicles are stolen and smuggled to other countries. With our global economy, communication networks, and transportation capabilities, vehicle theft is no longer just a local crime problem. The elimination of or reduction in border control between many countries has benefited transnational commerce and has eased traveling. Organized criminal groups have also capitalized on this opportunity. Vehicles stolen in one country can easily be driven to others with minimal possibility of detection and interdiction. This is particularly true in Europe, where the abolition of borders between 25 countries rendered the freedom of movement ideal for auto thieves. Densely populated areas near international land borders and seaports generally have the highest vehicle theft rates and, correspondingly, the lowest recovery rates.

According to the 2004 organized crime situation report by the Council of Europe [12], "As globalization facilitates the expansion of international trade in almost any sector, so does it facilitate transnational operations of criminal organizations on classical crime markets, such as trafficking in drugs, arms, vehicles, cigarettes and others". Economic disparity between affluent countries and neighboring developing countries is clearly a precipitator. Unfortunately, inadequate and ineffective data exchange exacerbates the problem and challenges authorities.

The following example demonstrates the pervasiveness of this crime and clearly establishes a nexus between organized crime, vehicle theft, and fraud. Border guards in Finland encounter a late model Mercedes, registered in the US, preparing to cross the border into Russia. Unfortunately, even though the guards are suspicious, they are unable to readily access US databases to determine whether the vehicle is stolen. Based upon their sovereign laws and the fact that they were unable to make inquiry into the vehicle's status, the guards were compelled to allow it to proceed without further delay.

In this particular scenario, the Mercedes was not yet reported stolen when it was encountered in Finland. Approximately three months later, the lien holder, Mercedes-Benz USA Credit, attempted to notify the buyer that payment had not been received. It was soon discovered that the vehicle was purchased under assumed identity using an innocent person's credit information. Interestingly, the buyer made two loan payments under the assumed identity even though the vehicle was already illegally exported. This was probably done to avoid any possible suspicion during transit.

A review of this case highlights the involvement of transnational organized criminal enterprises engaging in vehicle theft and fraud to generate huge profits. It is obvious that an individual or opportunist car thief could not accomplish such a complex transaction. First, a convincing identity theft had been perpetrated, which facilitated the purchase of a luxury automobile with minimal investment. Next, the vehicle was illegally exported from the United States to Europe. Finally, it was likely sold to an unsuspecting buyer.

Additionally, the dynamics of illicit markets such as these compromise legitimate business opportunities and government operations (by avoidance of taxation). It is difficult to establish a sound business decision that would justify legally exporting a vehicle from North America that was originally made in Europe and then shipping it back to Europe. Unfortunately, criminals are very adept at exploiting inefficiencies and weaknesses in the system. It may seem logical to assume that the flowing direction of stolen vehicles would simply follow the illegal trafficking of other goods. But in practice it is not quite the case, as Williams writes [13]: "Most of the markets have become global in scope and generally involve trafficking of illicit products from the developing world or states in transition to the developed world. The exceptions are arms and cars. Luxury motor vehicles in particular go from the countries of Western Europe to states in transition in Eastern Europe and the former Soviet Union and to developing states in Africa. Similarly cars stolen from the United States often end up in Central and South America. This is an interesting reversal of the direction of most illicit flows."

1.4 ENFORCEMENT AND PREVENTION STRATEGIES

1.4.1 Preventive Measures

Traditionally, vehicle theft has been considered a local crime issue. This view has dramatically changed. It is generally understood that vehicle theft must be addressed cooperatively and strategically. An effective anti-vehicle theft campaign needs to focus on several vital elements, such as efficient data exchange, coordinated law enforcement activities, aggressive prosecution, public awareness and community education, and vehicle security measures.

It is important to acknowledge that from a general perspective, vehicle theft is a preventable crime. Deterrence and crime prevention are always more cost effective than enforcement, investigation, and prosecution afterward. Ultimately, the public is responsible for taking reasonable precaution to protect its property. Owners should be informed about vulnerability and ways to avoid vehicle theft. The NICB recommends a "layered approach" to prevent vehicle theft [14]. Accordingly, the more layers of protection on a vehicle, the more difficult it is to steal.

The following four layers of protection are considered in the NICB's approach:

Layer 1: Common sense

- Never leave an unattended vehicle running.
- Remove keys from the ignition.
- Lock doors and close windows.
- Park in a well-lit location.

Layer 2: Warning devices and active anti-theft devices (see Chapter 8)

- Audible alarms
- Steering column collars, steering wheel/brake pedal locks
- Theft deterrent program decals
- Identifying marks and identification concealed in and on the vehicle
- VIN etching on glass and other components

Layer 3: Immobilizing devices and passive anti-theft devices

- Kill switches (electrical/fuel system disablers)
- Smart keys

Layer 4: Tracking devices

Efforts must be undertaken by manufacturers to enhance passive anti-theft systems and component parts marking. Passive anti-theft systems are critical because they require minimal effort to activate and are not dependent on manual application by the operator. Component parts marking is necessary to readily identify items stripped from stolen vehicles.

1.4.2 Investigative Measures

There are many enforcement strategies. Some are reliable and true investigative techniques, and others are new and innovative. Crime pattern targeting and aggressive high-intensity patrol activities prove effective (see Chapter 21). Covert undercover work, the use of informants, and "store-front" operations are also important activities in disrupting criminal organizations (see Chapter 18). Bringing together multiple disciplines, such as police, customs agencies, and insurance organizations, in the form of task forces is often well served in dealing with cross-jurisdictional international issues.

Bait cars, license plate reading cameras, and gamma ray scanners are three methods of technology used to combat vehicle theft (see Chapters 18 and 20). Bait cars are typically used in high theft areas and are equipped with tracking systems, audio/video recording devices, and electronic equipment capable of remotely disabling the vehicle. Before these systems were available, the police relied on hit and miss manpower-intensive surveillance. The use of bait vehicles has been well received by the police, prosecutors, public, and media. These are an efficient use of resources and often involve cooperative partnerships between the insurance industry, the police, and other organizations. For example, in the state of Arizona (US), the Automobile Theft Authority issues grants to police agencies to purchase complete bait car systems, and the insurance industry donates vehicles for use as bait cars [15]. Prosecution is simplified because of the strength of the audio/video recorded evidence. The theft rate in Arizona has decreased significantly since bait cars were deployed [16].

Digital cameras are being strategically deployed to record license plates on vehicles at a variety of locations, such as parking structures, critical infrastructure facilities, and international borders. Additionally, police are using mobile license plate reading cameras in a wide spectrum of environments. License plate reading cameras digitize the alpha-numeric characters, are linked to crime information computers, and rapidly identify wanted vehicles (see Chapter 20). License plate reading cameras are beneficial in both interdiction and generation of intelligence.

Gamma ray scanners are utilized to screen containers for contraband and are particularly useful at land border ports of entry and seaports (see Chapter 20). The sheer volume of vehicular traffic crossing land borders and of containers passing through seaports is overwhelming. Trained operators can identify disparities declared in containers and are capable of recognizing items of concern, including vehicles.

1.5 ORGANIZATIONS

1.5.1 Goals

Many associations and organizations around the world are interested in detection and repression of auto theft. Some of the major associations and organizations are presented in this section. It is not possible within the scope of this chapter to survey and present all these organizations. The reader is invited to check the websites of these organizations. They usually contain very pertinent information, from both preventive and repressive perspectives. Even if the association is not local to the investigator, he or she will find valuable information that will improve his or her knowledge of auto theft and its prevention.

1.5.2 Professional Associations

A/ The International Association of Auto Theft Investigators (IAATI, http://www.iaati.org)

The IAATI is the largest and most important professional association uniting auto theft investigators from around the world. The IAATI official definition is as follows [17]: "The International Association of Auto Theft Investigators (IAATI) was formed in 1952 in order to improve communication and coordination among the growing family of professional auto theft investigators. It has grown to 4,208 members representing over 35 countries and includes representatives of law enforcement agencies, as well as many others with a legitimate interest in auto theft investigation, prevention and education. We recognize that, just as law enforcement agencies cannot successfully function independent of one another, auto theft investigation requires the active participation of the private sector; therefore, our membership also includes the insurance industry, automobile manufacturers, car rental companies and, of course, the National Insurance Crime Bureau and its sister agencies in Canada and Europe."

The IAATI has more than 3,800 members worldwide and has regional chapters and international branches. The United States counts more than 2,900 members and is divided into five regional chapters: North Central, Northeast, South Central, Southeast, and Western. The IAATI has an Australasian Branch (formed in 1994 in Australia), a European Branch (formed in 1990), and a United Kingdom Branch (formed in 2001).

B/ North American Export Committee (NAEC, http://www.naec.ws)

The NAEC was formed in 1995 by representatives from the United States and Canada. The NAEC now also includes Mexico (see Chapter 18 for more detailed information). The NAEC official mission and vision are as follows [18, 19]: "The mission of the NAEC is to bring together those entities that share a common goal of combating the exportation of stolen vehicles and to facilitate contacts for the exchange of information and ideas to achieve that goal. The NAEC vision is to provide a model plan that can be implemented at every port to stop the exportation of stolen vehicles. This model includes verifying the validity of both vehicle and supporting documentation. It also includes methods of identifying vehicles hidden inside containers without interfering in the daily commerce of the ports."

1.5.3 Government-Sponsored Organizations

A/ Interpol (http://www.interpol.int)

Interpol is the world's largest international police organization. It was created in 1923, and today it includes 184 member countries [20]. It facilitates cross-border police cooperation and supports and assists all organizations, authorities, and services whose mission is to prevent or to combat international crime (see Chapter 22 for more detailed information).

B/ Europol (http://www.europol.eu.int)

The European Police Office (Europol) is defined as [21]: "the European Law Enforcement Organisation which aims at improving the effectiveness and co-operation of the competent authorities in the Member States in preventing and combating terrorism, unlawful drug trafficking and other serious forms of international organised crime." There are 25 Member States in the European Union.

C/ US Organizations

The following US governmental agencies (nonexhaustive list) are generally referred to as auto theft prevention authorities:

- Arizona Automobile Theft Authority (http://www.azwatchyourcar.com)
- Colorado Auto Theft Prevention Authority
- Illinois Motor Vehicle Theft Prevention Council (http://www.icjia.state.il.us/mv)
- Maryland Vehicle Theft Prevention Council (http://www.mdautotheft.org)
- Michigan Automobile Theft Prevention Authority
- New York Motor Vehicle Theft and Insurance Fraud Prevention (http://www.criminaljustice.state.ny.us/ofpa/mvtifpmain.htm)
- Pennsylvania Auto Theft Prevention Authority (http://www.watchyourcar.org)
- Texas Automobile Theft Prevention Authority (http://www.txwatchyourcar.com)
- Virginia State Police Help Eliminate Auto Theft (HEAT) (http://www.heatreward.com)

D/ Australia National Motor Vehicle Theft Reduction Council (NMVTRC, http://www.carsafe.com.au)

The NMVTRC is a not-for-profit association created by the joint initiative of all the Australian governments and the Australian insurance industry. Its mission is to drive down the high level of vehicle theft in Australia to benefit the economic and social well-being of the nation [22]. The association works with police, insurers, motor trades, vehicle manufacturers, registration authorities, and justice agencies.

1.5.4 Privately Sponsored Organizations

A/ Insurance Bureau of Canada (IBC, http://www.ibc.ca)

The IBC is the national trade association of nongovernmental property and casualty insurers. Member insurance companies provide about 90% of the home, car, and business insurance sold in Canada [23]. IBC Investigative Services works in cooperation with insurers, law enforcement agencies, and the Canadian Coalition Against Insurance Fraud to detect and prevent insurance crime and to gather evidence in aid of prosecuting offenders and securing restitution.

B/ National Insurance Crime Bureau (NICB, http://www.nicb.org)

The NICB is a not-for-profit organization supported by property/casualty insurance companies in the United States. Its goal is to facilitate the identification, detection, and prosecution of insurance criminals through a collaboration between insurers and law enforcement agencies (see Chapters 18 and 19 for more information) [24].

C/ Oficina Coordinadora de Riesgos Asegurados (OCRA, http://www.ocra.com.mx)

OCRA is comparable with the IBC and NICB, representing the majority of insurance companies in Mexico (see Chapter 18 for more information).

ACKNOWLEDGMENTS

The author would like to thank Eric Stauffer, Jean-François Chevalley, and Manu Poza for their input in the writing of this chapter.

BIBLIOGRAPHY

[1] Junghaus P. (2004) *Halte au Traffic*, JC Lattès, Paris, France.

[2] Interpol (2005) *Vehicle crime*, available at http://www.interpol.int/public/vehicle/default.asp, last access performed on April 2, 2005.

[3] Manitoba Public Insurance (2001) *Sister of slain police officer leads national committee to reduce auto theft*, available at http://www.mpi.mb.ca/english/newsroom/articles/2001/nr_P6116.html, last access performed on October 19, 2005.

[4] Europol (2004) *2004 European Union organised crime report*, Office for Official Publications of the European Communities, Luxembourg.

[5] Europol (2002) *An overview on motor vehicle crime from a European perspective*, available at http://www.europol.eu.int, last access performed on October 19, 2005.

[6] Federal Bureau of Investigation (2001) Section V—Analysis of motor vehicle theft using survival model. In: *Crime in the United States 2000 Uniform Crime Report*, US Department of Justice, Washington, DC.

[7] Federal Bureau of Investigation (2005) *Crime in the United States 2004 Uniform Crime Reports*, US Department of Justice, Washington, DC.

[8] Interpol (2004) *International Crime statistics—International motor vehicle theft statistics for 2003*, available at http://www.interpol.int/public/statistics/ics/default.asp, last access performed on August 22, 2004.

[9] National Highway Traffic Safety Administration (1998) *Auto theft and recovery: Effects of the anti car theft act of 1992 and the motor vehicle theft law enforcement act of 1984*, NHTSA report number DOT HS 808 761, available at http://www.nhtsa.dot.gov/cars/rules/regrev/evaluate/808761.html, last access performed on October 19, 2005.

[10] Bureau of Transportation Statistics (2005) *National Transportation Statistics 2005*, US Department of Transportation—Research and Innovative Technology Administration, Washington, DC.

[11] CCC Information Services (2005) *2004 Most stolen vehicle list points to thieves' needs for speed*, available at http://www.cccis.com, last access performed on October 19, 2005.

[12] Council of Europe Octopus Programme (2004) *Organised crime situation report 2004—Focus on the threat of cybercrime*, Department of Crime Problems, Directorate General of Legal Affairs, Council of Europe, Brussels, Belgium.

[13] Williams P (1999) Chapter 9—Emerging issues: Transnational crime and its control. In: *Global Report on Crime and Justice*, ed Newman G, Published for the United Nations Office for Drug Control and Crime Prevention, Center for International Crime Prevention, Oxford University Press, New York, NY.

[14] National Insurance Crime Bureau (2005) *Layered approach to protection*, available at http://www.nicb.org/public/newsroom/whereismycar/layeredapproach.cfm, last access performed on October 19, 2005.

[15] Arizona Automobile Theft Authority (2004) *2004 Annual Report*, Arizona Automobile Theft Authority, Phoenix, AZ.

[16] Scarborough S (2004) Bait cars reel in thieves, *The Arizona Republic*, March 28, 2004.

[17] International Association of Auto Theft Investigators (2005) *What is IAATI?*, available at http://www.iaati.org, last access performed on March 26, 2006.

[18] North American Export Committee (2005) *NAEC Mission*, available at http://www.naec.ws, last access performed on October 19, 2005.

[19] North American Export Committee (2005) *NAEC Vision*, available at http://www.naec.ws, last access performed on October 19, 2005.

[20] Interpol (2005) *Interpol member countries*, available at http://www.interpol.int, last access performed on October 19, 2005.

[21] Europol (2002) *Welcome to the European Police Office*, available at http://www.europol.eu.int, last access performed on October 19, 2005.

[22] National Motor Vehicle Theft Reduction Council (2005) *Who we are*, available at http://www.carsafe.com.au, last access performed on October 19, 2005.

[23] Insurance Bureau of Canada (2005) *About us: IBC profile and members*, available at http://www.ibc.ca, last access performed on October 19, 2005.

[24] National Insurance Crime Bureau (2005) *About NICB*, available at http://www.nicb.org, last access performed on October 19, 2005.

CHAPTER 2

VICTIM AND WITNESS INTERVIEWS AND COLLECTION OF CIRCUMSTANTIAL INFORMATION

Jean-François Chevalley
Manuel Poza

2.1 INTRODUCTION

This chapter presents the latest techniques in victim and witness interviews and collection of circumstantial information. Although the practice described is presently used in Switzerland, it is directly applicable or can be easily adapted for use in other countries. Even though laws and procedures concerning the application of criminal investigation vary greatly from country to country, the working methods of law enforcement personnel present some common grounds. Fortunately, some governments spend time and resources to fight auto theft, notably by creating special department or teams specialized in auto theft investigation. Other countries, may be less concerned by this particular criminal phenomenon or not having the same amount of available resources, do not place any emphasis on the battle against the theft of vehicles.

The goal of the interview of victims of and witnesses to an auto theft is to collect all the pertinent information related to the theft. This needs to be carried out using a systematic and comprehensive approach, which will lead to the opening of the investigation. In addition, circumstantial information surrounding not only the theft, but also the vehicle, can be obtained from sources other than the victim (or owner) and witnesses. Such sources include but are not limited to the car dealer, repair/body shop, manufacturer, insurance company, and department of motor vehicle (DMV) records. Research conducted later in the investigation might lead to the recovery of the vehicle, to the arrest of the perpetrator(s) of the theft, and/or even to the elimination of an organized crime group, from which more information may be obtained.

To properly announce the theft of a vehicle throughout the police department, or even among different law enforcement agencies, it is necessary to have a solid database, if possible, at the national level. This database must contain important information such as the license plate number, vehicle identification number (VIN), brand, model, year, and color of the vehicle. The strength of a database is only as good as the accuracy of the information placed in it. The theft report is the document used to input information into the database. Thus, the establishment of the theft report is a very important step that needs all the attention of its writer. In Switzerland, an electronic database is available at the national level and contains all the vehicles reported stolen in the country as well as all the Swiss-registered vehicles stolen in foreign countries. The access to this database is easy and rapid. In other countries, such as the United States, this type of information can be readily accessed from the terminal computer within a patrol car. This type of electronic tool allows for a control of vehicles at a very large scale and increases the chances of quickly identifying a stolen vehicle.

2.2 GENERAL APPROACH TO INTERVIEWS

2.2.1 Initial Report

Presently, vehicles are indispensable to the active life of many people, particularly in areas where public transportation is not readily available; the car is more than just comfort. When a car breaks down, it is a serious problem that can immobilize an individual. Theft is even worse because it can create a very serious inconvenience to the victim, who will need to change his or her schedule, report it to the law enforcement agency, file a claim with the insurance company, and, if the financial situation permits it, look for a replacement vehicle. In some instances, the theft of one's vehicle might prevent him or her from working, affecting income for the whole family.

Although auto theft is a worldwide problem, the manner in which these thefts are reported to the police and investigated by the police greatly varies. In most countries, a theft requires a filed complaint to the police for the insurance to create and process a claim. The time between the occurrence of the theft and the moment it is reported to the police can vary from a few seconds to several hours or even days if, for example, the owner was on vacation during the theft and discovered the crime only after returning.

The interview of the victim allows for the collection of information necessary to the creation of a police report, which officially recognizes the theft of the vehicle. The questioning of the victim can be performed at the crime scene, in a police station, or in a remote location. It is important to prioritize this interview and complete it as early as possible. In some instances, circumstances do not permit to interview the victim at the early stage of the investigation.

It is not necessary to have great interviewing skills to record the theft complaint. Agencies typically have template documents or forms already preprinted and designed to report this type of crime. Conversely, when circumstances allow for and once the initial crime report has been processed, an experienced police investigator should contact the victim and proceed to a more detailed interview, which could be written as an affidavit.

2.2.2 Main Interview

As in every case when dealing with a victim, the interview needs to be conducted in a courteous manner. Depending on the situation and the condition of the victim, who might feel this theft as an emotional and/or physical aggression, the investigator must show some comfort, thus creating a certain level of confidence. Simultaneously, the investigator should get a feel for the level of truthfulness of the statement provided by the victim. In cases where victims appear extremely shocked, it is difficult to imagine that they could be acting. However, it is important to always keep in mind that there are some "victims" with excellent simulating skills and the most criminal minds.

The investigator has to conduct the interview in a complete manner and collect all the necessary information. In some instances, it is necessary to insist in getting some information that is not easily revealed by the victims. Even if certain information does not appear useful at the early stage of the investigation, it can reveal itself to be extremely pertinent at a later date. Thus, every bit of information must be collected and recorded. Details provided by the victim(s) cannot be neglected at this stage and need to be precisely noted. Some of these details might play a crucial role in the identification of the criminals or in the exposure of the fraud. In the latter case, frequently the victim-perpetrator makes up an over-detailed story regarding the theft, believing it would render the statement more credible. When subsequent interviews are carried out, the victim-perpetrator will not remember all these details or will provide contradictory details. In opposition, a real victim typically provides a reasonable level of details and very likely remembers the same details during subsequent questionings. Thus, when too many details are provided and contradictions occur between interviews, it should be considered as suspicious.

In general, it is good practice to record the interview as an affidavit, meaning that the interviewee swears this statement to be true in front of a person legally authorized to administer an oath. This is particularly helpful when exposing falsely reported thefts.

2.2.3 Identities

The investigator or officer who records the theft report must ensure the identity of the victim(s) and the witness(es). In some instances, the victim(s) will have every reason in the world to provide a false identification to the police. For example, a victim might not be a victim but a perpetrator, should not have been present at the location and time of the theft, or has an open search warrant for an unrelated crime. There have also been many cases of impersonation for different and unbelievable reasons. Thus, it is always extremely important to perform an identity check.

Properly checking an identity is achieved by an official identification document such as a passport, identity card, or driver's license. It is necessary to note at minimum the first and last names, date of birth, social security number (or equivalent), and a valid address. An investigation might last several days, weeks, months, or even years. People can move during that time. Thus, it is crucial to have permanent data such as date of birth and social security number (or equivalent) that would allow finding these people.

In addition, it is necessary to record the identity not only of the witnesses to the crime, but also of the people accompanying the victim to and/or during the interview. These people might provide testimony that will become very important later in the investigation. People always talk or reveal secrets (in confidence) at some point. Information that appears to be insignificant at first might reveal itself as the key point of an investigation at a later time. For every subsequent interview, the identity of the victim or witness should be checked again, unless the interviewer can positively identify the interviewee due to prior encounter and identity check.

2.2.4 Quality of the Interview

The quality of the investigation often depends on the nature, amount, and accuracy of the information recorded by the interviewer.

A/ Listen But Never Suggest

At all times, it is very important to attentively listen to the victim and never suggest answers. If the interviewer suggests possible answers, it could jeopardize the genuineness of the interviewee's statement, which is not acceptable. For example, if the victim hesitates or does not frankly answer whether the vehicle was locked or not, it is critical not to suggest that it was locked. Nevertheless, it is important to help the victim to recall the memory by asking (but without suggesting) what particular gestures or actions were performed when he or she left the vehicle (e.g., talking on the cell phone, carrying groceries, looking for keys, etc.). Another example is if the victim cannot describe with accuracy where the vehicle was parked before the theft. The investigator should never suggest that it was parked in a particular location because, for example, it is a hot spot for auto theft. The investigator should rather pull out a map to bring the victim's memory back. It is also possible, and sometimes better, to visit the different possible locations with the victim.

B/ Objectivity Versus Subjectivity

It is extremely vital to always record what is observed and not what is perceived or assumed from one's observation. It is necessary to establish the written statement of the victim with the highest degree of objectivity. Subjective elements could have serious effects to the subsequent investigation. By illustration, one should never record that a thief must have defeated a vehicle's ignition because the victim presented all the keys. This is an assumption that can only be verified once the vehicle has been recovered and forensically examined. For example, a police officer wrote in a report that the thief broke the side window to enter the car before driving it away. In this particular case, the officer had no facts to rely on and just assumed this was the way it happened. This assumption will bring future readers of this report (investigator, judge, prosecutor, defense attorney, etc.) to believe that broken glass was found at the scene, collected, and even maybe examined, to have determined it was a side window. During the interview, the victim probably insisted on the fact that the car was locked and the police officer believed that it was appropriate to provide an explanation to this fact in the report. If the interviewer writes perceptions instead of observations, it could be detrimental not only to the outcome of the investigation itself, but also to the outcome of the subsequent trial, perhaps with dire consequences.

2.3 FORMS

To file the police report, different forms and documents are used by law enforcement agencies. Each form presents its particularities, advantages, and drawbacks. Because the laws and procedures vary greatly among countries or even states and cities, it is impossible to present a form that could be used universally by all police departments. However, each police department should have such forms available. They allow for the officer handling the filing process to make sure no element is left out in the initial process. It also offers a certain streamlining of the process, which increases its efficiency. At least one form designed to report a theft of a vehicle should be available within a police department. Also, a section of that form, or better a second form, should also be created to report the recovery of the vehicle.

Figure 2-1 presents an English translation of the form used by the State Police of Vaud *(Police Cantonale Vaudoise)*, Switzerland to record the initial complaint. This provides an idea of how the form is organized and the type of information presented. A second page, not shown, contains the narrative from the plaintiff.

The strength of the form mostly lies in the information placed on it. Thus, it is essential that the person filling the form does it with the highest level of precision and accuracy. Once the form is completed and forwarded to the interested services and divisions, including the crime record department, it is then forwarded to the district attorney's office. It is used as an official legal complaint as long as the victim agrees to pursue the complaint through the justice system. The form presented in Figure 2-1 has the advantage of being entirely designed around the theft and recovery of a vehicle. Indeed, it is also a document that contains the same general text fields as any other form used to report a crime, notably those dealing with the identity of the victim. There are two text fields, present at the end of the form, that contain the information regarding the restitution of the vehicle and the results of the research performed by the investigators.

2.4 COLLECTION OF INFORMATION

2.4.1 Principle

It is important to distinguish three types of information that need to be collected and recorded by the police officer and/or the investigator:

- Information regarding the vehicle;
- Information regarding the theft;
- Information regarding the recovery of the vehicle, if applicable.

Each part includes several detailed categories that are described in this section. The police officer establishing the report needs to complete the different fields based upon the information provided by the victim(s) and witness(es), and the documents presented. It is

CANTON DE VAUD POLICE DE SURETE	**AUTO THEFT COMPLAINT** Written on: 20 April 2004 at 16:30 Location: Lausanne, Vaud, Switzerland By: Sergeant Mario Scarpape	**PASSENGER CAR** **MSP**
☒ **State Police File** **File :** 040413598	Original to: ☒ Auto theft task force Copies to: ☒ Vehicle tracking section ☒ Department of motor vehicle ☒ Violent crime section ☐ ☐	**Stamp and transmission date :** **Appendix(es)** ☒ See page 2 for narrative

Investigative Judge of the District of Lausanne ☒ **TM** ☐ **JIC** **REQUISITION Inv. #** B587F6 To the Chief of State Police: Proceed to all necessary investigation and interviews regarding this theft complaint. This form attests that the following articles of law must be respected: CPP 169, 170, and 171. Date : Judge: (stamp)	**Has the identity been verified?** Yes **Name** Crapulo — **First name** John **DOB** 31 December 1973 — **Occupation** Baker **Origin** Chazulante-le-Goudet, France — **Home phone** 555-555-1000 **Address** Rue de Lavaux 34, CH-1009 Pully — **Office phone** 555-555-1001 **Act in the quality of** Plaintiff, victim **File penal complaint** yes ☒ no ☐ — **Civil action desired** yes ☒ no ☐ **Does complainant desire an interview with Investigative Judge?** yes ☒ no ☐ **Declare physical address in the State of Vaud (in case not a resident of State of Vaud) as:** n/a

Vehicle description: (at time of theft)	**Tag/plate number**	VD 220066	**Make/model**	Lancia A112 Abarth
	Type	Coupe	**Color**	Royal blue
	VIN	ZLA33125XF6597234	**DMV registration**	VD658921
	Year	1985	**Mileage**	130,500 km
	Estimated value (new)	CHF 15,000	**Other**	Left front fender damaged

Date of theft:	**Between** April 8, 2004 **at** 14:15 **and** April 8, 2004 **at** 14:45
Location of theft:	Train station parking lot B
Parking description:	Public — **Covered** ☒ yes ☐ no — **Controlled access** ☒ yes ☐ no
Was vehicle locked?	☒ yes ☐ no — **Key on ignition** ☐ yes ☒ no — **Key inside vehicle** ☐ yes ☒ no
Purchase of vehicle:	**CHF** 1,500 — **Mileage** 55,000 — **Date** June 1, 1999
Number of keys:	**In possession of** 3 keys — 2 **originals** — 1 **copies** **Received at purchase** 2 keys — 2 **originals** — 0 **copies**
Previous owner:	Garage du Crozet, Avenue de la Gare, Cossonay, Vaud, Switzerland
Accompanying/Witness:	None
Stolen objects:	One digital camera Canon 20D with lens was in vehicle at time of theft
Insurance:	Phenix Assurance, Lausanne — **Comprehensive** ☒ yes ☐ no — **Deduct.** CHF 500
Suspects/Perpetrators:	n/a
Modus operandi:	Broken glass present at scene
Signatures:	**Plaintiff** — **Secretary** — **Criminal investigator** **Additional information, see page 2 or appendixes to this report** ☒

Discovery date: **Discovery location:** **Technical examination:** **Vehicle's conditions:**	**Date** **Time** **By** **Doors locked** ☐ yes ☐ no — **Key on ignition** ☐ yes ☐ no **Engine** **Collision signs** ☐ yes ☐ no — **Mileage** **Damage**
Penal code violations: **Objects Inventory:** **Objects belonging to:**	
Information Broadcast:	**Date** 8 April 2004 — **Time** 16:20 — **Phone** ☒ — **Fax** ☒ — **Mail** ☐
Vehicle restituted:	**Date** — **Time** — **Signatures :** **To**
Results of vehicle search **See page 2** ☐	**Police station** — **Date** — **Signatures**

Figure 2-1

English translation of the form used to record auto thefts at the State Police of Vaud (Police Cantonale Vaudoise), Switzerland. This form is used to record the initial complaint. A second page (not shown here), as an annex to this form, contains the narrative from the plaintiff.

important when gathering the information and recording it to keep in mind that the police report is also an official legal document on which the subsequent penal procedure relies.

The collection of that information can be performed through different sources. The following list contains most of these resources but is not exhaustive:

- Victim(s)
- Victim's family, friends, and coworkers
- Witness(es)
- Suspect(s)
- Suspect's family, friends, and coworkers
- Department of motor vehicle
- Vehicle's insurance company
- Vehicle's dealer
- Vehicle's repair shop(s)
- Vehicle's body shop(s)
- Vehicle's manufacturer

2.4.2 Information Regarding the Vehicle

A/ Vehicle's Characteristics

The following information regarding the vehicle must be obtained:

- Tag or license plate number (registration)
- Make and model
- Color
- VIN
- Year
- Mileage
- Options/accessories
- Estimated value as new
- Condition of the vehicle (accident, etc.) before the theft
- Presence of tracking system on vehicle

These different parameters are usually straightforward to record. The mileage might be an element of great significance once the vehicle is recovered. Thus, if the owner does not recall the exact mileage, it is important to note an estimate. In all instances, some information such as VIN, tag or license plate number, and brand and model of the vehicle needs to be verified through the corresponding DMV. This is usually a simple process. Other information, such as the condition of the vehicle, its estimated value as new, and the presence of a tracking system, can be verified through repair and body shops, car dealer, manufacturer, insurance company, DMV's record, or tracking company. All this information should be very rapidly forwarded to the information and coordination unit of the police department or broadcasted through an all points bulletin to increase the chance of recovery of the vehicle. This is particularly true with vehicles equipped with a tracking system, where the tracking company should also be put on notice immediately, so a recovery effort can start with the greatest chance of success (see Chapter 20).

B/ Purchase of the Vehicle

The following elements should be recorded:

- Price
- Date
- Mileage at time of the purchase
- Type of purchase (credit, loan, cash)
- History of previous thefts

A history of previous thefts might reveal itself to be useful in cases of fraud. The price, date of purchase, and mileage at time of purchase are useful to make an estimate of the value of the vehicle at the time of the theft. Also, in case of insurance fraud, the investigator can determine whether the victim declared an excessive value in the claim.

C/ Seller and Previous Owner

Vehicle sales do not always go well. The relationship between the seller and the buyer can go bad after the sale. This is particularly true if the car sold proves to be a lemon or had previous hidden damage. If any data in this regard are available, it is necessary to put it in writing. In case of fraud, the former owner might also provide pertinent information regarding the condition of the vehicle.

D/ Keys

The number, types, and availability of keys for the vehicle are very valuable elements to record. It is important to determine how many original and copied keys the owner is in possession of and how many original and copied keys were provided at the time of the purchase. The number and function of keys provided with a new vehicle vary among manufacturers and individual models. The most pertinent information in this regard is obtained from the manufacturer or dealer. Additionally, law enforcement agencies can benefit from information resources such as the European vehicle identification database (EuVID). EuVID contains notably a catalog with description of original keys sold with new vehicles. When the vehicle is sold used, any number of original and copied keys can be provided. Again, the seller might provide more information in this matter.

This information might allow the investigator to start the investigation within the surroundings of the victim. It brings additional elements in case of insurance fraud. Often, the fraudulent owner contracts a third person, provides the keys to the vehicle, and asks to have the vehicle stolen. Any discrepancy between the number and types of keys received at the time of the purchase and the number and types of keys in possession of the owner at the time of the theft must be explained (see Chapter 10).

E/ Insurance

The type of insurance that was contracted for the vehicle needs to be known. Some people have comprehensive insurance, some have the minimum required liability insurance, and some do not have any insurance.

Vehicles in Switzerland have to carry civil responsibility insurance to be allowed to be driven on the road. Most countries have the same requirements; however, they are also many countries that do not require any insurance. Furthermore, among the countries requiring insurance, not all have an efficient communication channel between the insurance company and the DMV, forcing an uninsured vehicle to have its tag or license plate seized.

Information should be collected from insurance companies. They can provide very interesting facts, such as past due payments, several other claims with different vehicles, or a highly exaggerated claim report. When the vehicle benefits from comprehensive coverage, some owners hesitate less to declare the theft, because the claim might compensate them more (see Chapter 19).

2.4.3 Information Regarding the Theft

A/ Date and Time

This information allows for an estimate of the time span since the theft and the establishment of links in cases of serial auto thefts. It also allows for the investigator to evaluate the crime patterns and the crime's genuineness. Chapter 21 provides more information regarding crime mapping.

For example, if the vehicle was in a parking lot for a certain period of time or was stolen while the victim was on vacation, it is not possible for the victim to provide the officer or investigator with an accurate time or even date. In such cases, an indication of the period during which the theft could have happened needs to be written in the report. Law enforcement agencies have several means to establish the timeline regarding the movement of the vehicle such as border control, speed traps, traffic light radar, traffic surveillance cameras, toll surveillance cameras, or even traffic violation citations, but it is preferable to have a starting date and time from which to look to be more efficient.

The exact date and time of the theft are also meaningful when identifying insurance fraud. For example, a person filed in Switzerland a police report regarding the theft of his vehicle in Italy. The exact date and approximate time (sometime in the morning) was provided to the police. The investigation was able to determine that the person came back from Italy by plane, with a ticket that was purchased the night before the date of the vehicle's alleged theft.

B/ Location

It is essential to be very accurate when collecting information about the location of the theft. Some locations are known to be hot spots for auto theft, usually leaving less doubt about the veracity of the crime. Some other locations are not common for this type of crime. In any instance, it is possible that physical evidence is still present at the scene, such as broken glass, tire tracks, or a cigarette butt. An examination of the scene might reveal pertinent evidence, which might lead to the identification of the criminal(s). Unfortunately, it is rare that the police, even less the crime scene investigation unit, are dispatched to the site to corroborate the victim's statement and to search for evidence.

A certain amount of vehicles are stolen while located in a parking garage. In general, parking garages are locations that can be very interesting for the investigator. In this instance, it is important to verify if there are any data available regarding the movements in and out of the parking garage. The parking garage management should be contacted to obtain more information. Also, there are often surveillance cameras. These are important elements of evidence for the investigators, because they can provide a visual identification of the suspects. All possible surveillance cameras should be identified and the information contained in them obtained and reviewed. Finally, some parking garages offer valet parking. In this instance, it is necessary to contact all the employees that were in contact with the vehicle and determine their policy in regard to the locking of the vehicle and location of the keys.

When dealing with vehicles that were parked in private garages, it is particularly important to determine whether they were locked or not or if they were broken into. The determination of the modus operandi in this situation provides direction to the investigation. For example, if an investigation starts with the theft of a car that was locked in a private garage, it is important to consider different modus operandi. On one hand, the perpetrator, unless he or she is an acquaintance of the victim, has to monitor the garage to determine whether the vehicle is present or not. The perpetrator might have left traces of his or her presence, such as shoeprints and break-entry evidence when getting in the garage or cigarette butts around the street corner while surveilling the residence. Crime scene investigators should be able to detect and collect this evidence, which will be forwarded to the crime laboratory for examination. On the other hand, if the suspect is an acquaintance of the victim or is the "victim", he or she will have easy access to the premises, might have used a key, and tried to simulate a break-entry on the door. Different shoeprints and traces might be left that are not consistent with a stranger committing the crime. The investigation is directed by the information obtained from the observations made at the scene by the forensic specialists.

There have been several cases where the vehicle was a secondary target rather than a primary target. In some instances, burglars find the keys to the vehicle while sweeping the house. Hence, it provides them with a good opportunity to leave with the vehicle that was parked in the garage. This is a modus operandi often exhibited by Eastern European criminals. In these instances, it is essential to have a crime scene investigation unit processing the house or premises because they will likely contain many traces leading to the suspects and the possible recovery of the vehicle.

C/ Locking Condition

The locking condition of the vehicle is an important element to obtain. Also, the question of whether the key was inside the car (e.g., attached to the ignition) or not is pertinent. This information is important when establishing the modus operandi and if there was broken entry or not. This information can also influence the denial of the insurance claim by the insurer. It is also relevant to know it in case the vehicle is recovered because it may demonstrate insurance fraud.

D/ Accompanying Person and Witness

Often with carjacking, several people might be riding in the vehicle when it was stopped and overtaken. The testimony of these people is as crucial and pertinent as the testimony of the driver or owner of the vehicle. No detail should be neglected during the interviews of the passengers. Also, when witnesses such as bystanders are identified, it is imperative to take their deposition and not to neglect any of the details provided.

E/ Stolen Goods

Occasionally, victims report that many goods and property were stolen with the vehicle, such as jewelry, computer, or expensive audio/photo/video equipment. When the thieves are caught and interrogated, often the investigator realizes that the victim exaggerated in the declaration of the value of the stolen property to obtain a larger amount in compensation for the claim. However, it is hard to demonstrate such facts. In any instance, it is important to make an accurate inventory of the stolen goods declared by the victim. This is part of the police report and will help in the investigation later on.

F/ Potential Suspects

Victims can sometimes provide names of potential suspects in regard to the theft of their vehicle. Indeed, they may provide a description of the suspect. It is important to record as much detail as possible in this regard.

G/ Modus Operandi

It is necessary to gather all the information related to the means by which the car was stolen. It is possible, and advised, to write a narrative describing the event. The following example stresses the importance of gathering as much information as possible regarding the modus operandi and to record it in the initial report.

During the police report filing process, the owner of a restaurant gave some details about the circumstances surrounding the theft of his vehicle to the police officer. He stated that the keys went missing during the evening at his restaurant. Nothing was written in this regard in the police report. The investigators who were assigned to this case the next day were first dubious about the veracity of the theft. They started to cautiously investigate the owner of the restaurant unbeknownst to him. Finally, after a new interview of the victim, the investigators learned about the disappearance of the keys. They then turned toward the potential perpetrators among the personnel and clientele of the restaurant and found out that one of the employees actually committed the theft.

2.4.4 Information Regarding Vehicle Recovery

The first step in the discovery of a stolen vehicle is the protection of physical evidence that might be present in the vehicle and at the scene surrounding the vehicle. The vehicle should not be touched. It is important to have a crime scene investigator dispatched at the scene to process the vehicle, particularly when it is linked to a homicide or other violent crime (see Chapters 3 and 4). If the vehicle needs to be moved, a towing company of confidence should be called and nothing should be touched inside the vehicle.

A/ Date and Time

Whenever possible, the exact date and time of the recovery of the vehicle must be recorded. This information allows the investigator to determine the time span during which the vehicle was stolen. This could lead to the determination of the activities of the thieves. Also, if crime mapping is performed, this information is very pertinent.

B/ Person Who Discovered the Vehicle

The identity of the person who discovered the vehicle should be verified and recorded. This person's statement should be as accurate and complete as possible. Some elements might become very pertinent further in the investigation.

C/ Location of Discovery

As for the location of the theft, the location of recovery might present physical evidence left by the perpetrators (see Chapter 3). If it is possible to check this location, it might reveal pertinent information.

The access to the location of discovery is also very important to examine and record. For example, some locations are difficult to access or are not known to the public. This might mean intimate knowledge by the thief, which could help in the search for a suspect. Crime mapping also considers the location of discovery of the vehicle, which could help understanding the type of theft. The example presented in paragraph E (below) is a good representation of the importance of the location of discovery.

D/ Mileage

It is important to record the mileage at the time of the recovery to complete the information about the mileage recorded at the time of the theft report. It provides information to the investigator regarding the number of kilometers or miles driven during the theft. This could be very useful when a suspect is arrested, because it allows for the police to estimate the range of action of the suspect during the theft, which could lead to the identification of the route and other possible crimes. Nevertheless, if the vehicle is recovered very shortly after the theft, an abnormally great difference in the mileage might be very pertinent information, particularly with leased vehicles.

E/ Locking Condition

It is critical to record the locking condition of the vehicle. A locked vehicle will attract the attention of the investigator because a thief rarely locks a vehicle when abandoning it, unless he or she had the key in his or her possession. This could help in the determination of the authenticity of the theft.

When keys are found in the ignition, different questions might arise among the investigators. First, the victim may not have told the whole truth regarding the theft. Second, the victim may have left the keys in the vehicle before the theft and did not provide this information during the filing of the police report. For example, in the first case, frequently people driving under the influence of alcohol or drugs who become involved in an accident by themselves, try to mislead the police by reporting a theft the next day. They usually forget about the keys inside the vehicle after the accident, their priorities being not to get caught and to get home. A case revealed that a vehicle reported stolen in Memphis, Tennessee (US) by its owner was recovered in Chattanooga, Tennessee (US), about 560 kilometers (350 miles) from the location of the alleged theft. Interestingly, the vehicle was recovered in a water pond about three kilometer (two miles) behind the house of the owner. When the vehicle was pulled out of the water, the set of keys was still present in the ignition. One of the keys did open the house of the alleged victim, who always declared that no keys were lost or stolen, which was obviously true; the keys were simply forgotten in the ignition lock when the owner disposed of the vehicle. In other instances, it is also possible that the theft was committed by an acquaintance of the victim who had access to the keys.

F/ Condition of the Vehicle

It is important to describe the condition of the vehicle upon recovery. More information will be obtained by the forensic and technical examination of the vehicle performed at future date, but the basic and obvious condition of the vehicle must be recorded. Often, the vehicle is recovered damaged or destroyed. If it is established that the damage were not present prior to the theft, it is critical to catalog them in the report. It is also important to examine them and look for any physical evidence that would establish a link with the origin of the damage, which could link the vehicle to other crimes. Signs of forced entry, such as a broken door handle, broken side windows, the use of a Slim Jim, or the trunk popped open with a pry bar, must be accurately recorded.

Finally, all the damage not covered by an accident should be observed and described. This includes voluntary vandalism or violence that resulted in some damage. Usually, this damage is limited to the interior of the vehicle, such as lacerated seat upholstery, spray painted windows, or engine alterations.

G/ Inventory of Property and Goods Found in the Vehicle

It is useful to inventory all goods and property found inside a vehicle at the time of the recovery. In most cases, these objects are property of the victim, but in some instances, it is possible that some objects do not belong to the victim. This information is typically not available at the time of the recovery, and therefore all objects must be inventoried.

Foreign objects are identified when the vehicle is returned to its owner. These objects are often brought by the thief to commit the theft or to commit other crimes or they are personal objects that were simply forgotten. These objects might bear very important physical evidence and might lead to the identification of the perpetrator(s). They should not be neglected and should be collected as evidence for future forensic examination. Thieves also smoke, eat, or drink inside the vehicle, and items such as bottles and cigarette butts might bear DNA traces and/or fingerprints (see Chapter 4). A recent case occurred in Switzerland in which a simple earring was discovered in the trunk of a stolen-recovered vehicle. Because of the identification of this earring, a link was established with an organized crime group specializing in the burglary of luxury apartments.

ACKNOWLEDGMENTS

The authors would like to thank Eric Stauffer for translating this chapter into English.

CHAPTER 3

GENERAL CRIME SCENE CONSIDERATIONS AND DOCUMENTATION

Moira Johnson
Simone Reynolds

3.1 INTRODUCTION

Whether investigating crimes related to motor vehicles or crimes in general, crime scene investigators must perform their duties with the highest degree of skill and integrity. They must understand and competently apply an array of procedures and techniques related to detecting, assessing, recording, and collecting physical evidence located within a crime scene. For example, large visible items of physical evidence may simply be located within a scene using systematic searching procedures, whereas microscopic and latent evidence require certain physical and/or chemical enhancement techniques. Crime scene investigators must also possess relevant skills to ensure the scene is adequately recorded using handwritten notes, sketch plans, photography, and in some cases video [1–4]. The evidence they collect from a scene must be handled in a way that prevents loss and/or contamination and satisfies all legal requirements related to continuity and subsequent admissibility in a court of law. Furthermore, crime scene investigators, besides technical skills, must also possess particular personal attributes, including acute observation skills, high level of attention to detail, and an analytical ability to successfully examine, interpret, and present the physical evidence in a court of law.

3.2 NOTIFICATION AND INITIAL ASSESSMENT

Upon receiving a request for the services of forensic specialists at a crime scene, specific information is required to determine the resources needed to examine the scene. This information includes the type, size, and location of the incident, the time of occurrence, identification and location of victims and suspects, their condition, and what physical evidence is known to be present at the scene, such as blood, weapons, or vehicles. Crime scene personnel often use "what, where, when, who, why, and how" when obtaining this information. Additional information includes the name of the investigating police officer and the first officer at the scene. In some cases, it is necessary to clarify whether scene cordons have been established and a scene log recording the entry and exit of personnel has been initiated. Indeed, some crimes are multiple-scene incidents and require two or more scenes to be examined.

Incident types can range from a relatively minor burglary to a complex homicide involving multiple victims. The type of incident and the information available at the time of the initial notification generally determines the level of forensic response to a crime scene. For a minor incident, only one crime scene investigator may be required. However, for incidents identified as serious or where there are secondary or multiple scenes, there may be a need for two or more crime scene investigators and other forensic specialists. It may be necessary to allocate separate forensic teams to examine primary and secondary scenes to avoid cross-contamination between the scenes. A stolen-recovered motor vehicle is an incident type that may start with a relatively minor forensic response but that can escalate into a major forensic response if the vehicle is identified as a secondary scene or as part of a multiple-scene incident. For example, this may occur when a vehicle is used to flee the scene of an armed robbery, is involved in a homicide, or is implicated in a hit and run incident.

When determining the response requirements, the crime scene investigator should also consider the impact that the size and location of the scene has on the examination process. Because crime scenes start to deteriorate immediately after the incident has occurred, it is imperative to start the examination of the scene in the shortest time possible to avoid degradation or loss of physical evidence. Factors that may degrade physical evidence over time include weather and contamination due to poor scene security or interference by domestic animals or wildlife [5]. Large scenes and scenes located in busy urban or remote areas may require additional personnel and equipment to reduce examination times.

3.3 SCENE ATTENDANCE

In most cases, crime scene investigators are not the first respondents to attend an incident scene. Generally, uniform police or emergency services personnel, such as fire fighters and medical personnel, are the first people to arrive. The primary concern of emergency personnel is the preservation of life, property, and environment. Over recent years, a greater awareness of the importance of physical evidence has developed, and most emergency personnel do what they can to protect the scene and its associated evidence.

Responsibility for initiating scene preservation procedures falls on the first police officer(s) at the scene [2]. These procedures include setting up road closures, establishing cordons, and initiating scene logs. The first police officer(s) should also, whenever possible, gather information from witnesses, emergency personnel, and bystanders. This information is important for different reasons. First, it helps establish the outer perimeter of the scene, and second, it assists in determining the sequence of events and identifying victims and potential suspects.

Upon arrival at the scene, details including the date, time, location, names of forensic and other police personnel present, and the type of incident should be recorded by the crime scene investigator. In some cases, it may also be necessary to record environmental conditions such as weather, ambient temperature, wind direction, and initial odors present such as gasoline because these conditions may change before a detailed examination of the scene can be conducted.

The crime scene investigator should assess the scene perimeters and, if necessary, adjust the cordons to ensure all physical evidence is protected [6]. At major scenes it is beneficial to establish inner and outer perimeters. The crime or incident scene is included in the inner perimeter, which is restricted to essential personnel only such as forensic members and the investigating case officer. An outer perimeter is cordoned around the inner perimeter, which allows for a police command post to be established where briefings can be held and updates regarding the scene examination can be discussed. Also, this allows for effective scene control as bystanders and the media are kept at a reasonable distance from the scene outside the outer perimeter.

Liaison between the crime scene investigator, the first officer at the scene, and the investigating case officer is essential to confirm details already received during the initial notification and to obtain any new or updated information. Information regarding who has entered the scene and along which entry path and whether any items have been touched, moved, or removed from the scene must also be obtained. This includes identifying any actions undertaken by emergency personnel, such as forcing entry into premises or vehicles, breaking windows, or moving victims or deceased bodies in the course of their duties. The first police officer at the scene can also provide information regarding additional items within the scene, such as discarded medical items and extraneous physical evidence such as shoeprint or tire tracks created by emergency personnel and their vehicles.

Also, time has a great influence on the physical evidence. This is illustrated by the pattern in confusion described by Kind, as shown in Figure 3-1 [7]. The original shoeprints present at the scene on the top left diagram are the most important evidence. However, other shoeprints, tire tracks, animal prints, and other objects quickly obscure the original shoeprints as shown in the top right and bottom diagrams. These extra prints and objects could have either been left before the discovery of the scene or by the scene workers such as emergency personnel. Thus, it is essential for the crime scene investigator to interview first responders and to try to evaluate the pertinence of the different evidence found at the scene. The crime scene investigator should then conduct a preliminary examination of the scene to identify occupational health and safety hazards and to ensure that procedures such as wearing personal protective equipment (PPE) and clothing are followed to ensure a safe working environment.

In some cases it is necessary for the crime scene investigator to establish an entry and exit path for the remainder of the scene examination [2]. The entry and exit path area is extensively searched for any physical evidence and, when deemed to be clear, is marked and identified as the access path. A preliminary walk through the scene, preferably with the first officer at the scene, is required to plan for the scene examination. The planning stage of the scene examination includes identifying the examination techniques required

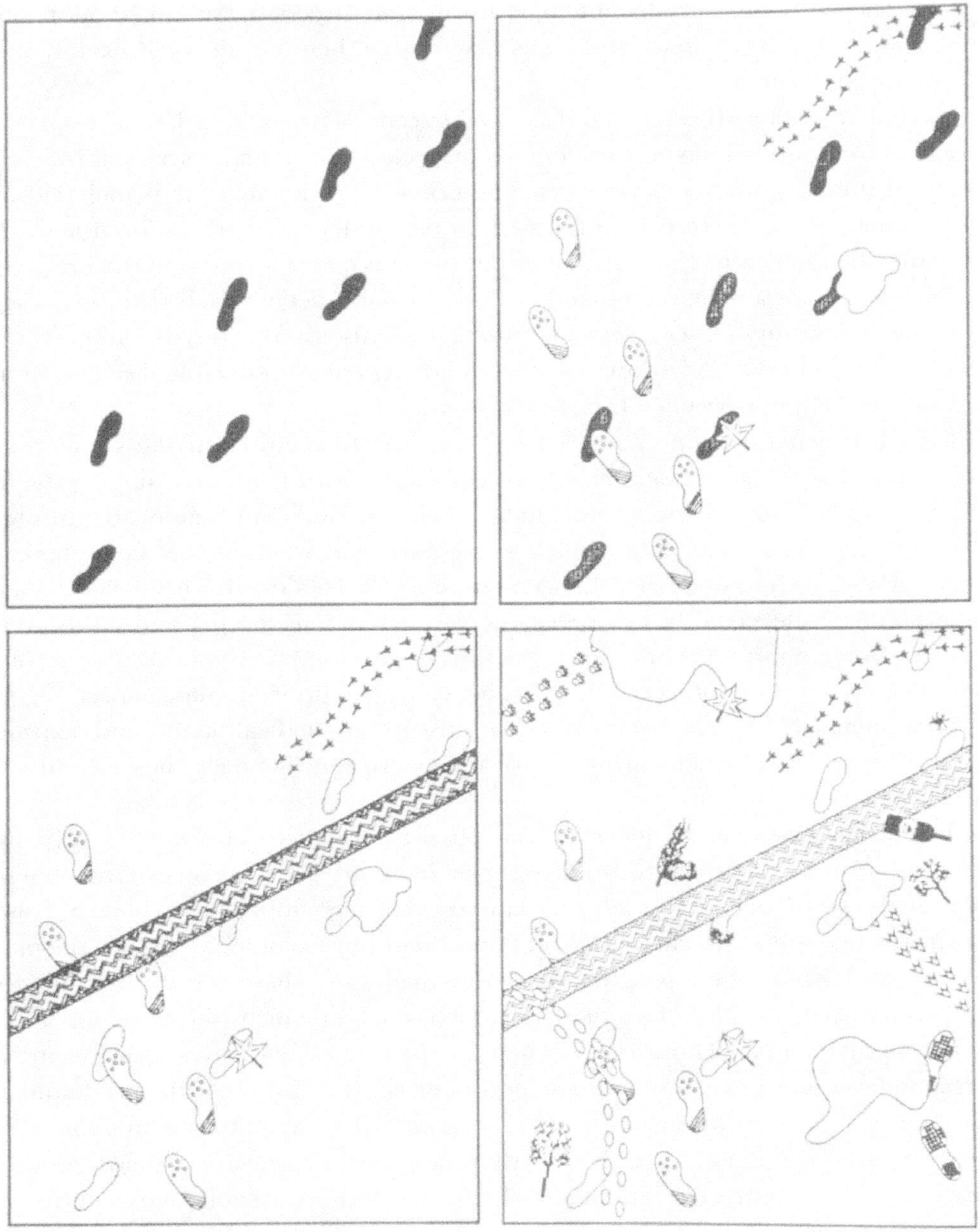

Figure 3-1

Illustration of how original physical evidence at a scene can soon be lost or obscured with extraneous material or marks. The evolution of the original shoeprints is shown from the diagram on top left to the one on bottom right. The crime scene officer should be able to identify relevant physical evidence by observing the condition of the scene upon arrival and asking pertinent questions during witness interviews. (Source: Kind SS. (1987) The scientific investigation of crime, Forensic Science Services, Harrogate, United Kingdom, p. 55. Reprinted with permission of Alan Douglas Kind.)

and the need for other forensic specialists such as fingerprint and firearm experts, other criminalists, and forensic pathologists. Involvement of external experts and urban utilities, such as mechanical, electrical, water, and gas personnel, may also be considered when necessary. The sequence of examinations, coordination of multiple forensic examinations, and prioritization of evidence collection are also determined.

3.4 SCENE EXAMINATION

3.4.1 Searching Techniques and Evidence Identification

Generally, scene examinations are conducted in a sequential order. A main objective of any examination is to locate and record physical evidence within the scene in situ. Further detailed examinations to identify, detect, and record latent or trace physical evidence are then conducted followed by the collection and preservation of physical evidence for further laboratory analysis or presentation in court.

At any crime scene, a methodical search of the scene is required to locate physical evidence. During the initial search, no items or physical evidence are moved before the scene has been completely recorded in situ. The scope of the search is determined by the initial information received and observations made by the crime scene investigator upon arrival at the scene. There are five common techniques that can be used to conduct the initial search of the scene: the line or strip, spiral, numbered grid, zone or quadrant, and wheel searching patterns (Figure 3-2) [2–6]. Selection of the appropriate search technique depends largely on the type, size, location, and complexity of the scene. There are a number of advantages and disadvantages associated with each technique, and it is the responsibility of the crime scene investigator to select a technique that guarantees thorough and systematic search.

The searching technique used at a scene involving a motor vehicle can incorporate a line search of the area surrounding the vehicle and a zone or quadrant search of the exterior and interior of the vehicle. The crime scene investigator should coordinate a line search of the ground leading up to and surrounding the vehicle for physical evidence such as shoeprints, dropped items around or under the vehicle, and extraneous tire tracks, possibly from a second vehicle leaving the scene. The exterior of the vehicle can be searched in zones such as front, back, sides, and undercarriage for evidence of paint transfer, scuff marks, blood stains, and damage to panels, headlights, rear lights, and tires. During the initial search of the vehicle interior, physical evidence clearly visible on the seats, floor wells, and luggage and engine compartments should be recorded in situ. The interior of the vehicle can then be separated into zones dividing the front and rear seats, floor wells, and luggage and engine compartments into individual areas to ensure that a systematic search of the vehicle is conducted (see also Chapters 4 and 16 for the different vehicle search procedures).

As physical evidence is located during the initial search, the crime scene investigator should consider identifying the locations with numbered markers. Markers are usually traffic cones or plastic triangles with a sequential set of numeric, or in some cases alpha,

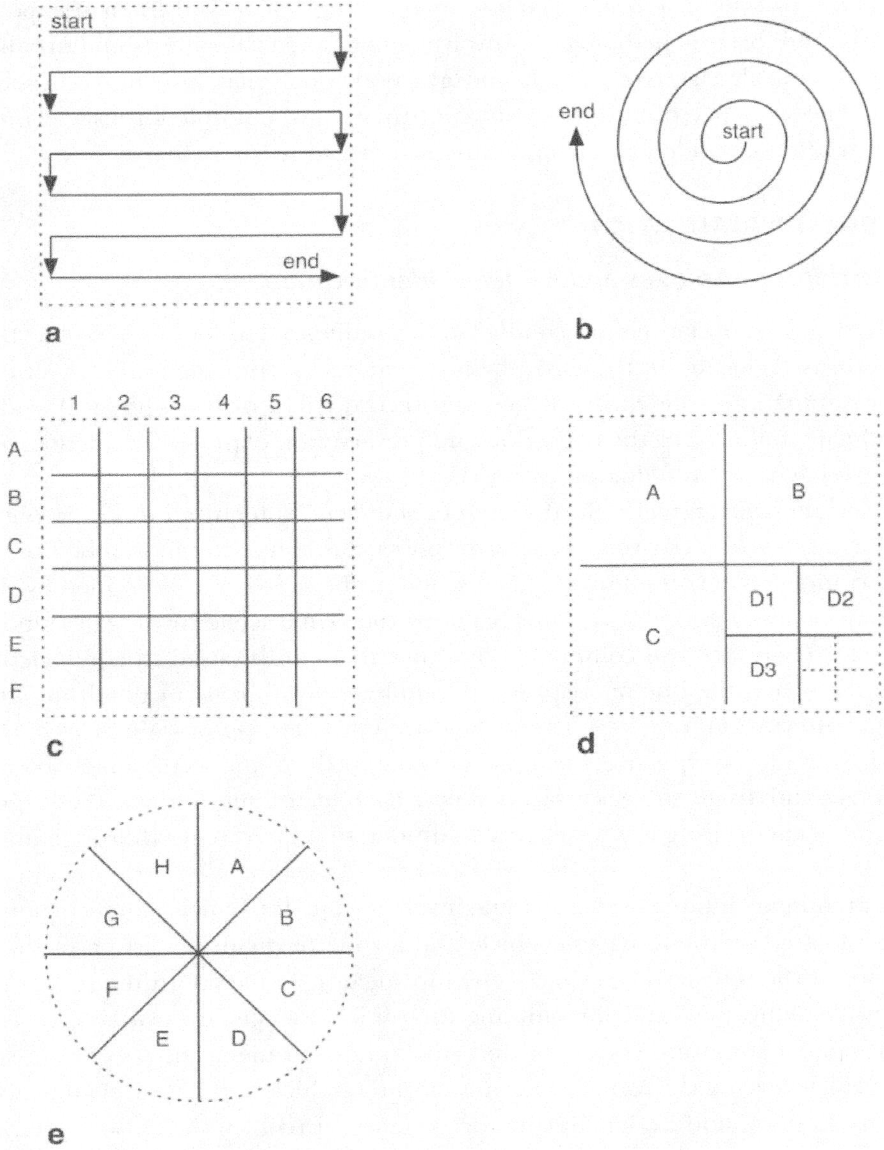

Figure 3-2

Different crime scene search pattern configurations. (a) Line or strip, (b) spiral, (c) numbered grid, (d) zone or quadrant, and (e) wheel.

characters printed on each side, as shown in Figure 3-3. It is good practice to match the number on the marker to the identification number of the evidence that it designates (see Subsection 3.4.6).

The markers are particularly useful for large scenes, because they assist in identifying the relative location of the physical evidence in the scene in overall and general photo-

Figure 3-3
Example of the use of numbered markers to identify physical evidence at a crime scene. A set of keys is shown as 1, a cell phone as 2, a shoeprint as 3, a candy wrapper on the floorboard as 4, and a pistol as 5.

graphs (also see examples in Figures 3-7 and 3-8). The number on the marker is included in the description of the evidence. This can assist in differentiating between multiple items of the same evidence type such as shoeprints, tire tracks, and fired cartridge cases. Numbered graph labels are also used to identify the location and position of latent evidence such as fingerprints or trace evidence such as fibers and hairs.

3.4.2 Note Recording

Written notes recorded at the time of examination are referred to as contemporaneous notes and may later be required for criminal or civil court proceedings. These notes should include a description of the scene and surrounding area, identifying it as rural, urban, or residential, and details of the location, description, and condition of physical evidence identified during the initial search of the scene [2].

When the scene involves a motor vehicle, the crime scene investigator should record at least the location and direction in which the vehicle is facing and the type, make, model, and color. Other details such as damage to door and luggage compartment locks, ignition and steering column, windows, and rubber seals and, in some instances, the temperature of the engine compartment must also be recorded. Accident damage to vehicle panels, headlights and rear lights, undercarriage, tires and wheels, and windows also needs to be noted. Examples of such notes are shown in Figure 3-4. The position of all doors, windows, engine and luggage compartment lids, gear shift lever, hand brake, and driver's seat should also be detailed in the written notes. The vehicle identification number (VIN) and the license plate number must be recorded. When the VIN plate has been removed or destroyed (as it is often the case with burned vehicles), it is necessary to record other identifying numbers such as engine number or secondary VINs (see Chapter 6).

3.4.3 Sketch Plan

A sketch plan of the scene is often used to supplement written notes because it allows for an overall visual perspective and shows the relationship between the items of evidence and their positions within the scene [3–6]. When drawing a sketch plan, it is important to include the location, date, time, and name of the person drawing the sketch. The sketch plan should reflect a northern aspect, which is recorded at the top of the page. The northern aspect should be identified with a standard northern arrow pointing toward the north relative to the diagram. If accurate measurements of the scene and the location of physical items within the scene are not recorded, the words "not to scale" should also be included on the sketch plan. An example of a sketch plan is shown in Figure 3-5.

There are some generally accepted symbols that can be used in the sketch plan to indicate permanent structures such as walls, doors, windows, kitchen sinks, and toilets [8]. Further, all moveable items, such as furniture and personal items that are recorded on the sketch plan, should be clearly identified at the time the sketch plan is drawn. In some cases, it may be useful to use different colored pens to identify physical evidence within the scene. In fire scenes or burnt out vehicles, it may be useful to indicate the area of origin with red hash marks and use red arrows to indicate fire direction. Subsequent fire and smoke damage can then be recorded in blue or green hash marks. A key or legend identifying the hash marks should also be included on the corner of the sketch plan if this system is used to ensure the plan is accurately interpreted at a later date. It may be useful to use generic vehicle diagram templates to record the location of damage, physical evidence, or areas where further forensic examinations were conducted on the vehicle, as shown in Figure 3-6.

In some cases, where the crime scene may be extensive or complex, it may be necessary for the crime scene investigator to obtain registered building plans, street maps, or maps detailing rural areas from the local council or a similar authority [5]. Where this occurs, the crime scene investigator should request several photocopies of the plan. This is beneficial because a single copy can be used as a working copy at the scene, whereas other copies are forwarded, with the sketch plan, to other forensic specialists to compile a computer-aided drawing plan, if required [2, 3].

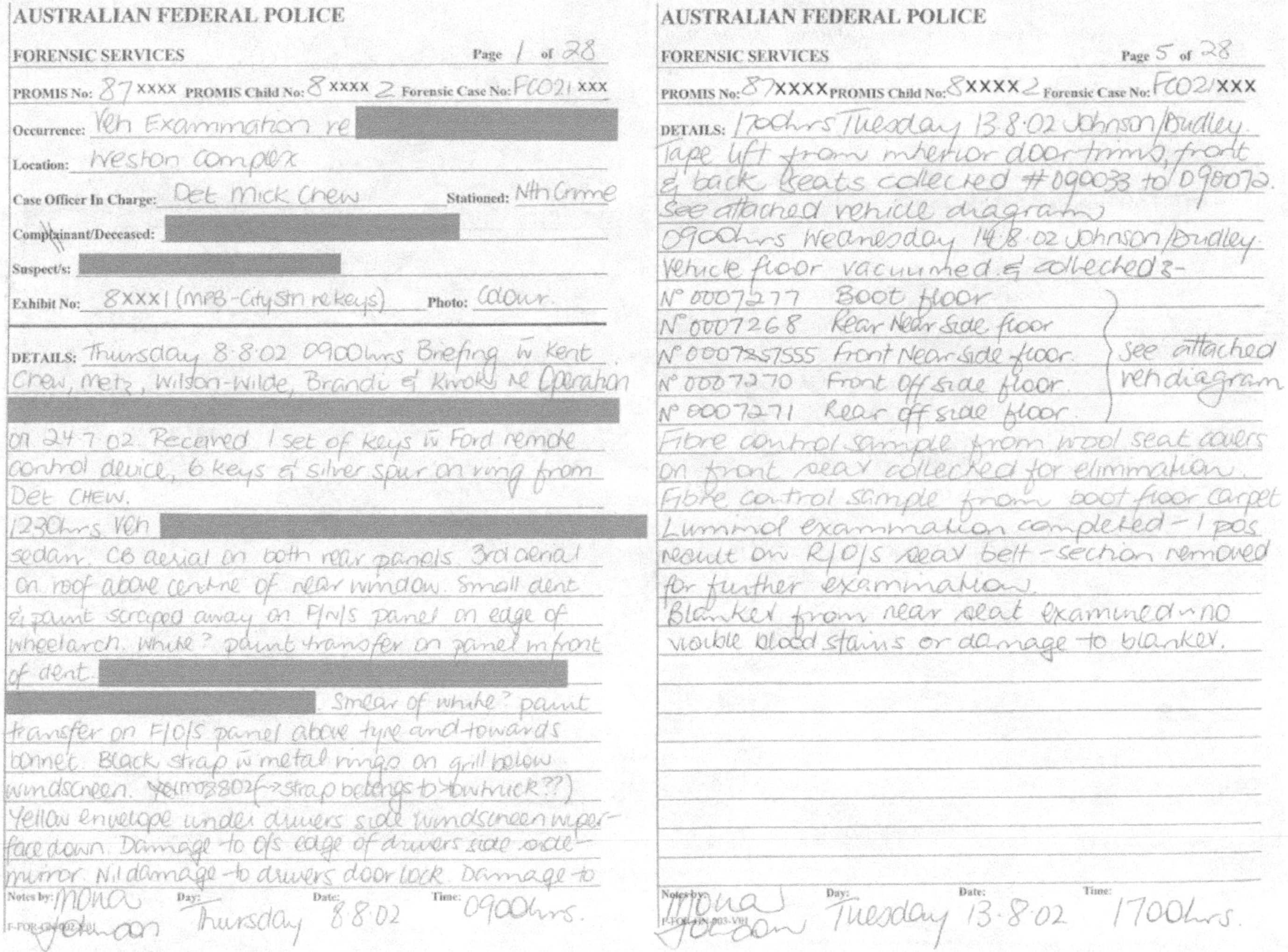

AUSTRALIAN FEDERAL POLICE

FORENSIC SERVICES — Page 1 of 28

PROMIS No: 87xxxx PROMIS Child No: 8xxxx2 Forensic Case No: FC021xxx

Occurrence: Veh Examination re [redacted]

Location: Weston Complex

Case Officer In Charge: Det Mick Chew Stationed: Nth Crime

Complainant/Deceased: [redacted]

Suspect/s: [redacted]

Exhibit No: 8xxx1 (MPB-City Stn re keys) Photo: Colour.

DETAILS: Thursday 8.8.02 0900hrs Briefing w Kent Chew, Metz, Wilson-Wilde, Brandi & Knoks re Operation [redacted] on 24.7.02. Received 1 set of keys w Ford remote control device, 6 keys & silver spur on ring from Det CHEW.
1230hrs Veh [redacted] sedan. CB aerial on both rear panels. 3rd aerial on roof above centre of rear window. Small dent & paint scraped away on F/N/S panel on edge of wheelarch. White? paint transfer on panel in front of dent. [redacted] Smear of white? paint transfer on F/O/S panel above tyre and towards bonnet. Black strap w metal rings on grill below windscreen. ~~Yelm08802~~ (→strap belongs to towtruck??) Yellow envelope under drivers side windscreen wiper - face down. Damage to O/S edge of drivers side side mirror. Nil damage to drivers door lock. Damage to

Notes by: Mona Johnson Day: Thursday Date: 8.8.02 Time: 0900hrs.

F-FOR-...

AUSTRALIAN FEDERAL POLICE

FORENSIC SERVICES — Page 5 of 28

PROMIS No: 87XXXX PROMIS Child No: 8XXXX2 Forensic Case No: FC021XXX

DETAILS: 1700hrs Tuesday 13.8.02 Johnson/Dudley. Tape lift from interior door trims, front & back seats collected #090033 to 090072. See attached vehicle diagram.
0900hrs Wednesday 14.8.02 Johnson/Dudley. Vehicle floor vacuumed & collected:-
N° 0007277 Boot floor
N° 0007268 Rear Near side floor
N° 0007~~2~~57555 Front Near side floor
N° 0007270 Front Off side floor
N° 0007271 Rear off side floor
} see attached veh diagram
Fibre control sample from wool seat covers on front seat collected for elimination.
Fibre control sample from boot floor carpet.
Luminol examination completed - 1 pos result on R/O/S seat belt - section removed for further examination.
Blanket from rear seat examined - no visible blood stains or damage to blanket.

Notes by: Mona Johnson Day: Tuesday Date: 13.8.02 Time: 1700hrs.

F-FOR-...003-VIII

a b

Figure 3-4

(a) Example of written notes recorded during a vehicle examination. (b) Notes completed in relation to further examinations conducted during a vehicle examination.

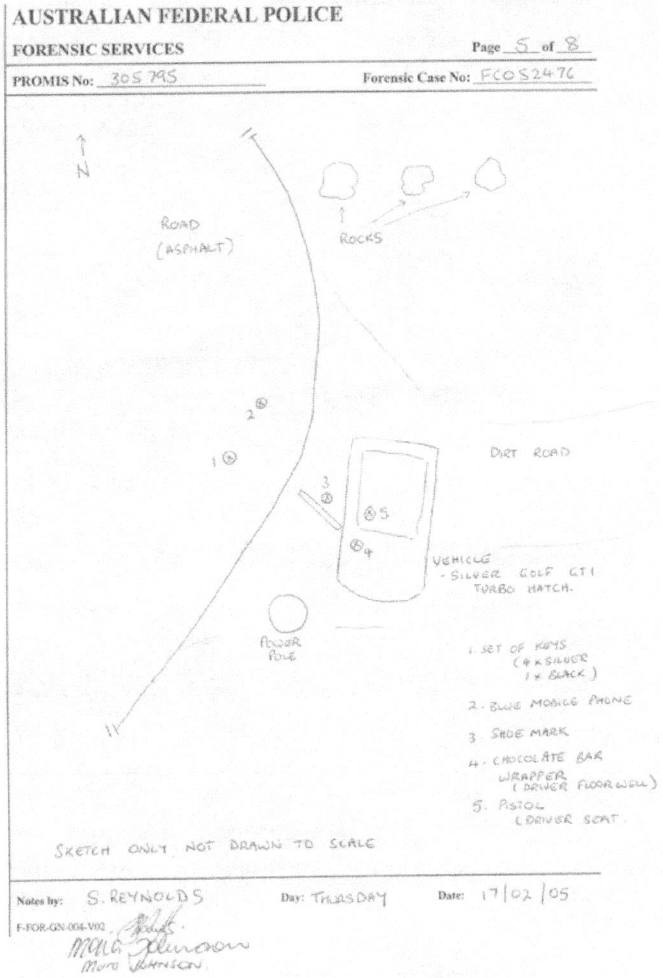

Figure 3-5

Example of a scene sketch plan involving a vehicle.

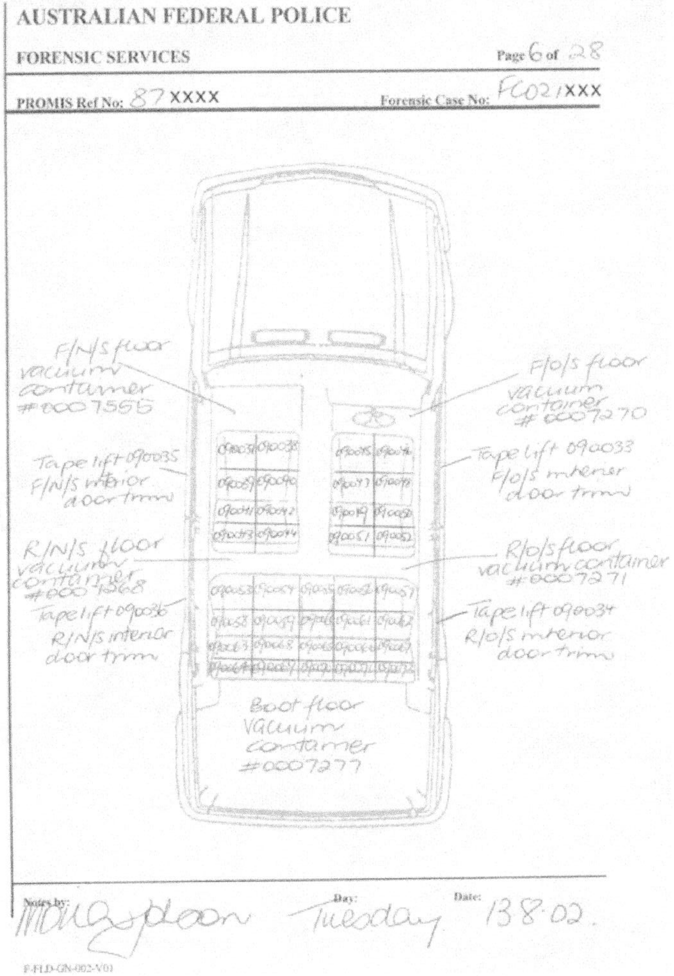

Figure 3-6

Vehicle diagram recording where further examinations such as vacuuming and tape lifts for trace evidence were conducted.

3.4.4 Photography

A systematic approach to photographing the scene and all the physical evidence in situ should be adopted once the initial search has been conducted, physical evidence identified, and initial notes and sketch plans recorded. In some cases, the photographs, or at least preliminary photographs, will be taken as the crime scene investigators penetrate the scene for the first time. This is particularly important if the survey of the scene requires the moving of some objects. This systematic approach to photographing the scene includes overall, mid-range, and close-up (macro) photographic techniques [2–4, 6].

Overall photographs should depict the area beyond the inner perimeter of the crime scene. If the scene is so large that one photograph will not suffice, panoramic photographs should be taken to allow the scene to be viewed in a continuous horizontal plane. With the current advances in digital photography, panoramic photographs can be taken using a specialized tripod that rotates 360 degrees and records the scene in one scan [9, 10]. The resulting image is stitched along one seam and allows for a navigational virtual reality view of the scene on a personal computer or laptop computer. The advantage of a panoramic view is the overall visualization it provides of the scene. However, a disadvantage of this technology is the distortion that occurs which does not allow for precise measurements to be taken directly from the photograph as with 1:1 photography using a macro lens.

When determining the position from which to take the overall scene photographs, the crime scene investigator should consider the location and positioning of the physical evidence and the numbered markers within the scene. This allows for consistency in perspective across all close-up, mid-range, and overall photographs. The numbers printed on the numbered markers placed next to the physical evidence items during the initial search should be visible in the overall, mid-range, and close-up photographs. Aerial photographs and photographs recorded from an elevated height may also assist in recording an overall perspective of the scene. Different views of the same scene are shown in Figures 3-7 and 3-8. Note how the perspective of the same scene changes depending on the view. In Figure 3-8, it is very easy to evaluate the positions of the physical evidence relatively to each other, which aids the viewer in understanding the dynamic of the crime scene.

In some cases, photographic techniques such as time exposures, painting with flash, oblique flash techniques, and scale photography may be required to effectively record the scene [11, 12]. Time exposure photography and painting with flash, as shown in Figure 3-9, are techniques that can be used at nighttime scene examinations such as fatal or hit and run motor vehicle accidents. Depending on available lighting conditions, painting with flash can provide extra lighting in darkened areas of the scene when recording overall scene photographs. Once the time exposure setting has been triggered, the crime scene investigator manually activates the flash unit at predetermined positions around the perimeter of the scene. This technique allows vehicles, debris, and tire tracks to be visible in long distance photographs, which would otherwise not be visible in normal flash photography.

Figure 3-7

Scene of a hit and run motor vehicle accident where numeric and alpha markers were used to identify the location of physical evidence such as scuff marks, tire tracks, and clothing items. (Photograph courtesy of AFP Forensic Services.) See Color Plate.

Figure 3-8

Aerial view of the scene depicted in Figure 3-7. Yellow markers and white outline indicate scuff marks caused by the movement of the victim on the ground. Black and white markers indicate tire tracks created by the backward and then forward movements of the vehicle. Markers 1, 14, and 16 identify dental stone of tire tracks. (Photograph courtesy of AFP Forensic Services.) See Color Plate.

a

b

c

Figure 3-9

(a) Example of nighttime photography performed without a flash by increasing the exposure time. Note how the streetlight is the only source of light and exposes the left part of the vehicle properly; however all the details on the right part are too dark to be properly recorded.

(b) Same photograph taken with the manual flash on camera. Note how it is possible to see slightly more details on the right side of the vehicle; however, it is now not possible to distinguish the presence of houses behind the fence.

(c) Same photograph using the painting with light technique, which results in a highlight of the dark areas. Note how it is possible to not only distinguish the background with the presence of houses, but also all the details on the right side of the vehicle with the suppression of shadows by painting the photograph with light from different directions.

Oblique flash techniques help prevent "flash back" or "hot spots" when photographing shiny or reflective surfaces such as vehicle panels and windows. The flash unit is removed from the bracket attached to the base of the camera but is still connected via the synchronization cord, as shown in Figure 3-10. When the flash unit is held at an oblique angle to the surface being photographed, the flash reflects off the surface at an oblique angle instead of reflecting directly back into the camera lens.

Scale photography techniques involve using a regular or macro lens and placing a numbered graph label, ruler, or tape measure beside the physical evidence such as tire tracks, as shown in Figure 3-11. This ensures that 1:1 photographs can be printed to scale at a

Figure 3-10

Demonstration of holding the flash unit at an oblique angle to avoid hot spots on reflective surfaces. In this instance, the crime scene officer is recording fingerprints found on the door handle of a stolen-recovered vehicle.

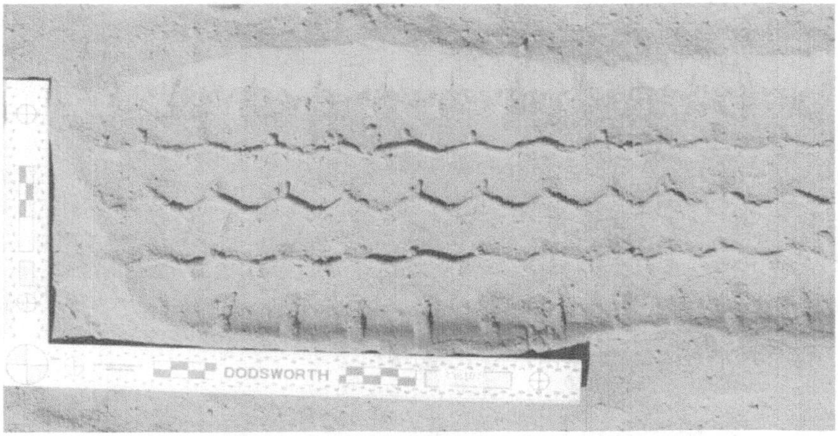

Figure 3-11

Photograph of a tire track with scale recorded at a scene of a burglary.

later time for comparative analysis. When taking these photographs, the scale must be completely flat and on the same plane as the object and the camera lens must also be parallel to the object. If the camera lens is not parallel to the object, the scale in the photograph will be distorted, making it very difficult to produce a 1:1 photograph. Scale photography is used when recording a wide range of physical evidence, including shoeprints, tire tracks, toolmarks, bite marks, fingerprints, assault injuries, and blood spatter patterns.

3.4.5 Further Scene Examinations

Once the scene and all the physical evidence has been recorded in situ, a more detailed search of the scene can be conducted. This search involves moving items within the scene, which can reveal the location of evidence previously concealed or hidden from clear view. Notes and sketch plans must be updated and additional photographs taken of the evidence in situ. Latent evidence, such as fingerprints or blood stains, may also be identified as a result of the application of physical and chemical enhancement techniques and must also be recorded in situ.

The crime scene investigator must describe the evidence items collected during the scene examination. Identifying features such as brand, color, size, and condition of the item should be recorded. Other details can include vehicle identification numbers, width of tire marks, length of shoeprints and toolmarks, and so forth. During this examination, it may be necessary to collect trace evidence such as hair, fiber, and paint transfers from the item such as clothing or weapons before packaging to prevent its loss or destruction during transport to the laboratory [2].

3.4.6 Evidence Collection

At this stage of the examination, exhibits are collected from the scene. This may involve collecting the whole item or taking swabs or scrapings from the item, such as blood stains or other residues for further laboratory examinations or analyses. The crime scene investigator should always ensure they are wearing disposable gloves when handling physical evidence items to prevent fingerprint and trace DNA contamination and to protect against infectious diseases. These gloves should be changed regularly during the collection process (i.e., between items), particularly when collecting or handling heavily soiled items and when collection instruments such as tweezers are not used. Nevertheless, coverall suits, such as Tyvek suits, may be used to prevent contamination and to protect the crime scene investigator.

Appropriate packaging, such as containers for cutting instruments or syringes, must be selected to ensure the protection of the physical evidence and to avoid injury to any personnel. Packaging such as paper bags as opposed to plastic bags for items bearing biological stains, such as blood stains, must be used to prevent degradation of the evidence due to excessive moisture leading to bacterial growth [2–4].

To commence a record of continuity related to the movement of the physical evidence from the scene to the laboratory, certain details regarding the item should be recorded on

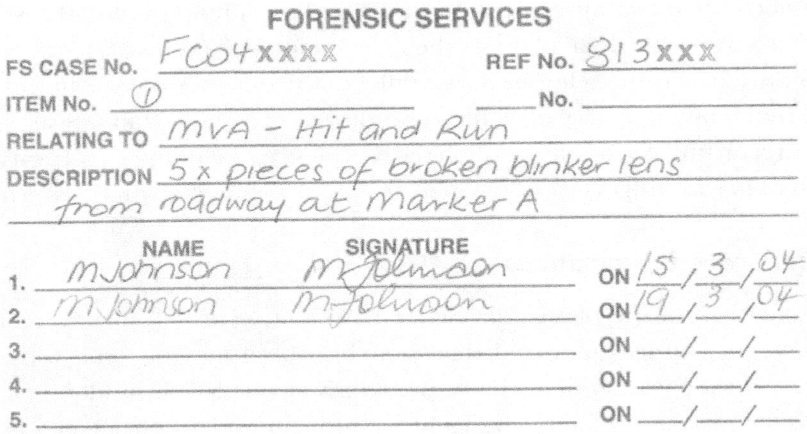

FORENSIC SERVICES

FS CASE No. FC04XXXX REF No. 813XXX

ITEM No. ① ______ No.

RELATING TO MVA - Hit and Run

DESCRIPTION 5 x pieces of broken blinker lens from roadway at Marker A

	NAME	SIGNATURE	
1.	M Johnson	M Johnson	ON 15/3/04
2.	M Johnson	M Johnson	ON 19/3/04
3.			ON __/__/__
4.			ON __/__/__
5.			ON __/__/__

Figure 3-12

Example of continuity label with collection details affixed to the packaging of each exhibit collected from a crime scene.

a packaging label, as shown in Figure 3-12. These details include the case reference, the item number, a description of the item, the location where the item was found, the time and date of collection, and the name and signature of the person collecting the item. After collection, the item is placed in an appropriate package and is sealed with tape along all opening edges to ensure the package is fully enclosed. Tamper-proof tape is then placed across the sealing tape and signed as intact, as shown in Figure 3-13.

During collection, each item is assigned a unique designator generally according to the sequence in which the items are collected. This designator may simply be numeric (i.e., 1, 2, 3, 4) for a single isolated scene or may be alpha-numeric if there are multiple scenes relating to the same incident that need to be examined simultaneously. For example, items collected from a hit and run scene may be assigned a number prefixed with the letter A (i.e., A1, A2, A3). Items collected during the examination of the vehicle involved in the incident, found some distance from the scene, may be assigned a number prefixed with the letter B. In circumstances where new evidence is generated from existing evidence, such as the collection of hairs, fibers, or swabs from an article of clothing, the typical numbering method should be as follows [13]:

Item 1 trousers
- Item 1.1 debris on front of trousers
  - Item 1.1.1 fibers from debris
- Item 1.2 debris on back of trousers
  - Item 1.2.1 fibers from debris
  - Item 1.2.2 cotton swab from debris
  - Item 1.2.3 glass particles from debris

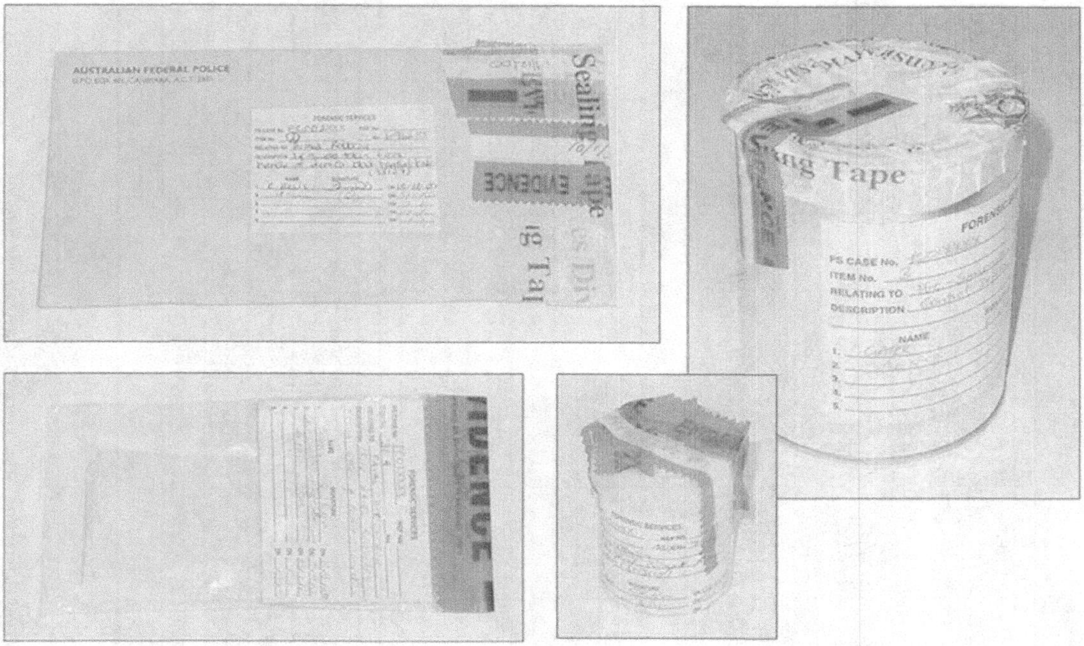

Figure 3-13
Sealed exhibits showing tamper-proof tape across the seal and the continuity label affixed to the packaging.

Upon returning to the laboratory, details of all items collected from the scene must be recorded in an official item register or exhibit book, as shown in Figure 3-14, along with the name, badge or personnel number, and signature of the crime scene investigator who collected the items. The items are then stored in a secure exhibit storage area while awaiting collection by laboratory staff for further examination and analysis. When the item is collected from the exhibit storage area by laboratory staff, details including their name, badge or personnel number, signature, and the name of their division or work area (e.g., fingerprints, firearms and toolmarks, serology) are recorded in the item register or exhibit book along with the date. The laboratory staff must also sign and date the continuity label on the item packaging. When the package is opened for further examination and analysis, the tamper-proof tape that was placed across the seal at the time of collection tears in multiple places and shows that the package has been opened. When the package is resealed, another signed section of the tamper-proof tape is placed across the seal and the continuity label is updated.

The details recorded on the package labels and in the item register or exhibit book establish a chain of custody and provide an audit trail that verifies the continuity of the physical evidence from the time of collection at the scene to its production in court or other judicial hearing. It should be emphasized here that a complete chain of custody

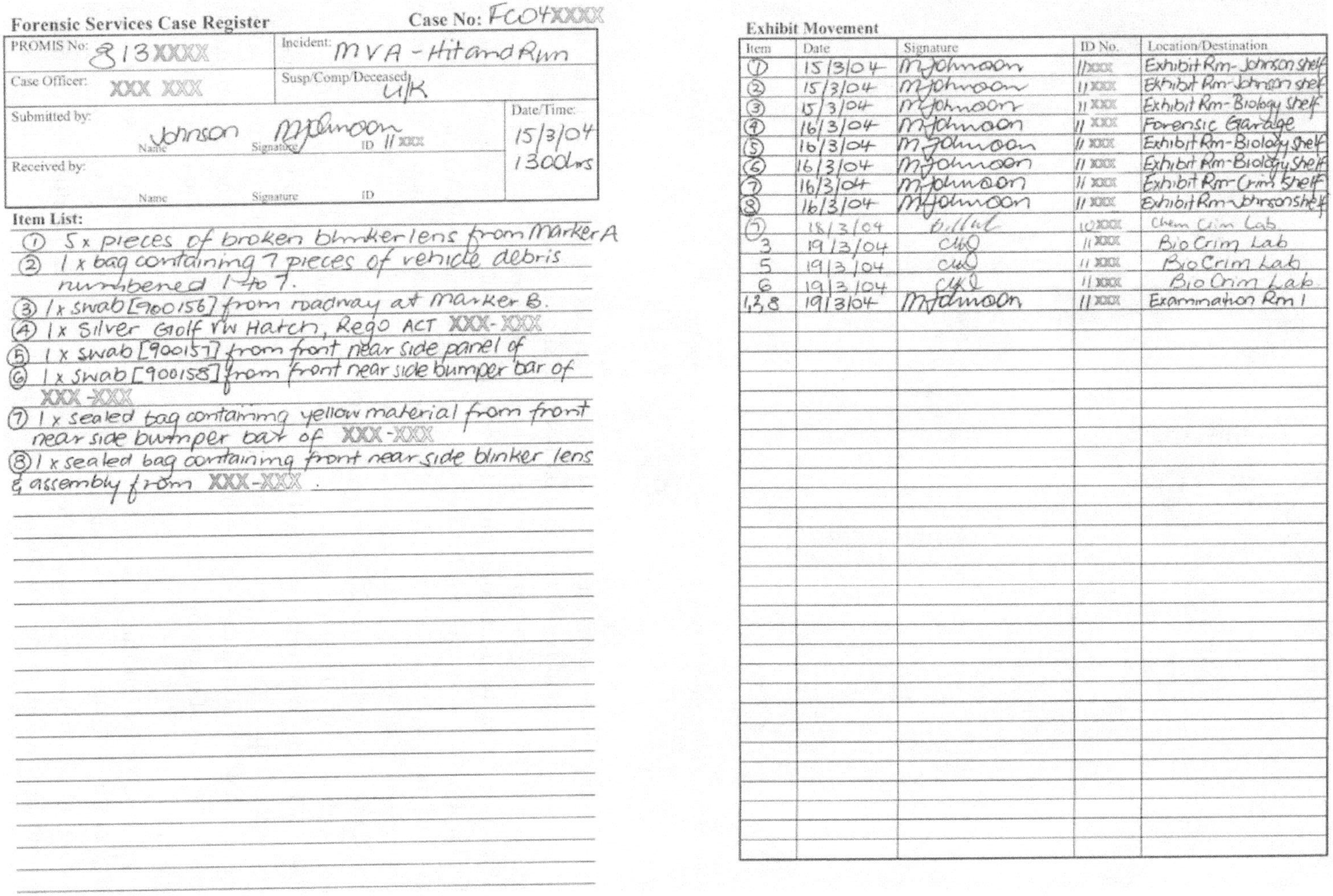

Forensic Services Case Register **Case No:** FC04XXXX

PROMIS No: 813XXXX | Incident: MVA - Hit and Run

Case Officer: XXX XXX | Susp/Comp/Deceased: U/K

Submitted by: Johnson (Name) MJohnson (Signature) ID 11XXX | Date/Time: 15/3/04 1300hrs

Received by: Name Signature ID

Item List:

① 5 x pieces of broken blinker lens from Marker A
② 1 x bag containing 7 pieces of vehicle debris numbered 1 to 7.
③ 1 x swab [900156] from roadway at Marker B.
④ 1 x Silver Golf VW Hatch, Rego ACT XXX-XXX
⑤ 1 x swab [900157] from front near side panel of
⑥ 1 x swab [900158] from front near side bumper bar of XXX-XXX
⑦ 1 x sealed bag containing yellow material from front near side bumper bar of XXX-XXX
⑧ 1 x sealed bag containing front near side blinker lens & assembly from XXX-XXX.

Exhibit Movement

Item	Date	Signature	ID No.	Location/Destination
①	15/3/04	MJohnson	11XXX	Exhibit Rm - Johnson shelf
②	15/3/04	MJohnson	11XXX	Exhibit Rm - Johnson shelf
③	15/3/04	MJohnson	11XXX	Exhibit Rm - Biology shelf
④	16/3/04	MJohnson	11XXX	Forensic Garage
⑤	16/3/04	MJohnson	11XXX	Exhibit Rm - Biology shelf
⑥	16/3/04	MJohnson	11XXX	Exhibit Rm - Biology shelf
⑦	16/3/04	MJohnson	11XXX	Exhibit Rm - Crim shelf
⑧	16/3/04	MJohnson	11XXX	Exhibit Rm - Johnson shelf
⑦	18/3/04	[illegible]	10XXX	Chem Crim Lab
3	19/3/04	CKO	11XXX	Bio Crim Lab
5	19/3/04	CKO	11XXX	Bio Crim Lab
6	19/3/04	CKO	11XXX	Bio Crim Lab
1,3,8	19/3/04	MJohnson	11XXX	Examination Rm 1

Figure 3-14

Example of exhibit details recorded in an official item register showing personnel details, exhibit description, and movement within the laboratory.

related to all evidence collected from a crime scene must be demonstrated through documentation to ensure the evidence is legally admissible in court. It is regarded by the court as a break in the chain of custody if any of the continuity details relating the evidence items are brought into question, for example, names, signatures, or dates that are missing from the label or item register, or if there is any indication that the items may have been handled by a person who does not have an official capacity in the examination process. This break in the chain of custody generally renders the evidence inadmissible in a legal proceeding.

3.4.7 Case Management

A case file should be initiated, and the electronic case management systems should be updated with all scene details. The case file should be identified with the forensic case number that allows for a chronological filing and archival system. The crime scene investigator's name, date of examination, and electronic case management number should also be recorded. The case file must contain all written notes, sketch plans, diagrams, and other associated paperwork (council plans, exhibit logs, and receipts, etc.). A copy of any digital photographs taken of the scene and physical evidence should also be included in the case file. As part of quality assurance procedures, the case file is submitted for administrative and technical audits by other qualified personnel within the section.

3.5 REPORT PREPARATION AND COURT PRESENTATION

The manner in which the scene examination and collection of physical evidence were conducted is crucial for the preparation of a report and presentation of evidence in court. It is thorough and methodical scene examination and exhibit-handling procedures that allow for the production of a detailed report and the presentation of credible evidence in court.

The report should contain a record of the crime scene investigators' qualifications, skills, and experiences and a declaration statement regarding the integrity of the author and the authenticity of the information contained within the report. Generally, the main body of the report should include all the details of the examination in chronological order, commencing with the time the crime scene investigator arrives at the scene. The report should then detail the movements of the crime scene investigator in and around the scene, the examination techniques used, and the evidence observed, including a description, their location, and details of who collected them from the scene [14]. The report should be written in simple terms, avoiding police or scientific jargon, and should be free of intellectual elitism. It is also important that the spelling, grammar, and format are thoroughly checked by another experienced senior crime scene investigator. Simple spelling or transcription errors can have a detrimental effect on the overall professional image of any forensic practitioner, no matter how legally or scientifically sound their findings are. The details in the report must be truthful, and the author of the report should have no direct interest in the outcome of the trial. It is the forensic practitioners' role to assist the jury. Opinions regarding the reconstruction of events or physical evidence interpretation are usually recorded toward the end of the report.

The main body of the report may be supplemented with attachments or appendices, including evidence movement logs or a glossary of terms. The report may also include attachments such as diagrams, sketches, scale plans, or maps, which can provide an easy visual means of indicating the location of evidence within a crime scene.

Photographic and video evidence can assist the court in visually orientating themselves within a crime scene and see the scene as it was when examined by the crime scene investigator. It also allows the court to view the physical evidence in context. The use of charts to display photographic evidence relating to comparisons is particularly useful because it allows the court to examine and assess class and individual characteristics. A comparison chart also enables the crime scene investigator to draw attention to various features related to a comparison, for example, comparing similarities between a forged VIN stamped into a vehicle to a set of dies seized during a search warrant on a suspects residence (see Chapters 6, 7, and 14).

Other new technologies may be used, including digital presentations, 3D digital scans, and virtual reality reconstructions. In most cases, special permission to use such media must be sought from the magistrate or judge and are generally only permitted if both the prosecution and defense counsel give consent. Furthermore, many of the newer presentation tools rely heavily on digital technologies, and in most countries there is still concern related to the security and integrity of digital evidence, particularly digital images. Laws concerning the use of digital evidence in a criminal proceeding are still in their infancy, and although many software companies are developing programs to protect the security of digital evidence, it will be some time before the court accepts it with total confidence.

Opinion evidence is admissible in court when the witness demonstrates suitable qualifications, knowledge, and experience to the court that their opinion will satisfactorily aid the jury in the search for the truth. Witnesses who fulfill this criterion are often referred to as expert witnesses. The crime scene investigator, as an expert witness, must explain his or her observations and examinations to a jury with clarity and confidence. Ultimately, the objective of the scene examination, laboratory analysis, and interpretation of the physical evidence is the presentation of the findings in a court of law [14].

ACKNOWLEDGMENTS

Many thanks to the following people for their encouragement and assistance in writing this chapter: Detective Senior Sergeant Fran Poole, New South Wales Police, Forensic Services Group; Elizabeth Craft and Jonathan Stenson, Imaging Section, AFP Forensic Services; and Samantha Johnson.

BIBLIOGRAPHY

[1] Cobb P. (1998) Forensic science. In: *Crime scene to court: The essentials of forensic science,* ed White P, The Royal Society of Chemistry, Cambridge, United Kingdom.

[2] Saferstein R. (2001) *Criminalistics: An introduction to forensic science,* 7th edition, Prentice Hall, Upper Saddle River, NJ.

[3] Lee HC, Palmbach TM, and Miller MT. (2001) *Henry Lee's crime scene handbook,* Elsevier Academic Press, San Diego, CA.

[4] Fisher BAJ. (1993) *Techniques of crime scene investigation,* 5th edition, CRC Press, Boca Raton, FL.

[5] Geberth VJ. (1996) *Practical homicide investigation tactics, procedures and forensic techniques,* 3rd edition, CRC Press, Boca Raton, FL.

[6] DeForest PR, Gaensslen RE, and Lee HC. (1983) *Forensic science: An introduction to criminalistics,* McGraw-Hill, New York, NY.

[7] Kind SS. (1987) *The scientific investigation of crime,* Forensic Science Services, Harrogate, United Kingdom.

[8] Reekie RF. (1969) *Draughtsmanship,* 2nd edition, Edward Arnold, London, United Kingdom.

[9] Wide X-stream (2005) http://www.wxs360.com, last access performed on May 27, 2005.

[10] Panoscan (2005) http://www.panoscan.com, last access performed on May 27, 2005.

[11] Blizter HL and Jacobia J. (2002) *Forensic digital imaging and photography,* Academic Press, London, England.

[12] Redsicker DR. (1991) *The practical methodology of forensic photography,* Elsevier, New York, NY.

[13] ASTM E1459-92 (reapproved 1998) *Standard guide for physical evidence labeling and related documentation,* ASTM International, West Conshohochen, PA.

[14] Rothwell T. (1998) Presentation of expert forensic evidence. In: *Crime scene to court: The essentials of forensic science,* ed White P, The Royal Society of Chemistry, Cambridge, United Kingdom.

CHAPTER 4

FORENSIC EXAMINATION OF STOLEN-RECOVERED VEHICLES

Part I: Technical Examination and General Forensic Traces

Marc Demierre

4.1 INTRODUCTION

4.1.1 Challenges

When a vehicle reported stolen is discovered, the forensic investigator must answer four basic questions:

I Has the vehicle really been stolen?
II How was it stolen?
III Who stole the vehicle?
IV Has the vehicle been used to commit other crimes?

The technical examination of the vehicle and the search, collection, and examination of forensic traces are performed to answer these questions.

The first question concerns the possibility of insurance fraud. If it is determined that the vehicle was actually not stolen, the investigator is dealing with a case of insurance fraud. This situation occurs quite frequently, and thus it is extremely important to determine the genuineness of the theft. The first question is actually answered once the examinations required to answer the second question have been performed. Additionally, the identification of the modus operandi used to steal the vehicle can help investigators to map crimes and to draw a list of potential suspects (see Chapter 21). To answer the first two questions, the examination first focuses on the determination of how the perpetrators entered the vehicle. Second, the modus operandi used to start and operate the vehicle is identified.

The third question concerns the identification of the perpetrator(s) and is a stepping stone in the investigation. This is performed by searching for and collecting traces left by the criminal(s) inside and outside the vehicle. Nevertheless, comparison traces must also be preserved from the vehicle. This is performed in case suspects are later arrested and traces from the vehicle are discovered on them.

Finally, the fourth question must be appropriately answered. As presented in Chapter 1, a vehicle can be stolen not only for its value (resale, export, or parts) but also to commit other criminal activities such as armed robberies, drive-by shootings, and burglaries, which could have much more serious consequences. Thus, the crime scene investigation of a recovered vehicle must not solely focus on the auto theft aspect of the crime but must also take into account all other possible crimes that could have been committed by the thief and/or his or her accomplices. It is extremely important for the crime scene officer to proceed to the detection and collection of any possible evidence that would demonstrate such crimes and possibly elucidate them. The technical examination and the search for forensic traces on stolen-recovered vehicles becomes of the greatest importance in an auto theft investigation and should be carefully conducted.

4.1.2 Crime Scene Considerations

As explained in Chapter 3, it is important to consider the recovered vehicle as any other crime scene. Thus, its protection and preservation must follow exactly the same procedures used with any other type of crime scene where trace evidence might be present. Additionally, the place where the motor vehicle has been found must be included in the crime scene examination. Criminals who exited the vehicle could have left traces on their fleeing path. This means that when the crime scene is secured and safe to enter, the motor vehicle is only accessed after a thorough examination of the surrounding area or after an access path has been delimited.

Wind, sun, rain, snow, and temperature can play key roles in the destruction of the evidence. The atmospheric conditions could urge an access path to speed up forensic examination of the vehicle and preservation of evidence. Also, it is important to specify that although the crime scene investigator often examines passenger vehicles, it is very likely that other types of transportation vehicles, such as commercial trucks and motorcycles, may be examined. The examination of these vehicles should follow the principles used with passenger cars and adapt them accordingly. Thus, this chapter refers to all types of vehicles as passenger cars.

4.1.3 Examination Facility

A main advantage offered by vehicles compared to "fixed" crime scenes is that they can be easily transported to a secure location. Therefore, it can be considered performing the forensic examination in a facility, which guarantees a proper preservation of the items of evidence and offers a better work place for the examiners. The transportation of the vehicle to the facility could engender the loss of some trace evidence and the contamination of some others. These losses and contaminations can be prevented with careful manipulations and preliminary examinations. For example, the tow truck driver usually penetrates the vehicle on the driver's seat to unlock the steering wheel, disengage the gear, and possibly

Figure 4-1
View of an inspection facility dedicated to the forensic examination of a vehicle. Although this facility is designed to accommodate two vehicles, only one vehicle is placed inside at a time to provide enough working space.

release the hand brake. This operation must be performed by the crime scene technician rather than of the tow truck driver. Also, one should touch the door handles with great care, because of the risk of losing pertinent fingerprints. Traces that would possibly be present on the external part of the vehicle must be detected and collected before moving the vehicle. If some traces are deemed safe for transportation, then their collection can be postponed until the vehicle is in the facility.

The facility used to examine vehicles must be large enough to comfortably work on one vehicle (Figure 4-1). A good rule of thumb is to have a facility designed to accommodate two vehicles but to have only one at a time in it. The facility must be outfitted with the minimum equipment for searching and detecting items of evidence. It should also be possible to completely darken the room to facilitate the use of luminol and other trace evidence search. Also, such places could offer a built-in superglue fuming chamber into which the whole vehicle can be placed (see Figures 4-18c and d).

4.1.4 Forensic Techniques

The techniques that can be used for the detection, recognition, and collection of evidence on and inside a vehicle include all the forensic techniques available to the crime scene officer. Sections 4.3 through 4.6 and 4.10 through 4.11 describe the detection and collection of the most common items of evidence encountered during the processing of a stolen-recovered vehicle. These items include but are not limited to:

- Fingerprints and other ridge skin impressions;
- Biological fluids and DNA;
- Microtraces (fibers, glass, paint, and soil);
- Toolmarks;
- Drugs;
- Gunshot residue;
- Explosives.

The detection of fingerprints and other ridge skin impressions, biological fluids, microtraces, and toolmarks are routinely performed on recovered vehicles. However, the detection and collection of drugs, gunshot residues, and explosives require particular techniques that apply only in a specific set of circumstances and are explained in Part II of this chapter.

4.2 DETERMINATION OF THE MODUS OPERANDI

4.2.1 Penetration of the Vehicle

There are several means of penetrating a vehicle. Of course, if the vehicle is not locked, a door or the trunk will simply be open. In some instances, convertible vehicles are left with the roof retracted, and thus no barriers of protection against entry are available. But most vehicles are locked, and the perpetrator must enter in some fashion.

A/ Window Breaking

The simplest manner consists of breaking a window, as shown in Figure 4-2. This technique is usually loud and apparent and so not really convenient to the auto thief. If someone is driving down the road with a broken window, it creates an increased risk of being caught by a police patrol. Consequently, this modus operandi is almost exclusively used by criminals who burglarize vehicles for their content rather than to steal them. In the rare instance

Figure 4-2

Window breaking technique to penetrate a vehicle. Notice how the window breaks into small pebble-like pieces, typical of tempered glass.

when a window is broken to steal the vehicle, the thief typically does not break the driver's window, because glass would fall on the seat, bringing discomfort and possible injuries during the operation of the vehicle; he or she usually chooses the window that is the least seen by passersby, the least obvious, and the most convenient (typically the small side glass on or near the rear doors).

Vehicle windows are made of safety glass, with the side and rear windows made of tempered glass and the windshield made of laminated glass. It is impossible to enter a vehicle by breaking the windshield, unless the whole unit is removed, because it is designed to hold together when broken. Thus, criminals typically do not tamper with the windshield. Conversely, tempered glass is designed to completely break into small pebble-like pieces. Therefore, the breakage of side or rear windows creates a large opening, with no large pieces left in the frame. Unfortunately, contrary to regular flat glass, it is not possible to determine from which side a piece of tempered glass has been broken. Additionally, when tempered glass breaks, it does so with great violence, which disperses glass fragments both inside and outside the vehicle in proportions that are most often independent of the side from which the impact was administered. Manufacturers have started to equip some vehicles with enhanced protective glass (EPG) or high security glass (HSG) [1, 2]. EPG and HSG are safety laminated glass, similar to the windshield but placed on side windows for different safety reasons. One benefit of such glass is its effect on auto theft deterrence.

The crime scene investigator must carefully note which window(s) is(are) broken. In addition, the investigator must attentively examine the frame into which the broken window was fitted. It is often possible to estimate the height of the window prior to its breaking by looking at the remaining small fragments stuck inside the frame, as shown in Figure 4-3.

Figure 4-3

Often, small pieces of glass remain stuck inside the window frame. They can be collected as comparison glass and can help the investigator to estimate the height of the window when it was broken. In this particular example, note the glass still present at the top of the frame and along the rear part of the frame. The window was rolled up at the time it was broken.

This can bring important information as to whether the window was rolled all the way up or not. For example, if it is determined that the window was rolled down, offering an opening sufficient enough to place one's arm through, there would be no reason for a thief to break it.

B/ Doorframe Bending

Another technique used to enter a locked vehicle is bending the top part of the doorframe away from the roof. The bending can be done by simply pulling on the door (some vehicles offer very little resistance) or by using a tool such as a lever or an air wedge. The opening created can be very small, just enough to place a tool that will move the locking button. Such tools are either metal wires or metal or plastic rods, as shown in Figure 4-4. In some instances, a simple coat hanger might do the trick. With older cars, the lock button is located near the rear part of the door next to the window and sticks out like a golf tee. In this case, a simple loop can grab the button and pull it. More modern vehicles have this button completely inside the door panel when locked, thus preventing this operation. It is also possible to create a much larger opening, which allows for the arm to reach and release the locking mechanism. An example of a bent doorframe is shown in Figure 4-5.

An alternative technique to bending the doorframe consists of trying to force down the door glass using a flathead screwdriver, or a similar tool, as a lever. In older vehicles, the window-rolling mechanism might not be very tight, and a small play might suffice to create an opening large enough to lodge a wire through. In such instances, it is possible that toolmarks will be left on the window frame where the tool was applied to force down the window. With some other vehicles, it is also possible to go through the vent window.

Figure 4-4

Typical tools used to open door by moving or pulling the lock button. The screwdriver is used to provide enough leverage to move the wire tool, which is used to reach the lock button.

Figure 4-5

View of the top portion of a doorframe that has been bent away from the roof to access the vehicle. (Photograph courtesy of Robert Mangine, North American Technical & Forensic Services.)

C/ Maneuvering the Lock Linkage

When the lock cylinder rotates, it acts on different rods, part of the lock linkage, which eventually activates the door latch. It is possible with many cars to directly push/pull or move the lock linkage to release the latch, thus bypassing the lock cylinder. This is the method typically used by professional automobile mechanics or locksmiths to open a vehicle without the keys. A lockout tool, typically called a Slim Jim, a multi-car opening tool (MCOT), or a wire tool, is inserted inside the door through the weather stripping between the window and the doorframe, as shown in Figures 4-6 and 4-7. The tool then grabs a

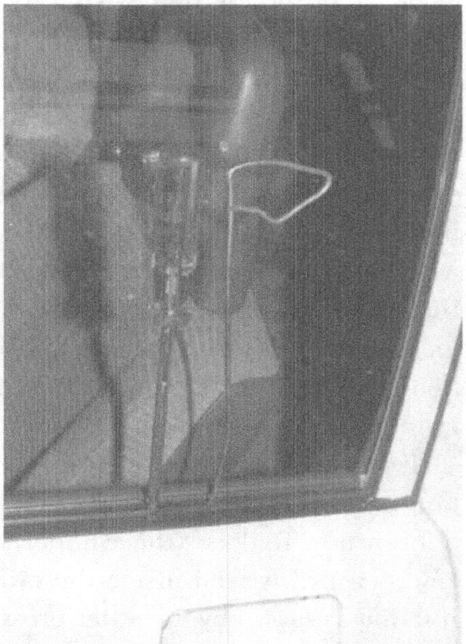

Figure 4-6

Example of placement of tools used to reach the lock linkage system. In this configuration, a screwdriver is used to create an opening between the weather stripping and the window and the wire tool is used to wiggle the linkage system to release the latch.

Figure 4-7

Close-up view of the configuration shown in Figure 4-6. Marks on the weather stripping can result from the use of such tools.

linkage rod and moves it in the proper direction, which releases the latch. In some instances, a wedge is used to hold the opening in the weather stripping to permit the proper manipulation of the lockout tool. This technique requires a certain experience and knowledge of each car's lock linkage system. Some modern vehicles are now designed to prevent this procedure from being accomplished.

When such a technique is used, there are often marks left on the weather stripping. The investigator must carefully inspect the weather stripping of each door to find any possible evidence of the use of a lockout tool.

D/ Lock Cylinder Defeating

This technique consists of either picking the lock cylinder or force rotating it. Both techniques applied to ignition lock cylinders are presented in more detail in Chapter 9. The same principle applies to door lock cylinders. Evidence of forced rotation is usually obvious but still requires the removal of the cylinder for internal inspection. An example of this is a case of a homicide investigation where a vehicle's owner reported his Porsche stolen. When the vehicle was recovered by the police, the observation of the driver's door handle and lock cylinder revealed that it had been force rotated. Figure 4-8 shows a general view of the door handle with the key and Figure 4-9 shows a detailed view of the dust cover of the cylinder. It is possible to easily notice the marks resulting from the forcing action.

It was not possible to identify the exact tool used to force the lock; however, the important point is to determine whether the forcing of the lock allowed for the opening of the door or not. To do so, the cylinder was taken out of its casing, as shown in Figure 4-10. The key is easily inserted inside the cylinder (Figure 4-11a); however, it does not allow for its rotation as shown by the wafer protruding in Figure 4-11b. The other wafers do not present any damage. This demonstrates that the lock was not forcibly rotated but only damaged by the insertion of a tool, which damaged one of the wafers. The owner of the vehicle tried to simulate the forcing of the door handle to support the theory of the vehicle's theft.

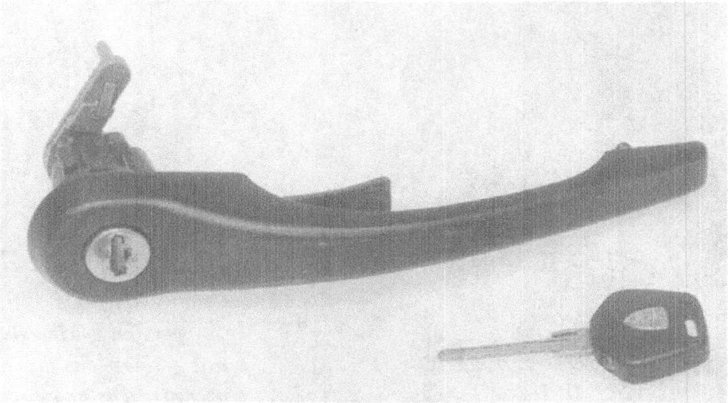

Figure 4-8

View of a door handle collected from a Porsche involved in a homicide. The vehicle was recovered with the handle presenting marks of mechanical forcing. External damage normally typical of the forced rotation of the cylinder was present.

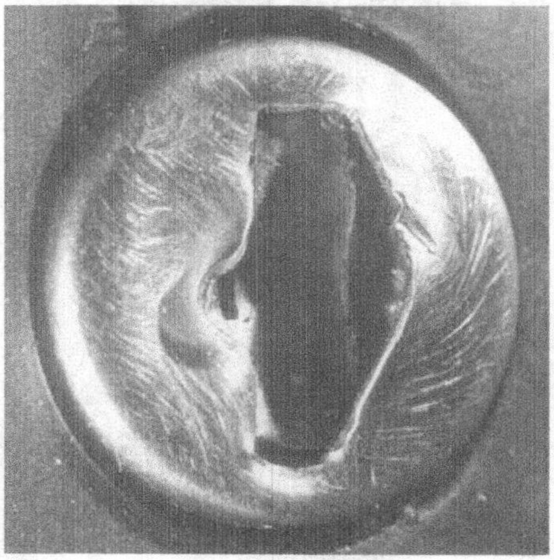

Figure 4-9

Close-up view of the dust cover of the lock shown in Figure 4-8. Note the toolmarks and the damage on the cover, which are usually indicative of a mechanical forcing of the lock.

E/ Doors and Trunk Forcing

The last technique consists of forcing the doors or the trunk using heavy-duty tools, such as pry bars or crowbars. This technique is rarely encountered, because it is virtually impossible to force open a modern vehicle's door in this fashion. Doors and door latches are designed to resist this type of attempt. Trunks are also very difficult to open, but not as hard as doors. This type of vehicle entry is usually very obvious and can be clearly docu-

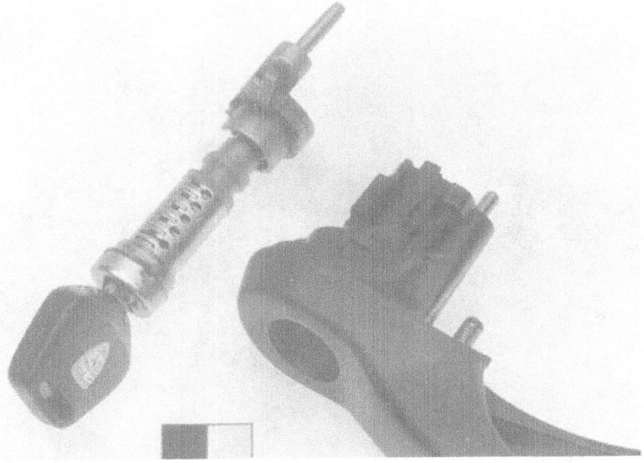

Figure 4-10

To determine whether the evidence of mechanical forcing on the lock allowed for its opening, it is necessary to take the cylinder out of its housing for laboratory examination.

a

b

Figure 4-11

(a) View of the wafers on the cylinder, showing no damage, except for the second wafer from the left. This cannot be consistent with a forced rotation of the cylinder, and therefore the mechanical forcing was not successful in opening the vehicle.

(b) When the key is inserted in the lock, all the wafers, except for the second from the left, correctly retract.

mented. Also, pertinent toolmarks and paint transfer usually result from such an operation. These must be photographed and collected by the crime scene investigator for possible linking with recovered tools. Figure 4-12 shows a trunk of a Lexus SC300 that was forced open. Notice the toolmarks on the trunk, the broken taillight, and the lock cylinder pushed inside the vehicle.

Figure 4-12

Result of the forcing of the trunk of a Lexus SC300. The lock cylinder was pushed inside its housing and the trunk was forced open using a tool. Notice the striated toolmarks on the trunk. (Photograph courtesy of Mark D. Culver, Key Fire Investigations, Inc.)

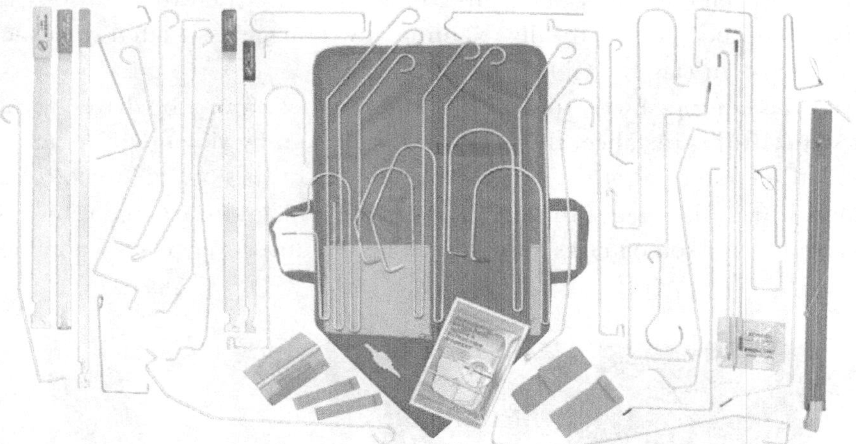

Figure 4-13

Example of a lockout toolkit (an auto opening kit Pro-lok Complete kit 2000 containing 47 pieces) used by professional locksmiths to penetrate vehicles. Auto thieves also use this kind of kit to illegally penetrate vehicles. (Source: Complete Kit 2000 (AKCOM00), http://www.pro-lok.com. Reprinted with permission of Pro-lok, Orange, California.) See Color Plate.

FI Auto Lockout Professional Toolkits

There are many auto lockout professional toolkits commercially available. These kits typically contain from a few pieces to several dozen pieces of material, which include Slim Jims, MCOT, other wire tools, long reach tools, and wedges [3, 4]. Figure 4-13 shows an example of such a toolkit.

It is also possible to find literature on how to unlock automobiles. Guides such as the *Car Opening Authority* and *The Best Damn Car Opening Manual Period!* are available to professional locksmiths and contain a plethora of information regarding the opening procedures of most vehicles [5, 6].

4.2.2 Starting the Vehicle

Once the criminal has penetrated the vehicle, he or she must defeat anti-theft systems and start the engine. More detailed information is provided on anti-theft systems in Chapter 8 and steering columns in Chapter 9. The investigator must be familiar with these systems to properly examine a vehicle to determine whether they have been defeated or not. Most modern cars are now equipped with electronic anti-theft systems and cannot be readily started without the proper equipment.

Older vehicles that are not equipped with any anti-theft systems are relatively easy to defeat. This usually requires taking apart the cover of the steering column to access the electrical wiring leading to the ignition lock. These wires are then cut or pulled from the connector and properly mingled together to establish contact. One last wire, energizing the starter's relay, is temporarily placed in contact with the two others to start the vehicle. The forensic examiner must carefully examine any damage associated with such a defeat of the ignition system (Figure 4-14).

Older vehicles equipped with the steering column lock cannot be driven without defeating this system. In this instance, the lock may be broken by violently forcing the steering wheel in one direction or another. The other solution consists of the forced rotation of the cylinder lock, which disengages the steering column lock and starts the vehicle simultaneously. This could leave some marks, particularly on the dust cover, as shown in Figure 4-15a.

Figure 4-14

View of the damage observed on the steering column of a vehicle reported stolen by its owner and recovered shortly after. The wire plug was pulled from the ignition switch (loose connector); however, no bridging between the wires was performed. It is not possible to start the engine of this vehicle with such a configuration. Additionally, the steering locking system was not defeated either. This case was eventually ruled as insurance fraud.

a b

Figure 4-15

(a) Example of fine marks left on the dust cover of a vehicle. An attempt to force-rotate the cylinder was carried out. (b) In some instances, the tool used to force-rotate the cylinder can break away and remnants will be left in the keyway.

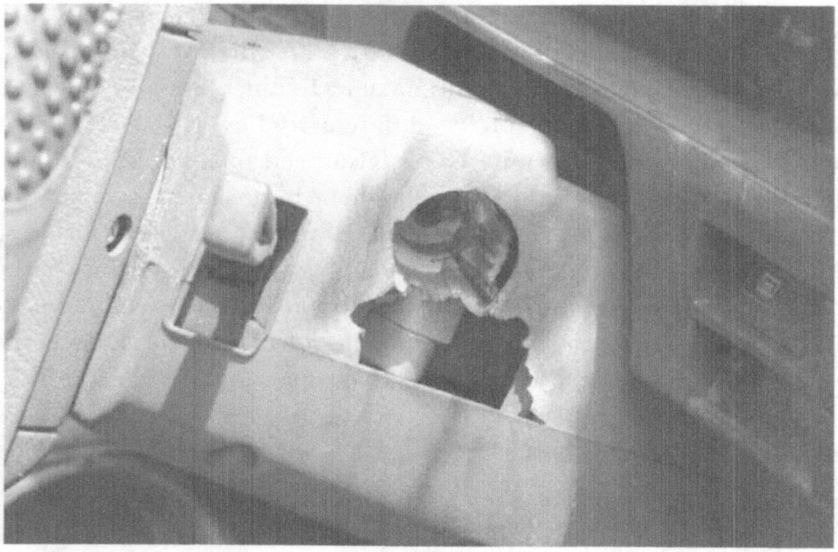

Figure 4-16

View of the steering column of a stolen-recovered vehicle where the cylinder was forcibly extracted. In some instances, it is possible to find the cylinder lying on the passenger's floorboard or rear floorboards. (Photograph courtesy of Eric Stauffer.)

In some instances, the tool used to attempt the forced rotation can break and a portion of it may be stuck in the keyway as shown in Figure 4-15b. This constitutes great evidence if the remainder of the tool is later found on a suspect. Finally, it is also possible to extract the lock cylinder, as shown in Figure 4-16. In such instances, it is often possible to find the cylinder lying on the passenger's floorboard or rear floorboards.

The technical examination of the vehicle is pursued with the search, collection, and examination of forensic traces.

4.3 FINGERPRINTS AND PALM PRINTS

4.3.1 General Considerations

Historically, fingerprints have been the first traces leading to the formal identification of perpetrators of crimes [7]. Today, it is still the most powerful means of identifying human beings. Thus, it is clear that the forensic examination of a stolen-recovered vehicle includes the search for fingerprints.

It is important to consider the possible collection of biological fluids before searching for fingerprints. The reason is that some techniques used for the development of fingerprints may compromise the integrity of the DNA evidence. Magnetic powder and multimetallic deposition are detrimental to the analysis of DNA [8]. Also, the use of ultraviolet (UV) light for more than 30 seconds on a given spot can degrade DNA and lead to irremediable consequences [9]. Roux et al. demonstrated that sticky side powder, black powder, DFO, ninhydrin with secondary treatment, cyanoacrylate with rhodamine, luminol, DAB, and amido black have little effect but require extra care when very little DNA is available to start with [10]. Other fingerprint enhancement techniques have not shown any effect on DNA analysis [11–14]. A summary of the different studies of the effect of fingerprint enhancement techniques on DNA analysis has also been published [15]. Therefore, as a general rule, collection of biological samples should be performed first on textured surfaces where the development of fingerprints is difficult. However, on smooth surfaces, the examiner should first carry out the development of fingerprints using great caution with the techniques that might have an effect on DNA. Additionally, the circumstances of the case must be taken into account to evaluate the pertinence of all evidence. Fingerprints in blood represent a particular situation, presented in the next section.

4.3.2 Fingerprint Search

When searching for fingerprints on the exterior portion of the vehicle (vehicle's body and windows), the entire area may be considered, not only the areas commonly touched by the regular user of the vehicle. It is also important to keep in mind that the body is subjected to different conditions that greatly affect the conservation of fingerprints, which can render the search for fingerprints very difficult, if not impossible. Environmental conditions (light, temperature, rain, wind) and contamination from the road such as water and dirt can completely preclude a fingerprint from being discovered.

4.3.3 Fingerprint Development

Most often, the development of fingerprints is performed with the use of luminescent or metallic powders, as shown in Figure 4-17a. Also, if the vehicle is wet, small particle reagent (SPR) is used (Figure 4-17b) (see also Chapter 13).

The whole vehicle can also be placed in a chamber and subjected to superglue fuming (cyanoacrylate). If the crime laboratory or crime scene unit is not equipped with such a special chamber, it is possible to build a fuming tent around the vehicle, as shown in Figure 4-18a and b. Some laboratories are equipped with a car-size superglue chamber as illustrated in Figures 4-18c and d.

For superglue fuming to work properly, airflow and humidity (about 80%) must be present in the tent [16]. This is achieved by adding one or more small fans inside the tent and by placing either enough open containers with water or one or more humidifiers. It is important to have a uniform distribution of the vapors inside the tent during the fuming process. The developed fingerprints can then be reinforced by the use of regular fingerprint powder or special dies [15]. Although the powder is easier to use, the application of dies typically offers a better result. In general, superglue fuming offers a much superior result in the development of fingerprints than regular brush powdering. However, its setup and realization are cumbersome and time consuming.

The interior portion of the vehicle does not suffer from outside conditions because it is usually protected. Of course, there are exceptions, such as with convertible vehicles retrieved with the roof retracted, vehicles recovered with all windows open, or vehicles found submerged in water (see Chapter 13). Except for the last scenario, traces inside these vehicles

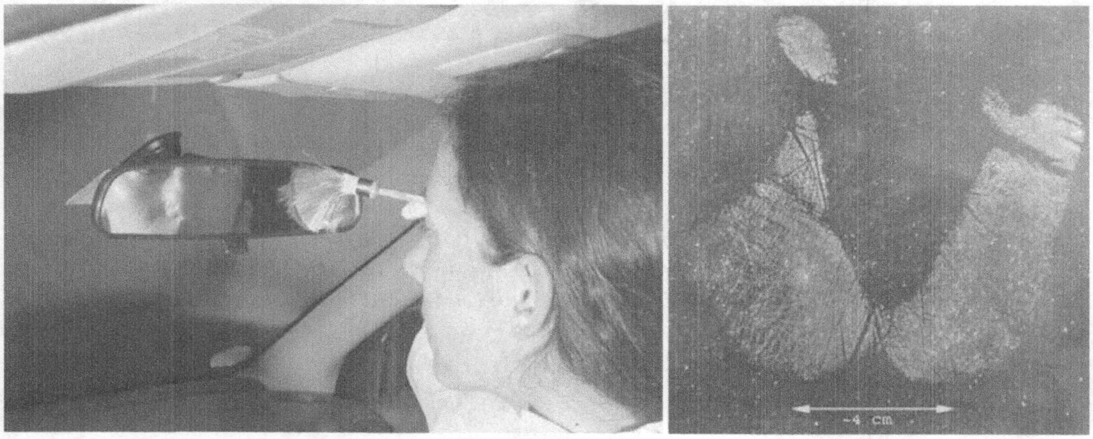

a b

Figure 4-17

(a) Development of fingerprints inside the vehicle by powdering smooth surfaces with argentoratum. (b) Use of small particle reagent on a wet surface on the vehicle's body panel. A palm print is discovered and observed with an alternate light source as the powder is fluorescent.

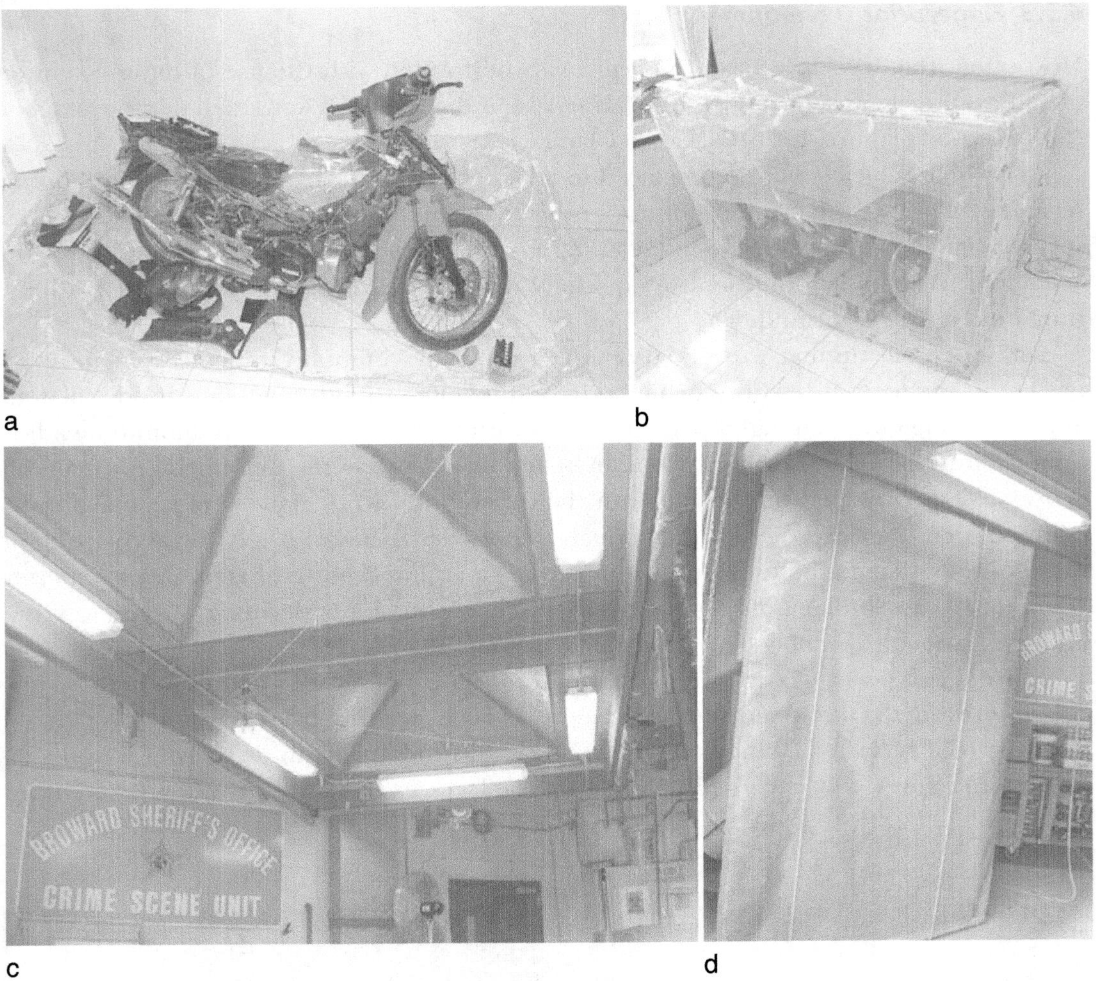

Figure 4-18

(a) Example of an improvised fuming chamber to subject a motor vehicle to vapors of cyanoacrylate (superglue fuming). This motorcycle was used by terrorists during the Bali bombing (see Chapter 17). Note how the different components of the vehicle (battery, fenders, etc.) were taken apart before the fuming to guarantee a maximum exposure to the vapors and the most efficient development of fingerprints. (b) The tent built up and the fingerprint development in process. (c) View of the car-size superglue chamber of the Broward Sheriff's Office in Fort-Lauderdale, Florida. The chamber is located in the vehicle's examination facility. (d) Curtains are brought down from the gigantic hood, which creates the chamber. (Photographs (a) and (b) courtesy of the Australian Federal Police. Photographs (c) and (d) courtesy of Sgt Stewart Mosher, Broward Sheriff's Office.)

still have a better degree of protection than traces present on the exterior of the vehicle. Unfortunately, the interior of a vehicle is rarely manufactured with surfaces facilitating the development of fingerprints. As a matter of fact, the plastic surfaces used on dashboard and other moldings are often rough and textured, rendering the use of powder very inefficient. Thus, the development of fingerprints should focus on the interior sides of windows, on the central mirror, and all other smooth surfaces that could have been touched. Some vehicles are equipped with some parts made of lacquered wood; these surfaces offer great results with powdering. In case of a vehicle used to smuggle illegal merchandise (see Chapter 16), the surfaces leading to the hidden space and the hidden space itself should be thoroughly examined for fingerprints.

It is also possible to fume the interior of a vehicle with superglue vapors. There are commercial field kits available to that effect. As with the tent, it is important to ensure a good distribution of the vapors as well as the proper humidity. When necessary, the forensic scientist should take apart components of the vehicle and bring them back to the laboratory, where they can be appropriately processed with the most advanced and appropriate techniques.

4.4 BIOLOGICAL AND DNA TRACES

4.4.1 General Considerations and Sampling

In the last few years, the importance of biological traces in the identification of human beings has dramatically increased. The discovery of DNA amplification by polymerase chain reaction allows for the analysis of very small amounts of DNA. Although the most commonly encountered biological traces during the examination of crime scenes are blood and semen, it is now possible to obtain DNA profiles from traces such as dandruff, other skin particles, and hair roots.

The best sampling technique consists of using a cotton swab slightly humidified with deionized water; the surface is then wiped with the swab onto which any biological traces collect (see Figure 4-19). When using larger cotton swabs, it is possible to humidify only one side, wipe the surface, and then wipe the surface again using the dry side for maximum efficiency. To prevent any possible contamination, the crime scene officer must wear gloves and mask during this operation. Gloves must be changed after each swabbing to prevent any cross-contamination. In addition, it is recommended that the crime laboratory obtain DNA profiles of all respondents to the scene and crime laboratory personnel that could have come in contact with the items of evidence to quickly pinpoint samples that may have been contaminated during evidence collection and to avoid "polluting" local and national DNA databases with irrelevant profiles.

4.4.2 Blood

The detection of blood outside and inside a vehicle can bring very pertinent information, not limited to the DNA profile of the person from whom the blood originates. As a matter

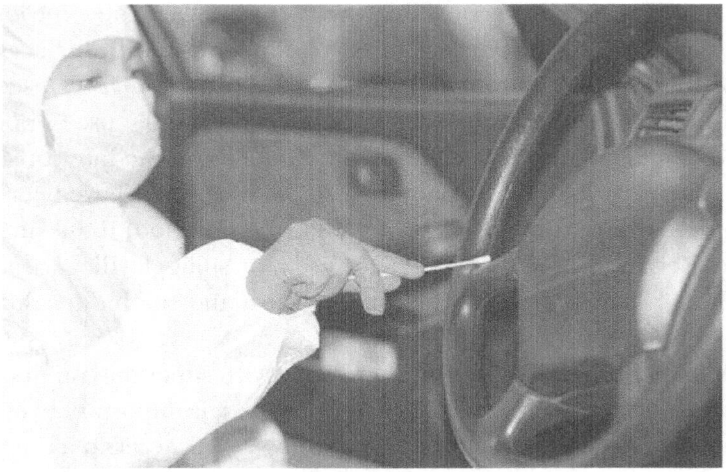

Figure 4-19
DNA is collected from different surfaces using a cotton swab slightly humidified with deionized water.

of fact, the spatial location and configuration of the traces can shed light regarding the activities that took place inside a vehicle and/or the involvement of different passengers inside the vehicle during the commission of a crime.

If blood is not visible, which could be due to a cleaning of the surface or simply to a poor contrast with the surface, it is possible to detect it by means of optical techniques. Optical enhancement techniques are generally nondestructive, and therefore their use is strongly recommended before proceeding with any chemical treatment. Although blood has a broad absorption spectrum in the entire light region (ultraviolet-visible-infrared), it exhibits a strong and narrow absorption maximum at 415 nanometers (nm). The enhancement of untreated bloodstains can be performed in either the absorption or the reflection modes. The absorption mode is recommended for lightly colored or luminescent surfaces, whereas the diffused reflection mode is recommended for dark or shiny surfaces [13].

When optical detection of blood fails, it is possible to use techniques involving colorimetric or chemiluminescent reactions. Luminol and leucocrystal violet (LCV) are two of the most powerful techniques presently used. These techniques are useful when bloodstains have been washed off or when the contrast with the background is too poor, as shown in Figure 4-20. It is important to keep in mind that LCV may jeopardize the subsequent DNA analysis of the bloodstains [17]. Conversely, luminol does not influence the DNA typing at all and, for this reason, is usually preferred [13]. Nevertheless, luminol is a chemiluminescence reaction, which usually offers the best contrast with all types of surface as it is observed in total darkness (Figure 4-20a through m).

Once a positive reaction has been obtained with a presumptive test, it is important to perform a confirmatory test to ascertain the presence of blood. Tests such as Hexagon OBTI or ABAcard HemaTrace are easily accessible and produce a quick

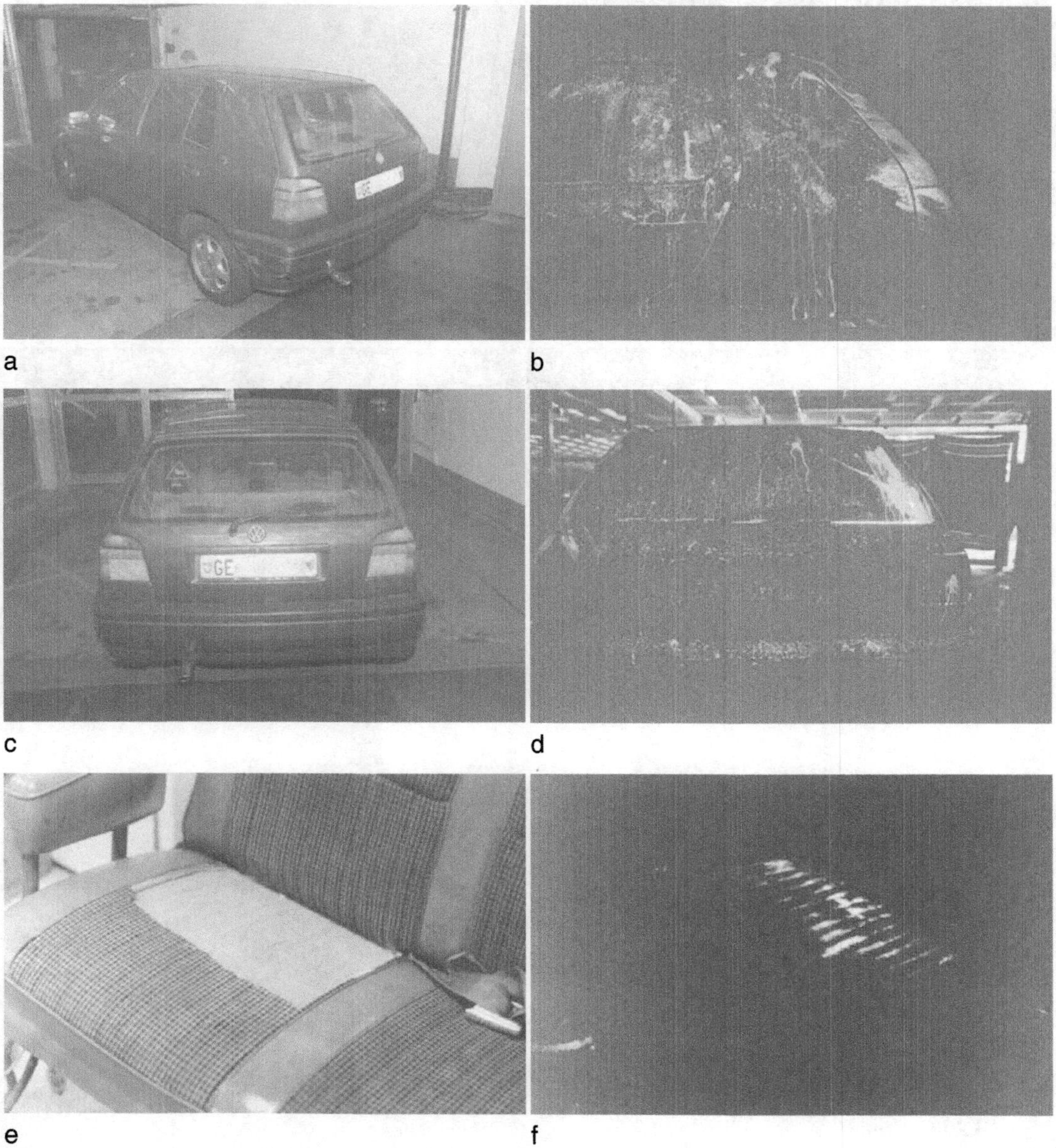

Figure 4-20

(a) This VW Golf was used in a homicide, and it was suspected that traces of blood were present on the body around the trunk. Unfortunately, due to the color of the vehicle, the bloodstains are not visible to the naked eye. (b) When luminol is applied to the vehicle, the bloodstains become readily visible. Then, it is possible to sample the different areas using the technique presented in Figure 4-19. (c) View of the same vehicle as in Figure 4-20a before application of luminol. (d) View after application of luminol. (e) View of a vehicle's rear seat whose top has been cut off for laboratory examination. No apparent stains are present. (f) View of the same seat as in Figure 4-20e after application of luminol. Blood is still present in the seat cushion. See Color Plate.

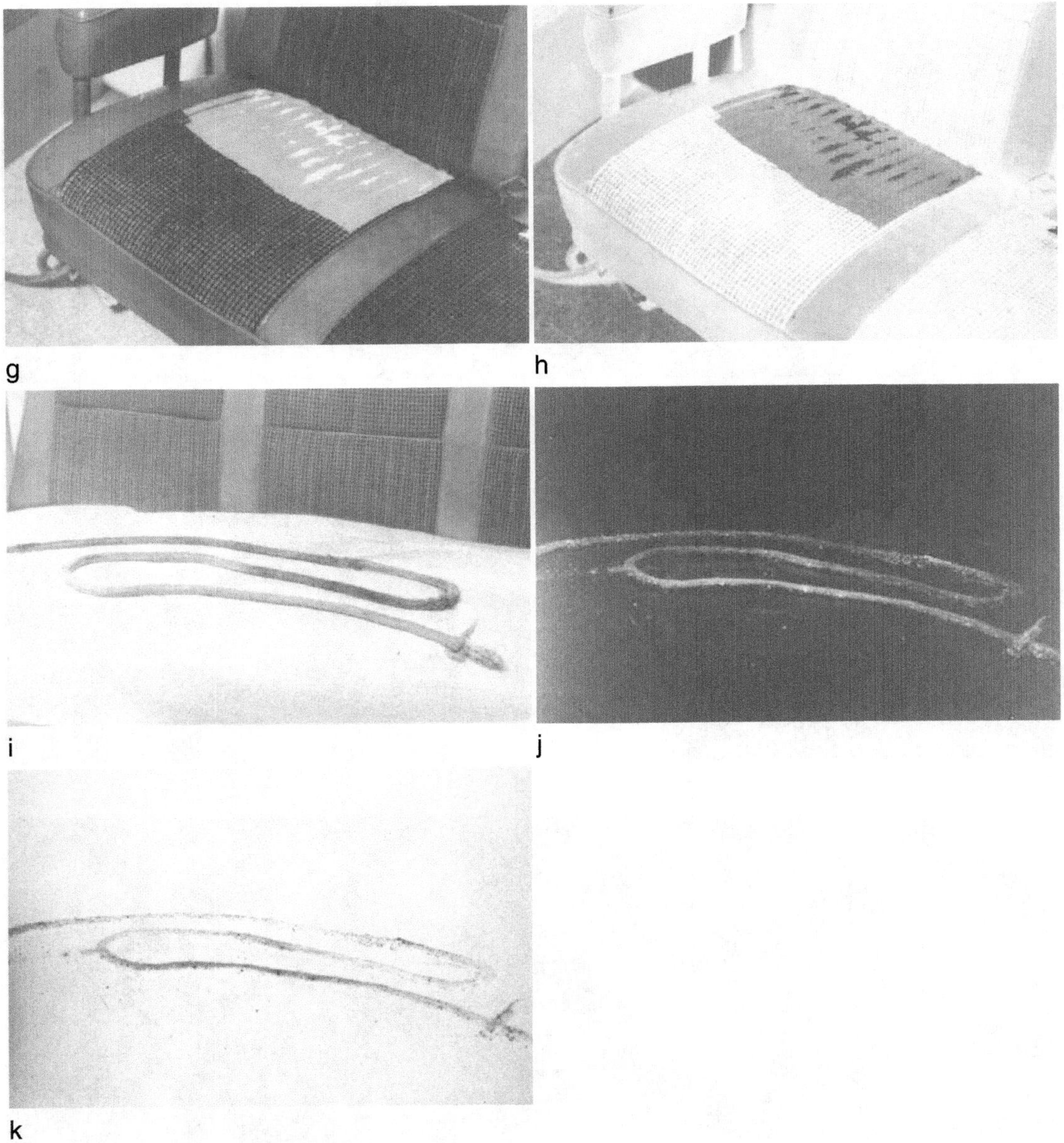

Figure 4-20 Continued.

(g) A particular photographic technique consisting of exposing the film to a short flash right after its exposition to the luminescence produced by the luminol provides a better visualization of the blood stains. Compare this image with Figure 4-20f. (h) By inverting the color of the image using computer software, it is possible to make the white-blue luminescence appear dark red, closer to the natural color of blood. This technique is particularly appreciated when presenting evidence to a jury. (i) This rope was found tied on the leg of a car seat and was allegedly used to tie a person who was bleeding. A naked eye examination did not reveal any blood traces. (j) Same view after application of luminol. (k) Same view after inverting the colors in computer software. See Color Plate.

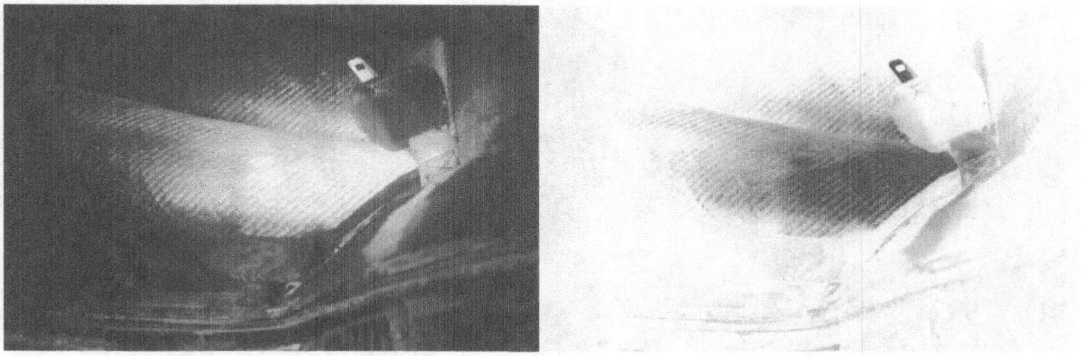

l m

Figure 4-20 Continued.

(l) View of a car seat after application of luminol, revealing important possible stains of blood. (m) Same view inverted with computer software. The color rendering provides a clearer picture of the bloodstain. See Color Plate. (Photographs (e) through (m) courtesy of Sgt. Stewart Mosher, Broward Sheriff's Office.)

confirmation of the presence of human blood [18, 19]. Figure 4-21 illustrates the use of the Hexagon OBTI test and a positive reaction for human blood.

In some instances (such as when a vehicle used by poachers is seized by the police), it might be necessary to search for animal blood. The presumptive tests are the same; however, the use of different commercially available confirmatory tests is recommended, because the confirmatory tests designed for human blood do not work with almost all animals. Thus, tests such as Sangur (Roche Diagnostics), Hemastix (Bayer), or Peroxtesmo KM (Macherey-Nagel) should be used.

4.4.3 Semen

The detection of semen in a vehicle might appear a priori useless. However, it becomes very pertinent in all cases where a (criminal) sexual act might have occurred in the vehicle.

Semen is not observed as easily as blood. This is partially because semen is almost colorless. Once dried, it leaves faint white-yellowish traces that can be tested using a presumptive test. The presumptive test used for semen involves the detection of acid phosphatase. It is also possible to detect semen using UV light or an alternate light source around 400 nm. In such case, semen may emit light around 560 nm. However, this situation is only rarely encountered as semen does not always fluoresce. The investigator should remember not to overexpose surfaces possibly containing biological traces with UV light because it rapidly contributes to the degradation of DNA.

Once a stain has been located, it can be collected by swabbing or cutting the surface. It is recommended that the surface be cut and the whole stain sampled because the chances of preserving a maximum amount of genetic material is greatly favored.

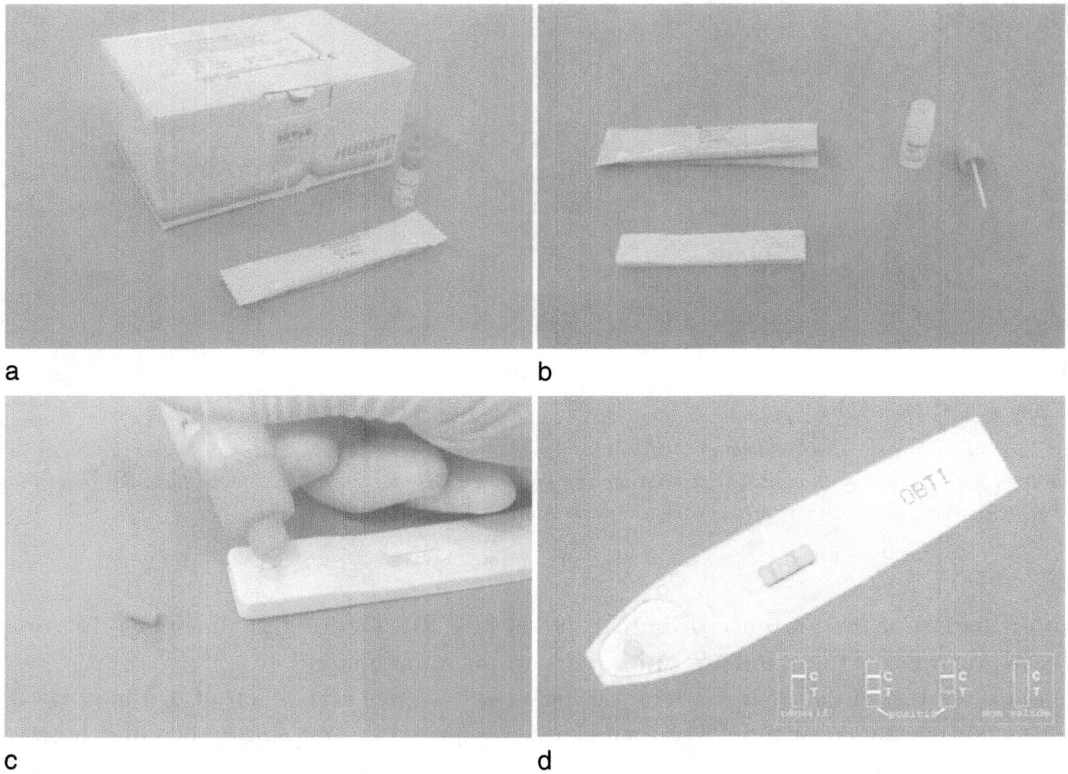

Figure 4-21

(a) The Hexagon OBTI kit used to confirm the presence of human blood. Similar kits are available for semen and are used in the same fashion. (b) The strip is removed from its package and the small bottle is opened. A portion of the trace is dissolved in the bottle with water. (c) After the bottle is vigorously shaken, a few drops are placed in the proper hole at the bottom of the strip. (d) The solution then moves through the strip by capillarity, and no band, one band, or two bands will appear in the viewer window. No band means that the test is invalid, one band means it is negative for human blood, and two bands mean it is positive for human blood.

It is also possible to screen a surface using a colorimetric presumptive test reacting to phosphatase acid. The screening can be direct by applying the reagent onto the surface. This brings the advantage of accurately locating the possible stains on the surface. However, if the surface is dark or purple in color, it might not provide sufficient contrast to observe the color change. In such instances, it is possible to perform indirect screening. A large filter-type paper is slightly humidified and pressed onto the surface for a couple of minutes. This transfers a small portion of any possible stains onto the filter paper, onto which the reagent is applied. This technique provides the good advantage of a clear contrast between the paper and the possible color change reaction. Because the direct technique does not interfere with the subsequent DNA analysis, it is preferred to the indirect technique whenever possible.

When a presumptive test provides a positive response, it is possible to confirm the presence of semen at the scene using prostate-specific antigen membrane test assays such as ABAcard p30 [20, 21]. The use of such tests is identical to the ones for blood presented in Figure 4-21.

4.4.4 Skin Contact Traces

Skin contact traces are the resulting traces left by the contact of skin onto a surface. Contrary to blood and semen traces, there are no techniques to detect and identify latent contact traces. One can only hope for the presence of such traces and swab the surface. Only laboratory analyses can reveal whether such a contact trace was present or not by detecting the presence of DNA. According to Wickenheiser, most collected skin contact traces belong to the last person who touched the surface [22]. Literature shows that steering wheels, among other objects, can bear sufficient quantities of DNA [23, 24]. Thus, these traces are of particular interest to the crime scene investigator because they might allow for the identification of the last person who drove the vehicle.

The collection of skin contact traces must be performed on all surfaces that could have been contacted by the thief(ves) during the theft or during the operation of the vehicle. The following locations should be swabbed (see Figure 4-22):

I The rearview mirror (both on the glass and on its frame): The rearview mirror is often adjusted by the driver unconsciously to properly operate the vehicle. When performing fingerprint detection with powder on the rearview mirror, the investigator should be careful not to use too much powder to preserve possible skin contact (DNA) traces.

II The gearshift lever knob: This is particularly important in vehicles equipped with manual transmission, because the knob is repeatedly touched by the driver during the vehicle's operation.

III The hand brake: Particular attention should be paid to the release button, which is often slightly textured and thus allows for a better transfer of skin cells.

IV The steering wheel: The operation of a stolen vehicle is usually performed under heavy stress, and thus the perspiration of the driver is highly increased resulting in more skin contact transfer on the steering wheel.

V Interior door and window handles and switches: The interior accessories are prioritized because they are protected compared with the exterior surfaces and because anyone could potentially touch exterior artifacts. To proceed to the most pertinent collection, different hypotheses regarding the number of people who traveled in the vehicle and their position must be taken into account.

VI The air bag: During the deployment of the air bag, the air inside the bag can reach temperatures up to 600°C, and the driver or passenger will violently contact it, making the deposition of skin contact traces very likely. In addition, the nature of the contact could result in a burn of the skin of the driver or passenger and, in some instances, in the deposition of blood. In cases of burn, if visible traces are present, it is also possible to compare the pattern present on the air bag to the pattern exhibited by the suspect or the suspect's clothing [25].

Figure 4-22

The different locations within a vehicle where skin contact traces typically occur and where possible DNA should be collected.

4.4.5 Hair and Dandruff

The search for and sampling of hair and dandruff must not be forgotten. Dandruffs are skin particles, and thus they contain DNA. The search for such traces is particularly pertinent around the headrests. The persistence of such traces can be extremely important. Therefore, it is very likely that a given portion of hair and dandruff particles found on the headrest belong to the regular driver of the vehicle. In such light, it is pertinent to sample DNA from the primary driver and regular passengers of the vehicle to eliminate legitimate DNA profiles among the traces. The collection of hair and dandruff is performed identically to the collection of fibers, as described in Subsection 4.5.2. Hair and dandruff must be placed in a paper bag, as is the proper practice with all DNA traces.

4.5 MICROTRACES

4.5.1 General Considerations

Microtraces are often neglected by crime scene investigators. This is probably because many crime laboratories do not favor these types of traces, for which the analysis is often labor intensive and requires complex analytical instrumentation. Nevertheless, the interpretation of the results is not necessarily straightforward, and such traces do not typically allow for the establishment of a direct and unique link with a possible suspect (see Chapter 5). Their inclusionary power is usually not as strong as traces such as fingerprints, toolmarks, and tire tracks.

However, depending on the circumstances of the case, it is extremely important to search for and collect microtraces. Microtraces, as well as impressed marks, can also become extremely pertinent in case of reconstruction of road accident fatalities, when the question of who was driving and riding in the vehicle must be answered [26, 27]. However, this chapter does not address this particular issue.

4.5.2 Fibers

Many studies have been performed regarding the transfer and persistence of fibers in automobile vehicles, and the reader is invited to refer to these publications for more information [28–30]. The detection and collection of fibers is usually a three-step procedure.

First, a visual observation is performed, both with ambient light and with an alternate light source, particularly with wavelengths in the UV range (254 to 366 nm). Many fibers are luminescent under UV light, which allows for their easy localization. When fibers are observed in this fashion, they can be precisely collected using tweezers and their exact location can be accurately recorded (Figure 4-23). Thus, it is possible to collect only fibers of interest or to collect different types of fibers in separate packages. This step is particularly relevant if the investigator knows what type of fibers he or she is looking for. Otherwise, the investigator looks for heteroclitic concentrations of fibers. Nevertheless, this step is the most important when searching for fibers from car seats on the clothing of the suspect. The fibers are placed in small paper bags or druggist folds, which prevent the unwanted effect of static electricity.

Second, a transparent adhesive tape is used to tape the surface from which fibers are to be collected (Figure 4-24). The total surface can be divided into different zones, and one piece of tape is used for each zone. This allows the analyst at the laboratory to estimate the distribution of the different types of fibers on the surface. It is also possible to perform 1:1

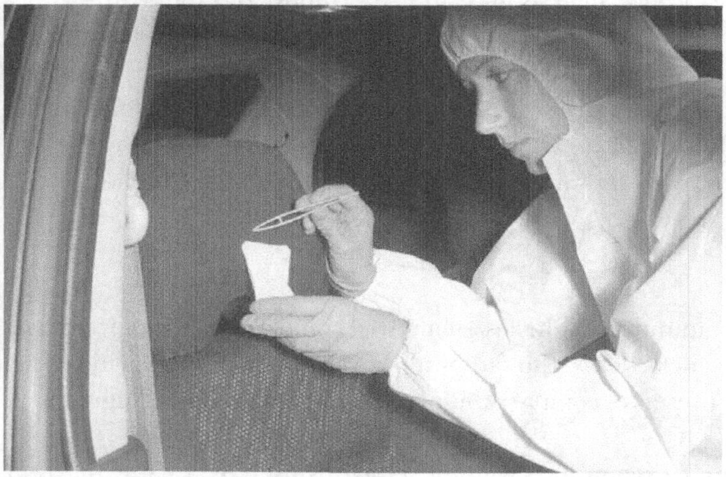

Figure 4-23

Collection of hair from a headrest using a pair of tweezers.

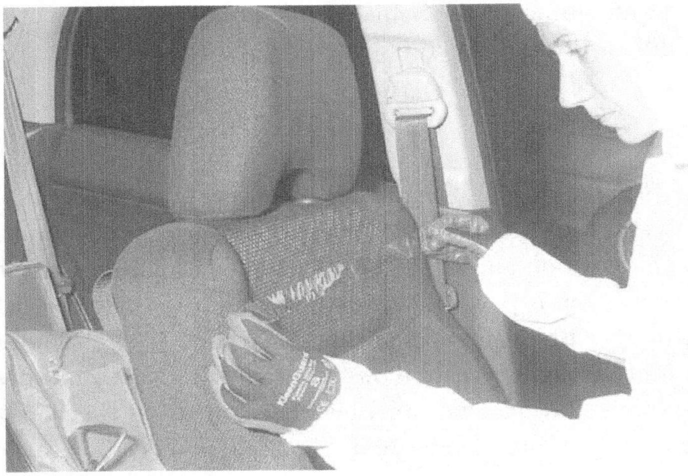

Figure 4-24
Collection of fibers using adhesive tape.

taping, which allows for a full reconstruction of the distribution of the fibers on the surface [31]. This technique presents the drawback of collecting the fibers from the substrate as well. Once removed from the surface, the adhesive sheet is then stuck on a transparent plastic sheet and placed in a paper bag.

The third step consists of vacuuming the surface from which fibers are to be collected using a vacuum cleaner equipped with a special filter (see Paragraph 4.11.5C). This step does not allow for the localization of the fibers on the surface because they are all collected on a small filter. The filter is then placed in a paper bag. This technique also collects any debris from the surface such as glass, bread crumbs, and dandruffs.

Fibers must be collected from the car seats, the seat belts, and, if applicable, the air bags. In every instance, comparison fibers from the substrate itself must also be collected in separate packages. A cross-transfer of fibers could have occurred: Not only the fibers from the suspect's clothing could be transferred on the car seats, but also the fibers from the car seats could be transferred onto the suspect's clothing (see Chapter 5).

4.5.3 Glass

Glass particles found in vehicles generally come from windows that were broken to penetrate the vehicle. Thus, the sampling of glass found in the vehicle is almost exclusively performed to preserve comparison material in case glass fragments are found on a suspect.

One particular exception is when a vehicle is used as a battering ram to break into a commercial facility. In this case, glass fragments from the building windows might also be present in the vehicle in addition to any potential glass fragments from the vehicle's broken windows. When dealing with such a crime scene, the investigator must collect comparison glass from all possible sources.

The collection of comparison glass should be performed preferentially on what remains of the window, such as the pieces left in the frame or from pieces located inside the vehicle. Multiple comparison pieces must be collected to ensure that the intravariation of the characteristics exhibited by a window can be estimated during the laboratory examination. Also, the investigator must very carefully observe any pieces of glass before moving them for possible shoeprints, tire tracks, fingerprints, or blood evidence. If such evidence is present on the glass, their examination must be prioritized because they do not influence the examination of the glass itself and usually lead to stronger inclusionary evidence.

Finally, glass fragments can be linked together by fracture assembly (physical match). Although this is typically not realizable with safety glass, due to the breakage pattern, it is readily feasible with regular flat glass if fragments large enough are still available. It is also possible with headlamp glass. Thus, when dealing with breaking entry into commercial facilities where a vehicle was used as a battering ram, the investigator must be very thorough in collecting glass fragments. If the vehicle is found at a later date, it is possible that it would bear glass fragments with a fracture pattern matching that from a portion of the store window, thus establishing a very strong link.

4.5.4 Paint

Paint evidence, and other protective coatings such as lacquer, enamel, and varnish, is sometimes recovered in forced-entry and hit and run cases. In some cases, it is possible to show conclusively that the paint came from a specific location if the chips are large enough and the edges can be fitted together in a jigsaw puzzle fashion (physical match). However, in most instances, only class characteristics are present.

Paint traces can usually be found in and around toolmarks left by tools that contacted a surface. Inversely, these tools could have collected paint from the surface of the vehicle. Thus, it is crucial to sample comparison paint from stolen-recovered vehicles when toolmarks are present. When collecting standard specimens of paint from automobiles, the specimen should be collected as close to the area of damage as possible to avoid the situation where an area further away from the location of interest was painted differently. When investigating the scene of a hit and run (such as if the vehicle was used as a battering ram) or the scene from where the vehicle was stolen, the investigator must vigilantly look for any possible paint traces left at the scene and collect them.

When collecting paint samples, it is best to place the paint into folded paper. Small plastic bags should be avoided because of the static electric charge that makes it extremely difficult to extract the small paint chips from the bag. Paint present on tools should not be removed. It is preferable to carefully wrap the end of the tool in such a way as to not dislodge the paint and submit the tool to the laboratory with the paint still intact.

4.5.5 Soil

Soil evidence may be encountered on a wide variety of forensic items. It may be found on shoes, clothing, and on the underside of and inside the vehicle. Soil evidence is useful in tying the suspect or a car to a location. Soil is a mixture of decayed and weathered rock and decomposed organic material known as humus. Soil contains a wide variety of minerals such as quartz, feldspar, and mica as well as other components such as partially decomposed leaves, pine needles, pollen grains, and other plant fragments. Thus, it is possible to differentiate soils from various locations by microscopic examination of the different components [32]. Known soil specimens from the crime scene are absolutely required for an analysis of the incriminated sample. Comparison samples should be collected from various locations at the crime scene; it is important not to dig deeper than an inch or so when collecting these specimens. The subsoil may have a significantly different composition than the topsoil, thus leading to confusing results.

Evidence specimens collected in a stolen car at different locations (carpets, tires, pedals, etc.) as well as on the suspect (shoes, clothing) should be carefully handled. The collection of the evidence is carried out by means of tweezers or spatulas (Figure 4-25). In certain instances, it may be possible to remove successive layers of the soil sample and reconstruct the activities of the subject based upon the different types of soil present.

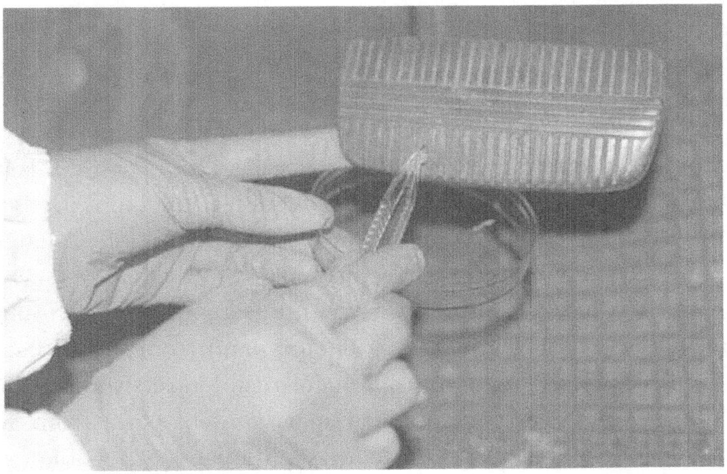

Figure 4-25

Collection of dirt from the brake pedal.

4.6 TOOLMARKS AND OTHER EVIDENCE

Toolmarks or marks of objects that have been used as tools are often found on stolen vehicles. Among the tools that leave identifiable marks are axes, knives, screwdrivers, chisels, crowbars, pliers, and cutters. Most of the time, the toolmarks are a result of the action of entering and/or starting the vehicle (see Figure 9-19).

The best places to find these toolmarks are on or around the different locks present on the vehicle (doors, trunk, or ignition). To collect these marks, it is best to take the entire lock back to the laboratory. This allows the toolmark examiner to have ample time to study the object and to make the most appropriate comparisons with any suspected tools. Casting or other methods of taking impressions of a toolmark should be used when it is not possible to collect the object onto which the marks are present. Whether the actual toolmarks are recovered or a cast is made, they should be accurately photographed. The tool may also have deposited traces in the form of paint, oil, or other contamination; these traces are sometimes just as valuable as the toolmark itself.

In some instances, toolmarks present on stolen vehicles can reveal intelligence regarding the modus operandi of the theft or the attitude of the perpetrators. Figure 4-26 is an example of toolmarks found on the roof of a stolen-recovered vehicle that were the results of vandalism.

All types of forensic traces can potentially be found in a vehicle. As an example, Figure 4-27 shows a shoeprint found on the cover of a stolen-recovered car. This kind of evidence must be collected and dealt with accordingly.

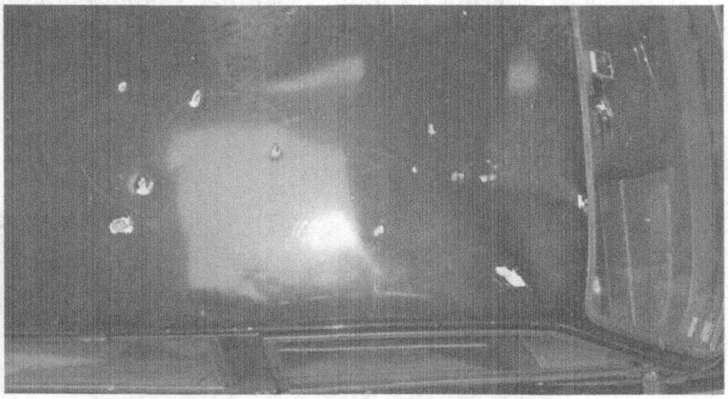

Figure 4-26

Toolmarks present on the roof of a stolen-recovered vehicle. They show the act of vandalism performed by the perpetrators.

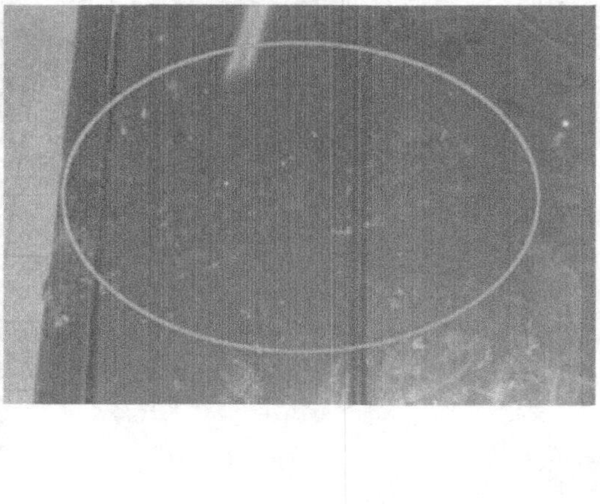

Figure 4-27

Example of a shoeprint found on a plastic cover of a stolen-recovered vehicle.

4.7 ABANDONED OBJECTS

Objects abandoned in a vehicle that do not belong to its legitimate owner/driver must be collected and submitted to the forensic laboratory for detection of ridge skin impressions and other possible biological traces. Figure 4-28 illustrates an example where the examination of several abandoned items in a stolen vehicle successfully led to the arrest of the two perpetrators. Usually, these objects can be tools, clothing, drinking cups and bottles, and cigarette butts. Regarding the latter, it is possible that many are abandoned in the vehicle and so it may not be feasible to analyze all of them at the laboratory. However, it is important to collect all of them. The laboratory will then determine which ones are analyzed, based upon their brand, where they are found in the vehicle, and other particular signs, such as the presence of lipstick. This type of evidence demonstrates the importance of collecting information regarding the habits of the primary driver and regular occupant(s). Also, cigarette butts can bring intelligence regarding the number of people that were present in the vehicle and their respective positions.

4.8 VEHICLES INVOLVED IN OTHER CRIMES

The last task of the crime scene officer is to determine whether a vehicle has been used to commit other crimes. This is performed by examining any potential damage present on the vehicle as well as any potential evidence present inside the vehicle. The damage present outside the vehicle

Figure 4-28

Objects found in a vehicle that was stolen and used to commit several crimes in France and Switzerland. The examination of these objects (pair of pants, T-shirt, and plastic bags) at the laboratory revealed DNA profiles and fingerprints, which lead to the identification of two suspects.

Figure 4-29

(a) This BMW M5 was used as an indirect battering ram to break into a jewelry store in Switzerland. Two wood beams were placed on top of a shopping cart and rested on top of the bumper of the vehicle on one side and on the door of the store on the other side. The vehicle then accelerated in reverse, pushing the door inside the store and creating access for the thieves. (b) Detailed view of the point of contact of the wood beams with the trunk of the vehicle. (c) View of the damage resulting from this operation.

can quickly provide intelligence regarding the type of activities that the vehicle underwent while in custody of the thieves. Figure 4-29 presents a vehicle that was used as an indirect battering ram. Figure 4-30 presents a vehicle that was used to pull a cable attached around an ATM. The damage created to the first vehicle is very specific to this type of modus operandi. Damage to the second vehicle is not expected to be significant, except in some instances.

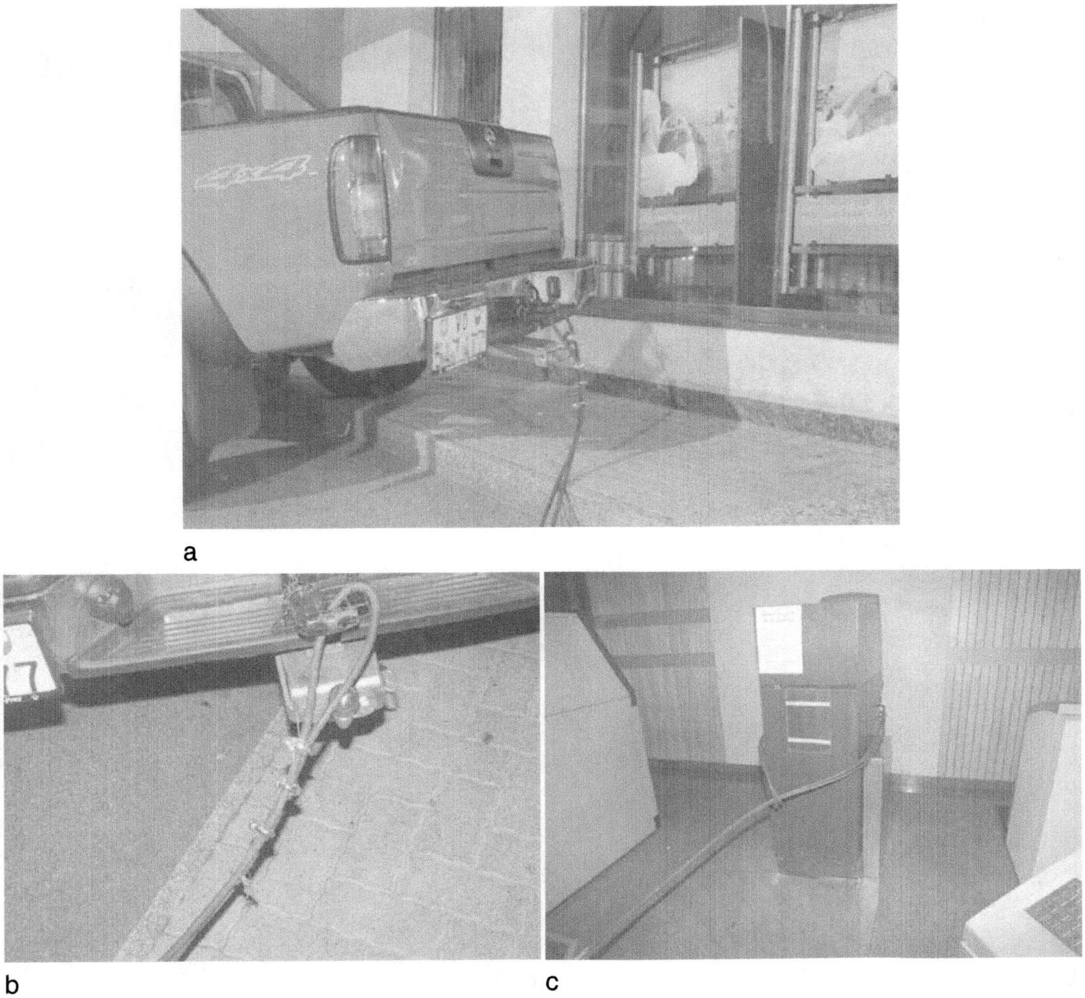

Figure 4-30

(a) This Nissan pick-up truck was used to pull an ATM from its solid foundations. (b) A steel cable was attached to the towing package of the truck. (c) The steel cable was then wrapped around the ATM and the truck accelerated forward to drag the ATM out of its foundations. These photographs were taken during a reconstruction.

ACKNOWLEDGMENTS

The author would like to thank every person who helped in the writing of this chapter. It would not have been possible to finish this writing without them.

BIBLIOGRAPHY

[1] Davis M. (2005) New glass deters thieves, *The Car Connection*, available at http://www.thecarconnection.com, last access performed on November 18, 2005.

[2] Memmer S. (2001) New developments in safety glass, *Edmunds.com,* available at http://www.autotrader.com, last access performed on November 18, 2005.

[3] Weinraub Enterprises (2001) *Pro-Lok professional locksmith,* available at http://www.pro-lok.com, last access performed on November 18, 2005.

[4] HPC (2005) *Car openers,* available at http://www.hpcworld.com, last access performed on November 18, 2005.

[5] HPC (2005) *Car opening authority* HPC Inc., Shiller Park, IL.

[6] Pro-Lok (2003) *The best damn car opening manual period!* Weinraub Enterprises, Orange, CA.

[7] Beavan C. (2001) *Fingerprints–The origins of crime detection and the murder case that launched forensic science,* Hyperion, New York, NY.

[8] Spear TF. (year unknown) *Summary of the impact of fingerprint reagents on the ability to obtain typing results using PCR-based DNA methods or conventional typing methods,* available at http://www.cacnews.org, last access performed on November 18, 2005.

[9] Andersen J and Bramble S. (1997) The effects of fingermark enhancement light sources on subsequent PCR-STR DNA analysis of fresh bloodstains, *Journal of Forensic Sciences,* **42**(2), pp 303–306.

[10] Roux C, Gill K, Sutton J, and Lennard C. (1999) A further study to investigate the effect of fingerprint enhancement techniques on the DNA analysis of bloodstains, *Journal of Forensic Identification,* **49**(4), pp 357–376.

[11] Stein C, Kyeck SH, and Henssge C. (1996) DNA typing of fingerprint reagent treated biological stains, *Journal of Forensic Sciences,* **41**(6), pp 1012–1017.

[12] Zamir A, Springer E, and Glattstein B. (2000) Fingerprints and DNA: STR typing of DNA extracted from adhesive tape after processing for fingerprints, *Journal of Forensic Sciences,* **45**(3), pp 687–688.

[13] Gross AM, Harris KA, and Kaldun GL. (1999) The effet of luminol on presumptive tests and DNA analysis using the polymerase chain reaction, *Journal of Forensic Sciences,* **44**(4), pp 837–840.

[14] Zamir A, Oz C, and Geller B. (2000) Threat mail and forensic science: DNA profiling from items of evidence after treatment with DFO, *Journal of Forensic Sciences,* **45**(2), pp 445–446.

[15] Champod C, Lennard C, Margot P, and Stoilovic M. (2004) *Fingerprints and other ridge skin impressions,* CRC Press, Boca Raton, FL.

[16] Wertheim PA. (1997) Atmospheric superglue method, *Minutiæ,* **44**(September-October), p 6.

[17] Grubwieser P, Thaler A, Kochl S, Teissl R, Rabi W, and Parson W. (2003) Systematic study on STR profiling on blood and saliva traces after visualization of fingerprint marks, *Journal of Forensic Sciences,* **48**(4), pp 733–741.

[18] Abacus Diagnostics (2005) *Human blood identification test,* available at http://www.abacusdiagnostics.com, last access performed on November 18, 2005.

[19] Bluestar forensic (2004) *Hexagon OBTI,* available at http://www.bluestar-forensic.com, last access performed on November 18, 2005.

[20] Abacus Diagnostics (2005) *Semen identification test,* available at http://www.abacusdiagnostics.com, last access performed on November 18, 2005.

[21] Hochmeister MN, Budowle B, Rudin O, Gehrig C, Borer UV, Thali M, and Dirnhofer R. (1999) Evaluation of prostate-specific antigen (PSA) membrane test assays for the forensic identification of seminal fluid, *Journal of Forensic Sciences,* **44**(5), pp 1057–1060.

[22] Wickenheiser RA. (2002) Trace DNA: A review, discussion of theory, and application of the transfer of trace quantities of DNA through skin contact, *Journal of Forensic Sciences,* **47**(3), pp 442–450.

[23] Ladd, C, Adamowicz MS, Bourke MT, Scherczinger CA, and Lee HC (1999) A systematic analysis of secondary DNA transfer, *Journal of Forensic Sciences,* **44**(6), pp 1270–1272.

[24] Hassler MY, Mihalovich JS, and Gibbons MM. (2001) Typing of DNA recovered from steering wheels: Detecting evidence of the non-habitual drive. Spring seminar of the California Association of Criminalistics, available at http://www.cacnews.org, last access performed on November 18, 2005.

[25] Schubert GD. (2005) Forensic value of pattern and particle transfers from deployed automotive airbag contact, *Journal of Forensic Sciences,* **50**(6), pp 1411–1416.

[26] Jochem G. (2004) Fiber-plastic fusions and related trace material in traffic accident investigation. In: *Trace evidence analysis: More cases in mute witnesses,* ed Houck MM. Elsevier Academic Press, Burlington, MA.

[27] Von Bremen A. (1990) The comparison of brake and accelerator pedals with marks on shoe soles, *Journal of Forensic Sciences,* **35**(1), pp 14–24.

[28] Roux C and Margot P. (1997) An attempt to assess the relevance of textile fibres recovered from car seats, *Science & Justice,* **37**(4), pp 225–230.

[29] Roux C and Margot P. (1997) Type population of textile fibres on car seats, *Science & Justice,* **37**(1), pp. 25–30.

[30] Roux C, Langdon S, Waight D, and Robertson J. (1999) The transfer and persistence of automotive carpet fibres on shoe soles, *Science & Justice,* **39**(4), pp 239–251.

[31] Nehse K. (2004) Using 1:1 taping to reconstruct a source. In: *Trace evidence analysis: More cases in mute witnesses,* ed Houck MM. Elsevier Academic Press, Burlington, MA.

[32] Murray RC. (1982) Chapter 13—Forensic examination of soil. In: *Forensic science handbook,* ed Saferstein R. Prentice Hall Regents, Englewood Cliffs, NJ.

Part II: Chemical Traces—Drugs, Explosives, and Gunshot Residue

Francesco Saverio Romolo

4.9 INTRODUCTION

Motor vehicles are often used in the commission of crimes and are found in criminal activities such as transporting drugs, explosives, and firearms. In these cases the forensic scientist can look for chemical traces due to a transfer from the source to the vehicle. This transfer is a phenomenon of capital importance in forensic science, generally known as the principle of exchange of Locard [1]. There are three major classes of chemical traces described in this chapter that could have some probative value in solving a crime: drugs, explosives, and gunshot residue. The detection of the presence of drugs, explosives, or gunshot residue can represent very important evidence in the investigation of a crime.

4.10 CHEMICAL TRACES

4.10.1 Illicit Drugs

Drugs and explosives are mainly solid materials. Particles can be transferred during any contact with a receiving surface or can fall on the floor of the car or in the cargo area of a truck. In the past, most drugs were made from plants, such as the coca bush for cocaine, opium poppies for heroin, and cannabis for hashish or marijuana. According to the United Nations Office on Drugs and Crime data from 95 countries, in 2003, 52% of the seizures were related to cannabis, 25% involved opiates, 10% involved amphetamines, and 7% involved cocaine [2]. There are two main forms in which cannabis is consumed: herbal cannabis and cannabis resin. The former comprises the flowering tops and leaves of the plant. Cannabis resin is popularly referred to as hashish and consists of the secretions of the plant created in the flowering phase of its development. Pure cocaine and pure heroin are white or colorless crystalline powders, but street heroin samples can be brown because of insufficient purification procedures. In the last decade, the most significant trend has been the increase in the number of seizures of amphetamine-type stimulants. Global amphetamine-type stimulant production is currently above 400 tons, three quarters of which is either methamphetamine or amphetamine and one quarter of which is "ecstasy" [2]. Amphetamine-type stimulants are generally found in solid form as powder or tablets. Physical aspect and physicochemical properties of the different compounds are important for their recognition and for their detection (Table 4-1) [3].

Table 4-1

Illicit Drugs: properties of pure compounds [3].

Compound	Formula	Molecular weight [g/mol]	Physical state	Melting point
Amphetamine	$C_9H_{13}N$	135	Liquid	—
Amphetamine phosphate	$C_9H_{13}N$, H_3PO_4	233	Solid	Decomposes at about 300°C
Amphetamine sulfate	$(C_9H_{13}N)_2$, H_2SO_4	368	Solid	Above 300°C with decomposition
Cocaine	$C_{17}H_{21}NO_4$	303	Solid	98°C
Cocaine hydrochloride	$C_{17}H_{21}NO_4$, HCl	339	Solid	About 195°C with decomposition
Heroin	$C_{21}H_{23}NO_5$	369	Solid	173°C
Heroin hydrochloride	$C_{21}H_{23}NO_5$, HCl, H_2O	423	Solid	229°C to 233°C
Methamphetamine	$C_{10}H_{15}N$	149	Liquid	—
Methamphetamine hydrochloride	$C_{10}H_{15}N$, HCl	185	Solid	170°C to 175°C
Methylenedioxymethamphetamine (MDMA, Ecstasy)	$C_{11}H_{15}NO_2$	193	Liquid	—
Methylenedioxymethamphetamine hydrochloride (MDMA, Ecstasy)	$C_{11}H_{15}NO_2$, HCl	230	Solid	147°C to 153°C

4.10.2 Explosives

Trace detection of explosives and drugs is generally dependent on their vapor pressure, that is, their ability to generate vapors at a given temperature in equilibrium above the solid (or liquid) phase. These data have been thoroughly studied for explosives and are included in Table 4-2, along with other physicochemical properties [4].

Most of the high explosives and all the inorganic compounds in black powder and pyrotechnical mixtures are solid. Nitroglycerine (NG) and ethyleneglycoldinitrate (EGDN) are liquid at room temperature but are always mixed with other compounds in commercial products. Nitrocellulose (NC) allows NG (and EGDN) to form a gelatinous mixture. One of the strongest commercial explosives is called blasting gelatin and consists of 92% to 94% NG gelatinized with 6% to 8% of a special type of NC, called guncotton [5]. In some applications, called extra dynamites or extra gelatin dynamites, a fuel/oxidizer mixture is added to the composition. Fuels such as sawdust or wood meal (fine sawdust) and oxidizers such as sodium nitrate or, more often, ammonium nitrate are typically used. NG can also be used in some smokeless powders. Smokeless powders are low explosives, used in firearms as propellants for projectiles. The composition of smokeless powders can be distinguished between single-base powders (NC), double-base powders (NC plus NG), or triple-base powders (NC, NG, nitroguanidine). In propellant manufacturing one or more additives are always used, such as dinitrotoluenes (DNT), diphenylamine(DPA), or ethylcentralite

Table 4-2

Explosives: properties of pure compounds [4].

Compound	Molecular formula	Molecular weight [g/mol]	Physical state	Melting point [°C]	Temperature of explosion [°C]	Vapor pressure at 20°C [Pa]
Ethyleneglycoldinitrate (EGDN)	$C_2H_4N_2O_6$	152	Liquid	—	237	5.1
Nitroglycerine (NG)	$C_3H_5N_3O_9$	227	Solid	13	270	0.03–0.2
Triacetone triperoxide (TATP)	$C_9H_{18}O_6$	222	Solid	94	227	0.4
2,4-Dinitrotoluene (2,4-DNT)	$C_7H_6N_2O_4$	182	Solid	69	270	2.5
2,4,6-Trinitrotoluene (TNT)	$C_7H_5N_3O_6$	227	Solid	81	288	0.001
Pentaerythritolte-tratrinitrate (PETN)	$C_5H_8N_4O_{12}$	316	Solid	141	210	1 to 8 $\times 10^{-6}$
Cyclotrimethylene-trinitramine (RDX)	$C_3H_6N_6O_6$	222	Solid	204	217	1 to 4 $\times 10^{-7}$

Table 4-3

Typical formulations of some smokeless powders produced in the United States [6].

	Composition [%]				
Type	NC	NG	DNT	DPA	EC
Single base M6	87	—	10	1	—
Single base MIO	98	—	—	1	—
Double base M2	77.45	19.5	—	—	0.6
Double base M5	81.95	15.0	—	—	0.6
Double base M8	52.15	43.0	—	—	0.6
Double base M21	53.0	31.0	—	—	2.0
Double base N5	50.0	34.9	—	—	—
Double base MDM	48.6	27.0	—	—	1.1

Totals might not reach 100% because only the components of forensic interest are presented.

(EC). Table 4-3 shows the main components in typical formulations of several smokeless powders produced in the United States [6].

These explosives can transfer traces of NG and EGDN through vapor due to their high vapor pressure, leaving traces behind without the necessity of a direct contact. Another explosive that can leave traces due to vapor transfer is triacetone triperoxide (TATP). It is an explosive in solid form that undergoes substantial sublimation [7]. There are other explosives containing volatile compounds or mixtures, such as ammonium nitrate fuel oil (ANFO) or emulsion explosives, where diesel or other hydrocarbons are present [5].

4.10.3 Gunshot Residue

Chemical traces from firearms are generally called gunshot residue (GSR), cartridge discharge residue, or firearms discharge residue. GSRs are produced when a firearm cartridge is shot. They are composed of burned and unburned particles from the propulsive charge, traces from all the cartridge components, such as the primer, the case, or the bullet, and from the firearm itself. These particles leave the firearm mainly through the muzzle, but a great amount of GSRs passes through the ejection port in semiautomatic and automatic weapons. Also, they can escape through narrow passages, such as the breech end of the barrel in revolvers. If GSRs are found on a specific surface, it was in contact with a GSR source (firearm, spent cartridge case, etc.) or was in the proximity of a firearm at the time of the shooting [8, 9].

4.11 CRIME SCENE EXAMINATION

4.11.1 Security Phase

When a motor vehicle is found and is suspected to be associated with a crime, it must be treated as a scene of crime, including its surrounding area. Before starting any activity, it is necessary to acquire information about the case, particularly for security reasons. A car used by terrorists can be booby-trapped, requiring the intervention of the improvised explosive device (IED) specialists. A truck transporting dangerous material (e.g., radioactive waste) requires the intervention of different specialists at the scene to guarantee the safety of the public and responders. During this preliminary activity, the access to the area surrounding the vehicle must be controlled. As long as a danger exists, forensic issues cannot be fulfilled if they are incompatible with security concerns. Forensic experts may start recording (e.g., video, photographs, sketches) from a safe place before taking control of the zone.

4.11.2 Forensic Phase A: Outside the Vehicle

The vehicle can be considered to be both a scene of crime and evidence. Evidence that is not protected from alteration may not provide useful information. In case of bad weather conditions or damage to the vehicle (e.g., broken windows), it can be protected using a tent or some plastic sheeting or can be transported to a special location for subsequent examination in optimal conditions. It is important to carefully consider the modifications brought to the vehicle due to protection activities. Protective plastic sheets should not be placed in contact with the vehicle without examining its external surface first. More severe destruction of evidence could occur when moving the vehicle. If someone has to penetrate the vehicle to move it, he or she should wear appropriate personal protective equipment (PPE) such as mask, scene suit, gloves, boots, and head cover to prevent any risk of contamination.

Although wearing disposable PPE reduces or eliminates the risk of depositing hairs, fibers, DNA, or trace material from a person's clothing inside the vehicle, it could also destroy evidence such as latent footwear marks or fingermarks. For this reason, any activity inside the motor vehicle during the examination phase must be carefully considered and avoided if unnecessary. Thus, any activity that can be performed properly at the scene without moving the vehicle should be done at the site. Also, before any internal examination of the vehicle, a thorough examination of the external surfaces of the vehicle must be conducted.

After examination, photography, and note taking of the outside of the motor vehicle, it might be necessary to proceed to some evidence sampling. General sampling procedures for chemical traces are described later in this chapter; however, a specific procedure for the exterior is adopted for shooting distance determination. The nitrite ions and smokeless powder residues around the bullet entrance hole can be transferred onto an adhesive sheet. After the transfer, lead and copper deposits around the hole can be visualized by spraying suitable reagents directly on the target. The adhesive sheet is examined later in the laboratory using the modified Griess test (MGT) after alkaline hydrolysis of the residues [10].

4.11.3 Forensic Phase B: Vehicle Entrance

The area surrounding the vehicle may be a public space where a full contamination control procedure cannot be reasonably adopted. Thus, the potential contamination of the inside of the motor vehicle must always be taken into account and its occurence must be reduced to a minimum. A contamination control procedure should include a decontamination zone where disposable PPE can be correctly suited on, blank samples can be taken, and crime scene equipment can be cleaned before being used in the vehicle. A blank sample should be taken on disposable gloves and jumpsuit after being suited on the crime scene officer and before entering the vehicle in every search for chemical traces. These blank samples should be analyzed with all other samples taken inside the vehicle. They help to ensure that analytical results were not due to contamination coming from the personnel attending the scene or the surrounding environment. If on-site analysis can be conducted, the apparatus should be set up between the decontamination area and the vehicle. In the same decontamination area, crime scene equipment can be decontaminated after use and disposable PPE can be discarded.

4.11.4 Forensic Phase C: On-Site Analysis

There are several instruments developed for airport and border control that can be transported to a scene. Some instruments can detect both contraband drugs and explosive vapors, and other apparatus are only explosive vapor detectors (EVD). Until the early 1970s, most EVD were based upon gas chromatography (GC) with electron capture detection (ECD). Because of the low selectivity of the electron capture detector, these instruments exhibited a high rate of false alarms. At that time, vapor detection systems based upon ion mobility spectrometry (IMS) started to appear on the market. IMS is a highly sensitive analytical technique able to detect a wide range of chemical compounds (both organic and inorganic) at trace levels in the gas phase. The potential use of IMS for the detection of contraband drugs was discovered almost since the advent of the technique [11]. After the explosion of Pan Am flight 103 over the Scottish town of Lockerbie, high throughput rugged detectors for airport security (based upon IMS) were developed to detect explosive particles rather than vapors [12]. Also, thousands of handheld IMS systems were used in the Gulf War in 1990 and 1991 to detect chemical warfare agents [13]. Thermedics Incorporated (Woburn, Massachusetts) developed a system for the detection of explosive traces mainly for airport security activities, commercialized under the name of EGIS. It is based upon high-speed GC combined with a highly selective and sensitive chemiluminescence detector (CD). It is probably the only well-functioning, but very expensive, instrument available on the market that is based upon the GC-CD technology [14]. There are several comparative studies about commercial on-site apparatus for both explosives and illict drugs [15–17]. All these systems can be handheld, or they include a sampling device that can trap traces onto a suitable substrate and a desorption heater that thermally desorbs traces into the analytical apparatus. In a forensic setting, it is important to understand that the systems described for on-site detection only produce indicative results that must be confirmed in the laboratory with specific analytical techniques. Figure 4-31 shows members the on-site IMS apparatus used by members of the Ecole des Sciences Criminelles of the University of Lausanne in Switzerland.

Figure 4-31

The on-site IMS apparatus used by the members of the Ecole des Sciences Criminelles of the University of Lausanne (Switzerland) close to a vehicle to be examined.

4.11.5 Forensic Phase D: Sampling

Initially, when examining the inside of a motor vehicle, attention should be paid to the collection of clearly visible and macroscopic items that can be recovered by hand with the use of tweezers or small brushes. It is not possible to proclaim a rule as to the order of collection of items that is suitable for all types of evidence. Activities inside the vehicle can damage latent fingermarks; however, the use of powders for such marks can create problems for the sampling of chemical traces. Specialists should carefully evaluate the case to decide the correct order of activities inside the car and should enter the vehicle only after well-thought out planning. The choice of a correct sampling technique is critical for the proper detection and identification of chemical traces and must be decided during the planning. The sampling method depends on several factors, including the nature of the compounds sought, the type and area of the surfaces under examination, and the analytical techniques used for the detection and confirmatory analysis. Chemical traces can be sampled by using one of the techniques described hereafter.

A/ Gas Trapping

When using the gas trapping technique, samples are collected by drawing a known volume of air through a trap. The trap can be a sampling tube containing a suitable material such as Tenax resin [18]. Samples can be desorbed and analyzed either on-site to get preliminary results or in the laboratory. Gas trapping can be used for explosives with high vapor pressure such as EGDN, NG, and TATP. This sampling technique is particularly useful when large volumes are examined, such as inside the cargo area of a truck. The main advantage of gas trapping is the minimal activity required inside the vehicle. Large amounts of air can be drawn from a little hole, without the need for the examiner to enter the volume to be sampled. The main disadvantage of this technique is its lack of sensitivity, particularly with explosives or drugs that exhibits very low vapor pressure.

B/ Swabbing

Swabbing is the method of choice to work on tables, floors, and smooth fabrics such as leather or plastic. It is difficult to evaluate which is the best swabbing system. Different materials for swabbing are often used: cotton balls, synthetic wool, filter paper, nonwoven cotton cloth, and Acrilan. An ideal swabbing system should efficiently remove the traces from the surface with as little coextracted interfering compounds as possible. Swabs are generally used to collect particulate matter before being thermally desorbed into an IMS system on-site or in another desorption system later in the laboratory, mainly for GC analysis. The sample is taken by firmly rubbing the area several times, using hands, tweezers, or a specifically designed swab holder. Figure 4-32 shows an example of a commercial swab prewetted by isopropanol and water being used to sample a surface [19].

If a wet swab is used, it must be ensured that the sampled compounds remain stable in the solvent used for the swab. For example, when in water solution, some explosives can be degraded by hydrolysis and bacterial activity. And, because the nature of the traces sampled

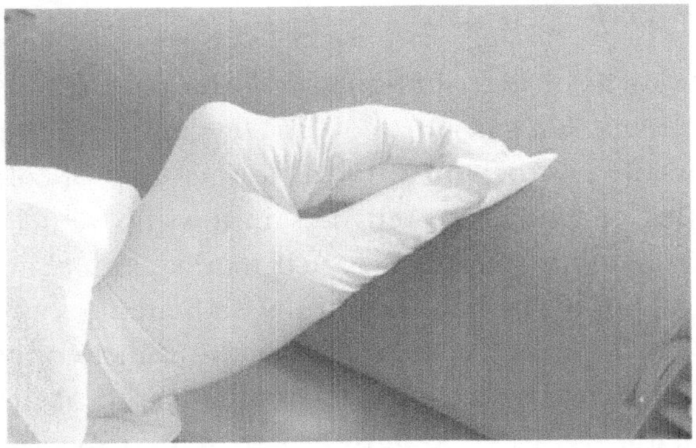

Figure 4-32

Example of swabbing of a vehicle's dashboard using a commercial swab prewetted with isopropanol and water.

with the swab is most often not known, it is difficult to evaluate which is the best solvent to use with a swabbing system. Nevertheless, there is no general solvent that would suit perfectly all sorts of chemical traces. They all present some advantages and drawbacks. The material of the swab and the solvent, if any, should be chosen after careful consideration of the traces sought and the analytical techniques to be used later in the laboratory. Prewetted single-sealed swabs should minimize the possibility of contamination.

C/ Vacuum Lifting

Vacuum lifting is generally preferred with fabrics. Particles are lifted by an air flow and trapped in a sampling device, generally a filter. It is important that the filter does not interfere with the compounds sampled. For example, drugs are generally vacuum lifted using a Teflon filter. Many on-site systems are equipped with a vacuuming apparatus. A disposable filter is inserted in the sampling unit of the on-site IMS apparatus, shown in Figure 4-31, before sampling (Figures 4-33 and 4-34). The sampling unit is then used inside the car (Figure 4-35). A specially equipped vacuum cleaner with disposable filters can also be used (Figure 4-36).

Typically, vacuuming is performed just above the surface to be examined. When using a vacuuming unit, care must be taken to ensure that the components transferring the traces to the filter are absolutely clean. It is best to use disposable sampling units. If analytical instruments are available on-site, it is possible to perform a test before sampling to ensure that the sampling device is clean. A filter disk put into a disposable syringe barrel attached to a vacuum pump can be used as a sampling unit. The Forensic Science Laboratory of Northern Ireland developed an efficient vacuuming system for the recovery of organic and inorganic cartridge discharge residue [20, 21]. The suction sampling apparatus consists of a 25-millimeter diameter in-line Delrin filter holder and a 25-millimeter diameter Fluoropore membrane filter. Traces can be trapped on the filter using flexible plastic tubing and a vacuum pump (Figure 4-37). Several filter holders and filters are commercially available. Their dimensions and chemical composition can be chosen, based upon the desired sampling capability and the subsequent analytical steps. Larger holders and filters allow for

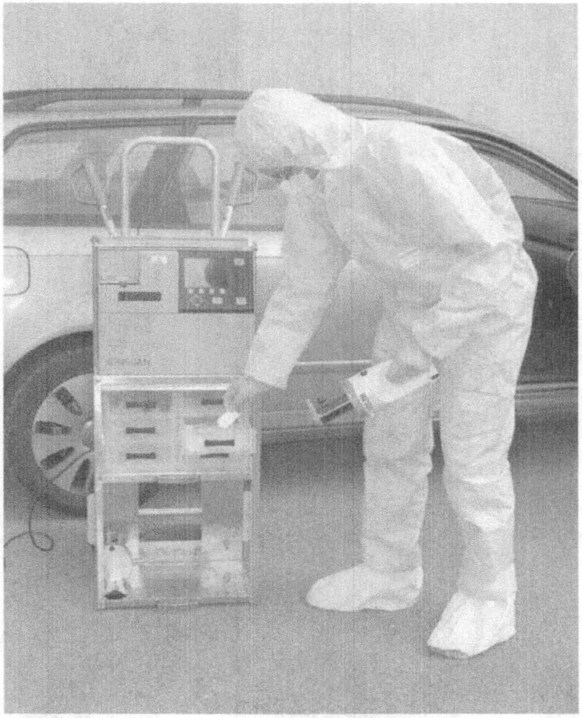

Figure 4-33

A crime scene officer with personal protection equipment at the scene of a stolen-recovered vehicle prepared to perform a vacuum lifting of chemical traces in the vehicle.

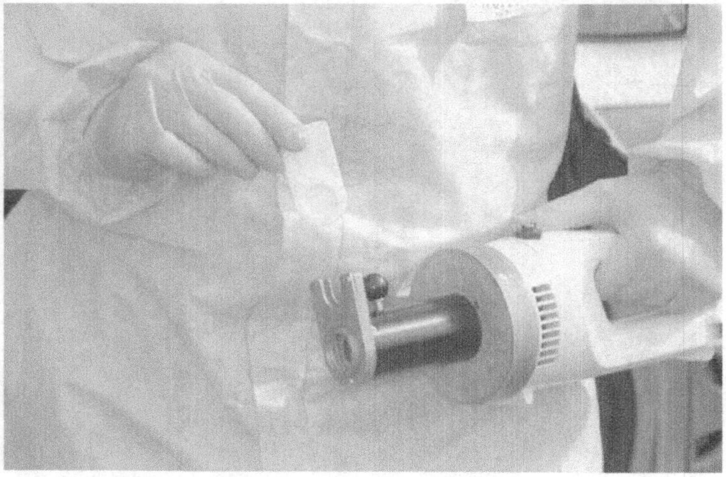

Figure 4-34

The disposable filter is inserted into the filter holder of the vacuuming apparatus.

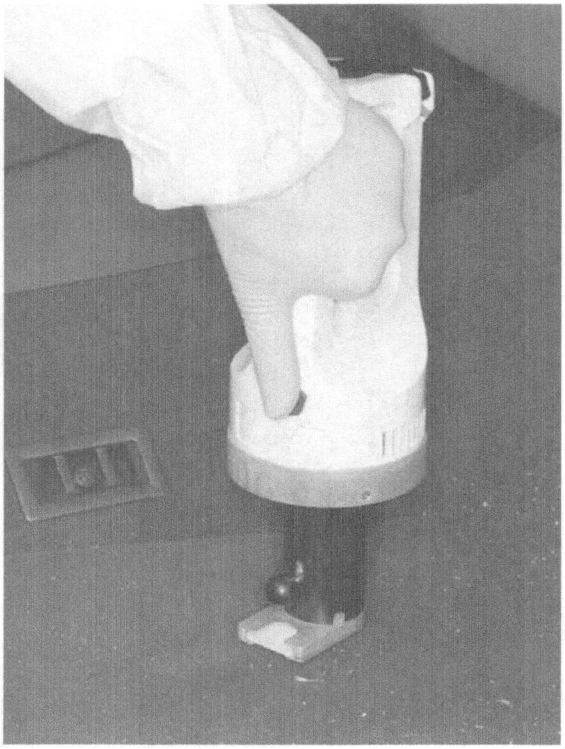

Figure 4-35
Vacuum lifting inside a car.

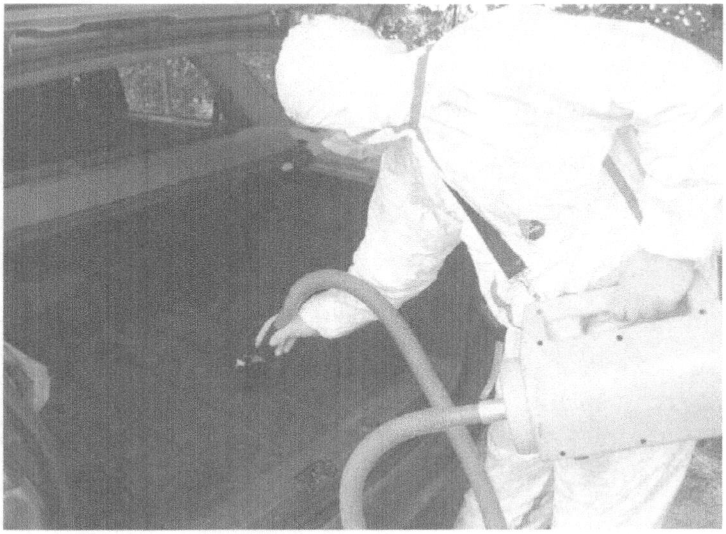

Figure 4-36
The specially equipped vacuum cleaner can also be used to vacuum lift samples of chemical traces inside a car.

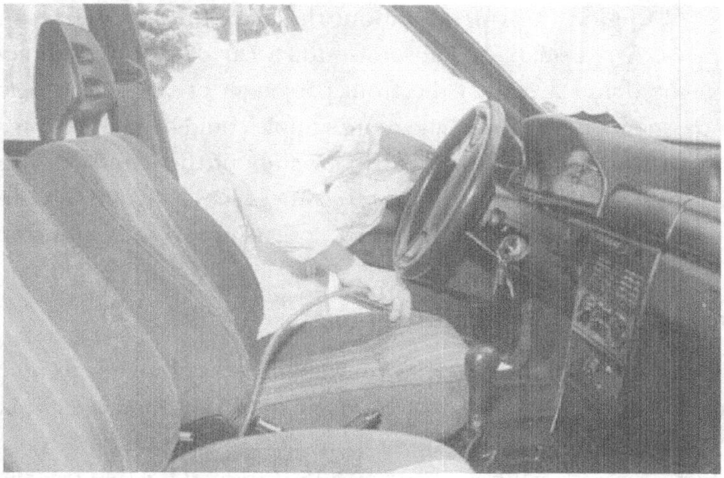

Figure 4-37

An expert sampling traces of explosives and gunshot residue using a Delrin filter holder, a Fluoropore membrane filter, a flexible tube, and a vacuum pump.

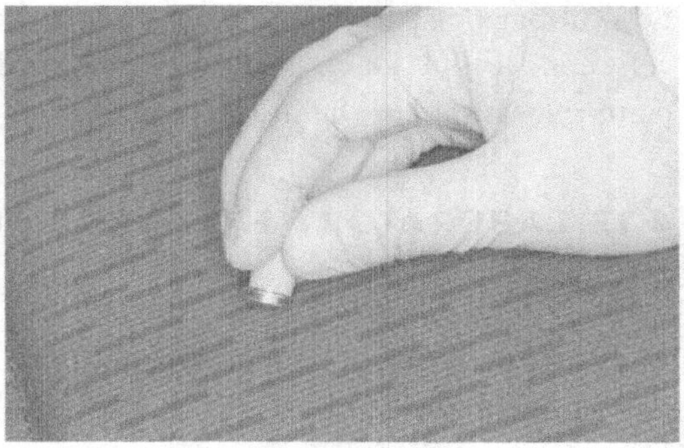

Figure 4-38

Tape lifting inside a car using small adhesive tabs.

larger samples to be collected. However, this can be problematic to extract and could yield larger amount of interfering material. Also, the material of the filters must be compatible with the solvents used in the extracting procedure.

D/ Tape Lifting

Collection of particles by adhesive tape lifting of surfaces is the favorite sampling approach for scanning electron microscope (SEM) analysis [8, 9]. The loss of stickiness restricts the size of area from which particles can be collected, and extraction procedures are more complex, due to the matrix effect of the adhesive. Figure 4-38 shows an example of tape lifting using a small adhesive tab performed on a fabric surface inside a car.

Before sampling, it is necessary to plan not only the sampling technique, but also the number of samples to be collected. When properly used, several samples can provide topographic data useful to gain information about the case, but too many samples can result in the loss of traces. A reasonable compromise for a car can be the choice of taking four samples from (1) the driver area, (2) the front passenger area, (3) the rear passenger side, and (4) the rear passenger driver side. One more sample could be taken from the space behind the rear seat and/or in the trunk. The information about the case should help develop the most appropriate strategy. For example, if an eyewitness sees a criminal hiding a pistol under the seats, samples should be taken from under the seats to verify this information.

4.11.6 Forensic Phase E: Packaging of Samples

Samples must be preserved immediately to prevent any loss of materials and any risk of contamination. The packaging chosen depends on the nature of the trace. Volatile compounds (e.g., EGDN, NG, and TATP) may evaporate if the containers are not airtight. In this instance, clean metal cans or nylon bags specifically designed for that use are recommended. Also, small particles can be lost from packages that are not properly sealed. Proper sealing and packaging prevent the possible transfer of foreign matter from outside during transport and storage too. This is an important step in the contamination control procedure. Containers can be sealed with sealing tape and then with evidence tape signed by the collector. Several bags or cans specifically designed to collect evidence are commercially available. A reference to the sketch done at the beginning of the forensic activity can easily help to associate the sample to the area sampled, thus recording topographical information.

4.12 LABORATORY EXAMINATION OF SAMPLES

4.12.1 Analysis

There are two levels of analysis for chemical traces that can be performed on samples: screening and confirmatory. Screening analyses are faster and cheaper than confirmatory analyses. However, screening analyses can give false-positive results and should solely be conducted to avoid more expensive and time-consuming analysis of negative samples. On-site analyses give only preliminary results that must be confirmed in the laboratory with more selective techniques.

Traces of drugs are generally confirmed using GC or high-performance liquid chromatography (HPLC) with mass spectrometry detection [22, 23]. For the confirmation of organic explosives, both GC-CD and HPLC with electrochemical detection were often used in the past. Now HPLC-MS systems are more easily found in forensic laboratories. Ionic chromatography and capillary electrophoresis are used for the analysis of inorganic components in explosive substances [24, 25].

GSRs are generally analyzed using a scanning electron microscope equipped with an energy dispersive x-ray analyzer (SEM-EDX) that can isolate and help identify individual GSR particles through both morphological and elemental characteristics. An extended use of lead-free ammunition in the future could give more importance to other analytical techniques for organic GSR detection and identification [19].

4.12.2 Interpretation of Results

When interpreting analytical results in a forensic context, the probability of the evidence supporting a criminal activity (i.e., some explosive or drug was transported in a car) should be weighed against the probability of the evidence supporting alternative explanations [26, 27]. Currently, only limited data are available to assess the likelihood that a car might have become contaminated with drugs or explosives without being involved in criminal activities. The most important "population study" about explosive traces is probably the survey published in 1996 by Crowson et al., carried out to determine the background levels of explosive residues in public places [28]. Samples were taken from 25 taxis (124 samples) and 10 buses (87 samples) and NG was never detected, whereas RDX was found in 3 samples taken from taxis. Another group of samples was taken from 21 police vehicles (120 samples), and NG traces were found in 8 of the 21 police cars.

Some tests simulating the transport of drugs or explosives in a vehicle can help rule out secondary or tertiary transfers and cross-contamination. Another aspect to consider during interpretation is the persistence of traces. During the investigation on the Mafia bombing attacks committed in Rome in 1993 (via Fauro, S. Giovanni in Laterano and S. Giorgio in Velabro), in Florence (via dei Georgofili, near the Uffizi Gallery), and in Milan (via Palestro), traces of the explosives used were found years later in the vehicles used to transport explosive and in places where explosive charges were prepared or hidden before the terrorist attacks.

When interpreting evidence from GSR, it is necessary to understand that although chemical analysis can identify traces from the discharge of a cartridge, they cannot distinguish whether the traces were deposited on the surfaces surrounding the weapon while shooting or if they are due to a transfer from a surface rich with GSRs (firearm, cartridge case, bullet hole, etc.) to the hands or clothes of someone [8, 9]. Only by carrying out a topographical analysis of the traces it is possible to discriminate between different hypotheses (e.g., leaving a pistol under the seat of a car or shooting from inside the car). Thus, in case of GSRs, it is primordial for the exact location of the sampling to be carefully documented.

ACKNOWLEDGMENTS

The author would like to thank the Institut de Police Scientifique, University of Lausanne (Switzerland), and RIS Carabinieri (Italy) for providing some of the illustrations.

BIBLIOGRAPHY

[1] Locard E. (1934) *La police et les méthodes scientifiques,* Les Editions Rieder, Paris, France.

[2] United Nations Office on Drugs and Crime (2005) *World drug report,* United Nations Publications, Geneva, Switzerland.

[3] Moffat AC, Osselton MD, and Widdop B. (2004) *Clarke's analysis of drugs and poisons,* 3rd edition, Pharmaceutical Press, London, United Kingdom.

[4] Oxley JC, Smith JL, Shinde K, and Moran J. (2005) Determination of the vapor density of triacetone triperoxide (TATP) using a gas chromatography headspace technique, *Propellants, Explosives, Pyrotechnics,* **30**(2), pp 127–130.

[5] Meyer R, Köhler J, and Homburg A. (2002) *Explosives,* 5th edition, Wiley-VCH, Weinheim, Germany.

[6] Urbanski T. (1984) *Chemistry and technology of explosives,* Volume 4, Pergamon Press, Oxford, United Kingdom.

[7] Tamiri T. (2000) Explosives: Analysis. In: *Encyclopedia of Forensic Sciences,* ed Siegel J, Knupfer G, and Saukko P, Academic Press, London, England, pp 729–745.

[8] Meng HH and Caddy B. (1997) Gunshot residue analysis—a review, *Journal of Forensic Sciences,* **42**(4), pp 553–570.

[9] Romolo FS and Margot P. (2001) Identification of gunshot residue: a critical review, *Forensic Science International,* **119**(2), pp 195–211.

[10] Glattstein B, Zeichner A, Vinokurov A, and Shoshani E. (2000) Improved method for shooting distance determination. Part 2—bullet holes in objects that cannot be processed in the laboratory, *Journal of Forensic Sciences,* **45**(5), pp 1000–1008.

[11] Karpas Z. (1989) Forensic science applications of ion mobility spectrometry, *Forensic Science Review,* **1**, pp 103–119.

[12] Ewing RG, Atkinson DA, Eiceman GA, and Ewing GJ. (2001) A critical review of ion mobility spectrometry for the detection of explosives and explosive related compounds, *Talanta,* **54**(3), pp 515–529.

[13] Eiceman GA. (2002) Ion-mobility spectrometry as a fast monitor of chemical composition, *Trends in Analytical Chemistry,* **21**(4), pp 259–275.

[14] Kolla P. (1997) The application of analytical methods to the detection of hidden explosives and explosives devices, *Angewandte Chemie International Edition in English,* **36**(8), pp 800–811.

[15] Rhykerd CL, Hannum DW, Murray DW, and Parmeter JE. (1999) *Guide for the selection of commercial explosives detection systems for law enforcement applications,* National Institute of Justice Guide 100-99, US Department of Justice, Washington, DC.

[16] Bruschini C. (2001) *Commercial systems for the direct detection of explosives (for explosive ordnance disposal tasks),* ExploStudy Final Report, Ecole Polytechnique Fédérale de Lausanne, Switzerland, available at http://diwww.epfl.ch/lami/detec/explostudy.html, last access performed on September 1st, 2005.

[17] Butler RF. (2002) *Mailroom scenario evaluation, final report,* National Institute of Justice, US Department of Justice, Washington, DC.

[18] Sigman ME, Ma C-Y, and Ilgner RH. (2001) *Performance evaluation of an in-injection port thermal desorption/gas chromatographic/negative ion chemical ionization mass spectrometric method for trace explosive vapor analysis,* Analytical Chemistry, **73**(4), pp 792–798.

[19] Romolo FS. (2004) *Organic gunshot residue from lead-free ammunition,* PhD thesis, Ecoles des sciences criminelles, Institut de police scientifique, University of Lausanne, Lausanne, Switzerland.

[20] Wallace JS and McKeown WJ. (1993) Sampling procedures for firearms and/or explosives residues, *Journal of the Forensic Science Society,* **33**(2), pp 107–116.

[21] Speers SJ, Doolan K, McQuillan J, and Wallace JS. (1994) Evaluation of improved methods for the recovery and detection of organic and inorganic cartridge discharge residues, *Journal of Chromatography A,* **674**(1–2), pp 319–327.

[22] Gough TA. (1991) *The analysis of drugs of abuse,* John Wiley and Sons, Chichester, United Kingdom.

[23] Cole MD and Caddy B. (1995) *The analysis of drugs of abuse: an instruction manual,* Ellis Horwood, Chichester, United Kingdom.

[24] Yinon J and Zitrin S. (1981) *The analysis of explosives,* Pergamon Press, Oxford, United Kingdom.

[25] Yinon J and Zitrin S. (1993) *Modern methods and applications in analysis of explosives,* John Wiley and Sons, Chichester, United Kingdom.

[26] Robertson B and Vignaux GA. (1995) *Interpreting evidence: evaluating forensic science in the courtroom,* John Wiley and Sons, Chichester, United Kingdom.

[27] Aitken CGG and Taroni F. (2004) *Statistics and the evaluation of evidence for forensic scientists,* Second edition, John Wiley and Sons, Chichester, United Kingdom.

[28] Crowson CA, Cullum HE, Hiley RW, and Lowe AM. (1996) A survey of high explosives traces in public places, *Journal of Forensic Sciences,* **41**(6), pp 980–989.

CHAPTER 5

TRACES AND THEIR EVIDENTIARY VALUE

Eric Stauffer

5.1 INTRODUCTION

The goals of forensic sciences are to:

- Determine whether a crime has been committed;
- Identify the victim(s) and perpetrator(s);
- Determine the crime's modus operandi.

These goals are reached by examining crime scenes, collecting items of evidence, examining these items at the laboratory, and properly interpreting the results. The establishment of links between the crime scene, the victim(s), and the perpetrator(s) is the most crucial part of the overall process. The demonstration that the perpetrator was at the scene of the crime is a necessary element in implicating that person in the direct commission of the crime. Indeed, finding items of evidence originating from the victim on a perpetrator (and/or the reverse) clearly demonstrates a contact (or inference thereof) between these two individuals. The principle behind the transfer of traces from a source (object or person) onto a recipient (person or object), which allows for the establishment of links, was first enunciated by French criminalist Dr. Edmond Locard. He stated that [1]: "Toute action de l'homme, et a fortiori, l'action violente qu'est un crime, ne peut pas se dérouler sans laisser quelque marque." This can be directly translated in English to "any action of an individual, and obviously, the violent action constituting a crime, cannot occur without leaving a trace." This principle, often referred to as Locard's principle of exchange, has been generalized to state that when two objects/people come into contact, traces from each object/person will be left onto the other one.[1] When such an ideal situation is available, it is said that a

[1] Locard wrote many books and articles into which other versions of the principle were enunciated. A famous one states that [2]: "Nul ne peut agir avec l'intensité que suppose l'action criminelle sans laisser des marques multiples de son passage, tantôt le malfaiteur a laissé sur les lieux des marques de son activité, tantôt, par une action inverse, il a emporté sur son corps ou sur ses vêtements les indices de son séjour ou de son geste." This can be directly translated in English to "No one can act with the intensity of a criminal activity without leaving multiple traces of his/her action. In some instances, the criminal leaves traces of his/her activity at the scene and in some other instances, in an inverse fashion, he/she takes on his/her body or clothing, traces from the scene or from the criminal act itself."

cross-exchange or cross-transfer occurred. Most often, the transfer occurs only in one direction, meaning that one object/person leaves traces onto another one, without taking any traces from it.

Figure 5-1 shows the links that may be established during the forensic investigation of a crime scene. Although in many instances the victim is present at the scene, he or she could also be moved from one scene to another. With regard to the investigation of auto theft, the vehicle can be considered as the victim in the figure. In this case, evidence can be used to link the offender to the vehicle and the vehicle to the crime scene(s) (such as the scene where it was stolen from, the scene where the vehicle has been driven to during its theft, or the scene where it was recovered).

Some traces can be used as primary links and some as secondary links. For example, a bloodstain is a primary link to a human being. To the contrary, a shoeprint is a secondary link to a human being. A shoeprint is a primary link to a shoe, which can then be linked to a human being (e.g., by finding it on a person or by finding DNA in the shoe). Another interesting aspect of linking different objects, people, and scenes exists in cases of repeated offenses. This aspect can be very useful when dealing with professional auto thieves, who steal vehicles on a regular basis. For example, if the fingerprints left on two a priori independent

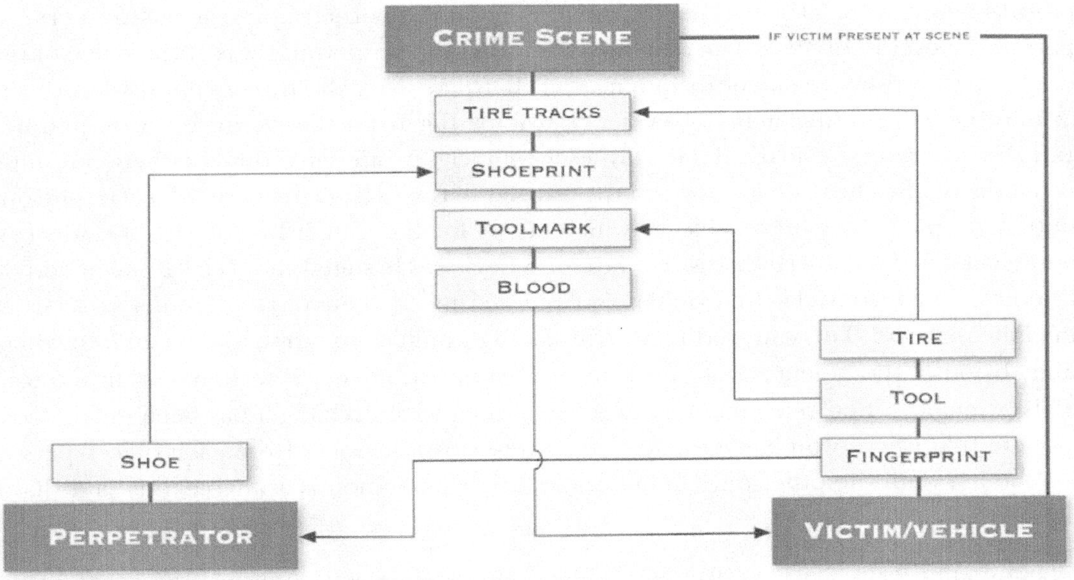

Figure 5-1

Schematic representation of different items of evidence used to establish links between crime scene(s), victim(s) (or vehicle[s]), and perpetrator. When the victim is found at the scene, a direct link is established as shown with the top right link. Thick lines with no arrows are direct links as the items or traces were present at the scene, on the vehicle (victim), or perpetrator. Thin lines with arrows represent the links established through the forensic comparison of the object/person with the trace(s).

vehicles are found to be from the same source, it is possible to demonstrate (infer) a link (common perpetrator) between these two vehicles (for more details on forensic mapping, see Chapter 21). Nevertheless, if two sets of tire tracks found on escape routes from the dumping sites of two a priori independent stolen vehicles originate from the same source, again it is possible to demonstrate a link (common escape vehicle) for the two thefts. Figure 5-2 illustrates this principle. These types of links across multiple criminal acts allow law enforcement to connect different crimes that were a priori independent but that are actually committed by the same offender or group of offenders. Additionally, this permits the information collected from each investigation, which might not be sufficient to resolve each crime independently, to be used jointly, which could bring the pieces of the puzzle together and lead to a successful outcome. This is particularly useful when dealing with organized crime.

The general concept of the examination of crime scenes has been presented in Chapter 3 and the particular application to vehicles in Chapter 4. The goal of this chapter is to present the different types of evidence (traces) that can be retrieved from a crime scene, more particularly from a vehicle. The forensic investigator performing stolen-recovered vehicle examination should be aware of what types of evidence can be recovered and what forensic value each type of evidence could bring to the investigation. Each type of com-

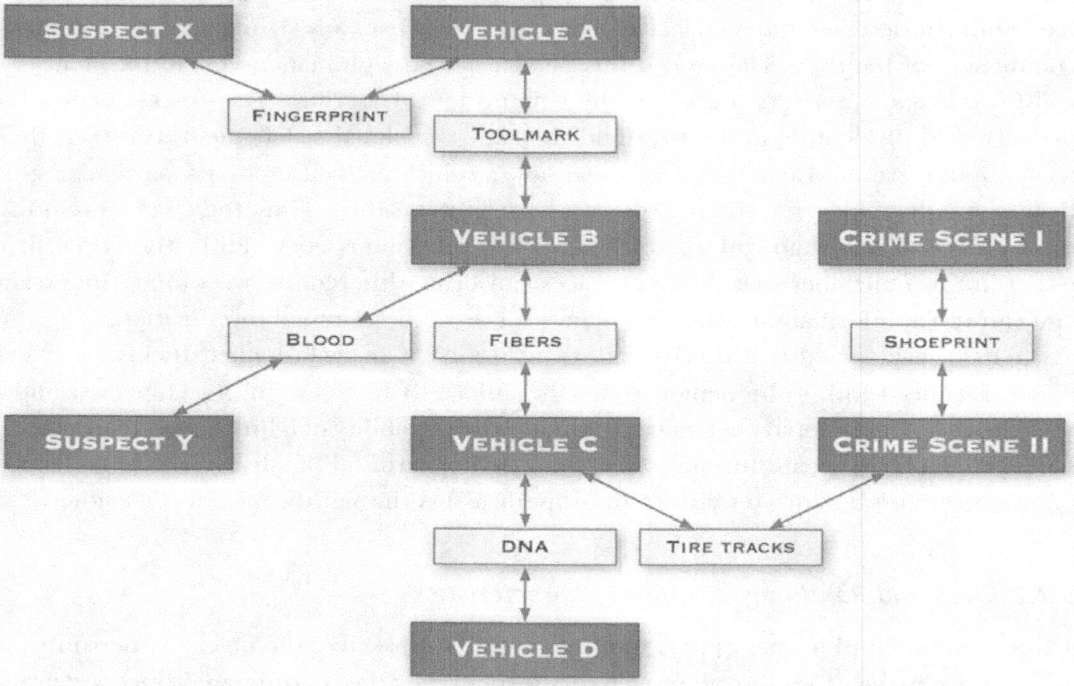

Figure 5-2

Schematic representation of the establishment of links between crime scenes, vehicles, and perpetrators with common traces and evidence.

monly encountered evidence is briefly presented along with its origin, characteristics, and forensic/evidentiary value. As stated previously, the exploitation of these items of evidence can bring a tremendous amount of intelligence to an auto theft investigation, which can ultimately result in the identification of criminals and the repression of this heinous crime.

5.2 TRACE PROPERTIES

5.2.1 Evidence Functions

Virtually, any traces or items can become evidence in a criminal investigation. However, there are some common types of traces such as fingerprints, DNA, and toolmarks that are of particular forensic interest due to their inherent nature and the circumstances under which they are left at crime scenes. An item of evidence must substantiate one of the three goals of forensic sciences; otherwise, it will bear no pertinence in the scope of the forensic investigation. Thus, there are items of evidence that help determine whether a crime has been committed, some that identify the victim(s)/perpetrator(s), and some that determine how the crime has been committed. Some evidence also brings answers or clues to more than one question. For example, a fingerprint left on the rearview mirror of a stolen-recovered vehicle furthers the primary goal of identifying the person from whom it originates. If this person has legitimate access to the vehicle, it will not necessarily bear any significance regarding the commission of the theft. However, if this person has no legitimate access to the vehicle, it will provide some answers regarding the commission of a crime. Tire tracks further the primary goal of identifying the tire (and then the vehicle) that left them. However, their location and spatial orientation can reveal how a vehicle maneuvered at a particular spot, from where the vehicle was coming, and to where it was heading. Thus, these same tire tracks can also reveal important information regarding the modus operandi of the crime. It is important to remember that different traces can bring different answers to a crime scene and can reveal information regarding many aspects of the criminal investigation.

Items of evidence are also classified as inclusionary or exclusionary. Inclusionary evidence permits a link to be demonstrated (or inferred) between an object/person and a trace, whereas exclusionary evidence eliminates the possibility of a link between an object/person and a trace. In any instance, the circumstances around the discovery and collection of evidence must be known in order to properly assess the significance of the evidence.

5.2.2 Class and Randomly-Acquired Characteristics

The examination of a trace is performed to identify its source: the object or person from which it originates. To do so, characteristics displayed by the incriminated trace are evaluated and compared with the characteristics exhibited by the source itself or by a comparison trace generated by the source. If these characteristics match and are uniquely provided or exhibited by this source to the exclusion of all other sources, it is possible to infer a unique link between the incriminated trace and the source. However, some of these characteristics are not always unique to one source and can be shared between many sources and, thus, many traces of different origins. In forensic sciences, it is common to consider two types of characteristics: class and randomly-acquired.

A/ Class Characteristics

Class characteristics are traits that define a group of individuals or objects collectively. They comprise the set of characteristics that are representative of a certain class. Thus, class characteristics are shared among many individuals or objects. With objects, these characteristics are usually due to the manufacturing process or to the primary material used to fabricate the object. For example, the size of a shoe is a class characteristic, because many shoes share the same size.

B/ Randomly-Acquired Characteristics

Randomly-acquired characteristics are traits that define and identify an individual or item as unique, exclusive to all other individuals or items. In general, randomly-acquired characteristics are created based upon the random history of an item or individual. These characteristics are also often referred to as accidental characteristics. For example, the cuts created on the sole of a shoe by random walking are individual characteristics, because their location, shape, and dimensions are unique to that particular sole and no other.

5.2.3 Requirements

Not all traces left by an object or person are suitable for establishing links with their source. Indeed, not all features exhibited by the trace are suitable, either. Champod identified five conditions that should or must be respected for a trace or trace's features to be reliable in identifying its source: distinguishability, high intersource to intrasource variance, known variance in time, normalization (standardization of examination), and independence [3]. From a practical perspective, it is also important to take into account the cost and time necessary to perform the examination, detectability, and recordability of such features.

Ideally, the features exhibited by a trace are individual (they are not generated by any other sources) and immutable (they never change). However, because nothing is ever ideal, the study and interpretation of traces in forensic sciences can become quite a complicated science.

5.2.4 Evidence Strength

In an ideal world, the crime scene investigator has unlimited time, resources, materials, and personnel available to perform the examination of a stolen vehicle or any other crime-related vehicles. Unfortunately, these circumstances do not often exist. Thus, in some instances, the crime scene investigator must prioritize his or her search of a vehicle and the collection of some specific evidence to the detriment of some others. This is the reason why it is crucial for the investigator to be very familiar with the different types of evidence and their potential strengths. As a good rule, all pertinent evidence must be collected and if the pertinence of a trace cannot be estimated at the scene, it should be also collected. It is always better to collect useless evidence than to not collect useful evidence.

The evidentiary strength of a trace can only be fully appreciated and evaluated once placed in the context of its discovery and within the circumstances of the crime. However, without taking into account the circumstances surrounding the trace and the crime, some traces are usually stronger than others in the establishment of the link between them and their source of origin. The reason lies in their individualization power, i.e. how unique the link established between the incriminated trace and the putative source is.

If a trace bears class characteristics only, it will not be possible to distinguish it from other objects sharing the same set of class characteristics. It can only be correlated to a group of sources. Thus, such traces are called "class evidence" or "corroborative evidence." The link established between the trace and the source is not unique. However, in some instances, it is possible to further qualify this link by integrating the frequency of the characteristics observed within a known population. The interpretation of the strength of evidence, and particularly of class evidence, requires accurate population study as well as a mathematical and statistical approach. There are books dedicated on this particular topic [4–6].

Conversely, some types of evidence allow the criminalist to infer the identification of a unique source from which they originate. These types of evidence can lead to individualization because they display randomly-acquired characteristics in sufficient quantity and quality to individualize their source [3]. In this situation, when a link is established between the incriminated trace and the source, it allows for the exclusion of all other sources as the origin.

Table 5-1 presents different class evidence and evidence leading to individualization that are commonly encountered in forensic sciences.

It is important to specify that when in presence of evidence leading to individualization, it does not necessarily mean that it is possible to establish a unique link between the trace and the source in all instances. As a matter of fact, randomly-acquired characteristics must be present in sufficient quality and quantity and must be observed, recorded, and interpreted to establish this link. This is not feasible with every trace leading to individualization.

Although this type of evidence appears at first sight as more powerful inclusionary evidence than class evidence, it is always important to place evidence in the right context to determine its strength in establishing links. Without taking into account possible circumstances, a toolmark has potentially more value in identifying an object than does paint. However, this could change for a given case when the circumstances around which the evidence was discovered are taken into account.

Finally, the strength of a trace does not only lie in its inclusionary power, but also in its exclusionary power. Class evidence is as powerful as evidence leading to individualization

Table 5-1

Different class evidence and evidence leading to individualization commonly encountered in forensic sciences.

Class evidence	Evidence leading to individualization
Paint	*Fingerprints*
Fibers	*Footprint*
Glass	*Ear print*
Hair	*Lip print*
DNA	Shoeprint
Drugs	Tire tracks
Gunshot residues	Toolmarks (including firearms)
Ignitable liquid residues	*Handwriting*
Explosives	*Signature*

This list is nonexhaustive. Traces in italic are used to identify human beings, other traces are used to identify objects.

because the exclusion of common source stands firm. Thus, although class evidence cannot normally establish a unique link between a trace and a source, it can bring very pertinent information in excluding a suspect or an object as being the source of the trace.

5.2.5 Comparison Process

Paul Kirk, a famous American criminalist, stated that [7]: "The real aim of all forensic science is to establish individuality, or to approach it as closely as the present state of the science allows." He added that [7]: "Criminalistics is the science of individualization". To establish individuality and, therefore, to enable the inference of a unique link between the trace and its putative source, the forensic scientist proceeds by comparing characteristics exhibited by both the trace and the source.

The practice first consists of comparing the class characteristics. The presence of matching class characteristics between the trace and the possible source establishes that the trace may have originated from this source. However, it does not establish that the trace did originate from the source, because class characteristics are shared among more than one source. The strength of the link can be qualified accordingly. Terms such as "possible," "probable," "likely," and "very likely" should be used in reporting the link between the source and the origin. The presence of unexplained nonmatching class characteristics between the trace and the source allows for the exclusion of a common origin. The absence of some class characteristics does not necessarily preclude a common origin.

Once the class characteristics of the source and the trace are deemed matching, it is possible to observe and compare randomly-acquired characteristics. Similar principles apply here. The presence of matching randomly-acquired characteristics between the trace and the source allows one to infer that the trace originates from the source. The presence of unexplained nonmatching randomly-acquired characteristics between the trace and the source leads to the exclusion of a common origin.

5.3 EVIDENCE LEADING TO INDIVIDUALIZATION

5.3.1 Fingerprints and Other Ridge Skin Impressions

Fingerprints have been used in forensic sciences to identify human beings for about 100 years. The first identification processes were performed to identify recidivists. Fingerprints are very valuable traces because they are readily left by a person touching an object, fairly easily detected and collected at the scene, immutable, and finally, they exhibit features that are unique to the person from whom they originate.

Fingerprints form on a finger's skin surfaces early in the embryonic development, and the ridge configuration is created in a random manner, mainly due to the stress (mental and physical) of the pregnancy. Fingerprints are patterns of friction ridge skin. There are other similar patterns on the human body, such as on the palms and on the feet.

In a forensic setting, it is possible to find fingerprints under three different configurations:

I Latent: When the fingerprint is not visible to the naked eye with white light. Special physicochemical techniques can be applied to reveal and enhance such fingerprints. This is the most common manner in which fingerprints are found at crime scenes.

II Patent: When the fingerprint is visible to the naked eye with white light. These traces could be either negative (removal of material such as a finger touching a sooty area) or positive (deposit of material such as a bloody finger touching a surface). Even if the fingerprint is visible, it is still possible to apply physicochemical techniques to improve the observation of impression.

III Plastic: When the fingerprint is a 3D impression, such as in mastic. When dealing with such types of fingerprint, it is possible to obtain a cast of the fingerprint.

Fingerprints display class and randomly-acquired characteristics. Class characteristics are represented by the general shape of the fingerprint. There are three main types of general shapes: arches, loops, and whorls. These general shapes are found in proportions of approximately 5%, 65%, and 35%, respectively. Also, there are some subclasses such as ridge tracing (for whorls) and ridge counting (for loops). Fingerprint classification systems are based upon all these class characteristics. Randomly-acquired characteristics are represented by minutiae as well as ridge and pore structures. Minutiae are the ridge characteristics such as bifurcations, or ridge endings that constitute the friction ridge pattern of the fingerprint.

The comparison process is performed by matching the general shapes of the incriminated and comparison fingerprints and a sufficient amount of minutiae (of sufficient quality). When a certain number of matching minutiae in sufficient quality is present between the incriminated and comparison prints, it can be inferred that the origin of the fingerprint is uniquely attributed to the source. The process is identical (although class characteristics vary accordingly) with all other friction ridge skin impressions.

Fingerprints are the most powerful trace used to identify human beings. In addition, one does not necessarily need to know a possible suspect to identify the source of a fingerprint. The Automatic Fingerprint Identification System (AFIS) is a database readily available in many countries [8]. It contains fingerprints from convicted offenders, crime scenes, and, according to the legislation of the country or state, other classes of people, such as suspects or foreign nationals. Thus, if a fingerprint is discovered at a crime scene, it is possible to search the database for a potential suspect.

5.3.2 Shoeprints

A shoeprint is the impression from a shoe onto a surface. Three types of shoeprints are usually considered (similarly to fingerprints): latent, patent, and depressed. Latent and patent shoeprints are 2D, whereas depressed shoeprints are 3D. Shoeprints are an extremely useful type of evidence used to link shoes to crime scenes. Some thieves understand that they can be identified with fingerprints and thus, use gloves or pay more attention to what they touch to reduce the chances of leaving fingerprints at crime scenes. However, it is difficult to avoid walking and leaving shoeprints behind. Some burglars wear socks over shoes to prevent the deposition of shoeprints.

A shoeprint is used to link it to a putative shoe. It does not establish a direct link with an individual, conversely to fingerprints, because shoes can be worn by different people. However, when shoes are found on a suspect, it is likely that the suspect will have some involvement in the crime or at least some knowledge of the whereabouts of the shoes at the time of the crime. A link between a shoe and a person can also be established by the presence of DNA in the shoe. Also, shoeprint evidence is not limited to the mere identification of its source. When several shoeprints are available as part of a walking action, it is possible to estimate the height of the individual and his or her gait characteristics [9]. Furthermore, based upon the spatial location and configuration of the shoeprints, it is possible to identify the activities that took place at a crime scene (modus operandi). This information could be very pertinent in a particular case.

Shoeprints display class and randomly-acquired characteristics. Class characteristics comprise the general pattern of the sole, its size, and some manufacturing characteristics. Randomly-acquired characteristics include accidentally-acquired characteristics such as cuts, indentations, removal of material, and regular wear and tear. These characteristics are dependent on the history of the shoe, which is typically random. No two shoes have undergone the same exact conditions, thus rendering the accidentally-acquired characteristics totally unique to one shoe only.

The process of identification starts by comparing the class characteristics between the trace and the putative shoe (or usually a shoeprint made with the putative shoe). In case of a match, the examiner pursues the comparison with the randomly-acquired characteristics. It is important to remember that, contrary to fingerprints, shoes are not immutable and randomly-acquired characteristics vary with the shoe's use. New characteristics can appear and old ones can disappear with regular wear and tear. This phenomenon must be taken into account when performing the examination. When a certain number of randomly-acquired characteristics in sufficient quality are present between the print and a putative source, the origin of the shoeprint can be uniquely attributed to the examined shoe.

When no comparison material is available, it is possible for the examiner to attempt to determine the brand and type of shoe using published references and electronic databases [10, 11]. Also, some police departments now have a database of shoeprints collected from crime scenes and impressions of shoes seized from burglars and other criminals [12–15]. This database is the equivalent of AFIS for shoeprints. This brings one more piece of information to the investigator, which could lead to the arrest of a suspect.

5.3.3 Tire Tracks

A tire track is an impression from a tire onto a surface. Tire tracks are treated as shoeprints and can be dealt with almost identically. Most tire tracks are found in the depressed (3D) configurations; however, it is possible to find latent and patent (2D) tire tracks.

As with shoeprints it is possible to establish a link between the trace (tire tracks) and the source (tire). This does not establish a direct link with a vehicle, per se, even though wheels are not changed as often as shoes are with human beings. Thus, a vehicle bearing the tire that made the tire tracks at a crime scene has a high probability of being involved in the criminal activity from which the tire track was extracted. Tire tracks also bear more significance than the mere identification of the tire. Passenger cars have four tires, and thus multiple tracks can be readily available at a scene of crime. Besides their direction of travel and the maneuvers made by the vehicle at the scene, which help in the determination of the modus operandi, the spatial location of the tracks relative to each other can be of extreme pertinence. Data such as the circumference of the tire, as well as track width, wheel base, and turning radius of the vehicle can be extracted. This allows the investigator to refine the search for a suspect vehicle or to eliminate or incriminate a vehicle already in custody.

Tire (or a tire track made with a putative tire) tracks display class and randomly-acquired characteristics. Class characteristics comprise the general pattern of the tire thread, its size, and some manufacturing characteristics. Randomly-acquired characteristics include accidental characteristics such as cuts, indentations, removal of material, and regular wear and tear. As with shoeprints, these characteristics are acquired based upon the random history of the tire, and because no two tires undergo the exact same conditions, these characteristics are unique.

The process of identification starts by comparing the class characteristics between the trace and the putative tire. In the case of a match, the examiner pursues the comparison with the randomly-acquired characteristics. Similarly to shoeprints, these characteristics present on tire tracks vary with time, and this must be taken into account during the comparison process. When a certain number of these characteristics in sufficient quality are present between the trace and a putative source, it is possible to infer that the tire track originates from that unique tire only.

When no comparison material is available, it is possible for the examiner to attempt to determine the make and model of the tire using published references and electronic databases [14–16]. In addition, some information regarding the tire size could be extracted from the trace, which aids in the identification of its make and model. Once a make and model of the tire is identified, it is also possible to determine which vehicles are (originally) equipped with such tires and to cross-reference those to the dimensions of the vehicle extracted from the tracks.

5.3.4 Toolmarks

A toolmark is the pattern left by the forcible contact of a tool onto a surface. This can include a variety of different types of impressions, such as those created by a pair of pliers or a hammer, but also those created by the barrel of a firearm onto a bullet, by the die during the stamping of a serial number, or by the teeth of a person during the biting of a fruit.

In general, two main types of toolmarks are considered: striated and impressed. Striated marks are the result of a lateral movement between the two contacting surfaces, such as dragging or sliding action. This motion results in a mark displaying a series of parallel striations. Impressed marks are the result of the penetration of one surface onto another without lateral movement between the two surfaces. This leaves a mark as a negative mirror image of the strongest surface onto the weakest surface.

Toolmarks are used to link a trace (mark) to the source (tool) that created it. These can be very strong evidence linking the tool to a crime scene. Nevertheless, evidence present on the tool, such as the location where it was discovered (at a suspect's home, for example), fingerprints, or DNA, can establish further links. Finally, the study of toolmarks at crime scenes can reveal information regarding the modus operandi of the crime. For example, the toolmarks left on the door handle, door lock, or space between the door and window of a vehicle can reveal how the car was broken into. Indeed, the marks left on the ignition lock cylinder can reveal the type of tools used to forcibly remove it.

Toolmarks exhibit both class characteristics and randomly-acquired characteristics. Class characteristics comprise the general type of mark (striated or impressed), its general shape, and its dimensions. As such, a series of identical screwdrivers leaves the same class characteristics when creating striated toolmarks under the same conditions. Toolmarks also include randomly-acquired characteristics, which are the result of accidental characteristics such as cuts, broken pieces, and regular wear and tear. One tool can leave different types of toolmarks depending on different parameters. Thus, in general, it is considered that the nature and quality of toolmarks depend on the tool itself, the hardness of both surfaces, the force applied, and the relative motion of one surface over the other.

The process of identification starts by comparing the class characteristics between the trace and the putative tool (or usually a toolmark made with the putative tool). In the case of a match, the examiner pursues the comparison with the randomly-acquired characteristics. Similarly to shoeprints and tire tracks, the characteristics present on toolmarks vary with time and this must be taken into account during the comparison process. When a certain number of these characteristics in sufficient quality are present between the trace and a putative source, it is possible to infer that the toolmarks originate from that unique tool only.

Some crime laboratories keep a database or a file with the casts from all the toolmarks collected from crime scenes. It is important for the investigator to communicate clearly with the laboratory to fully benefit from the laboratory's capabilities. When no comparison material is available, the investigator should check with the database or file to see whether any of these toolmarks could match the incriminated mark [17].

When dealing with bullets and cartridge casings (and other elements of ammunition), the study of the class characteristics of toolmarks can reveal very pertinent information regarding the make and model of firearms that could have fired them. When dealing with rifled barrels, the number, width, orientation, and angle of the lands and grooves left on the bullet by the barrel are class characteristics specific to a certain make and model or to a group of makes and models of firearms. When combined with the caliber of the bullet, it is possible, by referring to the database called General Rifling Characteristics (GRC) managed by the Federal Bureau of Investigation, to determine which makes and models of firearms exhibit these characteristics [18]. This can bring important leads in an investigation. Nevertheless, the shapes and relative positions of the extractor, ejector, firing pin, and bolt head marks are also class characteristics (specific to one or more makes/models of firearm) left onto the cartridge casings. Again, combined with the caliber, by using the database, it is possible to determine the makes and models of firearms sharing these characteristics [18].

5.4 CLASS EVIDENCE

5.4.1 Body Fluids and DNA

Virtually all secretions (solid or liquid) originating from a human body contain some levels of DNA. Typical samples often used to recover DNA include semen, blood, saliva, vaginal secretions, urine, dandruff, and hair. DNA stands for deoxyribonucleic acid and is the genetic material of all human beings that constitutes the chromosomes. There are two types of DNA used in forensic sciences: nuclear DNA, found in the nucleus of most cells of the human body, and mitochondrial DNA (mtDNA), found in the mitochondria. One inherits half of his or her nuclear DNA material from the mother and the other half from the father. The mitochondrial DNA is inherited exclusively from the mother.

DNA is a very long polymeric molecule where nucleotides comprise each link of the chain. There are four types of nucleotides constituting the molecule: adenosine, thymine, cytosine, and guanine. The sequence of some of these nucleotides acts as a code for the genes, which determine the characteristics of the human being. However, these coding nucleotides only count for about 3% of the total nuclear DNA, leaving about 97% of the nuclear DNA noncoding. As of today, science does not believe noncoding DNA has any other purpose but to fill the gaps in the chain. Among this noncoding DNA, there are sequences of nucleotides that repeat themselves, such as if the DNA was stuttering. These repetitions of DNA sequences are the characteristics that are analyzed by the forensic laboratory to identify human beings. A site where a repetition occurs is called a locus. For each locus, a different range of number of repetitions (called alleles) can be found. Because one inherits half of his or her nuclear DNA from each parent, two alleles are present for each locus (which can be identical).

The analysis of DNA is technically complicated but consists of determining the different alleles for a given number of known loci. DNA is a very powerful tool in identifying individuals; however, it does not present the uniqueness exhibited by fingerprints. DNA does not have randomly-acquired characteristics; it only bears class characteristics in the form of different alleles within different loci. Fortunately, the frequency of occurrence of each allele for each locus is known for different given populations. Additionally, each locus is independent from the others. Thus, once the alleles of the different loci are known, it is possible to express a frequency of occurrence of such a profile within the population. For example, one profile might be encountered at a frequency of 1 in 1,000,000 people, whereas another profile might be encountered at a frequency of 1 in 60,000,000,000 people. Thus, the rarity of the DNA sample greatly varies from one profile to another, in some instances, making the link as close to unique as possible. It is always important to remember that identical twins share the exact same DNA, independently of the frequency of occurrence of the DNA in the population.

DNA is extremely useful, and it became even more useful with the development of a database capable of storing DNA profiles. The combined DNA index system (CODIS) is the database equivalent to AFIS for fingerprints [19]. Thus, it is not necessary anymore to have a known suspect to make a sample of DNA useful to the investigation. It is possible to submit the sample of DNA to the CODIS database, which, depending on the jurisdiction's law and statutes, contains convicted offender's profiles and DNA traces found at crime scenes, and to determine whether a matching DNA profile is present or not.

5.4.2 Paint

Paint is a substance used as a coating to protect and/or to decorate a surface. Traces of paint can be transferred when a painted object contacts another (painted or not) object or surface with enough violence to separate the paint from the original object. This type of situation is often encountered in burglary and hit and run investigations. For example, the forcing of a door with a pry bar usually leaves toolmarks on the door jam and, possibly, paint from the pry bar. Conversely, the paint from the door jam might also be transferred to the pry bar. This paint can then be used to establish a possible link with its source. The rare cases of cross-exchange of paints from both objects are ideal, because they better support the hypothesis of contact.

Paints can present varying compositions according to their use but, in general, are composed of a binder (or vehicle), pigments, extenders (or fillers), and a solvent. When vehicles are painted more than one layer is deposited onto the body's surface. Usually, there is a primer, a filler, a coat, and then a topcoat (transparent finish). There could be more layers depending on the type of paint and the painting process. In addition, some vehicles are repainted after an accident or for simple decorative reasons, resulting in the modification of some layers and the addition of new ones. Paint traces left can be deposited on an object under two different conditions. When the contact of both surfaces is almost parallel and in opposite directions, the mechanical friction results in a paint smear. When the contact involves a mechanical shock rather than friction, paint chips can be detached from the surface and transferred to an object.

Paint presents only class characteristics. These characteristics are present in the form of color, physical appearance, and chemical (organic, mineral, and elemental) composition. When multiple layers are present, the layer sequence is also a class characteristic in and of itself. This increases the probative value of the link. Without taking into account the circumstances of the case, it is not possible to identify a paint trace as originating from only one source. The class characteristics exhibited are shared among many sources. There is one exception to this rule: when the link between an incriminated paint chip and a source is demonstrated by fracture assembly (also called physical match). In this instance, the assembly process only takes into account the fracture pattern, independently of the paint chemistry. This fracture presents a random pattern, which is unique to the source and to the trace.

The international forensic automotive paint data query database (PDQ) is a useful tool in determining the make and model of vehicle based upon the paint trace's chemical and color information [20]. Thus, there is not necessarily a need for a comparison paint from a suspect vehicle to lead an investigation. The information gathered from a paint chip found at a crime scene could be searched against the PDQ to determine which vehicles are (originally) manufactured with such a paint. The PDQ contains information on more than 13,000 vehicles; however, it is not an exhaustive database.

5.4.3 Fibers

A fiber is a thread typically used to make a textile. Natural fibers can be of animal, mineral, or vegetable origins. The most commonly encountered fiber is cotton. Also, within the textile industry, synthetic fibers now represent a very important part of the market, and nylon dominates it. Fibers can become very useful in the forensic investigation because they are widely available and are readily transferred onto different surfaces following Locard's principle of exchange. For example, when an individual sits in the driver's seat of a vehicle, a transfer of fibers from the seat occurs onto the clothing of that person. Conversely, a transfer of fibers from that person occurs onto the seat of the vehicle [21]. Additionally, transfer of fibers from the floor mat of a vehicle can occur to the occupants' shoe soles

[22]. The crime scene investigator should keep that in mind when processing a vehicle and collect floor mat fibers as comparison samples.

Fibers do not exhibit randomly-acquired characteristics. As with paint and other mass-produced materials, they bear class characteristics such as refractive index, color, other optical properties, physical appearance, and chemical (mostly organic) composition. Thus, it is not possible to identify an incriminated fiber, or a series of incriminated fibers, to one unique source. However, when analyzing several independent class characteristics, it is possible to reduce the number of putative sources and to increase the strength of the link between the incriminated trace and the source. In some instances, the link can be strengthened very close to uniqueness.

Hairs are fibers of animal origin. They can be very valuable in a forensic investigation. For example, Swiss poachers are often caught by analyzing hairs found in the trunk of their cars. They use their cars to transport illegally killed wild boars and other wildlife. It is possible to readily determine the species of a hair based upon the microscopic observation of its scales, medulla, and other characteristics [23]. Finally, when the hair bears a root, it is also possible to proceed to DNA analysis.

5.4.4 Glass

Glass is a product of inorganic materials that solidified but did not crystallize. Glass is extremely prevalent in modern everyday life. It is mainly composed of silicon dioxide (SiO_2) and contains other chemicals depending on its type and use. When a window or a glass item is broken, very small fragments are ejected to significant distances (up to several meters away). These fragments can collect on clothes, hair, and shoe soles. When fragments of glass are found, it is interesting, from a forensic perspective, to compare them with a source window to determine whether a link can be established [24].

Glass only presents class characteristics. These characteristics are present in the form of color, patterns, density, refractive index, fluorescence, and chemical (mineral and elemental) composition. Without taking into account the circumstances of the case, it is not possible to identify a glass fragment as originating from only one source: the class characteristics exhibited are shared among many sources. There is one exception to this rule: when the link between an incriminated glass fragment and a source is demonstrated by fracture assembly (physical match). In this instance, the assembly process only takes into account the fracture pattern, independently of the glass physical and chemical properties. This fracture presents random patterns, which are unique to the source and to the trace.

When the refractive index of glass is determined, it is possible to know the frequency of occurrence of such a refractive index within the population. Thus, this allows the criminalist to attribute a weight to the rarity of the particular refractive index, which may strengthen the link between the trace and the source.

Finally, it is always important to keep in mind that glass can contain other traces such as fingerprints, shoeprints, and blood. Thus, it is very pertinent to carefully observe glass not to miss this type of evidence.

BIBLIOGRAPHY

[1] Locard E. (1934) *La police et les méthodes scientifiques,* Les éditions Rieder, Paris, France.

[2] Locard E. (1920) *L'enquête criminelle et les méthodes scientifiques,* Flammarion, Paris, France.

[3] Champod C. (2000) Identification/Individualization: Overview and meaning of ID. In: *Encyclopedia of Forensic Sciences,* ed Siegel J, Knupfer G, and Saukko P, Academic Press, London, England, pp 1077–1084.

[4] Aitken C and Taroni F. (2004) *Statistics and the evaluation of evidence for forensic scientists,* 2nd edition, John Wiley and Sons, Chichester, England.

[5] Buckleton J, Triggs CM, and Walsh SJ. (2004) *Forensic DNA evidence interpretation,* CRC Press, Boca Raton, FL.

[6] Robertson B and Vignaux GA. (1995) *Interpreting evidence: evaluating forensic science in the courtroom,* John Wiley & Sons, Chichester, England.

[7] Kirk PL. (1963) The ontogeny of criminalistics, *Journal of Criminal Law, Criminology and Police Science,* **54,** pp 235–238.

[8] Komarinski P. (2005) *Automated fingerprint identification systems (AFIS),* Elsevier Academic Press, Burlington, MA.

[9] Bodziak WJ. (2000) *Footwear impression evidence: Detection, recovery, and examination,* 2nd edition, CRC Press, Boca Raton, FL.

[10] Ashley W. (1996) What shoe was that? The use of computerised image database to assist in identification, *Forensic Science International,* **82**(1), pp 7–20.

[11] Foster + Freeman (2003) SoleMate, Product note, 3.

[12] Girod A. (1996) Computerized classification of the shoeprints of burglar's soles, *Forensic Science International,* **82**(1), pp 59–65.

[13] Geradts Z and Keijzer J. (1996) The image-database REBEZO for shoeprints with developments on automatic classification of shoe outsole designs, *Forensic Science International,* **82**(1), pp 21–31.

[14] Foster + Freeman. (2005) *SICAR 6—Tyre mark and shoe print evidence management system,* available at http://www.fosterfreeman.co.uk, last access performed on November 22, 2005.

[15] STAMP (2004) *Shoeprint tyreprint acquisition and matching program,* available at http://www.stampmatch.com, last access performed on November 22, 2005.

[16] Foster + Freeman (2005) *TreadMate - A reference database of vehicle tyres and tyre tread patterns to assist the identification of vehicles from crime scene data,* available at http://www.fosterfreeman.co.uk, last access performed on November 22, 2005.

[17] Geradts Z, Keijzer J, and Keereweer I. (1994) A new approach to automatic comparison of striation marks, *Journal of Forensic Sciences,* **39**(4), pp 974–980.

[18] Schehl SA. (2000) Firearms and toolmarks in the FBI laboratory—Part 2, *Forensic Science Communications,* **2**(2).

[19] Federal Bureau of Investigation (2000) *The FBI's combined DNA index system program,* US Department of Justice, Washington, DC.

[20] Royal Canadian Mounted Police (2005) *RCMP fact sheets: Pain data query (PDQ),* http://www.rcmp.ca/factsheets/fact_pdq_e.htm, last access performed on November 22, 2005.

[21] Roux C and Margot P. (1997) An attempt to assess the relevance of textile fibres recovered from car seats, *Science & Justice*, **37**(4), pp 225–230.

[22] Roux C, Langdon S, Waight D, and Robertson J. (1999) The transfer and persistence of automotive carpet fibres on shoe soles, *Science & Justice*, **39**(4), pp 239–251.

[23] Debrot S, Fivaz G, Mermod C, and Weber JM. (1982) *Atlas des poils de mammifères d'Europe*, Institut de zoologie, Université de Neuchâtel, Neuchâtel, Switzerland.

[24] Curran JM, Hicks TN, and Buckleton JS. (2000) *Forensic interpretation of glass evidence*, CRC Press, Boca Raton, FL.

CHAPTER 6

VEHICLE IDENTIFICATION

William T. Smylie

6.1 INTRODUCTION

When a vehicle is at the center of a crime or is involved in a criminal activity, the vehicle's true identity may be called into question. This question might arise because the vehicle's true identity has been almost completely obliterated by being burned, submerged, or dismantled or because it has been purposely altered.

A vehicle's true identity is the factory original identity of a car or truck as it left the final assembly plant, pertaining principally to its vehicle identification number (VIN) but also including to a lesser extent its model year, make, model/series, body style, exterior color, engine, and drivetrain information. Given mass production of thousands of the same model, similarly trimmed and equipped "cookie cutter" vehicles each year, the only feature truly unique to any specific car is its VIN. Once assigned and affixed by the manufacturer, this unique grouping of numbers and letters that make up the particular VIN belongs to that vehicle exclusively. Although there are usually some components, large and small, that are part of the vehicle that have separate serial numbers of their own, the VIN is the unique identifying number that represents the vehicle as a whole. In most cases, the original VIN affixed to a vehicle consistently follows it during an entire service lifetime through periodic license plate updates, transfers of ownership, changes in title and registration from state to state, and eventually even possible replacement of some of the vehicle's original major components due to extensive wear, damage, or even theft. The VIN is the principal identifying number used by virtually all government vehicle titling and registration authorities for issuing ownership and related documents for modern motor vehicles.

6.2 VEHICLE IDENTIFICATION NUMBER FORMAT

6.2.1 General Structure

The modern 17-digit VIN, as shown in Figure 6-1, not only uniquely identifies a particular vehicle but also contains information that describes the vehicle itself [1]. The first three characters taken as a whole are referred to as the world manufacturer identifier (WMI) section. The second section, called the vehicle descriptor section (VDS), is composed of the next five digits in North America. This is followed by the ninth digit, which is called

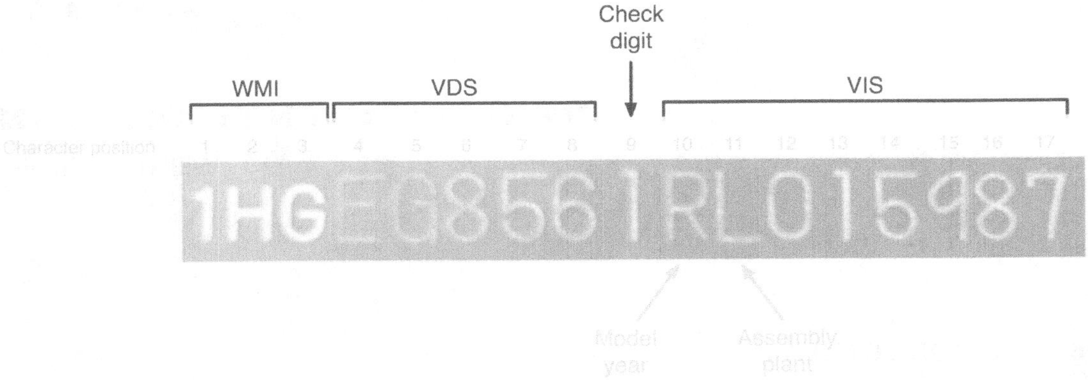

Figure 6-1

General structure of a typical 17-digit VIN. This particular example is a US assembled Honda passenger car.

the check digit. Finally, the last eight characters are defined as the vehicle identifier section or vehicle indicator section (VIS).

In the United States, the structure and content of the VIN is described in the *Code of Federal Regulations*, Title 49, Transportation [1]. It should be noted that the Society of Automotive Engineers (SAE) has been designated by the International Organization for Standardization (ISO) to assign the WMI portion of the VIN [2]. The National Highway Traffic Safety Administration (NHTSA), which is responsible for the application of VIN regulations in the United States, also follows the SAE WMI code system [3]. Several standards are available from the SAE regarding the VIN structure [4–8].

ISO has several standards dedicated to the VIN, which are adopted by the European Union. ISO Standard 3779 describes the VIN content and structure [9]. The VIN structure is almost identical to the one described above for North American cars. The VIN is 17 digits long and contains the three main sections WMI, VDS, and VIS; however, the presence of a check digit is not cited in the standard. Thus, the VDS ranges from characters 4 through 9 instead of 4 through 8. It is also stated that if a manufacturer does not use all the spaces inside the VDS, these spaces shall be filled with alphabetic or numeric characters of the manufacturer's choice. ISO also publishes standards regarding the WMI and the world parts manufacturer identifier (WPMI) [2, 10].

It should be noted that both in the ISO standards and in the *Code of Federal Regulations*, the list of characters that can be used in VINs is limited to the following [1, 9]:

A B C D E F G H J K L M N P R S T U V W X Y Z 1 2 3 4 5 6 7 8 9 0

This limitation has been designed to avoid confusion between letters and figures that can appear similar. Thus, the letters I, O, and Q do not appear in a VIN, so they are not confused with the numbers 1 and 0.

6.2.2 *World Manufacturer Identifier (WMI)*

This section identifies the manufacturer of the vehicle. The first character, sometimes supplemented by the second character, designates the nation of origin of the vehicle such as the United States, Canada, or Japan. Table 6-1 presents a list of some countries of origin with their respective code.

Although the first character indicates the country of origin, it is important to keep in mind that most major vehicle manufacturers today have a global presence. For example, Honda vehicles, which are generally seen as from a Japanese-based automobile company, are actually assembled in a number of different countries. It is not uncommon to see Honda VINs beginning with 1 or 5 for US-built cars, 2 for Canadian-built units, 3 for those assembled in Mexico, and even S for those assembled in the United Kingdom, in addition to the expected J symbol for Japan. Because this situation is common to a number of other global manufacturers, the

Table 6-1

Partial list of country codes used in the WMI.

	Code	Country
Africa A–H	AA-AH	South Africa
Asia J–R	J	Japan
	KL-KR	South Korea
	L	China
	MA-ME	India
	MF-MK	Indonesia
	ML-MR	Thailand
	PL-PR	Malaysia
Europe S–Z	SA-SM	United Kingdom
	SN-ST, W	Germany
	SU-SZ	Poland
	TR-TV	Hungary
	VA-VE	Austria
	VF-VR	France
	VS-VW	Spain
	YF-YK	Finland
	YS-YW	Sweden
	ZA-ZR	Italy
North America 1–5	1, 4, 5	United States
	2	Canada
	3A-3W	Mexico
Oceania 6–7	6A-6W	Australia
	7A-7E	New Zealand
South America 8–9	8A-8E	Argentina
	8X-82	Venezuela
	9A-9E, 93-99	Brazil
	9F-9J	Colombia

best indicator as to where a particular vehicle was actually assembled would be the plant character located at the 11th character of the VIN, as described in Subsection 6.2.5.

The second digit of the WMI identifies the corporate manufacturer with the third character indicating vehicle make and type on some models. Note that if the manufacturer produces less than 500 vehicles per year, the third character of the WMI is usually a 9. Table 6-2 presents a list of some common car manufacturers and their respective WMI codes. Note that a particular car manufacturer can have several WMIs because they manufacture

Table 6-2
Partial list of WMI codes.

WMI	Manufacturer	WMI	Manufacturer
JA	Isuzu	1F	Ford Motor Company
JH	Honda	1G	General Motors
JM	Mazda	1GC	Chevrolet
JN	Nissan	1GM	Pontiac
JS	Suzuki	1HG	Honda
JT	Toyota	1L	Lincoln
KL	Daewoo	1M	Mercury
KMH	Hyundai	1N	Nissan
KN	Kia	1VW	Volkswagen
SAL	Land Rover	1YV	Mazda
SAJ	Jaguar	2F	Ford Motor Company
SCC	Lotus	2M	Mercury
TRU	Audi	2G	General Motors
VF1	Renault	2G1	Chevrolet
VF3	Peugeot	2G1	Pontiac
VF7	Citroën	2HM	Hyundai
VSS	Seat	3F	Ford Motor Company
WAU	Audi	3G	General Motors
WBA	BMW	3VW	Volkswagen
WBS	BMW	4F	Mazda
WDB	Mercedes-Benz	4M	Mercury
WMW	Mini	4S	Subaru-Isuzu
WP0	Porsche	4US	BMW
WVW	Volkswagen	5L	Lincoln
WV2	Volkswagen	6F	Ford Motor Company
YK1	Saab	6H	General Motors Holden
YS3	Saab	6MM	Mitsubishi
YV1	Volvo	6T1	Toyota
ZAM	Maserati	9BW	Volkswagen
ZAR	Alfa Romeo		
ZFA	Fiat		
ZFF	Ferrari		
ZLA	Lancia		

cars in different countries (such as Volkswagen in Germany, United States, Mexico, and Brazil) or because they manufacture a great number of cars within a country (such as General Motors [GM] or Ford Motor Company).

6.2.3 Vehicle Descriptor Section (VDS)

In North America, the next five-digit segment of the VIN (positions 4 through 8) is called the VDS. In Europe and other countries following the ISO standard, the VDS ranges from position 4 through 9, and there is no check digit [9]. The VDS contains characters that describe certain attributes of the vehicle, which may include such things as restraint system specifications, intended market information, model/series, body style, transmission, and engine type/displacement. On sport utility vehicles (SUVs) and light trucks, this section may also contain information on gross vehicle weight rating (GVWR), brake system, chassis type, cab type, and drivetrain. Although the VDS section appears in the same place within the VIN on all different makes of vehicles, there is some disparity between the various manufacturers as to the type of information contained in it. Some manufacturers outside the United States do not use all characters of the VDS and thus fill the spaces with a character of their choice [9].

6.2.4 Check Digit

In the North American VIN format, the ninth digit position is called the check digit. The check digit is a built-in security code that verifies that the makeup of the VIN as a whole conforms to a certain mathematical formula. It helps to determine whether the VIN is legitimate. Simply put, individual characters in the VIN have a preassigned numeric equivalent value. Further, each character position has an assigned numeric "weight" that depends on its location in the 17-digit field. Table 6-3 shows how to determine the check digit. In this process, numerals maintain their numeric value and letters are assigned numeric values according to Table 6-4 [1].

Table 6-3

Table used to calculate the check digit.

Position	1	2	3	4	5	6	7	8	Check digit	10	11	12	13	14	15	16	17
VIN #	□	□	□	□	□	□	□	□		□	□	□	□	□	□	□	□
Assigned value									0								
× Weight factor									0								
= Product	**8**	**7**	**6**	**5**	**4**	**3**	**2**	**10**	0	**9**	**8**	**7**	**6**	**5**	**4**	**3**	**2**
Sum of products =					Divided by 11 =					Remainder = □ = Check digit							

Table 6-4

Numeral values assigned to letters in the calculation of the check digit.

Assigned numerical values for letters			
A = 1	G = 7	N = 5	V = 5
B = 2	H = 8	P = 7	W = 6
C = 3	J = 1	R = 9	X = 7
D = 4	K = 2	S = 2	Y = 8
E = 5	L = 3	T = 3	Z = 9
F = 6	M = 4	U = 4	

Table 6-5

Example of the calculation of the check digit of a VIN using the formula of Table 6-3.

Position	1	2	3	4	5	6	7	8	Check digit	10	11	12	13	14	15	16	17
VIN #	1	G	3	A	J	5	5	M	4	S	6	3	0	5	6	5	6
Assigned value	1	7	3	1	1	5	5	4	0	2	6	3	0	5	6	5	6
× Weight factor	8	7	6	5	4	3	2	10	0	9	8	7	6	5	4	3	2
= Product	**8**	**49**	**18**	**5**	**4**	**15**	**10**	**40**	**0**	**18**	**48**	**21**	**0**	**25**	**24**	**15**	**12**
Sum of products = 312						Divided by 11 = 28 r 4				Remainder = 4				= Check digit			

First, the VIN is written in the proper cells (see Table 6-5). Then, the numeral values (if the character is a number) and the assigned numeral values (determined per Table 6-4 if the character is a letter) are filled. The assigned value for each VIN position is then multiplied by the weight factor to obtain the product of each character. The products are then summed and the total is divided by 11. The numerical remainder of the division is the check digit. If the remainder is 10, the letter X is used to designate the check digit. The correct numeric remainder, 0–9, or the letter X should appear at position 9 of the VIN.

In the early days of the 17-digit VIN, it was not uncommon to see a hard-working auto theft investigator, pencil in hand, hunched over a check digit key worksheet calculating the validity of a suspicious VIN; a cumbersome process to say the least. Today, almost all systems for checking a VIN such as the National Crime Information Center (NCIC) of the Federal Bureau of Investigation, the National Insurance Crime Bureau (NICB) on-line program, and the NICB VIN Assist programs automatically check to determine whether the VIN conforms to check digit standards. These databases and programs are covered in more detail later in the chapter.

6.2.5 Vehicle Indicator Section (VIS)

The next and final portion of the VIN concerns the eight characters occupying positions 10 to 17 called the vehicle indicator section, also known as the vehicle identifier section, or VIS. The character in position 10 denotes the vehicle model year, character in position 11 identifies the final assembly plant of the vehicle, and characters in position 12 to 17, the final six digits, indicate the sequential production number or the number order in which the vehicle was produced on the assembly line.

The 10th character identifies the model year using a single digit as shown in Table 6-6 [1, 9]. The first column (in italic) shows the early year codes attributed in the ISO Standard 3779. VINs from this era were not yet required to be 17-character long and there was no uniform location for the year designation within them.

A simple way to identify the vehicle's year when viewing a VIN is to remember that letter A corresponds to 1980, although the use of letter A was optional for manufacturers in 1980, because the 17-digit VIN was not yet widely in use. The letter B corresponds to 1981 and so on in ascending order up through the alphabet. As stated previously, the letters I, O, and Q are skipped. In addition, the letters U and Z are skipped in the year code only, because they are sometimes mistaken for the letter V and the number 2, respectively. The letters U and Z may appear in areas of the VIN other than as the year designator.

Hence, the 2000 model year designator for vehicles is the letter Y. From model year 2001 through 2009, the numbers 1 through 9 are used in ascending order. Model years 2010 through 2030 will be represented by letters A through Y, again in ascending year/letter order (still skipping letters I, O, Q, U, and Z). Model years 2031 through 2039 will be represented in order by numbers 1 through 9.

Table 6-6

List of vehicle model year codes.

		1980	A	1990	L	2000	Y	2010	A
1971	*1*	1981	B	1991	M	2001	1	2011	B
1972	*2*	1982	C	1992	N	2002	2	2012	C
1973	*3*	1983	D	1993	P	2003	3	2013	D
1974	*4*	1984	E	1994	R	2004	4	2014	E
1975	*5*	1985	F	1995	S	2005	5	2015	F
1976	*6*	1986	G	1996	T	2006	6	2016	G
1977	*7*	1987	H	1997	V	2007	7	2017	H
1978	*8*	1988	J	1998	W	2008	8	2018	J
1979	*9*	1989	K	1999	X	2009	9	2019	K

Note that the first column is in italic to emphasize that although these codes are attributed as is by ISO and were used on older vehicles, they are not commonly encountered.

6.2.6 Information Resources

A/ NICB

One of the most important resources of information for vehicle identification is the NICB. In partnership with the Insurance Services Office, the NICB is the official custodian of assembly and shipping records for vehicles produced for sale in the United States and Canada. The NICB also maintains records of vehicle secondary number locations from the manufacturers to assist investigators in vehicle examination. Nevertheless, the NICB collects much more than just vehicle manufacturer's information. Current and purged vehicle theft records that have been entered into the NCIB database are available as well as theft records on vehicles stolen in Canada. In addition, the NICB has access to vehicle salvage records, vehicle impound records, vehicle insurance claim data, information on vehicles presented to US Customs for export, and some information on vehicles stolen in Mexico. Although all this information is available to law enforcement agencies over the phone, most of it can be accessed online by subscribing to the NICB online service. Finally, the NICB also maintains a network of local agents around the United States who may be contacted for investigative assistance. The NICB website is found at http://www.nicb.org. The Canadian counterpart to the NICB is the Insurance Bureau of Canada (IBC), with its main office located in Toronto, Ontario; its website is http://www.ibc.ca.

B/ Websites Dedicated to VIN Interpretation

- http://www.vehicleidentificationnumber.com offers a comprehensive guide to VIN decoding by makes and models. The site is not the easiest to navigate, but it is worth the effort.
- http://www.analogx.com/contents/vinview.htm offers free but limited VIN decoding.
- http://www.angelfire.com/ca/TORONTO/VIN/VINcode.html offers an extensive source of information on the principles behind VINs and their decoding. Also, the site contains some very interesting information regarding some make's VDS characteristics.
- https://vintelligence.polk.com offers paying VIN decoding (around USD 1 per VIN) with comprehensive information.
- http://www.autoinsurancetips.com/vin_number.htm is a good guide to VIN decoding with some data of older vehicles (back to the 1950s).
- http://www.greatoldcars.com/decoding.htm offers the paying VIN decoding of classic cars.
- http://en.wikipedia.org is a free encyclopedia that contains several definitions and information regarding VINs. This source might be very helpful, but the information is to be taken with precaution.

C/ Commercial Databases

- XML VIN Decorder of DataOne Software at http://dataonesoftware.com
- VINaudit from VINtek at http://www.vintek.com
- VinPOWER at http://www.vinpower.com

D/ Automobile Manufacturers

Another source that can be extremely valuable is the vehicle manufacturers themselves and their dealerships. The authorized dealerships of the various automobile makers have direct access to internal computer databases that contain a large amount of information concerning their vehicles. Ford Motor Company's "OASIS" system is a good example. These company databases can be queried locally and can provide a complete service history, including dates, times, names, and locations for any VIN in their system. If the information provided by the dealer is not satisfactory, it is always possible to directly contact the manufacturer. Contact information may be obtained through a dealership or by checking on the Internet for their official website. An impressive list of manufacturer and vehicle parts supplier websites is listed at http://www.automotive-esources.com along with other interesting information.

E/ Other Internet Resources

The following are some resources that can provide helpful information:

- http://www.aama.com is the website of the American Automobile Manufacturers' Association.
- http://www.autoalliance.org is the website of the Alliance of Automobile Manufacturers marketing vehicles in the United States.
- http://www.aiam.org is the website of the Association of International Automobile Manufacturers. This is a site with information from the import manufacturers.
- http://www.cvma.ca is the website of the Canadian Vehicle Manufacturers' Association.
- http://www.nhtsa.dot.gov is the website of the NHTSA. This is a large site that includes information on VIN standards and vehicle safety standards.
- http://www.aamva.org is the website of the American Association of Motor Vehicle Administrators. This is another large site that includes links to the various state motor vehicle departments.
- http://www.paintscratch.com is a site that provides automotive paint code information along with color samples.
- http://www.carfaxonline.com is a site for the vehicle history service provider Carfax. The service is open to members. Law enforcement agencies are granted membership at no cost.

F/ Reference Material

A good set of reference material is important. The following are recommended:

- The annual editions of the *NICB Passenger Vehicle Identification Manual* and the periodically printed released editions of the *NICB Commercial Vehicle and Off-Road Equipment Identification Manual* [11, 12]. Both are available by contacting NICB.
- Any of the books from the *Cars and Parts* magazine "Matching Number Series" publications. This includes all three volumes of the *Catalog of American Car ID Numbers* spanning vehicle years 1950 to 1979 [13–15]. These books are an invaluable resource when called upon to examine older vehicles and are available from Amos Press Incorporated.

- All three volumes of the *Standard Catalog of American Cars* covering 1805 to 1942, 1946 to 1975, and 1976 to 1999 [16–18]. Also see the *Standard Catalog of American Light Duty Trucks* and the *Standard Catalog of Imported Cars* [19, 20]. These are all large well-illustrated guides that are very helpful. They are available from Krause Publications.
- The book *Auto Dictionary* written by John Edwards and published in 1993 by H P Books [21]. This is basically a dictionary of automotive terminology, including some slang language. This is a helpful guide to those people who are just starting out in vehicle theft investigation.
- *Understanding Automotive Specifications and Data,* written by James Flammang and published by Tab Books in 1986 is another helpful book for those new to the business [22].

6.2.7 Examples

The following are a few examples of actual VINs, broken down by segment and interpreted. This gives the reader an opportunity to practice VIN interpretation.

A/ First Example

1G1	ND52T	1	VY116615
WMI	*VDS*	✓	*VIS*

WMI 1 = USA, G = General Motors Corp, 1 = Chevrolet division.

VDS ND = Base level series Malibu, 5 = 4-door sedan, 2 = Restraint system dual front seat air bags and manual safety belts, T = 2.4-liter inline 4-cylinder fuel-injected engine.

Check digit 1

VIS V = 1997 model year, Y = Final assembly point Wilmington, Delaware, 116615 = sequential production number.

B/ Second Example

2MH	HM79V	7	4X676128
WMI	*VDS*	✓	*VIS*

WMI 2 = Canada, M = Mercury Division of Ford Motor Company, H = Incomplete vehicle, this passenger car was sent directly to an aftermarket supplier by the manufacturer to have a power-operated moon roof installed.

VDS H = Restraint system of shoulder/lap belts with dual front air bags and front seat side air bags, M79 = Marauder 4-door sedan, V = 4.6-liter dual overhead cam V8 engine.

Check Digit 7

VIS 4 = 2004 model year, X = Assembled in St. Thomas, Ontario, Canada, 676128 = Sequential production number.

C/ Third Example

JT8	BH22F	4	T0048456
WMI	*VDS*	✓	*VIS*

WMI J = Japan, T = Toyota Motor Corp, 8 = Lexus division passenger car.

VDS BH2 = LS400 4-door sedan with a 4.0-liter V8 engine, 2F = Restraint system including dual air bags.

Check Digit 4

VIS T = 1996 model year, 0 in this position on a 1996 Lexus LS400 = Final assembly plant of Tahara, Japan, 048456 = Sequential production number.

D/ Fourth Example

WDB	JF65H	2	XA899788
WMI	*VDS*	✓	*VIS*

WMI W = Germany (originally for West Germany), DB = Daimler-Benz, the maker of the Mercedes-Benz vehicles.

VDS JF65 = In combination indicates the E series 320W 4-door sedan, H = Restraint system of shoulder/lap belts for all five passenger positions with front and side curtain air bags.

Check Digit 2

VIS X = 1999 model year, A = Final assembly plant in Sindelfingen, Germany, 899788 = Sequential production number.

6.3 VIN PLATE LOCATIONS, TYPES, AND ATTACHMENTS

6.3.1 VIN Locations

Before model year 1954, vehicles were variably identified by their engine number, frame number, chassis number, or body number. From approximately model years 1954 to 1967, vehicles were identified using an overall serial/VIN number that could have been of almost any configuration and placed practically anywhere on the body. Starting in model year 1968, all passenger vehicles manufactured for sale in North America were required to have a VIN attached in a location viewable from the outside of the vehicle. In the United States, according to the *Code of Federal Regulations,* the VIN should be readable without moving any part of the vehicle through the vehicle glazing under daylight lighting conditions [1]. The ISO Standard 4030 describes that the VIN shall be located on the right side of the vehicle and if possible on the front half, or when the VIN must be readable from the outside, it shall be located inside the passenger compartment adjacent to the windscreen pillar [23]. In addition, the VIN shall be located in an easily visible position and in a manner that precludes obliteration or alteration.

With some notable exceptions, manufacturers generally place this public VIN plate around the top of the driver's side instrument panel area viewable at the base of the windshield, as shown in Figure 6-2. The most notable exceptions were the 1968 model year Ford passenger vehicles where the VIN was placed on the top passenger side instrument panel and BMW and Ferrari vehicles of the late 1960s and early 1970s where the VIN plate was placed on top of the steering column, viewable through the windshield, just forward of the steering wheel.

Figure 6-2

Placement of public VIN plate typically mounted behind the windshield glass, clearly viewable from outside the vehicle. Generally found at the base of the windshield, with exceptions found on the windshield pillar or at the base of the windshield toward the center.

6.3.2 General Plates

Since 1981 there has been some uniformity in the industry as to the VIN content and its general area of placement. However, the VIN plate's physical appearance and its attachment vary greatly. The ISO standard offers the possibility of having a VIN directly marked on an integral part of the vehicle or marked on a plate that is then permanently affixed onto the vehicle [23].

In the first few years after the 17-digit VIN was adopted, most VIN plates consisted of small rectangular strips of metal (stainless steel or aluminum) with the VIN characters stamped in a raised fashion or imprinted on the surface. VIN plates were generally black or silver in color, although some earlier GM public VIN plates were painted to match the instrument top panel color and a few imports have used some other colors. The plates were attached to the vehicle with rivets of several different types; GM, Ford, and Chrysler Corp used the six-petal "rosette" style rivet made of stainless steel or aluminum almost exclusively in either a black or silver color. This rosette style is shown in Figures 6-3 and 6-4.

Most of the imports used a round-headed "pop" style rivet made of stainless steel, aluminum, or plastic, as seen in Figure 6-5. A few imports also used sheet metal screws to secure the VIN, as with some older Audi models. On many vehicles the rivet heads are

Figure 6-3

Typical GM passenger vehicle VIN plate showing the six-petal rosette-style rivets. Also, note the barcode and the GM box logos. This particular example is from a Canadian-assembled Chevrolet.

Figure 6-4

Typical Ford passenger car VIN plate showing the six-petal rosette-style rivets. Also, note the air bag symbol and the Ford's trademark oval logos on the plate background (can be observed on the left side of the plate).

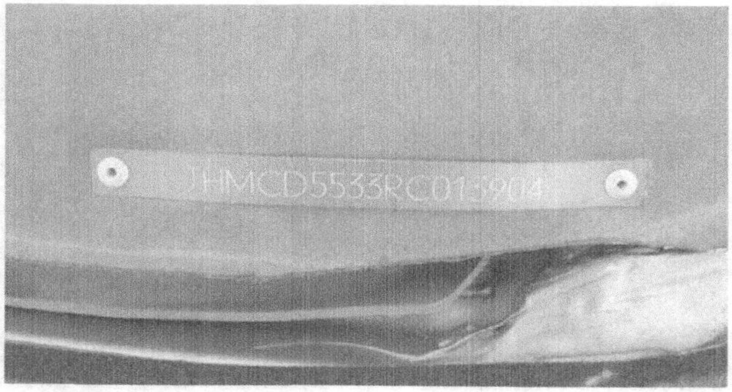

Figure 6-5

Typical Honda VIN plate of the 1990s with round head rivets. Many later model Honda VIN plates are attached with five-petal rosette rivets.

visible, on some others the heads are partially obscured by dashboard trim, whereas others have the rivets mounted completely out of sight underneath the instrument top panel. In the mid-1980s most of the Toyota family vehicles switched from a pop rivet style fastener for their VIN plates to a five-petal rosette-type rivet that looks distinctly different from the rosettes used by the domestic big three: GM, Ford, and Chrysler. This five-petal rivet style is shown in Figure 6-6. In recent years, Honda vehicle lines have adopted a distinctive-looking rosette style rivet of their own for VIN plates.

Various company names and logos found their way onto the VIN plate background surfaces during the 1980s and early 1990s with the eye-readable number. Simultaneously, the trend to use a bar code label on the VIN plate, as illustrated in Figure 6-3, started. More and more, VIN plates have appeared with laser-cut characters replacing the older raised or stamped style. Some manufacturers such as Mercedes and BMW have gotten away from using metal VIN plates and rivets to a great extent and have developed high-quality Mylar and acrylic VIN labels attached to the vehicle with adhesive materials, as illustrated in Figure 6-7.

Figure 6-6

Typical Toyota VIN plate attached with five-petal rosette rivets used on a number of models. Note the display of upside-down Toyota lettering in background (visible on the left side of the plate).

Figure 6-7

Mylar or plastic label VIN attached with adhesive to a plate mounted with rivets. This particular example is from a Mercedes-Benz built Sprinter cargo van marketed in the United States through Dodge and Freightliner dealerships.

6.3.3 Safety Certification Label

In addition to the public VIN plate, there are a few other places where the full VIN may easily be found in a vehicle. Since 1970, vehicles manufactured for distribution in the United States are required to display a Federal Motor Vehicle Safety Certification Label, commonly called the safety standards emblem [24].

Initially, this label typically contained language certifying that the particular vehicle met current safety requirements along with the full VIN. In more recent years, although this label still displays the full VIN (sometimes supplemented by a VIN bar code) along with the required safety certification/theft prevention verbiage, it may now also contain additional vehicle information including such items as assembly dates (month/year), passenger capacity, country of origin, GVWR, vehicle weight distribution, axle ratio, transmission code, paint color/trim codes, and original equipment manufacturer (OEM) tire specifications and inflation data. An example from a Ford vehicle is presented in Figure 6-8.

These safety standards emblems are usually made of paper with a clear overlay of a soft plastic film material and are attached with adhesive on almost all domestic vehicles and many imports. These softer labels are designed to self-destruct upon removal, with most leaving behind a portion of the adhesive layer from the label as a distinctive marking on the vehicle, as illustrated in Figure 6-9.

Some import models use metal plates attached with a combination of rivets and adhesive. Most of the time the emblems are attached to the vehicle somewhere on the drivers' side door striker face or the B pillar front door jamb area. As always, when dealing with vehicle identification, there are exceptions. Sometimes the emblem is found in the driver's side rear door opening, such as with some Honda family models, on the driver's door hinge post, or under the hood on or near the upper firewall. Although these mounting systems contain excellent safeguards against unlawful removal of the emblem, they are not infallible. Professional vehicle theft criminals have devised thermal methods and other tech-

Figure 6-8

Typical example of a paper Federal Motor Vehicle Safety Certification label showing the full 17-digit VIN and other features of the vehicle. This particular Ford example also displays the VIN in barcode form. These labels are usually found in the driver's door opening area either on the door striker face or on the B pillar.

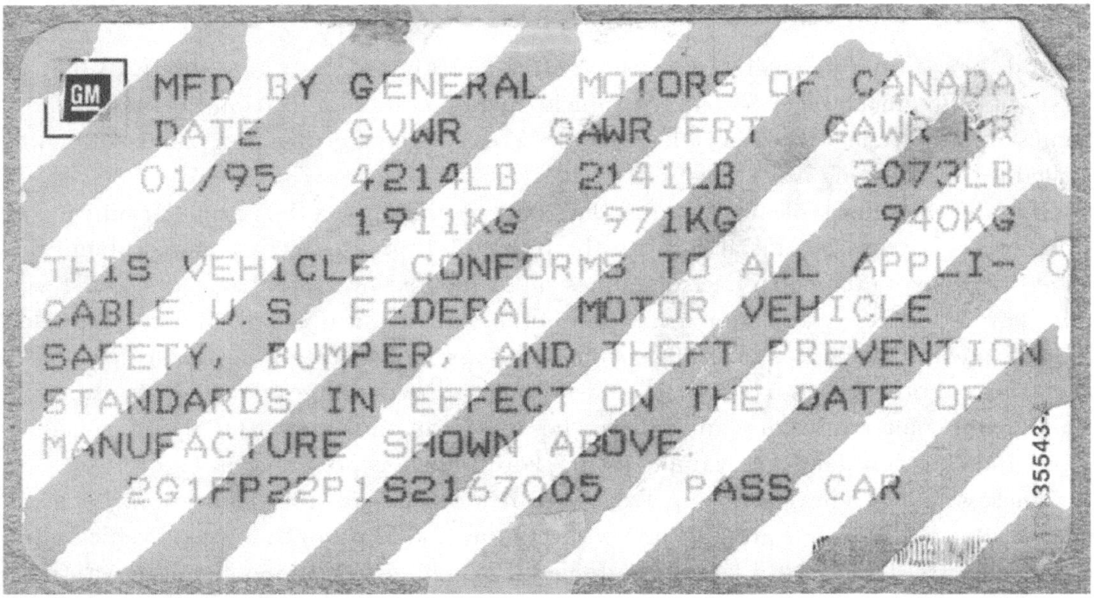

Figure 6-9

Example of tamper evident feature on GM Federal Motor Vehicle Safety Certification emblem. When removed, this label partially self-destructs, creating the distinctive diagonal striped pattern and leaving a similar striping on the vehicle from which it was removed.

niques to remove these emblems almost completely undamaged, allowing for placement on another vehicle. It is nearly impossible to remove a factory-installed certification label completely intact by simply pulling on it. Any certification label that comes off a vehicle this easily should immediately be considered suspicious.

6.3.4 Anti-Theft Label

In the United States, the Federal Motor Vehicle Theft Law Enforcement Act of 1984 provided that a number of major vehicle components be marked with a VIN-bearing theft prevention label. Pursuant to the theft prevention standards, these labels are required to be affixed to the major body parts of new vehicles designated to be "high-theft line" models from the latter half of 1986 forward [25]. These labels are generally found on the doors, lids (hood, trunk, hatch, tailgate), front fenders, both rear quarters, and both bumpers. Figure 6-10 is a diagram of a vehicle showing the different parts that are required to be marked as defined in the *Code of Federal Regulations* [25]. The engines and transmissions are also required to be marked with the VIN (or at least a portion of it) but not necessarily by a label.

The list of vehicles covered by these requirements changes somewhat annually. Most new passenger vehicles the investigator might encounter are fully labeled; some are partially covered with only the engine and transmission being marked. Some car lines are listed as "exempt" from the labeling requirements on the annual NHTSA list because of factory instal-

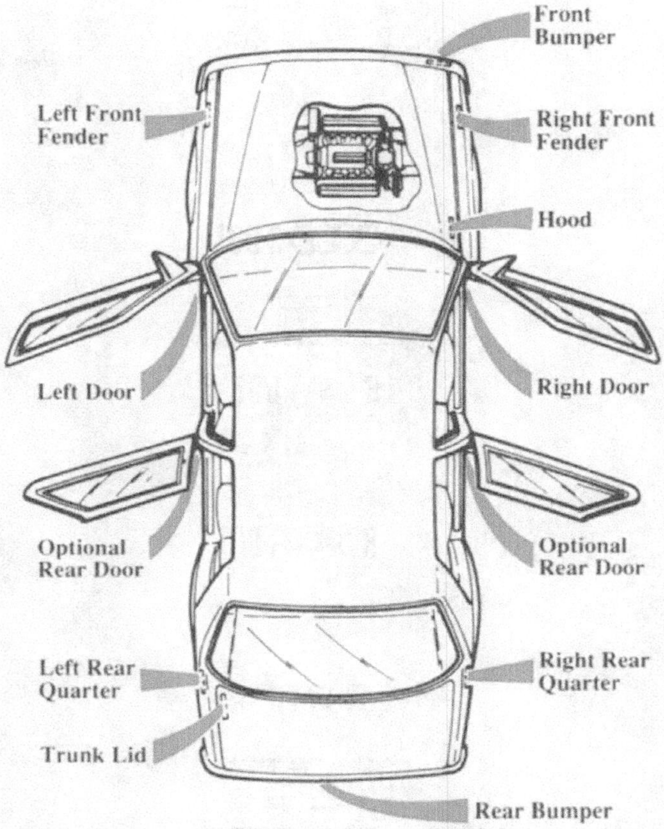

Figure 6-10

Locations of where anti-theft labels are affixed. (Source: 3M Security Systems Division (2001) Automotive Security Labels Verification Procedures, p. 3. Reprinted with permission of 3M Security Systems Division.)

lation of anti-theft devices deemed by NHTSA to be as effective as standard parts markings. These vehicles are sometimes referred to as having been granted the "black box" exemption. The latest list of car lines and components covered by the theft prevention standards may be found in the annual editions of the *NICB Passenger Vehicle Identification Manual* [11]. The most up-to-date vehicle parts marking list may also be obtained from the NHTSA website at http://www.nhtsa.gov in the "vehicles and equipment" section, under the "vehicles-related thefts" topic; the appendix "high theft vehicle lines" contains the listing [26].

Most of the theft prevention labels on body parts consist of a small rectangular emblem containing some kind of manufacturer identification on the border area with the full VIN printed in the middle. Figure 6-11 presents some examples of such labels manufactured by 3M for different car manufacturers. Each car manufacturer has two label types: a label to be affixed on the parts as in Figure 6-10 and a label with the mention R DOT that is affixed on replacement parts. These labels are bonded to the body part surface with a strong adhesive and are almost impossible to remove without causing some kind of noticeable damage to them. Most of the labels leave a latent image of its shape (footprint), viewable with ultraviolet light, on the body surface when removed (see Figure 6-37).

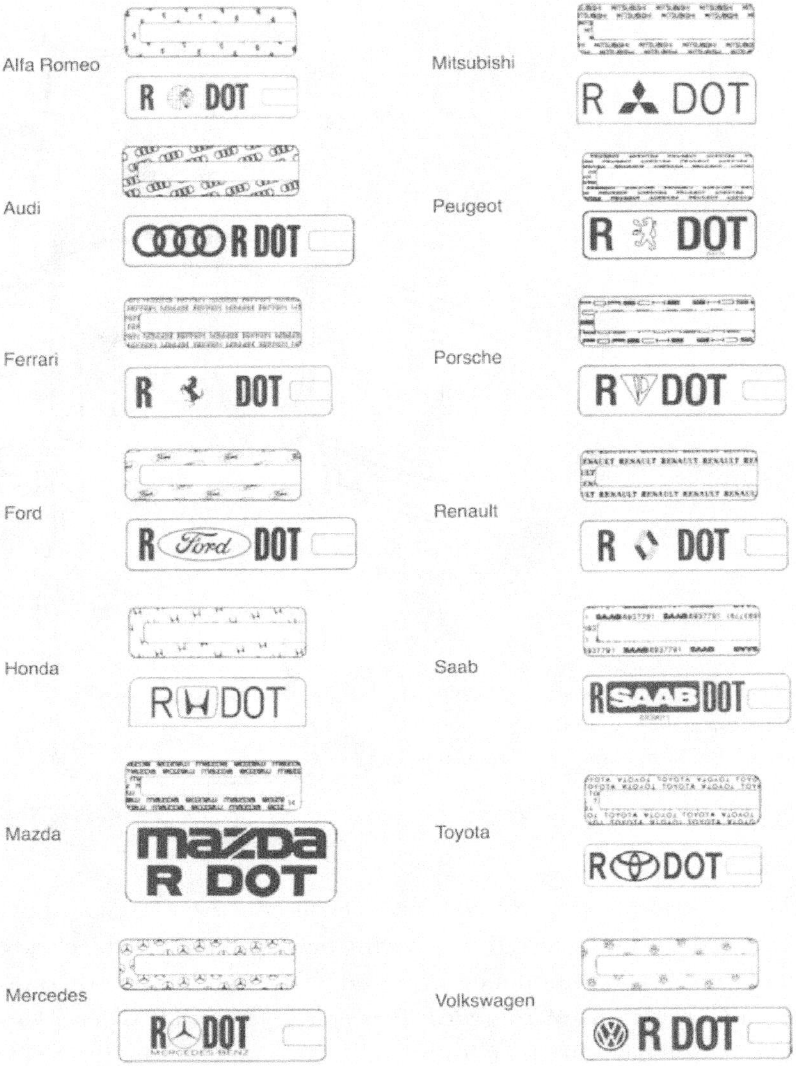

Figure 6-11

Examples of anti-theft labels manufactured by 3M for different manufacturers. The labels written R DOT are used for replacement parts. (Source: 3M Security Systems Division (2001) Automotive Security Labels Verification Procedures, pp. 9–11. Reprinted with permission of 3M Security Systems Division. Copyright 3M. All rights reserved. Reprinted with permission.)

6.3.5 Stamping of Other Parts

The engine and transmission versions of the theft prevention label are not actually labels on the majority of vehicles. Usually the VIN, or a portion of it, is stamped directly onto a smooth machined area or an "as-cast" surface of the engine block and transmission housing. Figures 6-12 and 6-13 show examples of such stamping. This VIN can also be attached with a plate as illustrated in Figure 6-14.

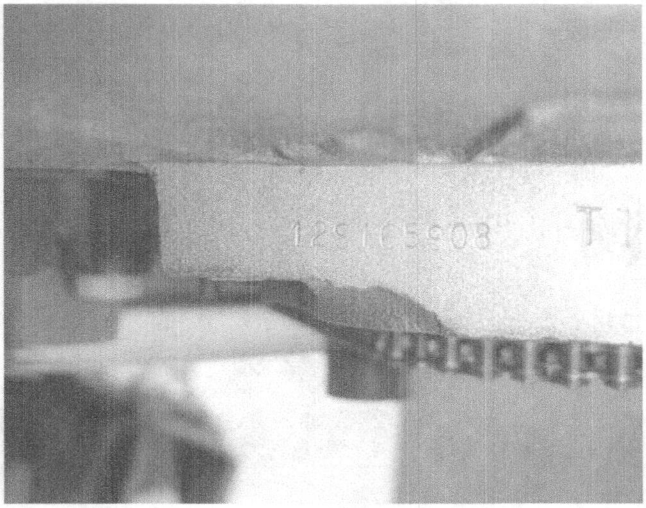

Figure 6-12

Stamped engine number VIN derivative on GM V8 engine. As seen in this example, these numbers are not always deeply stamped.

Figure 6-13

Typical stamped VIN derivative transmission number. This particular GM example from the 1990s is stamped in a style peculiar to GM at that time.

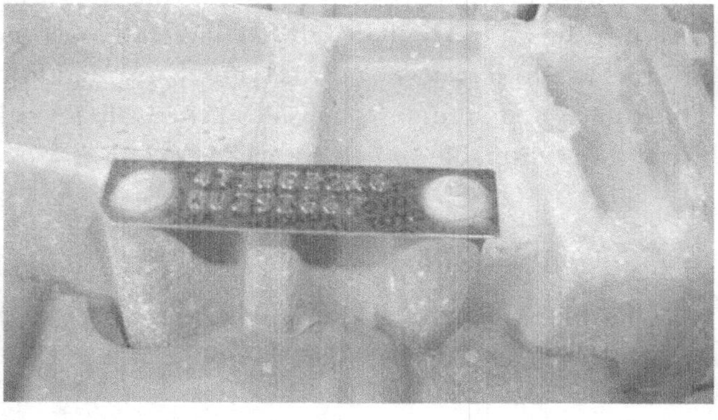

Figure 6-14

Engine number plate containing full VIN. Note the Toyota logos (visible on the right side of plate) and the drive nail mounting mechanism.

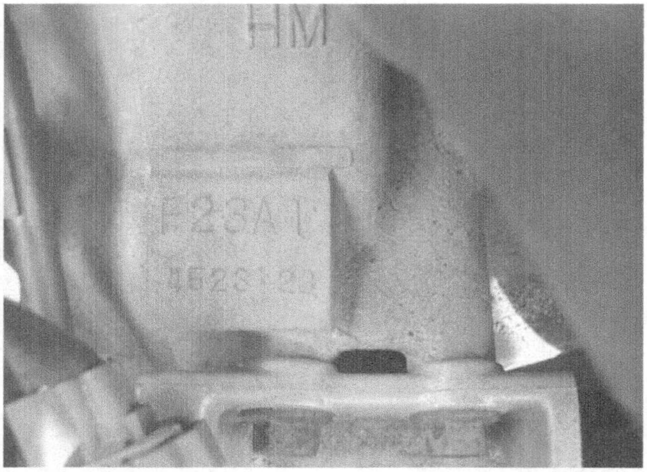

Figure 6-15

Example of a non VIN-based engine serial number, which may be cross-referenced back to vehicle VIN. This particular example was present on a Honda engine in addition to a VIN-based engine number tag similar to the one shown in Figure 6-14.

On most imports it is not unusual to find the entire 17-character number placed on these components, as seen in Figure 6-14. On most domestic vehicles a partial VIN is displayed, containing at least the last eight characters of the VIN (the VIS segment) and possibly prefixed with a make identifier character on GM models, as depicted in Figures 6-12 and 6-13. The stamping styles vary from machine stamps to dot matrix characters. It is important to remember that the VIN-based theft prevention identification numbers placed on engines and transmissions are separate and distinct from the serial numbers (non VIN-based) placed on these components by many manufacturers, such as seen in Figure 6-15. However, most non VIN-based serial numbers on engines and transmissions may be cross-referenced back to the VIN of the originally assembled vehicle.

A third place the full VIN might be found is on what is sometimes called a service parts identification label. Although these emblems are seen predominately on General Motors vehicles, they may also be found on some units from Chrysler Corporation as well. The emblems are usually rectangular in shape and several inches in length. They are attached with an adhesive and generally found on the underside of the trunk lid (GM), under the hood, inside the center console compartment, or inside the glove compartment, as seen in Figure 6-16. The VIN, 1G5CS18R4G0521160, on this particular example is printed out in full in the top left corner on the emblem in addition to some coded parts information. These parts code numbers may also prove valuable in the process of determining the identity of a questionable vehicle.

6.3.6 Window Etching

Another place a VIN marking might be found is etched into the lower face of the vehicle window glass near the bottom. Although not placed on vehicles directly by the manufac-

Figure 6-16

GM service parts identification emblem. This particular example from a GMC vehicle contains the full 17-digit VIN in the upper left portion.

Figure 6-17

VIN etching on vehicle glass. This particular example is noteworthy because it contains both the entire VIN and a telephone number to call for further information.

turer, these types of VIN markings are normally the product of aftermarket companies or local law enforcement crime prevention programs and are still a valuable tool in vehicle identification. A surprisingly large number of vehicles viewed at random may have these numbers placed on them, especially those of the Toyota family. Another type of aftermarket window etching that can sometimes be found is that of a coded membership number and toll free telephone number placed on the glass. This coded number can usually be cross-referenced to the VIN of the vehicle it was placed on by calling the listed phone number, provided that the company in possession of the records is still in business and willing to cooperate. The example in Figure 6-17 shows the full VIN and a service contact telephone number.

6.3.7 Other VIN Locations

Another possible location of a full or partial VIN would be on a sheet of paper or computer punch card, which lists coded vehicle assembly information. Such a list is sometimes called a "broadcast sheet," "Auto-Tel," or build sheet. An example is shown in Figure 6-18.

Many of these sheets are folded up and resemble the packing lists commonly found attached to the cartons of new major appliances. Others are smaller rectangular-shaped pieces of computer paper or resemble IBM keypunch cards. These sheets usually have the entire VIN printed somewhere. These sheets could be located almost anywhere in the vehicle, some of the most common locations being under the front seat cushion areas, taped inside the doors, taped or wadded up inside the fenders, behind/under the rear seats, or somewhere inside the trunk area. The placement of these items seems haphazard, and their presence in the vehicle is never guaranteed.

6.4 SECONDARY AND CONFIDENTIAL VINS

6.4.1 Principle

Secondary VINs are those placed on a vehicle, by the manufacturer, that are not readily viewable when casually observing the vehicle from the outside. The actual secondary VIN, sometimes referred to as confidential VIN, represents an extremely reliable source of a vehicle's true identity. Some examples of what may be considered to be secondary VINs are the VIN-based numbering placed on engines and transmissions that were discussed in the previous section. Although these numbers are quite valuable from an investigative standpoint, they now mainly serve to satisfy the major parts marking requirements and to match these particular components with their original vehicle; they are not intended to represent the vehicle identity as a whole.

6.4.2 Secondary (Nonconfidential) VIN

The actual secondary VIN placed on a vehicle is not always located in a spot one would consider to be particularly "confidential." On some imported passenger vehicles, the secondary numbering often consists of the entire VIN being stamped onto the face of the upper firewall area clearly visible when the hood is fully opened. This is illustrated in Figure 6-19.

Frequently, these numbers present the stamped VIN characters in a font style unique to a particular manufacturer. Occasionally, there may be minor character and length differences within the composition of the secondary number compared with the number on the public VIN plate, but the last six digits (sequential portion) should match. As mentioned

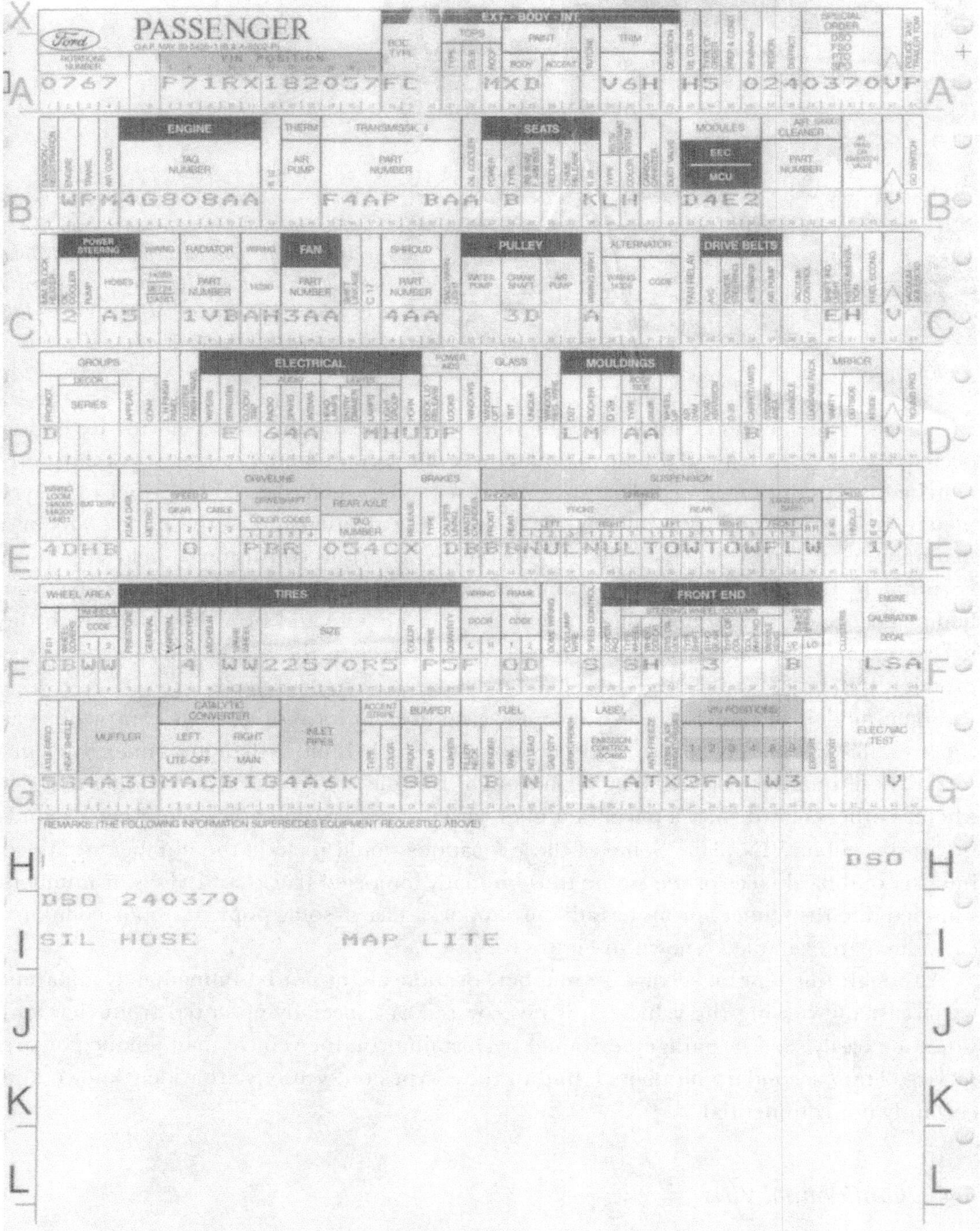

PASSENGER

A 0767 P71RX182057FC MXD V6H H5 0240370VP A

B WFM4G808AA F4AP BAA B KLH D4E2 V B

C 2 A5 1VBAH3AA 4AA 3D A EH V C

D D E 64A MHUDF LM AA B F V D

E 4DHB D PBR 054CX DBBNULNULTOWTOWFLW 1V E

F CBWW 4 WW22570R5 P5F OD S SH 3 B LSA F

G 554A3GMACBIG4A6K SS B N KLATX2FALW3 V G

REMARKS: (THE FOLLOWING INFORMATION SUPERSEDES EQUIPMENT REQUESTED ABOVE)

H DSO H

DSO 240370

I SIL HOSE MAP LITE I

J J

K K

L L

Figure 6-18

Typical large size build, broadcast, or "Auto-Tel" sheet. Although this particular Ford example shows only the VIS preceded by three numbers from the VDS of the VIN in the upper left-hand corner along with other information, many of these sheets display the entire VIN. Fewer of these documents seem to be placed in later model vehicles.

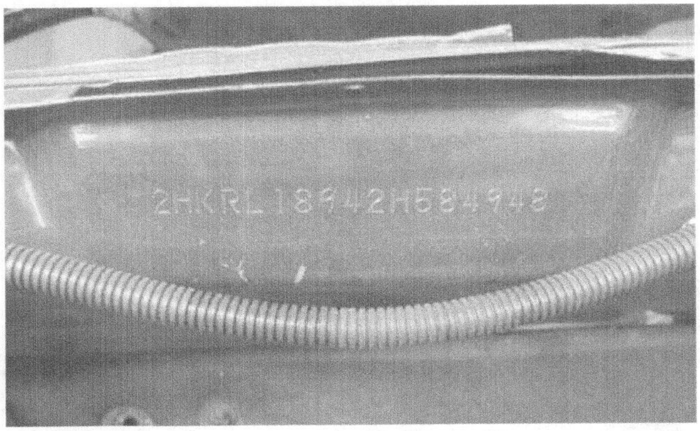

Figure 6-19

Typical secondary VIN stamped onto the upper portion of the firewall in the engine compartment. This particular example from a Canadian-assembled Honda passenger vehicle contains the full 17-digit VIN exactly as it is displayed on the public VIN plate.

earlier, there are always exceptions to the rules on vehicle numbering. A glaring exception occurred a few years back when the manufacturer of a certain line of pick-up trucks abruptly switched their customary VIN derivative based secondary/confidential identifier to a completely different secondary number with no apparent relationship to the public VIN at a manufacturing plant of theirs. This initially created a serious confusion in field situations. Fortunately, these occurrences are comparatively rare, and this particular manufacturer had acted responsibly by notifying the NICB of the change and providing them with the necessary cross-referenced production records to correlate these odd numbers with the proper VINs.

Some models of vehicles bear a secondary number in locations not as conspicuous as the firewall VINs but are readily noticed when the vehicle is more than just cursorily examined. These numbers vary in length and might include the whole VIN on some models, whereas others may display a partial number with just the last eight characters corresponding to the public VIN plate. Some of these locations would include the numbers stamped into the outboard faces of the frame rails on many imported trucks and SUVs or numbers stamped into the fender aprons or radiator support areas on some popular imported luxury cars. Such an example is shown in Figure 6-20.

Although this type of secondary number location might not be immediately apparent when casually viewing the vehicle, it is not covered or especially obscured from view and would be easily seen by garage personnel performing routine vehicle maintenance or tire service. Thus, secondary numbers found in these types of locations are widely known and certainly not confidential.

6.4.3 Confidential VINs

What are considered to be confidential VINs are those identification numbers placed on the vehicle by the manufacturer in a more or less hidden fashion and in a location that is not widely known. The viewing of these numbers might require some minor dismantling

Figure 6-20

Typical frame rail face stamped secondary VIN. This example from a US-assembled Mercedes-Benz SUV contains the entire 17-digit VIN prefixed and suffixed with the manufacturer's logo.

Figure 6-21

Typical frame top or bottom surface secondary confidential VIN. This example, stamped in dot style, shows a VIN derivative consisting of the letter X denoting the year, the letter S denoting the plant, followed by the six-character sequential production number.

of the vehicle or the use of an inspection mirror. These types of numbers could be located on the top surfaces of frame rails, such as presented in Figure 6-21. They might also be found on the floor pan underneath carpeting and trim or placed somewhere on the various sills, struts, aprons, flanges, and support members typical of vehicles with unitized body construction, as demonstrated with the example in Figure 6-22.

For the most part, these numbers are stamped into the metal surface with the characters in block, italics, or dot matrix style. Again, there are always exceptions; for example, the confidential VIN on some mid-size and smaller GM passenger vehicles dating back to the late 1980s consisted of a hidden riveted metal plate that was almost an exact duplicate of the public VIN plate. Although these GM plates contained the whole VIN, many confidential numbers may likely be a shorter VIN-based derivative usually consisting of the last eight characters of the VIN (the VIS segment) with possibly a star-like or asterisk-shaped symbol before and after the digits. Sometimes there may be one or more other characters that are symbols for make and model information preceding the eight VIS section digits. Many

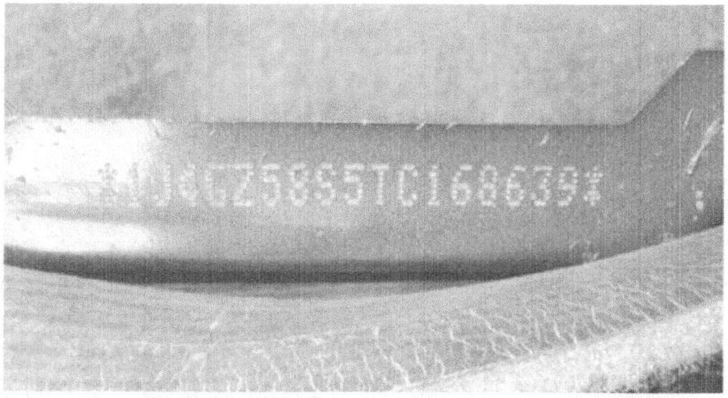

Figure 6-22

Typical secondary confidential VIN stamped into a structural member on a vehicle with unitized body constriction. This particular example from a Jeep vehicle shows the full VIN stamped in dot style with an asterisk stamped before and after the VIN. Viewing this example involved pulling back a section of carpeting.

manufacturers have a procedure in place when an error is made in the stamping of a secondary or confidential VIN. In such instances, the errant number is stamped over with the character X and the correct number is then stamped directly adjacent to the errant number. However, these mistakes seem to be a relatively rare occurrence.

For obvious security reasons, a discussion of the locations of confidential VINs on any particular makes, years, or models of motor vehicles is not presented in this book. In the United States, the best source for information on the location of the secondary or confidential numbering on a particular vehicle is the NICB. An NICB Investigative Assistant may be contacted by telephone by calling (800) 447-6282 and following the prompts. Another NICB resource for this type of information is to contact the local NICB field agent in one's area directly.

6.5 PARTICULAR VINS

6.5.1 Gray Market VIN

There are some glaring exceptions to all the advancements in vehicle numbering that need to be discussed. These exceptions concern gray market vehicles, kit assembled cars, and rebuilt motor vehicles.

Loosely defined, the term "gray market" is generally used to describe vehicles produced in a country, then imported into another country, but that were not engineered by the manufacturer to be marketed in the latter country. In the United States, although the means by which these vehicles reach the nation vary, they usually do not meet US requirements in one or more areas related to safety equipment, emissions, collision/impact standards, theft prevention, and of course vehicle numbering. Nevertheless, although vehicles intended for sale in Canada are not considered as gray market vehicles in the United States, they do present minor differences when compared with US market specifications.

Depending on the geographic region, dealing with gray market vehicle examinations may be a relatively common occurrence or just a rarely encountered oddity. Although gray market vehicle models from manufacturers in varied places such as Russia, France, Italy,

Germany, The Netherlands, The United Kingdom, South Africa, China, Japan, Sweden, Mexico, and Central and South America have been found to have reached the United States, the vast majority of these vehicles seem to be Mercedes-Benz, BMW, and, to a lesser extent, Porsche automobiles imported from Europe. Although many of these cars share a strong family resemblance with their US market counterparts, there are some definite differences. Aside from the occasional obvious right-handed steering/driving controls, the headlights, bumpers, model designations, gauges (usually metric), glass markings, and informational emblems (not worded in English) may all be different. Although the manufacturers in a number of countries that are members of the European Union comply with the VIN configuration that contains 17 characters as per ISO standard, there might not always be a factory-installed public VIN plate or driver's door area VIN emblem present. And although these VINs may be 17 digits long, they may present some differences in their content. These numbers may not display a model year designator or check digit, and the letter Z may be found to have been used several times within the number as a "filler character" to attain the 17-digit VIN length. For example, the gray market, sometimes referred to as "European-style," Porsche VIN WP0ZZZ93ZES000621 includes a year designator E in the 10th position indicating 1984 but has Z character fillers for some model information designations and for the check digit. In another example, although the gray market Mercedes-Benz passenger vehicle VIN WDB1260441A242521 from the mid-1980s does not display the Z filler characters, it lacks a valid check digit and year designator, as shown in Figure 6-23.

The primary factory VIN location on these vehicles is most likely on a plate somewhere under the hood with a stamped number probably in a place similar to where the secondary number would be found on a US market model of the same make. Although there are procedures in place under US law wherein a gray market vehicle may be brought into functional compliance through a process known as "federalizing," the vehicle's VIN configuration is not changed to comply. The original non-US VIN still remains the primary identification number. The US laws covering the federalizing process include provisions requiring a public VIN plate and a certification label, but these are aftermarket add-ons and usually do not look like factory original VIN items as depicted by the plate in Figure 6-23. Authenticating

Figure 6-23

Gray market VIN stamped into an add-on aftermarket plate attached to a gray market Mercedes-Benz from the mid-1980s. Even though it contains 17 digits, the VIN does not follow the US rules presented in Section 6.2.

the VIN and history of a questionable gray market vehicle is often a challenging undertaking on even the mainstream makes and especially so when dealing with the more obscure marques where no comparable US model or domestic distribution network exist. Interpol may be helpful in tracing these vehicles (see Chapter 22). The entire process surrounding gray market vehicles, their nonstandard VIN configurations coupled with the inevitable gaps that exist between US mandated procedures, and the various states' laws governing the titling of these cars, make this an area ripe for abuse.

6.5.2 Kit Assembled Cars

Another VIN exception area that needs to be covered concerns vehicles assembled from kits. This class of vehicle covers the spectrum from conveyances where the finished product is almost laughably crude to some very sophisticated well-assembled automobiles. Although these vehicles tend to be somewhat "individual" in appearance and equipment, they usually all share one trait: The finished product is generally based upon the platform of some kind of new, used, or salvage mass-produced vehicle model. There are companies that actually custom-build the kit car for the customer using new components, whereas others provide detailed written construction plans, a basic body, and some interior components with the buyer left to supply the vehicle's rolling chassis, engine, drivetrain, and related running gear.

In most cases the kit platform is an item supplied by the customer. The finished product probably uses the original VIN from this "donor" vehicle as its identification with appropriate notations added to the DMV title file indicating "kit", "assembled from parts" (ASPT), or "homemade" status. Of course, this varies depending on laws applicable in individual states or countries, with some jurisdictions supplying the kit customer with a state-issued VIN to identify their creation. In other cases, the kit supplier furnishes the customer with a manufacturer's certificate of origin (MCO) or a manufacturer's statement of origin (MSO) bearing a company-generated vehicle serial number.

Although most kit car companies are legitimate, the opportunities for abuse at the consumer end of these transactions are readily apparent. When called upon to examine any vehicle with a status declared as kit, assembled from parts, homemade, rebuilt from salvage, state-issued VIN, or any similar terminology, care must be taken to systematically establish the identity of each uniquely numbered component making up that vehicle. The examination could reveal that the entire car is actually a stolen vehicle disguised as a kit car. One or more of the major components making up such a vehicle could be from a parted-out stolen car, or the platform of the donor car for the kit might in fact be a late model vehicle that is already currently titled and the object of a substantial outstanding bank lien (fraud).

6.5.3 Rebuilt Vehicles

Much of the information on kit cars is applicable to rebuilt vehicles. These are usually motor vehicles that are declared as salvage by an insurance company or large fleet because of collision, fire, wind, or extensive water damage due to storms or flooding or vehicles stripped of a significant portion of parts. In most states a standard is established that, depending on the extent to which a particular vehicle is wrecked, flooded, burned, or stripped, it may be branded as "rebuildable" or "unrebuildable" on its title. Just as with examining kit cars, the investigation of a vehicle declared to be rebuilt should be conducted in a systematic fashion with care taken to uniquely identify as many of the components of the finished product as possible. Most states require submission of a detailed inventory of documents (receipts and bills of sale) for parts and labor services from the parties involved in the vehicle rebuilding before issuing a "clean" title. Having access to copies of these documents and any available insurance company vehicle examination reports or photographs as to what parts were originally damaged or missing are invaluable for comparison when physically examining the alleged "rebuilt."

6.6 OTHER VEHICLE MARKINGS

6.6.1 Air Bags

There are additional more obscure markings on motor vehicles that may prove to be valuable in the identification process. Some of the more prominent markings are found on air bag restraint devices. Personal experience has shown that besides those illegally altered, every passenger vehicle air bag device displays some type of serial numbering and date code markings such as the ones presented in Figures 6-24 and 6-25.

Most air bag serial numbers can be cross-referenced back to the VIN of the vehicle it was originally assembled with. In the United States, some of these serial numbers are available for immediate "on-line" cross-referencing through the NICB, whereas others may have to be cross-referenced through the manufacturer by contacting the NICB manufacturers' liaison desk. On a few vehicle models the air bag serial numbers are readily viewable by just

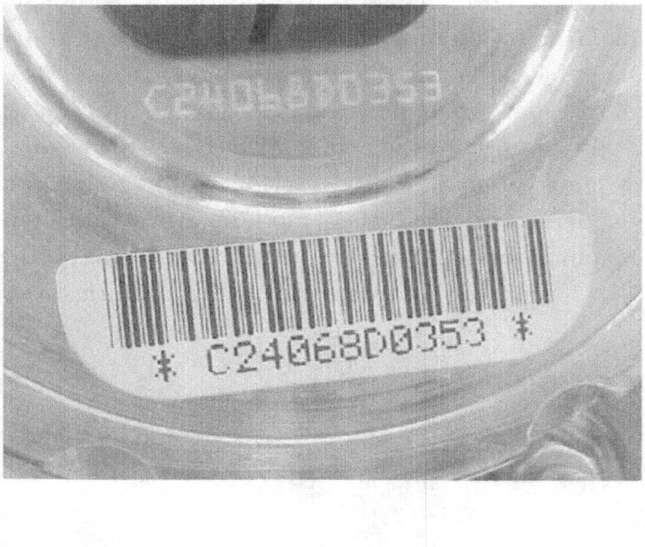

Figure 6-24

Example of numbering on the back of a driver's side air bag restraint device.

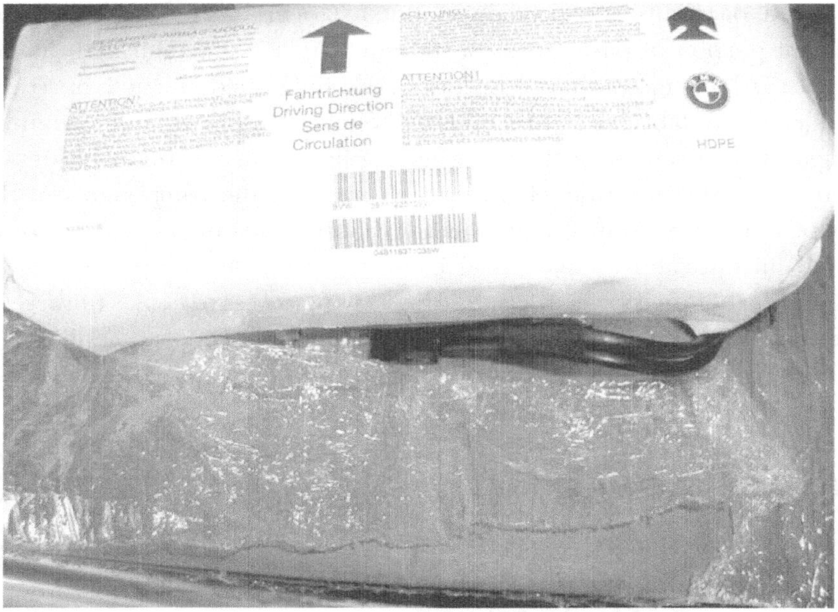

Figure 6-25
Example of numbering on a passenger's side air bag restraint device.

looking under the base of the steering wheel hub or behind the glove compartment box; unfortunately, on most, gaining access to these numbers involves some dismantling work. Caution must be exercised when examining vehicle air bags under these circumstances. These are powerful devices designed to violently deploy for the protection of properly secured occupants in the event of a collision. The activation of an air bag during its examination could be lethal to the investigator. Although there is a set of precautions that can be taken to lessen the risks, the best advice is to have an auto theft investigator with considerable experience handling air bags to assist. Also, it is always possible to take the vehicle to the appropriate dealership.

In addition to the actual air bags, there are other components within the deployment system that may display traceable serial numbers and date information, including the impact sensors and control modules. Any inquiries most likely have to be done through the manufacturer. Although experience with successfully tracing these items back to a specific VIN has proven spotty at best, the date information has frequently proven helpful.

6.6.2 Other Parts

When opening the hood, doors, trunk lid, tailgate, console, glove box, or spare tire compartment or peering beneath the seats, dashboard, and under the vehicle itself, it is

Figure 6-26

A vehicle body plate from Chrysler, located under the hood and containing the full 17-digit VIN on the bottom line in addition to other coded information on vehicle color, trim, and so on.

nearly impossible not to notice that modern motor vehicles have a lot of emblems and tags on them. Under the hood or in the trunk, these emblems or tags may be found on the underside of the lids themselves, on the engine, affixed to the transmission/transaxle, or on the various "bolt-on" components attached to the engine. Often informational emblems are located above the radiator area. The vehicle color code and trim plate tags may be found attached to the firewall or placed on inner fenders, as illustrated in Figure 6-26.

Close inspection may reveal a part number emblem placed on a trunk lid surface or down in the spare tire well. Underneath the car, they are often found on the frame rails (if applicable), floor stamping, transmission pan, fuel tank/lines, axle housings, and wrapped around clusters of wiring. In the passenger's compartment, a number of interesting markings may be found on the seat belt/shoulder harness straps (sometimes when fully extended), such as shown in Figure 6-27, inside some seat belt buckle assemblies, on the rear surfaces of seat cushion foam/padding and door upholstery panels, possibly on the back of the radio, and on some of the instrument panel gauges, attached to various wire clusters, and heater/AC duct work.

Many of these emblems may contain eye-readable characters in addition to a bar code. Although most of these emblems are part numbers, the part number may be exclusive to a specific year of vehicle or contain some coded date information. Most, but not all, year information contained on items like these probably refers to the calendar year the part was manufactured. For example, the emblem attached to the strap material of a seat belt/shoulder harness assembly installed in a 2004 model year vehicle might very well show a 2003 year because that could reasonably be the calendar year the belt/harness assembly was produced. A good rule of thumb on date codes is that unless there is evidence that the vehicle has undergone some fairly extensive repairs or modifications since leaving the assembly plant, the calendar year referred to on emblems should be the same as the vehicle model year or one year prior, not newer, nor too old.

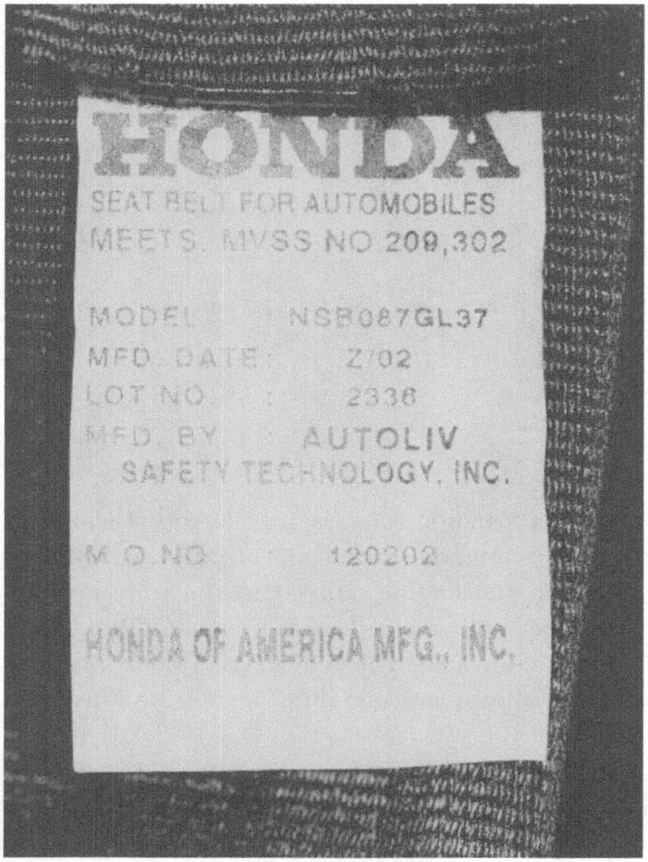

Figure 6-27

Seat belt certification label containing calendar year of manufacture (2002). This particular Honda example is partially sewn onto the safety belt strap material; other labels may be completely sewn (all sides) or directly printed onto the safety belt strap material itself.

6.6.3 Engine Emission Control Label

One particular informational emblem that does refer to the model year is the vehicle engine emission control label located under the hood. This label is generally mounted in plain sight on the area just above the top of the radiator facing upward, clearly visible on the underside of the hood or on the upper sides of the front shock towers. Occasionally, the labels are attached to the engine valve cover(s). This label is basically the manufacturer's certification that the engine installed in the vehicle conforms to the emissions standards applicable to that model year vehicle. It displays that year in a four-digit form (e.g., 2004) on the emblem.

Other engine information, including the engine family or group referred to by an alpha or numeric designation, is also listed. The first character of the designation gives the model year just as it would be listed at the 10th digit on the public VIN plate. Using the 2004 Mercury VIN (underlining added for emphasis) 2MHHM79V74X676128 broken down earlier as an example, the engine emissions control label, as shown in Figure 6-28, lists 2004 as the model year applicable to this vehicle. It also shows the engine group code as 4FMXV04.6VH5 (the first digit in the code indicating the model year).

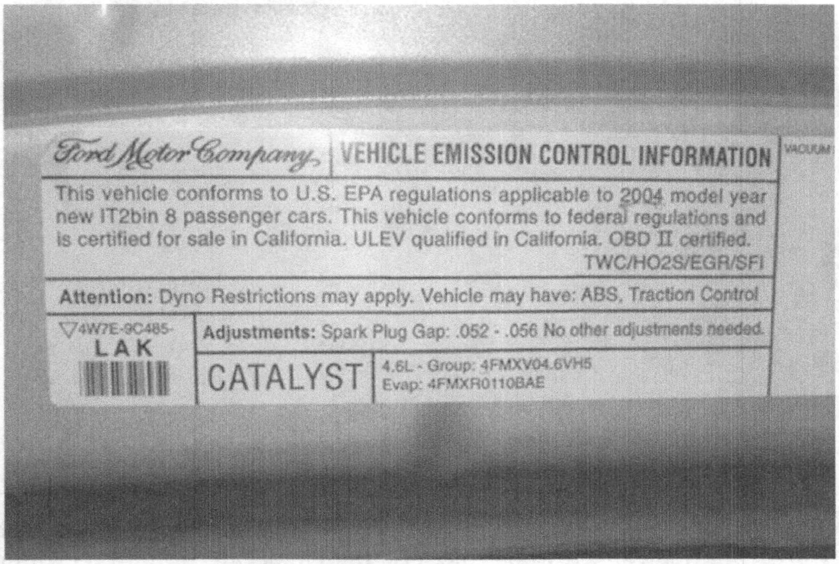

Figure 6-28

Typical engine emission control certification label, which is usually prominently affixed under the hood. This example of a 2004 Mercury shows the full model year in the second line of the label and again as a single digit (4) at the beginning of the engine group code number at the bottom right of the label.

6.6.4 Other Coded Date Information

Coded date information can sometimes be found on hoses, belts, and glass, including the mirrors. On most vehicles the calendar year in which the glass was manufactured is contained in the brand name and certification inscription area usually listed as a single standalone digit among the other characters. On some imported vehicles the calendar month and year of the window glass are contained in a coded form of small dots, with the position of certain dots in relationship to other characters inscribed on each particular pane determining the date information.

Moving on to the engine block (including items attached to it) and transmission, there are a number of interesting date-coded elements. In addition to the presence of VIN-based identification numbers and the possible existence of the separate component serial numbers discussed earlier, several raised (not stamped) numbers on the actual block and transmission housing surfaces may be present. These are generally referred to as casting numbers and normally denote certain information concerning engine family, displacement, and application/usage (truck, heavy duty, high performance, etc.) among other information. An example is presented in Figure 6-29.

On the block, there may be another raised number indicating the casting date in code that is not as neatly or clearly marked as the casting number. There is frequently a small round symbol near this date code called a casting clock that consists of raised dots (usually

Figure 6-29

Typical casting number area. This example is from a GM V8 engine block and shows the casting number (left), date-coded information, and a casting clock (right).

12) arranged in a circle with a pointer in the middle, as also shown in Figure 6-29. A similar arrangement sometimes seen on engines is a round symbol containing 12 equally sized raised pieces resembling a sliced whole pie. Casting clocks are not always round in shape. The date codes are usually fairly easy to decipher, because they are based upon common sense. The format possibly includes an alpha character denoting the month (A = January, B = February, etc. with the letter I probably skipped) and the year represented by one (no decade indicator) or two digits.

Some engine casting date codes may include a character denoting the day or week within a particular month. Casting numbers and date codes are sometimes seen on intake manifolds (not to be confused with the engine cylinder firing order), exhaust manifolds, engine heads, and water pump castings. The same casting numbers and symbols may also be found in similar formats on the transmission/transaxle housings. One may find date codes and other information stamped into surfaces, on labels, and on tags or data plates attached to transmissions, alternators, air conditioning compressors, power steering pumps, fuel pumps, fuel injector assembly housings (or carburetors on older models), and starter motors. Occasionally, stamped serial numbers are located on one or more of these items that might be traceable, particularly on some Honda family models.

6.6.5 Other Serial Numbers

It is also possible to notice stamped numbers (non-VIN derivative) on larger painted sheet metal parts under the hood on many vehicles. Although most of these are generic part identifiers, on some import models these could be body or radiator numbers that may be traceable to the VIN. These types of identifiers are usually larger and more clearly stamped than part numbers and might require moving some weather stripping/sealing material or plastic trim a couple of inches to view, as demonstrated in Figure 6-30. On some vehicle models, numbering found on lock cylinders and keys may be helpful in identification.

Another possible source of vehicle numbering deals with electronic data. Originally limited to the higher tech models, today more vehicles are equipped with electronic com-

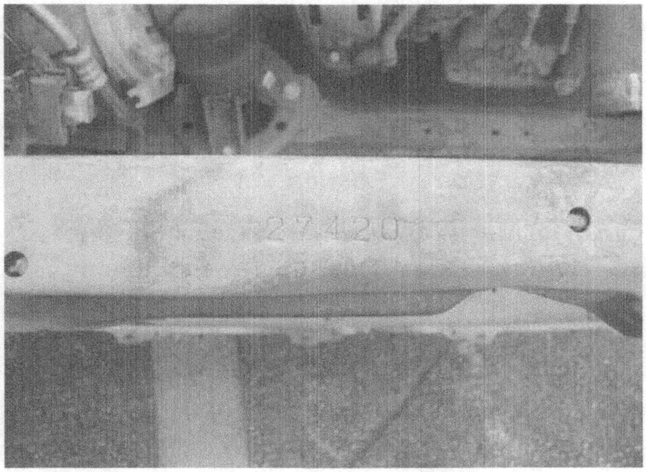

Figure 6-30

Typical stamped radiator/body number, non-VIN based, commonly found on many Nissan and Toyota vehicles. When used in conjunction with other particulars about the vehicle, this number may possibly be cross-referenced to the original VIN.

ponents that provide the VIN, mileage, and other data when interrogated using the proper equipment. The appropriate dealer for the make should be contacted regarding any request on this kind of service. Some vehicles, such as the 2005 Volvo S50, can also simply display their VIN by accessing the vehicle's settings through the radio display.

6.7 VIN CHANGING (OR RE-VINING, RINGING, OR TAGGING)

6.7.1 Principle

Tampering with the original/factory identification of a vehicle is almost always done to obscure the original identity. How much skill or finesse goes into this process depends on what result the offender is trying to accomplish. If the only purpose is to destroy the vehicle beyond recognition to facilitate an automobile insurance fraud or to render the donor vehicle for a salvage switch unidentifiable, the vehicle could simply be crudely dismantled, have all the identification numbers burned or roughly chiseled off, and subsequently be disposed of by being crushed as scrap, burned, or dumped into a lake or canal. Because the offenders in this scenario are only concerned with destroying the original vehicle identity and not thereafter substituting a new vehicle identity, they do not care how much damage is done in the obliteration process. The only expertise required here is a good knowledge of the location of the unique/traceable numbering on the vehicle and the ability to wield an air chisel, metal saw, or cutting torch to destroy them.

True skill and craftsmanship are required when the offenders need to carefully remove as much of the original identification numbering as possible and replace it as imperceptibly as possible with new identification. Regardless of which of the following techniques is chosen to change a vehicle's identity, the offenders must remove and/or change more than just the public VIN plate to make the substitution of the new vehicle identity a success. They must also contend with all other VIN bearing emblems, secondary and confidential numbers, and traceable component serial numbers discussed earlier in this chapter. The skill level and thoroughness displayed in the number replacement and alteration work done by the professional vehicle theft offender determine how difficult identifying a given vehicle will be.

6.7.2 Alteration of Existing VIN

Changing the identity of the vehicle may be done in several ways. The first involves altering the existing VIN markings to create a different number. The offender might change one or more characters on the original VIN plate such as converting the numeral 3 to an 8; a numeral 1 to a 4, 6, 7, 9, the letter L, or any combination of these characters; a 5 to the letter S; or the number 2 to the letter Z; which are the more popular alterations (see Figures 17-6 and 17-7). In addition to the obvious dilemma of probably throwing off the check digit calculation on a 17-digit VIN, this technique also tends to interrupt the font style of the VIN as a whole, causes variations in the color/ finish of the plate surface, and generally disrupts the original spacing between the altered and unaltered characters. Unless very skillfully done, these types of alterations are among the easiest to detect.

6.7.3 VIN Plate Switching

The next identification changing method involves using the original factory VIN plate and emblems from another vehicle. The "new" VIN will probably be obtained from a salvage vehicle of similar make, year, and model as the stolen vehicle being renumbered. This type of re-VIN is commonly referred to as a "salvage switch" or "salvage retag." The more skillful VIN retag offenders will be careful to use a salvage VIN from a vehicle as near in color and equipment specifications to the stolen car as they can both locate and mount the plates and emblems as close to factory specifications as possible. They may even transfer the engine, transmission, frame rails (if applicable), and speedometer/odometer cluster from the salvage vehicle to make certain the component numbers and mileage match up properly. Detecting a skillfully performed salvage switch in the field involves being able to recognize irregularities in the VIN plate such as scratches, bending, discoloration/excess weathering, and characters that are inappropriate descriptors for the vehicle model being examined, such as having a station wagon VIN on a convertible. It is also important to detect any anomaly in the plate mounting area such as damaged, scratched, or nonfactory rivets, tool scratches on the soft trim areas around the plate, the telltale presence of glue around the rivet heads, plate and adjacent trim pieces, or a mounting position slightly off-center from the original VIN. Such an example is presented in Figure 6-31.

On many vehicle models the factory placement of the VIN plate is in such a recessed location that it would require the offender to temporarily raise or remove the windshield to accomplish the plate switch. This disruption of the original glass mounting usually leaves behind some evidence of tampering on the windshield gasket (seal) or on trim pieces and mounting hardware such as toolmarks or excess adhesive. Many of the listed irregularities

Figure 6-31

Example of a poorly mounted VIN replacement plate. Although a good factory plate from another vehicle was used in this example, the offenders smeared a large amount of adhesive onto the rivets, plate surface, and even the surrounding soft trim areas.

Figure 6-32

Example of an excellently done counterfeit VIN plate from a successful stolen vehicle cloning operation. This example displays outstanding color and font reproduction quality, including manufacturer's background logos (Chrysler).

to watch for when examining the VIN plate are also applicable to the safety standards and other emblems.

6.7.4 Counterfeit VIN Plate

Another method of changing vehicle identity is through the use of counterfeit VIN plates and emblems. This category of VIN alteration is one of contrasts. Some of the best illegal VIN alteration workmanship seen are counterfeit plates used in stolen vehicle "cloning" operations that are so nearly identical to factory/original plates that the investigators involved often must grudgingly acknowledge a certain amount of respect for the level of skill required to produce them. An example of a Chrysler counterfeit VIN plate is shown in Figure 6-32.

On the other end of the counterfeit VIN spectrum are such laughably amateurish "plates" as those made with a plastic label-making kit and glued in place over the factory number plate, VIN's ink stenciled onto tin can material attached into place with tape, and VINs hand stamped into pieces of beverage can metal as seen in Figure 6-33, and attached with screws, wire, or even Velcro. Obviously, these latter counterfeits are simple to detect even with a casual examination, and the vehicle is usually easily identified.

Detecting more professionally made and installed counterfeit plates really depends on the accurate knowledge of the characteristics exhibited by legitimate factory VIN plate for

Figure 6-33

This is an example of an amateur attempt to manufacture a counterfeit VIN plate. This particular specimen, stamped into metal from a commercial beverage can, was mounted on the stolen vehicle with a combination of glue and Velcro.

the particular make and model in question. This includes the plate material (aluminum, stainless steel, plastic), color, sheen (flat, glossy), background symbols and logos, character font and style, character spacing, location and placement of the number as a whole on the plate (centered, offset either vertically or horizontally), location of a bar code (if applicable), and the color, style, and any markings of the attaching rivets. It is also important to know the exact factory placement location on the vehicle and whether or not it is mounted perfectly flat and leveled. The best way to examine a questioned VIN of this type is by use of a magnifying glass to carefully compare it with an original factory VIN of the same year, make, model, and plant. Many of the same comparison techniques should be used to examine the other numbering emblems on the vehicle.

6.8 HANDS-ON VEHICLE EXAMINATION

6.8.1 Equipment Needed

To perform a thorough vehicle examination safely and with minimum damage to the vehicle, the tools listed in Table 6-7 should be considered basic equipment for an auto theft investigator's toolkit.

6.8.2 Examination Site and Preliminary Data Collection

The optimal setting for a vehicle examination is a clean, spacious, well-lit, climate-controlled building with convenient access to power tools, lifting apparatus, cleaning agents, and computer links to all the pertinent databases close at hand. Unfortunately, the opportunity to examine questioned vehicles under these circumstances does not come along very often. Most likely, examinations take place outdoors either at the scene where the vehicle is first encountered or at an impound facility after it has been towed.

Generally, on-scene field examinations should probably be limited to making a quick identification if possible or simply to locating enough numbering irregularities to permit

Table 6-7

List of the basic toolkit equipment that should be in possession of the auto theft investigator when checking a vehicle's identity.

Small flashlight	Set of dental picks
Magnifying glass	Assorted screwdriver types and sizes
Varying sizes inspection mirrors	Disposable rubber gloves
Small magnet or "rot spotter" tool	Safety glasses
Small rubber ink eraser	Portable barcode scanner
Fine grit sandpaper	Portable ultraviolet light
Spray grease cleaning product	Fiber-optic viewing device
Clean shop rags	Fingerprint ink or powder
Set of small hand tools, e.g., wrenches, sockets, metric and SAE	Clear latent fingerprint lifting tape and latent print card
Small needle-nose pliers	Casting material for toolmarks

impounding the vehicle for further investigation. There is usually too much activity going on at the scene, including the possible presence of vehicle theft offenders watching every move made by the investigator during the identification process, to do a thorough examination. Examining vehicles in the privacy of a secure well-equipped impound facility is certainly preferable to doing it in the field. Processing in such a fashion also allows time to conduct a preliminary investigation on the vehicle's VIN history and production details before examining it more closely. This can include a 50-state DMV registration check to see whether the vehicle is registered elsewhere. It would also be advisable to check with the NICB for the shipping/assembly data and a VIN Assist printout of the specifications of the VIN in question. The NICB should also be asked for any prior records on the VIN such as salvage, damage claim history, previous export activity, any foreign theft records available, and any secondary number locations for the appropriate model of vehicle before the examination.

After obtaining as much vehicle history, assembly data, and secondary number location information on the vehicle to be examined as was initially available from the NICB, it is time to go through the vehicle examination process. Although nearly every step that can be taken in a vehicle examination is covered for instructional purposes in the following procedures, many of these steps might be unnecessary in an actual inspection depending on the circumstances.

6.8.3 Examination Process

A/ Determination of Make and Model

First, it is necessary to initially walk around the vehicle and to note its general condition. It is important to make certain of the exact make and model. Although this sounds pretty basic, there have been cases where the make and model of a vehicle have been disguised. For example, a renumbered Chevrolet truck could easily be disguised as the equivalent

GMC model, a minivan marked as Dodge Caravan or Plymouth Voyager may actually be a stolen Chrysler Town and Country model, or an Isuzu Rodeo may really be a VIN-switched Honda Passport. Thus, it is crucial not to take the apparent make for granted and to perform a detailed observation.

B/ Public VIN Plate

Second, the public VIN plate is checked. Using either the VIN Assist printout or the NICB vehicle identification manual, the investigator makes sure that the number passes check digit requirements and that it describes the vehicle to which it is attached. During the examination of the plate, its attachment is carefully scrutinized. One needs to make sure it is securely attached with the proper type of fasteners and that it looks like a known good number plate for that model and year vehicle. If the plate seems to be loose, it may be possible to attempt to gently lift and move it with one of the dental pick tools. A good original plate should remain tight. It should be noted if the plate is bent up, scratched, or markedly more or less weathered than the area around it.

C/ Stamped Firewall VIN

If a stamped number looks as if it has been tampered with, a small magnet (or "rot spotter" tool) should be run over it and the surrounding surface; the magnetic pull should be strong and uniform. The stamped number surfaces and surrounding areas should be picked with the dental probe tools; there should be no presence of anything but steel and a thin layer of primer covered with finish paint. No evidence of lead/solder, fiberglass, or plastic body filler, such as Bondo, should be present. An example of a VIN stamped in body filler is shown in Figure 6-34.

The possibility that an overlay of a false number might have been skillfully set into place over the original VIN or that the original number could have been cut out completely and carefully replaced with another must always be taken into account. It is critical to look directly at the back surface of the area where the number is stamped when possible or with the use of an inspection mirror if space is restricted. This might reveal a different VIN, as demonstrated in Figure 6-35.

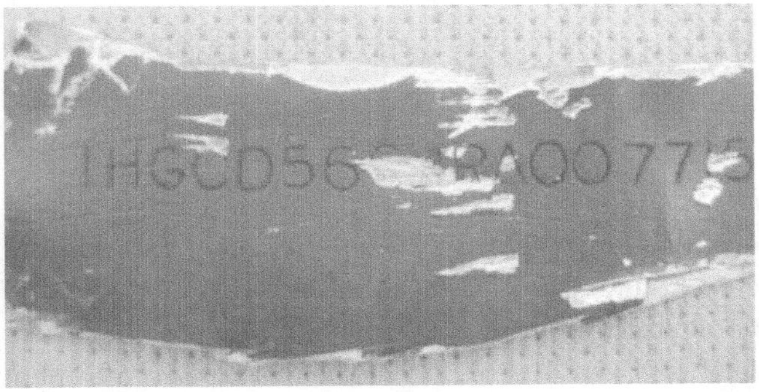

Figure 6-34

Counterfeit firewall secondary VIN stamped into plastic body filler.

Figure 6-35

Rear surface of firewall VIN as shown in Figure 6-34. Note shadow image of true VIN revealed on surface after light sanding.

It is also imperative to look for welds, burn marks, body filler material, and other signs of tampering in the area around the VIN. In some instances, with a little cleaning or very light sanding, it is possible to read the stamped number from the rear on the sheet metal; obviously, it should be the same as the number displayed on the front.

D/ Safety Certification Label

The number on the public plate should be compared with the VIN on the safety standards emblem, and they should match. Also, the other information discussed earlier in the chapter that may appear on this emblem, such as color/trim, country of origin, date of manufacture, and drivetrain specifications, should be noted. The information here should be consistent with the particulars of the vehicle being examined.

The investigator should also try to pull gently on one corner of the emblem. It should not peel off easily or cleanly. If a bar code is present in either place, it should be scanned with a reader. Although the scanner readout might contain an additional digit or two at the beginning or end, the same 17 characters of the VIN should appear in the exact same order as they do on the plate or emblem. If the number on the safety standards emblem is questionable, it should be very lightly rubbed with an ink eraser. This should have no effect on a good number.

E/ Anti-Theft Label

On vehicle models required to be equipped with theft prevention emblems on the major components, it is necessary to make sure not only that all of them are present, but also that the VIN on them matches the public plate and safety standards emblem. If a 3M-type label viewer is available, it should be used to view the label. These labels have a retroreflection security feature, which presents a different pattern than when viewed under regular light. An example of such a security feature is shown in Figure 6-36. When gently pulling on one of the corners of the emblem, it should remain tightly bonded to the surface. If the number on any of these emblems appears to be questionable, the same ink eraser procedure aforementioned should be used and the number should be unaffected. If the area around the theft prevention emblem can be sufficiently darkened, the use of a portable black light should reveal any latent footprints from previous emblems. An example of such footprint is shown in Figure 6-37. The label is also designed to break away when removed.

Figure 6-36

Retroreflection security feature present on 3M anti-theft label. (Source: 3M Security Systems Division (1996) Automotive Security Labeling System, p. 1. Reprinted with permission of 3M Security Systems Division.) See Color Plate.

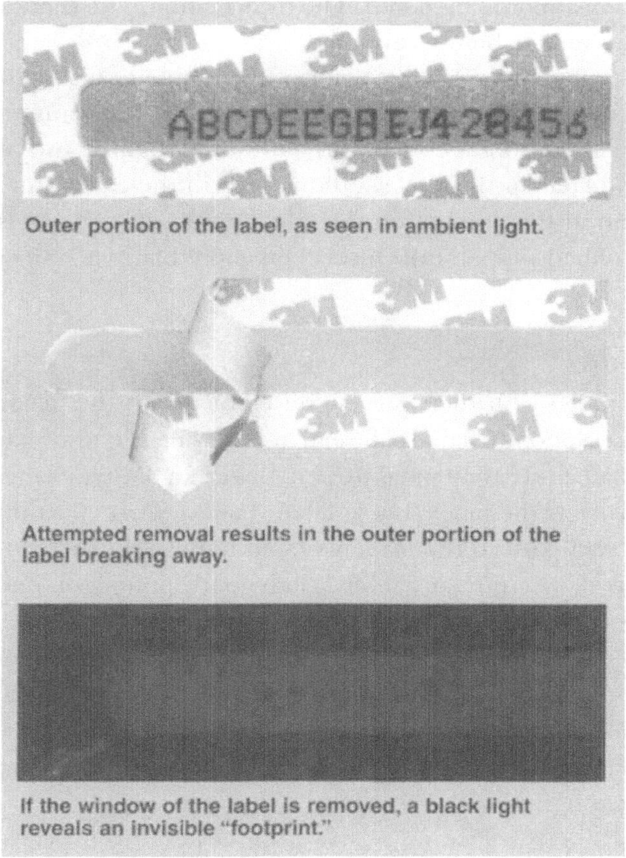

Figure 6-37

Illustration of the breaking away of a 3M anti-theft label when removed from vehicle and illustration of the footprint left after removal of the label and seen under ultraviolet light. (Source: 3M Security Systems Division (1996) Automotive Security Labeling System, p. 2. Reprinted with permission of 3M Security Systems Division.) See Color Plate.

F/ Other Markings

If a VIN-bearing part emblem is present, its VIN should be compared with the VIN on the vehicle. Any other spelled out or coded information on this emblem that can be deciphered should be consistent with traits of the vehicle under scrutiny. Any date information on the seat belt straps and buckles should be noted. The information from any body paint or trim code number tags found should be recorded.

The air bag restraint devices and related components should be checked for serial numbers during the examination. If the vehicle being examined is a model with a stamped secondary number that is right out in the open, such as on the firewall or shock tower, it is important to make sure that the number matches with at least the VIS segment of the numbers displayed on the plate and emblems.

It should be attempted to locate the VIN-based numbering on the engine and transmission. If these components have not been replaced, the numbers should be consistent with the public VIN. Simultaneously, any non VIN-based serial numbers present on these components should be identified and recorded exactly as they appear on the block or housing to be cross-referenced through the NICB or the manufacturer. The possibility of having any body serial numbers stamped into sheet metal areas should always be taken into account. These numbers and their locations should be noted along with any information, such as date-related casting numbers or other markings, for possible cross-reference at a later time. While under the hood inspecting the engine, it should be verified that the proper model year for the vehicle is listed on the emissions certification label and that the engine group/family code number displayed is consistent with the vehicle model year and specifications of the installed power plant. At any time during the examination of the vehicle, the investigator must be alert for the presence of the broadcast sheet, Auto-Tel, or build sheet-like items or other tags and emblems that might contain the entire VIN or a portion of it.

G/ Window Etching

Any etched VIN markings present on the various glass panes should be verified. When present these should also match. If the window etching is the type with a code and toll free telephone number, that information should be noted for possible future use. While checking the glass for etching, it is important to note any date information marked on the various panes by the manufacturer.

H/ Secondary VIN

If the vehicle in question has the secondary VIN placed on the frame, an attempt should be made to locate it. A frame secondary number located on the outboard face of the rail should be fairly evident. Again, the number should look like the frame number on a known good example of the same make, model, and year of vehicle. Because the lower areas of most vehicles are somewhat dirty, it may be necessary to clean off the dirt by spraying brake cleaner (using the proper gloves and safety glasses) and a rag to view the number clearly. In addition, many vehicle frames are coated with rust inhibitor or sound deadening material over the painted surface finish that may need to be removed with spray to view. Once cleaned up and the numbers apparent, the last eight characters of the frame number should match the public VIN. By running a small magnet across the number and the surrounding area, it is possible to determine whether the magnetic pull is strong and uniform. Using a magnifying glass, the number as a whole and each individual character separately are delicately inspected. This should reveal most evidence of tampering such as altered characters, number misalignment, spacing irregularities, grinding, or restamping.

A large area of the frame ahead of and behind the number should be cleaned and checked with a magnifying glass for evidence of welds, grinding, or other irregularities, which might indicate a new number area or length of rail that has been inserted. Touching the number surface area with the fingertips allows the investigator to get a feel of the surface. If the surface feels lower around the number as a whole or around any of the individual characters, it might be due to grinding or heavy polishing. Indeed, visual and touch inspection should be performed to determine whether the number area is raised up significantly above the surrounding surface. This could indicate that a false number overlay may have been placed over the original number or the original number was possibly welded over and restamped. Using the dental probe tools, the number surface and its surrounding area are gently picked to determine whether there is any evidence of foreign material there, such as plastic body filler or metal substitute products like "J-B Weld." With a rag and the spray cleaner, a spot on the frame a foot or more from the number should be heavily test cleaned to determine how resistant the frame surface finish in general is to removal; the finish around the number area should be just as resistant. If the finish on and around the numbers comes off much easier than it does from the other parts of the frame, the number has probably been tampered with and repainted. If the frame rail configuration is such that it is possible to view the rear surface of the number area, it is pertinent to clean that off as well to check for welds and other signs of tampering.

The process of checking secondary and confidential VINs on vehicles with numbers located on the top surface of the rail is identical to most of the steps listed above, except many of these numbers need to be viewed indirectly. Once one has a general idea of where the number stamps should be located on the vehicle, it is usually possible to find them by feeling the rail top with the fingertips. On SUVs and trucks with high ground clearance, this can probably be done by lying down next to the vehicle. On regular passenger cars that operation would be nearly impossible. In either case, it is much easier to locate and read frame top numbers while standing with the vehicle up off the ground on a lift or service ramp. Once the vehicle is safely raised and the numbers located on the top frame surface, it should be possible to read them by using the inspection mirror in combination with the beam from a flashlight. The result is shown in Figure 6-38. One may have to clean up the number somewhat by using spray cleaner or a very light sanding and then trying several different angles with the flashlight in conjunction with the inspection mirror to read the number clearly.

The use of a fiber-optic viewing device can be very useful to observe the number. Viewing frame top numbers using these devices is not always successful due to illumination issues

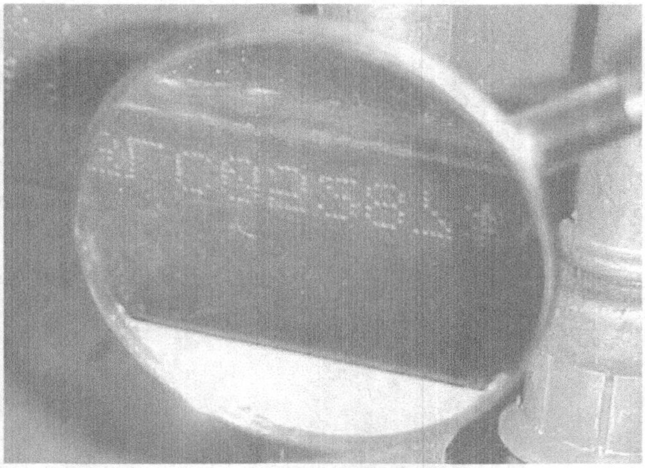

Figure 6-38

Typical frame top secondary/confidential VIN as viewed with an inspection mirror. This particular example, on a Ford Motor Company SUV frame rail, is a dot stamped derivative containing the year, plant, and sequential production number followed by an asterisk.

and space restrictions above the frame. A comparatively large amount of open space above the frame numbers is best for viewing. However, the mirror and flashlight method may end up being the better choice. Fiberoptics are best used to view numbers such as those on engines and transmissions where the area is not so constricted and light can be shone directly on the number.

In the event that a number is located but cannot be accurately read due to space restrictions or other factors, an ink lift might be the best method. This is done by cleaning up the number area as much as possible followed by smearing a very small amount of fingerprint ink uniformly onto the number surface with a fingertip or swab and carefully placing the appropriate size piece of clear latent print lifting tape over the entire inked number portion and surrounding area. Leaving a small portion of each end of the tape turned up (unsecured) greatly facilitates its removal. An example of such practice is shown in Figure 6-39.

After the tape is smoothed down firmly onto the entire number without sliding, it is removed as carefully as possible. The inked tape is placed onto a latent fingerprint card and should have a fairly readable lift of the number. This process may be repeated a few times to get a good and clear lift. This procedure may also be performed using fingerprint powder instead of ink, but the results are not as reliable because of difficulty in accurately controlling dispersal of the powder. Another variation on the number print process is lifting a cast of the number with some putty-like substance, such as casting material used by a toolmarks examiner. As with almost everything else when dealing with hands-on vehicle identification, these skills become more proficient with time.

A more extreme measure that can be taken to view a frame number, if other methods have failed, would be to loosen or remove the body bolts that secure the vehicle body to the frame and partially jack the body up a sufficient distance off the frame rails to allow the number to be clearly viewed. If the body portion covering the inaccessible frame top

Figure 6-39

Proper placement of tape for ink lift of frame secondary VIN image. The tape is left turned up on each end to facilitate proper removal.

number is a pick-up truck bed or similarly mounted specialty truck body, it might be possible to remove the bed/specialty body entirely. If the vehicle is of little or no value such as a wrecked, heavily stripped, burned, or submerged vehicle, cutting a viewing slot through the top surface of the floor pan, trunk pan, or rocker panel structure with a die grinder, air chisel, or torch directly above the place on the frame where the number is located might also be considered. It is crucial not to damage the number during the cutting process. Another last resort method involves the careful cutout of a small portion of the frame top surface where the number is stamped to view the number directly. These extreme methods allow for sufficient direct access to the frame top number area to permit taking photographs and conducting obliterated number restorations (as described in Chapter 7) if necessary. The procedure involving cutting the number portion out of the frame altogether has the added advantage of permitting any photographic, number restoration, or toolmarks comparison work to be performed at the crime laboratory.

Practically all the inspection methods discussed for the examination of secondary and confidential numbers stamped into the firewall, shock towers, or frame rail locations are applicable to examine secondary numbers stamped in other vehicle areas as well. With a few notable exceptions, these number locations are generally on vehicles with unitized body construction and may be hidden inside the fenders; stamped into the floor pan, trunk pan, and cargo areas; or placed on various sills, risers, and other structural members almost

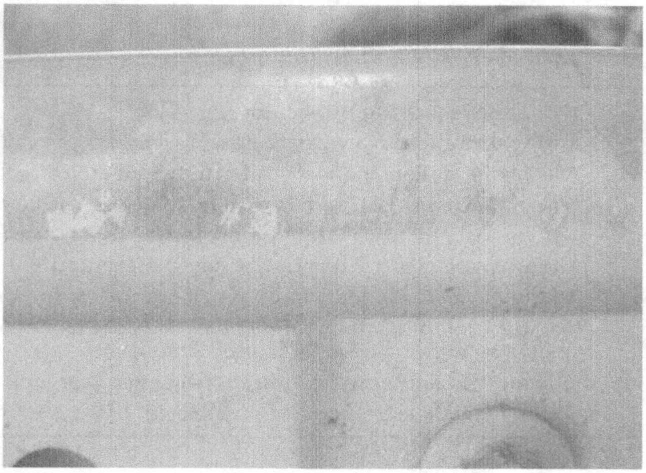

Figure 6-40

Body area where the secondary/confidential VIN is stamped on a unibody GM passenger vehicle. The number is obscured by a coating of weather sealing adhesive and finish paint.

Figure 6-41

The same body area as depicted in Figure 6-40 after cleaning with automotive brake cleaner spray revealing the presence of the secondary/confidential VIN derivative containing the make, year, plant, and six-digit sequential number.

anywhere on the vehicle. Locating some numbers might require loosening a fender, removing portions of the instrument panel, pulling up floor covering, removing seating and upholstery panels, moving insulation/weather-proofing material, or cleaning off adhesives or coatings. An example of the last scenario is shown in Figures 6-40 and 6-41. Even after locating some of these numbers, it may still require an inspection mirror to view them.

6.8.4 Summary

It is very important to have the right tools available to do the job safely and with minimal damage to the vehicle. As with all the other secondary numbers discussed, the best way to spot any irregularities is to know what a factory number looks like in that location on a

Table 6-8
Quick reference checklist used to verify the identity of a vehicle.

√	VIN passes check digit standards
√	VIN specifications match attributes of vehicle
√	VIN plate location and means of attachment are correct for make
√	VIN plate comparison with known good plate of same make, model, year, and plant
√	VIN on plate matches VIN on safety standards emblem
√	VIN on plate matches number on anti-theft emblems if applicable
√	VIN in obvious secondary locations (shock towers/firewall) consistent with public VIN
√	Any date information on safety belts, shoulder harnesses, and vehicle glass (including mirrors), should be either the model year or the previous year
√	Engine emission control certification emblem year indicators same as model year
√	Color indicator codes consistent with color of vehicle

known unmolested vehicle of the same year, make, model, and assembly plant. As mentioned earlier in the chapter, it might be pertinent to consider having the vehicle's electronic control systems interrogated by a dealer for any information of value.

Table 6-8 presents a summary checklist of the basic steps of the examination of a vehicle's identity. Having conducted the vehicle inspection, the vehicle's displayed identity may have been determined to be correct and unaltered. Otherwise, research on the information noted during the examination needs to be performed. If obliterated numbers were found during the examination, arrangements need to be made to attempt to have the numbers restored on the vehicle or have them removed and taken to the laboratory for restoration. If different whole or partial VINs were discovered during the inspection, these need to be checked through the appropriate databases for stolen status and run through the NICB for proper build-up, shipping, and history. All non VIN-based component serial numbers such as those on the engine, transmission, and air bags—or any other unique numbering not related to the VIN—that were discovered during the examination need to be cross-referenced through the NICB as well. Any date information or color and trim code data recorded during the inspection should be compared with the details of the vehicle. Normally, this type of research results in the successful identification of the vehicle in most cases. In cases where these steps are not sufficient to identify the vehicle or determine its status, there are other services available from the NICB, such as off-line/purged file searches and suspect vehicle runs. Also, requests for more detailed vehicle assembly information direct from the manufacturer in the form of traceability/build sheets can be made. Finally, it is always possible to request that an NICB agent go over the vehicle to assist with its identification.

BIBLIOGRAPHY

[1] National Highway Traffic Safety Administration (2004) Vehicle identification number requirements, *Code of Federal Regulations,* Title 49, Volume 5, Chapter V, Parts 565, pp 157–162.

[2] International organization for standardization (1996) *ISO/TR 8357:1996 Road vehicles—Instructions for the implementation of the assignment of world manufacturer identifier (WMI) codes for vehicle identification number (VIN) systems and for world parts manufacturer identifier (WPMI) codes,* Geneva, Switzerland.

[3] Society of Automotive Engineers (2005) *WMI/VIN Information,* available at http://www.sae.org/standardsdev/groundvehicle/vin.htm, last access performed on July 6, 2005.

[4] Society of Automotive Engineers (1981) *SAE Standard J272—Vehicle identification number systems,* Warrendale, PA.

[5] Society of Automotive Engineers (1981) *SAE Standard J273—Passenger car vehicle identification number system,* Warrendale, PA.

[6] Society of Automotive Engineers (1981) *SAE Standard J853—Vehicle identification numbers,* Warrendale, PA.

[7] Society of Automotive Engineers (1981) *SAE Standard J1044—World manufacturer identifier,* Warrendale, PA.

[8] Society of Automotive Engineers (2002) *SAE Standard J1108—Truck and truck tractor vehicle identification number systems,* Warrendale, PA.

[9] International organization for standardization (1983) *ISO 3779:1983 Road vehicles—Vehicle identification number (VIN)—Content and structure,* Geneva, Switzerland.

[10] International organization for standardization (1983) *ISO 3780:1983 Road vehicles—World manufacturer identifier (WMI) code,* Geneva, Switzerland.

[11] National Insurance Crime Bureau (2005) *NICB passenger vehicle identification manual,* Palos Hills, IL.

[12] National Insurance Crime Bureau (2005) *NICB commercial vehicle and off-road equipment identification manual,* Palos Hills, IL.

[13] Cars & Parts Magazine (1993) *Catalog of American car ID numbers 1950–59,* Amos Press, Sidney, OH.

[14] Cars & Parts Magazine (1991) *Catalog of American car ID numbers 1960–1969,* Amos Press, Sidney, OH.

[15] Cars & Parts Magazine (1991) *Catalog of American car ID numbers 1970–79,* Amos Press, Sidney, OH.

[16] Kimes BR, Clark HA, Dunwoodie R, and Marvin K. (1996) *Standard catalog of American cars 1805–1942,* 4th edition, Krause Publications, Iola, WY.

[17] Gunnell JA. (2002) *Standard catalog of American cars 1946–1975,* 4th edition, Krause Publications, Iola, WY.

[18] Flammang JM and Rowalke R. (1999) *Standard catalog of American cars 1976–1999,* 3rd edition, Krause Publications, Iola, WY.

[19] Lenzke JT. (2001) *Standard catalog of American light-duty trucks: pickups, panels, vans all models 1896–2000,* 3rd edition, Krause Publications, Iola, WY.

[20] Covello M. (2002) *Standard catalog of imported cars 1946–2002,* 2nd edition, Krause Publications, Iola, WY.

[21] Edwards J and Lawlor J. (1993) *Auto dictionary,* HP Books, Los Angeles, CA.

[22] Flammang JM. (1986) *Understanding automotive specifications and data,* Tab Books, Blue Ridge Summit, PA.

[23] International organization for standardization (1983) *ISO 4030:1983 Road vehicles—Vehicle identification number (VIN)—Location and attachment,* Geneva, Switzerland.

[24] National Highway Traffic Safety Administration (2004) Certification, *Code of Federal Regulations,* Title 49, Volume 5, Chapter V, Parts 567, pp 167–174.

[25] National Highway Traffic Safety Administration (2004) Federal motor vehicle theft prevention standard, *Code of Federal Regulations,* Title 49, Volume 5, Chapter V, Parts 541, pp 109–118.

[26] Kratzke SR. (2004) Department of Transportation—National Highway Traffic Safety Administration—Federal motor vehicle theft prevention standard: Final listing of model year 2005 high-theft vehicle lines, *Federal Register of March 3, 2004,* **69**(42), pp 9964–9969.

CHAPTER 7

RESTORATION OF SERIAL NUMBERS

Horst Katterwe

7.1 INTRODUCTION

The restoration of markings (serial numbers) is an important forensic discipline that includes the science and technology of materials. It deals with aspects of solid-state physics, chemistry, metallurgy, and engineering [1–6]. Markings such as serial numbers, letter codes, and label codes are applied to distinguish various items and to sign items in commercial use. In criminal cases, they are removed to conceal the item's true identity.

This chapter presents methods used to recover the markings in metals or polymers, even when erased or blotted out. Today, a wide range of restoration techniques are used in forensic science laboratories worldwide. Various etching methods, the application of thermal energy, the use of cavitation by induced oscillation, and other techniques are discussed in detail. These technical and scientific processes facilitate the reappearance of the markings, and this is illustrated with selected examples. Finally, a practical section is offered at the end of the chapter to provide practical remarks and suggestions in the application of serial number restoration.

7.2 SERIAL NUMBERING METHODS

Serial numbers can be applied in various manners depending on the substrate, the base material, and its environment. Table 7-1 gives an overview of important methods used in the automotive industry. Examples of marked surfaces are shown in Figures 7-1 to 7-8.

Several other marking methods and procedures are used today, such as electrochemical marking, embossing, offset printing, hot foil stamping, and laser printing. Nevertheless, the marking industry changes rapidly and new applications are often unveiled. The future will determine the way newly developed marking methods are used in the automotive, firearm, and jewelry industry.

7.3 TECHNIQUES USED TO OBLITERATE NUMBERS

The obliteration of a stamp or serial number is the defacing of its indentation by different means to the point that it cannot be recognized or visually identified. Traditionally, methods used to remove serial numbers involve physical abrasion of the substrate until the number

Table 7-1

Overview of important marking methods used in the automotive industry.

Method	Remarks
Die stamping (cold/hot working)	There are 2 types of press marking, also called conventional stamping: • Cold working for metals (Figure 7-1) • Hot working for plastics (Figure 7-2). This technique is suitable for use by machines or by hand (hammer punches). The stamping tool is pressed into the work piece vertically. The machines can be operated manually or pneumatically and exert a very high impact force. The amount of pressure applied to achieve a certain intrusion depth during the marking process depends on the hardness of the surface and base material. To achieve a given mark depth, an increase in hardness of the metal requires an increase of the amount of pressure. The stamped shapes may take one of three forms: 1. Sharp faced 2. Flat faced 3. Open faced Universal usage is ensured by the ability to accept a wide variety of easily changeable standard or special tools. Polymer substrates are not suitable to the described cold-working methods. However, suitable impression characters are achieved by using a heated die immediately stamped into a plastic surface.
Stylus/pin marking	Cold spinning operation is performed by a portable and nonportable dot marker and is used for vehicle identification number (VIN) marking. The VIN is directly inserted into the vehicle chassis using a dot marking head mounted in a special clamping support. It can be interfaced with computer (controlled) systems. A carbide tipped stylus indents most materials, including hard metals, and on flat or uneven surfaces usually by computer controlled marking. By comparison, the conventional stamping process requires a single stroke to impress a number. The pin stamping demands multiple strokes (Figure 7-3). The physical principles are the same as in die stamping or rolling; the material is being deformed beyond its elastic limit resulting in a permanent mark.
Roll marking	Cold presswork method is used to mark the periphery of cylindrical or solid circular work pieces. Shock absorbers and brake disks are marked in roll marking stations.
Type wheel marking	Cold presswork method (Figure 7-4) where vehicle markings are performed in a hydraulic press with preprogrammed or with computer-controlled typewheels.
Engraving	A chip cutting operation (Figure 7-5) where the substrate is cut away by a tiny spinning head leaves marks such as the serial number. Engraved dies can have a male or female embossing and can be applied by hand stamps. In manual operated and automatic engravers, the standard pantograph ratios lie between 1:2 to 1:7.
Scribe marking	A combination between cold spinning and engraving method, producing only microchips due to the micromachining cutting (Figure 7-6). Scribe markers for marking the VIN are often mounted on a robot arm. These are fully automated units. Sometimes there is an integrated camera system in the head of the marking machine for imaging the marked VIN.
Laser beam marking	Metals and polymers are marked by an intense laser beam through computer-controlled systems (Figures 7-7 and 7-8). The amount of heat imported from the laser beam is able to alter the structure in the material (steel or plastic). This phenomenon is called the heat-affected zone. Carbon dioxide (CO_2) or yttrium aluminum garnet (YAG) laser systems are used to produce VIN labels in metals and plastics. These marking processes are very fast, computer controlled, and applicable in automatic production lines.

Figure 7-1

Marking by stamping. Cold working (conventional stamping) in a stainless austenitic chromium-nickel steel alloy (ductile material). Depth of mark, 260µm.

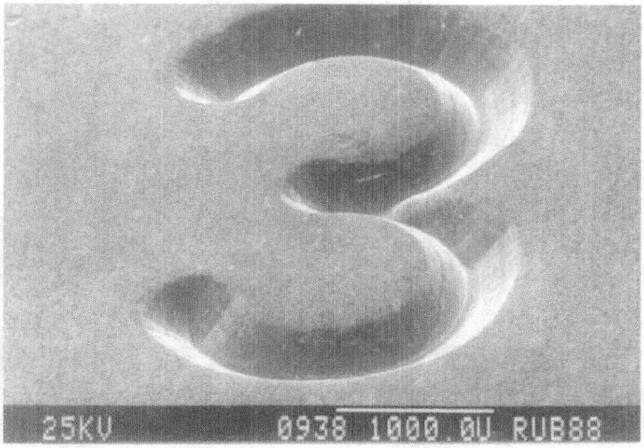

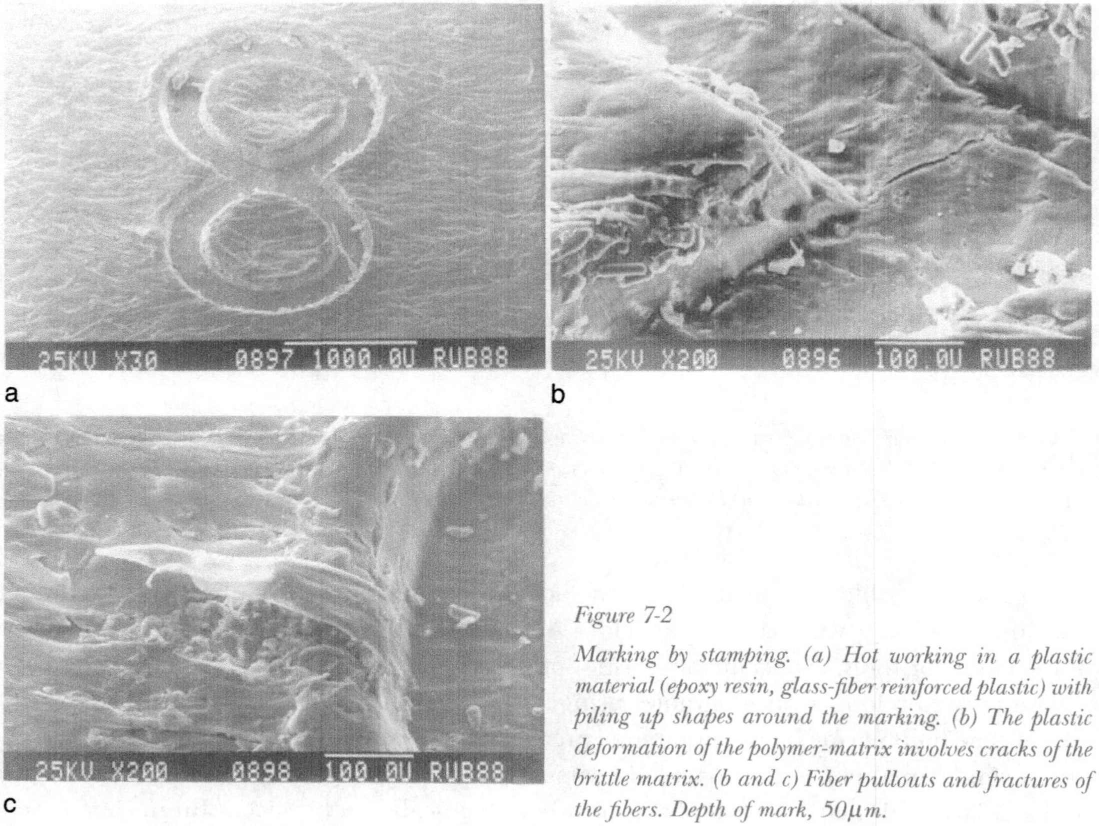

Figure 7-2

Marking by stamping. (a) Hot working in a plastic material (epoxy resin, glass-fiber reinforced plastic) with piling up shapes around the marking. (b) The plastic deformation of the polymer-matrix involves cracks of the brittle matrix. (b and c) Fiber pullouts and fractures of the fibers. Depth of mark, 50µm.

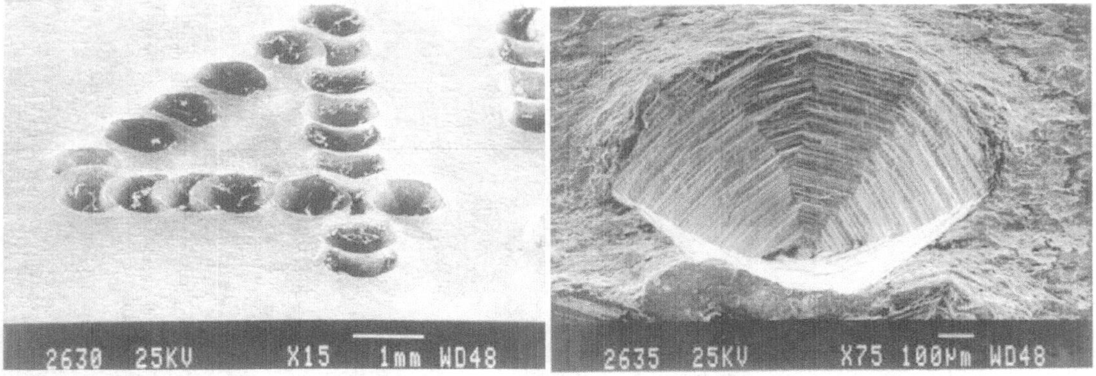

a b

Figure 7-3

(a) Stylus marking in a low alloy sheet steel. (b) With piling up shapes and big depth effect. Depth of mark, 320μm.

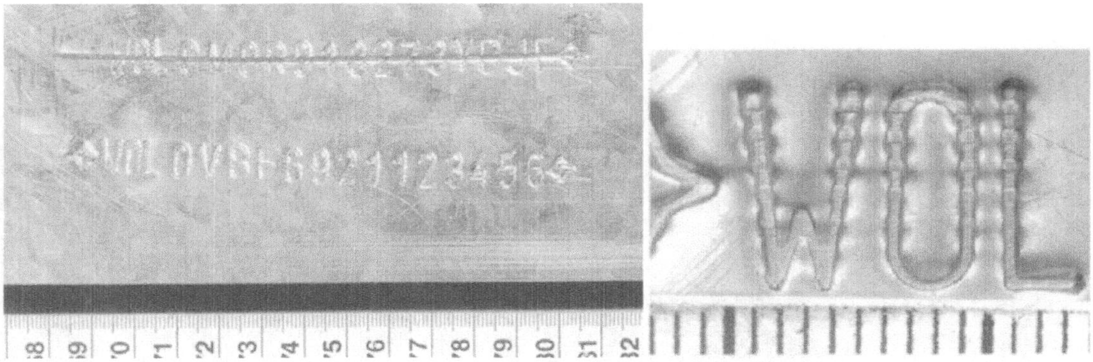

a b

Figure 7-4

Example of type wheel marking with the vehicle identification number (VIN) of an Opel Omega in galvanized low alloy steel. (a) Two VINs are shown. As the first one was marked by mistake at the factory, it was crossed out and the new correct VIN was marked. (b) Details of the mark at higher magnification. Depth of mark, 450μm.

is no longer visible. These methods include filing, grinding (usually with a grinding machine), sanding with sand paper/cloth or emery paper/cloth, or scraping with sharp implements or tools, especially on plastic substrates. Numbers can also be obliterated (so that they are no longer decipherable) using a sharp object, such as a center punch or cold chisel. The hammering on the numbered area takes place until it is damaged beyond legibility. A similar manner is the so-called overstamping. Alternatively, the use of heat or fire (particularly with plastics), acid, base, or other chemicals can be successful in the process of obliteration.

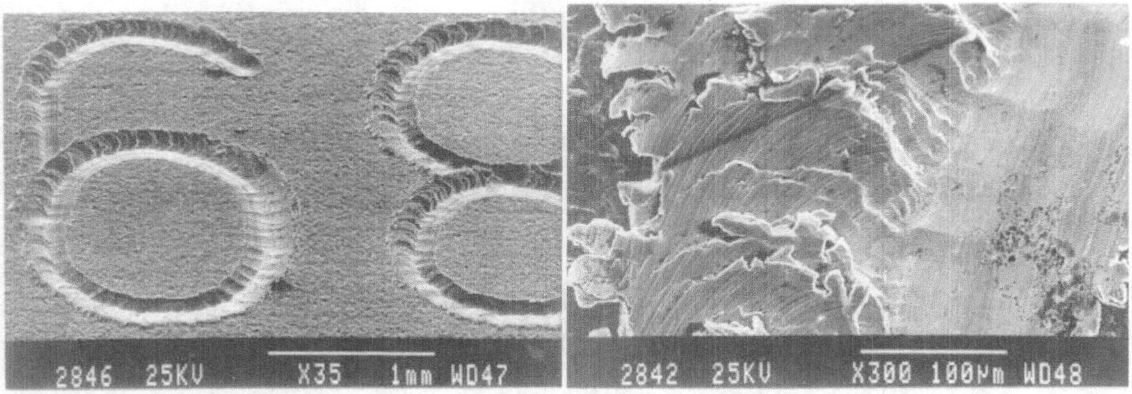

a b

Figure 7-5

(a) Engraving procedure (chip cutting operation) in a low alloy steel. (b) The direction of the working process (graver) and flat spangles are clearly visible. Depth of mark, 32μm.

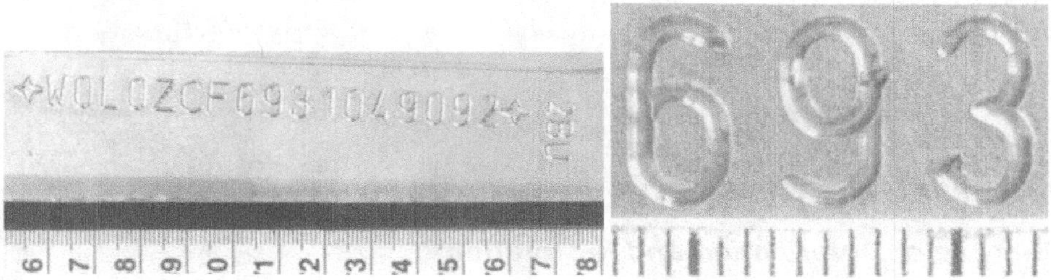

a b

Figure 7-6

(a) Scribe marking of the vehicle identification number of an Opel Astra in a low alloy steel. (b) Detail view at higher magnification. Depth of mark, 200μm.

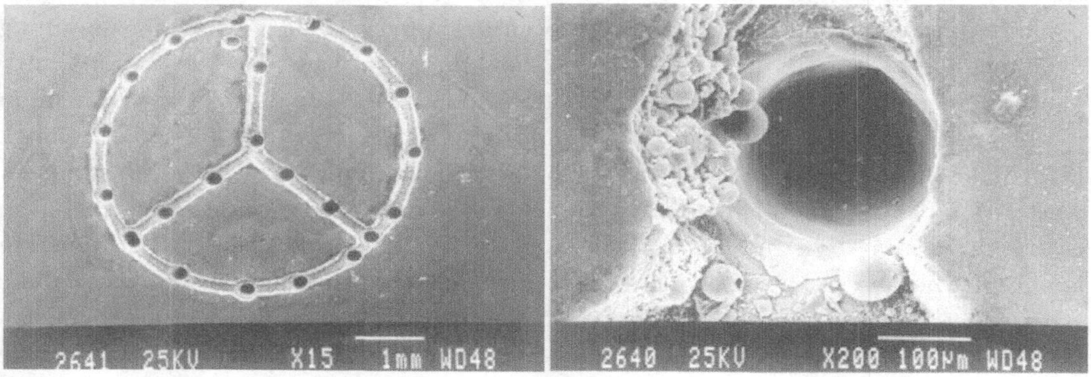

a b

Figure 7-7

(a) Laser beam marking in a galvanized steel sheet, painted with synthetic resin varnish (rutile [TiO_2] pigments). (b) The continuous laser beam is superimposed by pulsated emission: hole formation in the metal (depth: 440μm) and evaporation of the varnish.

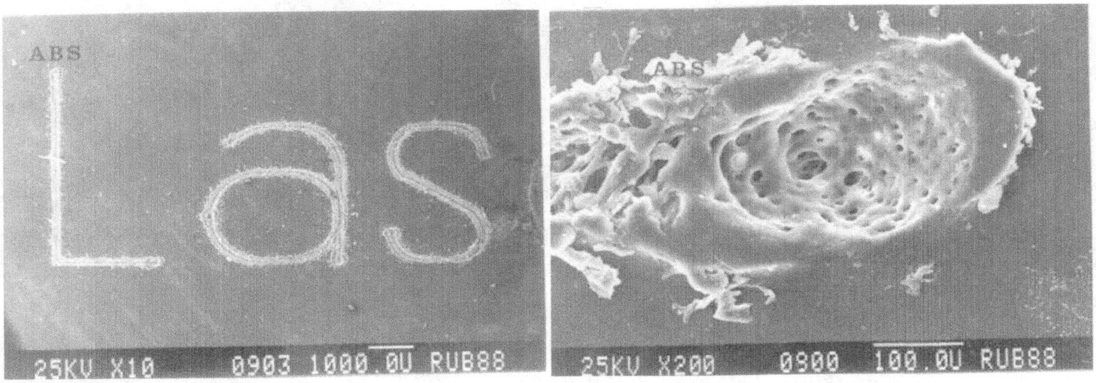

a b

Figure 7-8

Laser beam marking in an ABS-polymer material. (a) The material is melted and evaporated by the continuously emitted laser beam. (b) A "cratered landscape" is created. Depth of mark, about 32μm.

7.4 MATERIAL DEFORMATION AND CAUSED EFFECTS

7.4.1 Principle of Deformation and Restoration

A study of the behavior of the material—metals and plastics—during deformation is necessary to understand the fundamentals of restoration processes [5–7]. It is important to understand that restoration of serial numbers is only possible when the material onto which they were present (and have been eradicated from) still contains a physical deformation due to their original presence. When a die stamps a piece of metal, deformation of the metal beyond the vision of the naked eye occurs. Figure 7-9a shows two zones resulting from the deformation of the material due to the stamping process: the elastic (nonpermanent) and the plastic (permanent) deformation zones. If the zone of plastic deformation is still present after the obliteration, it should be possible to restore the serial number. Figure 7-9b shows an example where the criminal would partially grind or sand the number. This partial obliteration might appear complete to the untrained eye, as some of the material resulting from the grinding accumulates in the small hole and covers it. This situation is optimal, because a cleaning of the surface would allow the examiner to recover the stamped number. If the criminal pushes the obliteration further, until the stamped impression has completely disappeared to his or her eye, the cross section appears as in Figure 7-9c. In such instances, no mark is left from the stamped number on the metal. However, the zone of plastic deformation is still present. Thus, the restoration of the mark is still possible using the different techniques described in this chapter. Finally, if the obliteration goes much deeper in the material, as shown in Figure 7-9d, all physical and permanent deformations related to the original marking disappear and no restoration technique will

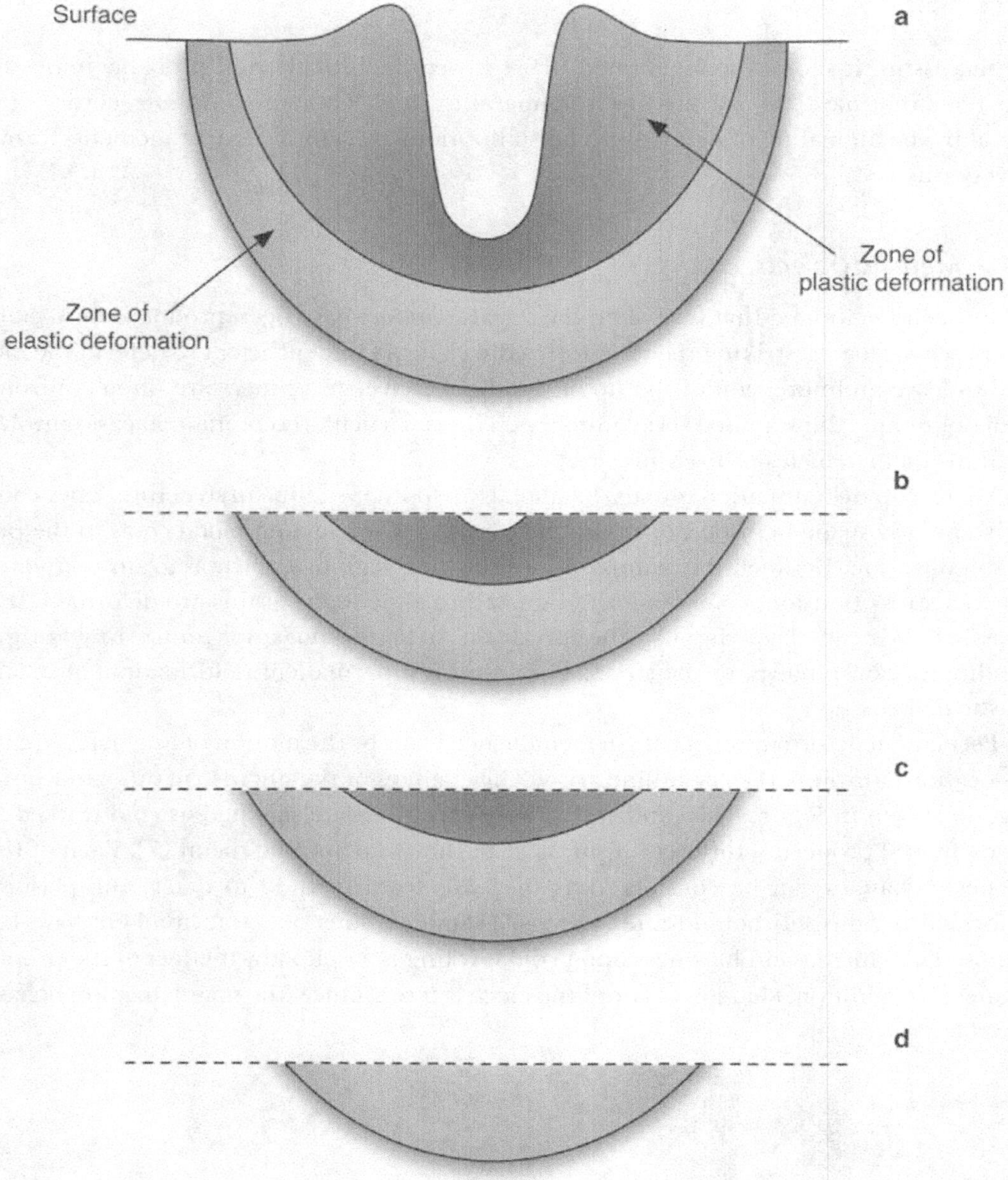

Figure 7-9

Cross section of a stamped number in metal.
(a) Two zones of deformation can be observed: the plastic zone, which is permanent, and the elastic zone.
(b) In some instances, the material produced by the grinding or sanding process accumulates in the cavity and hides it. In such instances, the criminal might stop the process, believing the has disappeared, while a very small portion of the cavity is still left and easily observed after cleaning of the surface.
(c) If the number is obliterated until it completely disappears to the naked eye, the plastic deformation zone is still present, allowing the examiner to recover the mark using different restoration techniques.
(d) If the stamp is obliterated beyond its plastic deformation zone (down to the elastic deformation zone), no restoration is possible. (Diagram courtesy of Eric Stauffer and adapted from "Coupe transversale d'un numéro estampé dans du métal," aide-mémoire de criminalistique, Institut de Police Scientifique et de Criminologie, Université de Lausanne, courtesy of Professor Pierre Margot.)

be successful. It should be mentioned that recovery is difficult, and probably impossible, for items that have been heated to a temperature high enough to cause recovery of the metal by the annealing of defects or recrystallization (i.e., atomic rearrangements forming new grains).

7.4.2 Metallic Objects

It has been mentioned that very often the serial numbers have been produced by a "stamping process," that is, striking the item with a die using a force sufficient to deform the metal and to leave an impression of the tip of the die. Forensic examiners are often confronted with obliterated die-stamped serial numbers. This is particularly common in cases involving theft of motor vehicles or firearms.

Microscopic examination reveals that metals are polycrystalline in structure. They consist of irregularly shaped crystals, or grains, which form when molten metal cools to the point of solidification. Between the grains are interlocking regions known as grain boundaries. When a stress (tension, compression) is applied to a metal, its grains are deformed. If the stress exceeds the elastic limit of the metal, the structure does not return to its original condition upon removal of the stress. The result is a permanent deformation, also called plastic deformation.

Permanent deformation occurs in metallic materials by the motion of line defects, called dislocations, through the crystalline array. The changes of the microstructure after punching are shown in Figures 7-10 and 7-11. These crystals are designated as cold worked and have physical properties different from those of the nondeformed metal [7]. Each of these property changes can be considered as the basis for a method to detect the plastically deformed regions left behind after the visible indentations of serial numbers have been removed. An increase in hardness upon cold working is a well-known effect of the property changes. In addition, changes occur in the electrical resistance, the magnetization behavior

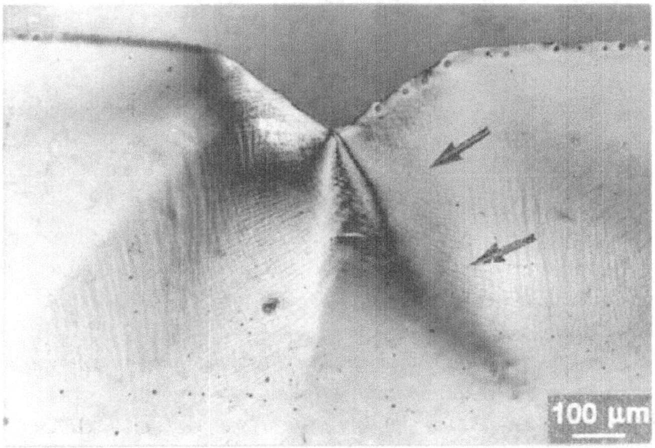

Figure 7-10

A metallographic cross section of a serial number (mark located at top of photograph). Material, Al 99.99; etchant: hydrofluoric acid 1%; time, 30 seconds. The arrows show the extension of the plastic zone: slip bands inside the crystallites indicate a high local deformation.

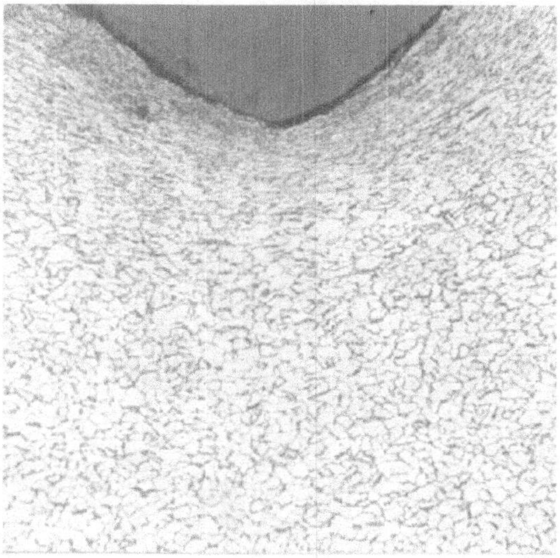

Figure 7-11

The metallographic microstructure (low alloy steel) of the serial number. After deformation, the shapes (textures) of the grains are changed, as it can be seen from the top of the photograph (bottom of the mark) to the bottom of the photograph (undisturbed metal).

of ferromagnetic alloys, the electronic work function, the chemical potential, the thermal conductivity, and the x-ray diffraction pattern.

The amount of plastic flow and the depth to which the plastic region extends below the indentation depend on the shape of the die and, as expected, the depth of the indentation. Blunt dies produce plastic flow to a greater depth (in relation to the depth of the indentation) than sharper v-shaped dies. On average, removal is expected to occur at the same depth below the indentation, and restoration techniques work on the deformation of the metal. Therefore, it is preferable for manufacturers to use blunter dies in marking serial numbers because more cold-worked material would be left behind, easing the recovery process.

When the metal has been deformed plastically, then there are highly distorted regions left, which possess high energy. The number and the amount of stored energy in these regions depends on the amount of deformation. Upon heating, these regions readily form nuclei of new grains, a process known as recrystallization [5–7]. Upon heating to higher temperatures, grain growth occurs around the new nuclei. Hence, the final crystal size depends on the amount of straining and the temperature reached. There is a critical amount of straining that gives the largest crystal size. Further straining would give more nuclei and hence smaller grains. This effect of recrystallization is applied in forensic science for measuring the depth of deformation (plastically deformed affected zone) after punching (Figure 7-12).

7.4.3 Organic Solids (Plastics)

In the last few years many experiments were carried out to reveal erased numbers in technically significant polymers [1, 8–12]. In contrast to the metallic materials, polymer materials consist of long chains forming a more or less dense network of chemical or physical cross-links, which as a whole is isotropic. In this disordered state, the entropy of the system

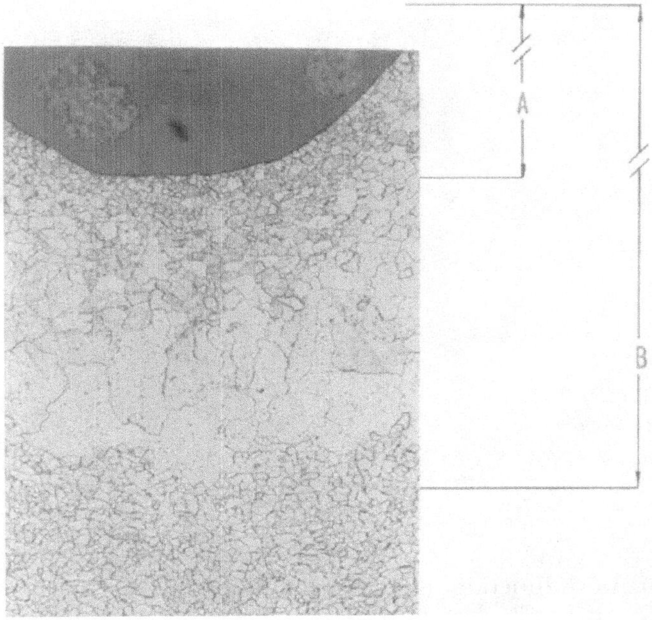

Figure 7-12

Estimation of the depth of possible restoration on a serial number in low alloy steel using the recrystallization effect. A represents the depth of the stamp mark. B shows the estimated restoration depth.

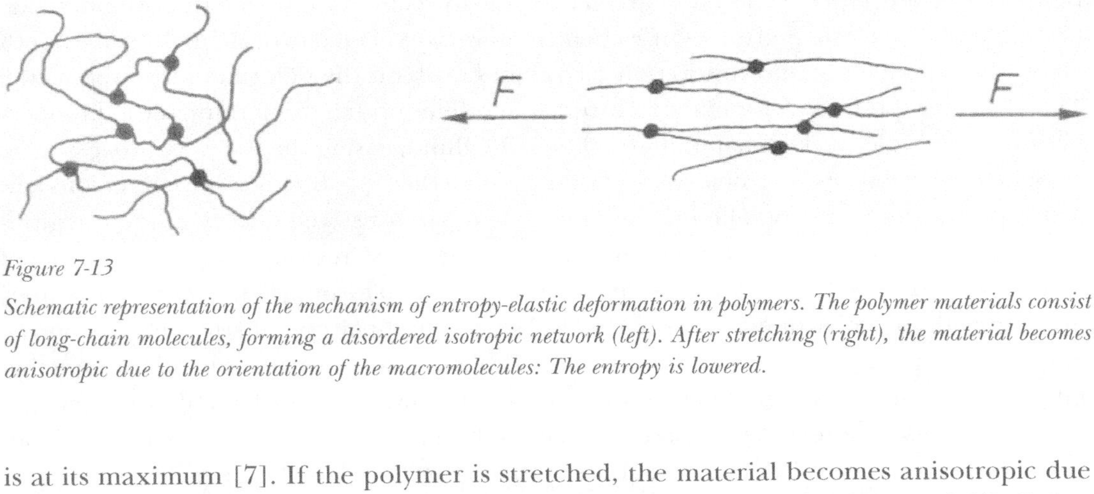

Figure 7-13

Schematic representation of the mechanism of entropy-elastic deformation in polymers. The polymer materials consist of long-chain molecules, forming a disordered isotropic network (left). After stretching (right), the material becomes anisotropic due to the orientation of the macromolecules: The entropy is lowered.

is at its maximum [7]. If the polymer is stretched, the material becomes anisotropic due to the orientation of the macromolecules and the entropy is lowered (Figure 7-13). Below the so-called glass temperature—a characteristic temperature for each polymer—this orientation of the molecules remains stable. Above the glass temperature, the Brownian motion forces the molecules back into the statistically coiled conformation. Therefore, when a warm stamp penetrates the surface of the polymer, leaving a marking, the macromolecules around the marking become oriented: The entropy is decreased by the marking process [7].

7.5 RESTORATION METHODS FOR METALLIC OBJECTS

7.5.1 Principle

There are many methods used to restore erased numbers in metals. The most important procedures, both destructive and nondestructive, are described in this section. Table 7-2 gives an overview of the different destructive and of nondestructive procedures.

Table 7-2

Destructive and nondestructive restoration procedures for metals.

Method	Status	Type
Chemical etching	Validated	Destructive
Electrolytic etching	Validated	Destructive
Heat treatment	Validated	Destructive
Ultrasonic cavitation	Validated	Destructive
Magnetic particle procedure	Validated	Nondestructive
Hardness profile measurements	Validated	Nondestructive
Relief polishing	Validated	Nondestructive
X-rays (transmission)	Under development	Nondestructive
X-rays (reflection)	Under development	Nondestructive
Scanning acoustic microscopy	Under development	Nondestructive
Electron channeling contrast	Under development	Nondestructive

7.5.2 Sample Preparation Techniques

Before any restoration method can be applied, the surface of the samples must be prepared. The preparation method consists of a series of steps during which material is removed mechanically from the sample surface by means of successively finer abrasives. Progressively, finer grades of waterproof abrasive paper (known as "wet and dry" paper) are used to gradually smooth out the surface. Some practitioners use grinding or sanding machines to smooth the surface, but this runs the risk of uncontrolled heating of the area, as well as having less control over the depth of the abrasion. For those reasons, hand sanding is generally preferred. The final surface, after using 320-, 400-, 600-, and 1,200-grit papers, has a mirror-like "metallographic" finish.

7.5.3 Destructive Restoration Procedures for Metals

A/ Chemical Etching

This is a major method used to recover obliterated serial numbers in metal substrates [1, 2, 4]. The purpose of this process is to create a visible contrast between the damaged and undamaged regions of the substrate. This contrast is the result of differential reflection or scattering of light from damaged area compared with that of the undamaged area.

After preparation, the real chemical etching procedure follows. This method originates from metallographic examinations in which etching solutions are applied to cross sections of metals to observe the crystal structures under reflected light. For recovering obliterated numbers, the method relies on the phenomenon that the rate of reaction of the applied chemicals with the substrate can differ between the damaged and undamaged areas. The damaged area usually has a different electrochemical potential than the undamaged surroundings and can therefore be attacked selectively or at least at a different rate. This greater reactivity is the basis for the recovery of serial numbers through etching.

The chemicals used depend on the composition of the metal. They range from simple alkaline solutions to more complicated mixtures, which result in an oxidation/reduction reaction with the metal. Care must be taken when using etching solutions, because they can be corrosive and/or toxic.

The solutions are generally applied by wiping them across the surface of the substrate with a cotton or cloth swab. This process is repeated several times, and the surface is observed closely between applications. Any visible characters should be noted immediately, because it is quite common for different parts of the character sequence to appear at different times during the restoration and/or to disappear shortly after becoming visible. Another method of application, particularly useful for engine block numbers on horizontal surfaces, is to form a frame of modeling clay around the area and then pour the etching solution into the area to a depth of two to five mm.

Many years of empirical testing have resulted in lists of reagents suitable for particular metal studies. Various reagents suitable for serial number recovery and the procedures to be followed have been discussed in the literature [13, 14]. It is important to note that different acidic solutions are generally necessary for different metals or alloys to obtain the best results possible. There are numerous formulations for chemicals that will etch different metals. For practical purposes, the most commonly used are described in Table 7-3.

Many markings are stamped in steel. The acidic solutions, which are widely used for steels, are aqueous solutions of hydrochloric acid and copper chloride, which sometimes contain an alcohol. By swabbing of the surface with this solution, a restoration is possible within a time ranging from seconds to minutes. This etchant forms etch pits and dissolves the plastically deformed regions more rapidly; the numbers become visible because of a difference in light reflectivity, as seen in Figure 7-14.

B / Electrolytic Etching

A modification of the standard chemical etching technique is electrolytic etching (electropolishing). This requires an instrument such as the one shown in Figure 7-15. It involves the addition of an electrical current in the etching process. A power supply (DC) is used, with the first plug connected to the body of the substrate and the other to the swab [4].

Table 7-3

Common metal etching formulations.

Substrate	Technique	Details
Cast-iron and steel	Fry's reagent	90 g copper chloride ($CuCl_2$) 120 ml hydrochloric acid (HCl) 100 ml water
Stainless steel	Acidified ferric chloride	5 g iron chloride ($FeCl_3$) 50 ml hydrochloric acid (HCl) 100 ml water
Aluminum alloys	Dilute sodium hydroxide	10 g sodium hydroxide (NaOH) 90 g water
Brass and copper	Acidified ferric chloride	19 g iron chloride ($FeCl_3$) 6 ml hydrochloric acid (HCl) 100 ml water

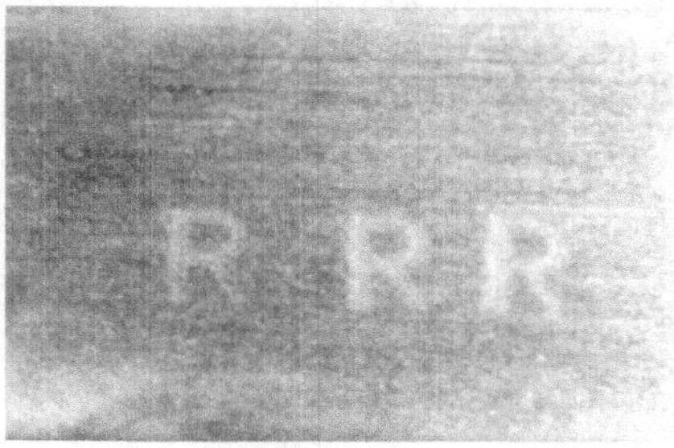

Figure 7-14

Characters on stainless steel that have been erased and restored after treatment with Fry's reagent.

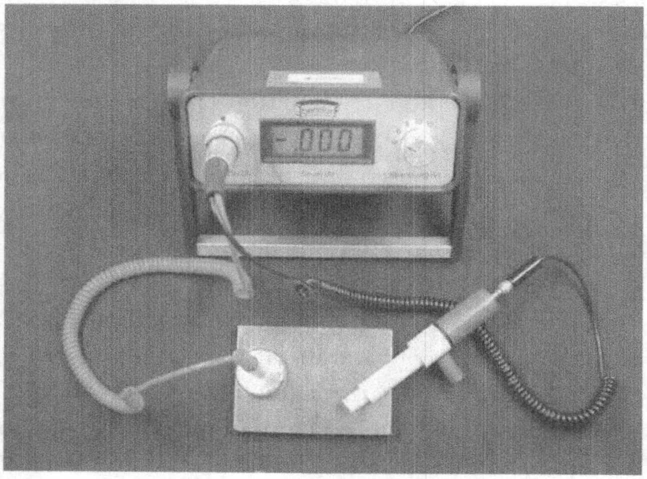

Figure 7-15

Electropolishing instrument with DC power supply, anode, and cathode including the cotton swab.

In this process, the metal to be etched is used as the anode in an electrolytic bath of some dilute acid or etching solution, such as (a) 900 ml ethanol and 100 ml hydrochloric acid or (b) 950 ml glacial acetic acid and 50 ml perchloric acetic acid. The cathode consists of a wire, which holds a cotton swab kept wet with the solution. The metal surface, which has been smoothed and polished, is gently swabbed over the area to be etched with the wet cotton cathode. Then, starting with a value below the decomposition potential of the solution, a voltage is applied across the electrodes. There is an optimum voltage (and amperage) that will give the best result. The optimum conditions must be determined by experiments. The method has found favor in some forensic laboratories, but opinions seem to be divided. The question that arises from the etching methods is whether the chosen etching solution is the best one for recovery. Therefore, several recovery procedures based upon different property changes are the basis of research programs at the Forensic Science Institute of the Bundeskriminalamt (BKA): heat treatment, magnetic particle method, ultrasonic cavitation, and so forth [1].

C/ Heat Treatment

This is a successful technique, particularly when applied to restoring obliterated serial numbers on cast-iron substrates. The success relies on the residual stresses present below the stamped area. The visualization technique differs from that of etching. In this method, heat is applied directly to the obliterated area (e.g., by means of local manual heating of the material surface by propane gas burner) until the metal glows in a light cherry red color. This results in the release of the residual tensile stresses and allows the deformed area to bulge above the surroundings (recrystallization effect). The thermal energy relaxes the upper areas of the material, and the stored elastic energy of lower microstructure areas pushes up the upper relaxed areas (arching of marks), as illustrated in Figures 7-16 and 7-17. After heating, the area is lightly rubbed with abrasive paper, which removes any soot or oxide layer from the raised characters, showing good contrast to the dark surroundings [1, 5, 15].

D/ Ultrasonic Cavitation

Ultrasonic cavitation or cavitation erosion is an erosive wear mechanism known of hydrodynamic processes like ship propellers or hydraulic pipes. The characteristics of cavitation are fatigue wear caused by implosions of bubbles, which are generated by raising and dropping pressure (hydrodynamic process). The bubbles are of high energy and have—after implosion—the capability to destroy metal surfaces. The implosion of the hollow space (cavity) results in a highly energetic water beam (microjet) directed at the surface of the material. When using cavitation as a method for restoration, the erosion of the material surface and the incubation time of the fatigue wear process are decisive. Strain hardening of the material under the stamp causes a longer incubation period.

Recovery of obliterated serial numbers in metal objects is possible if they are placed into a water bath that is excited by an ultrasonic sound frequency. It appears that the abrasion

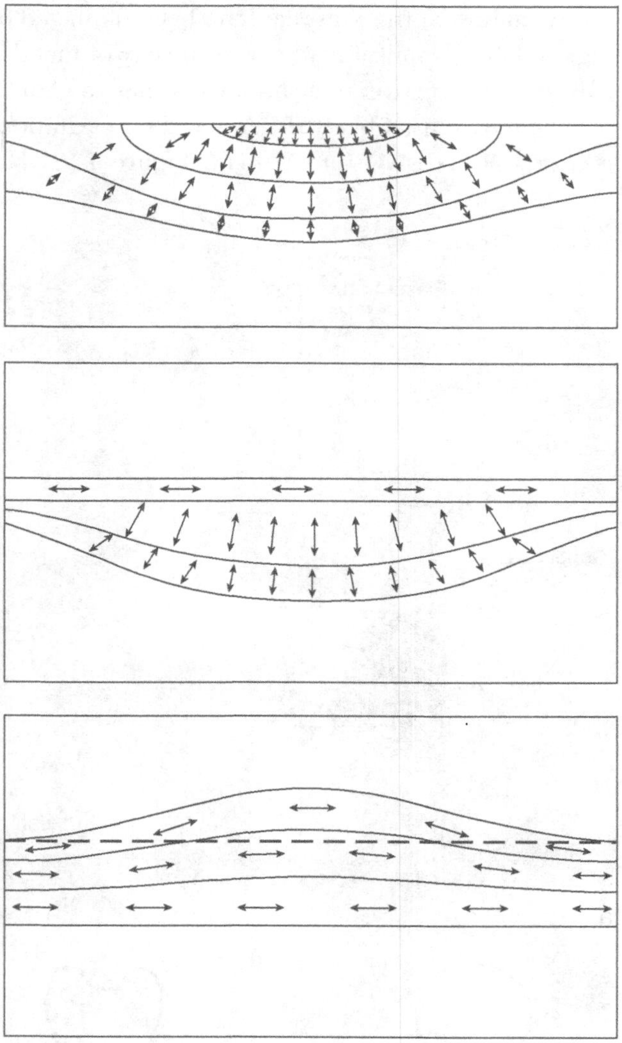

Figure 7-16

Relaxation model (recrystallization effect). Internal mechanical stress of the deformed material (top); thermal energy relaxes the upper areas of the material (middle); the stored elastic energy of lower microstructure areas pushes up the upper relaxed areas: arching of marks (bottom).

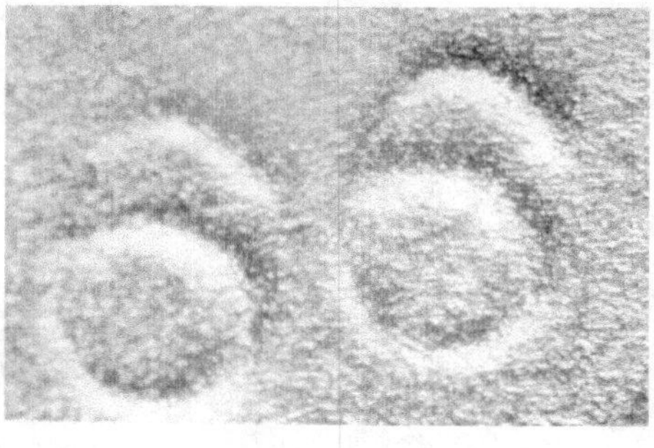

Figure 7-17

Result of a successful heat treatment procedure (arching of marks) with a piece of pearlitic gray cast iron (graphite flakes in a pearlite matrix).

occurs mainly at the sites that have been damaged by stamping. This method has the advantage of being applicable to a wide variety of metals and does not require the use of chemicals. It is destructive, in a similar manner to chemical etching.

The device used by the BKA to induce cavitation in water consists of a power supply that converts 50 Hz (AC) into 20 kHz (Figure 7-18) [1, 2, 5, 15]. Polished test specimens are

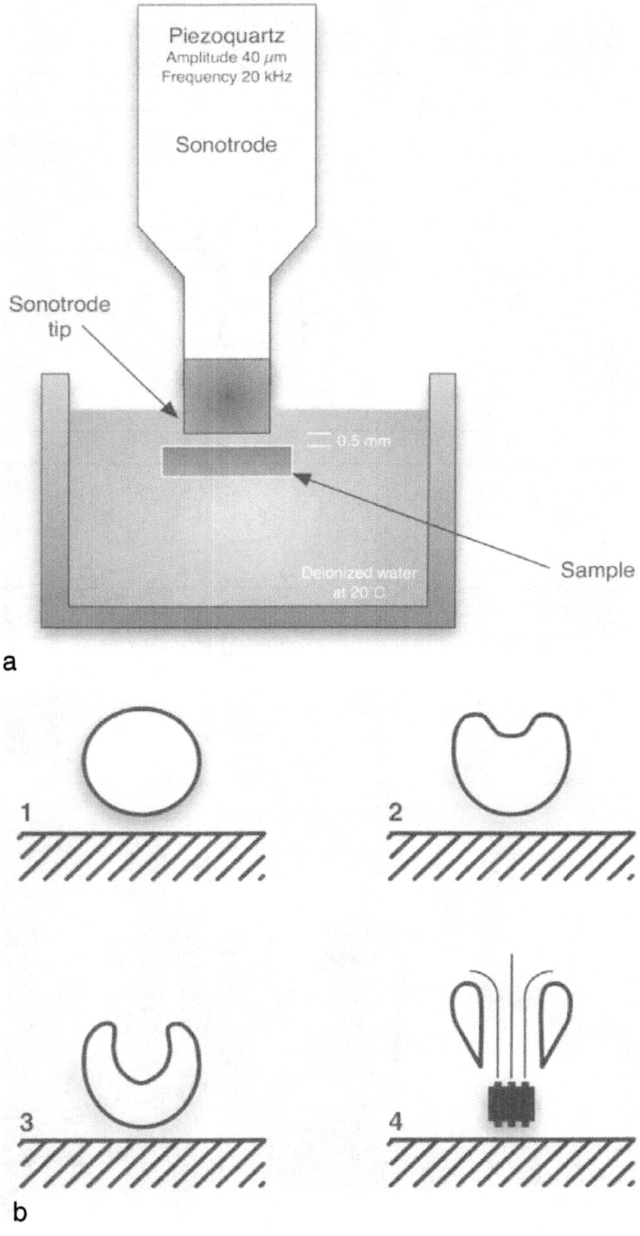

Figure 7-18

(a) Schematic representation of the laboratory device used for restoration by ultrasonic cavitation. It produces vibration-induced cavitation. (b) Fatigue wear processes by implosions of cavitation bubbles forming microjets. (Diagrams courtesy of Eric Stauffer.)

positioned directly below the horn tip to receive the full abrasive action of the cavitation bubbles (Figure 7-19). This serial number restoration technique works on a diversity of materials, and it is dependent on differences in the metals due to the mechanical deformation produced by the stamping process.

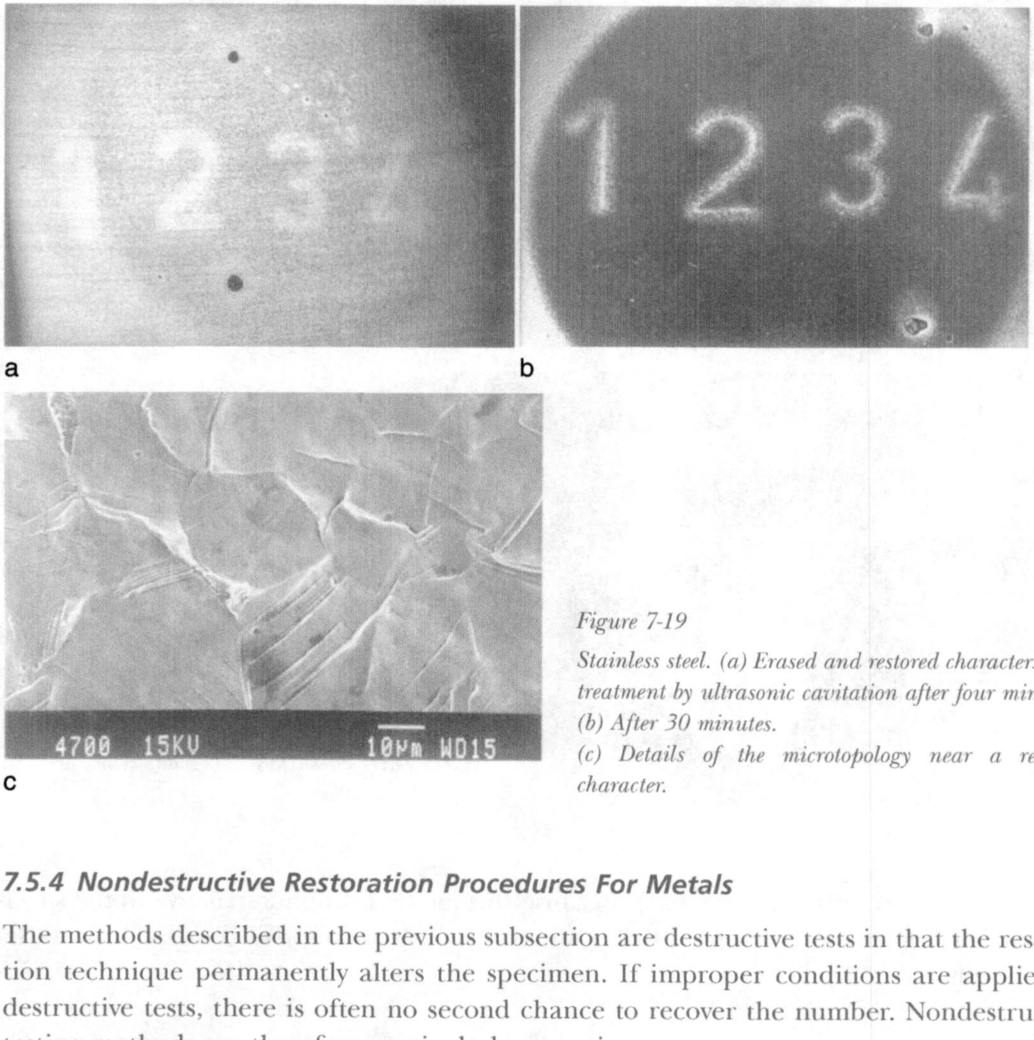

Figure 7-19

Stainless steel. (a) Erased and restored characters after treatment by ultrasonic cavitation after four minutes. (b) After 30 minutes. (c) Details of the microtopology near a restored character.

7.5.4 Nondestructive Restoration Procedures For Metals

The methods described in the previous subsection are destructive tests in that the restoration technique permanently alters the specimen. If improper conditions are applied in destructive tests, there is often no second chance to recover the number. Nondestructive testing methods are therefore particularly attractive.

A/ Magnetic Particle Method

As with other restoration methods, the specimen is first polished. It is then placed between the contact plates of a magnetic testing unit (Figure 7-20). The specimen is magnetized and then sprayed with fine magnetic particles, which outline the obliterated number if the

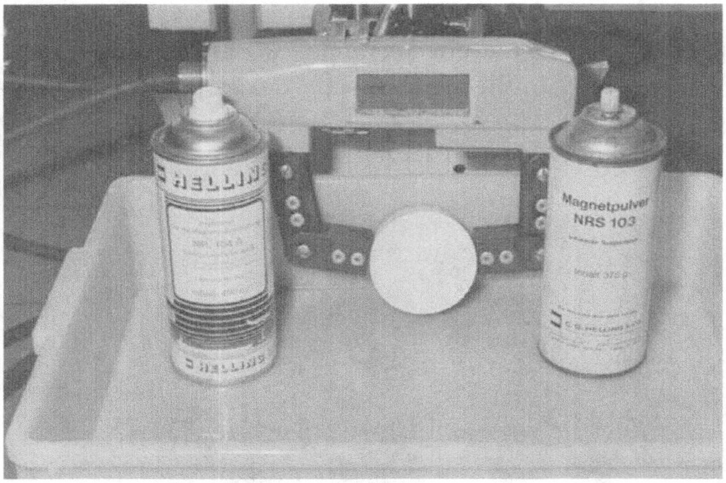

Figure 7-20

Laboratory device using magnetic particle method. See Color Plate.

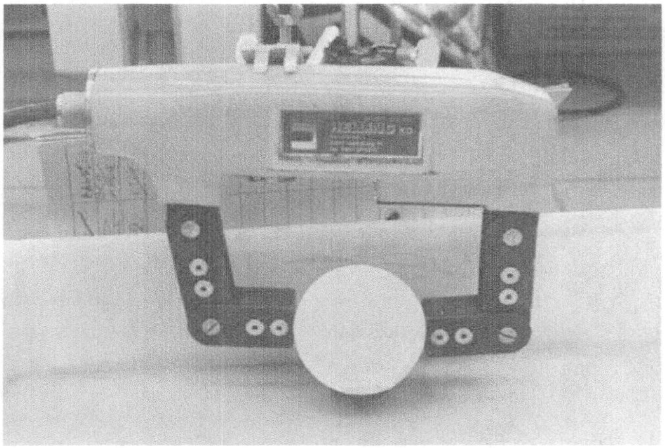

Figure 7-21

Magnetic metal (ferritic steel). Erased and restored characters after magnetization and spraying with fine magnetic particles. For contrast enhancement, it may be useful to paint the surface with white colors first. See Color Plate.

restoration is successful (Figure 7-21). Because this method is nondestructive to the specimen surface, it can be attempted first without affecting subsequent restoration work by other methods [1, 2, 4].

B/ Hardness Testing

The increase in hardness of a metal upon cold working is well documented [7]. To measure the hardness, an indenter (such as a Vickers diamond pyramide) is pushed into the sample using a predefined load. After reaching the peak load, the load is held at a constant value for a while. During the indentation procedure, the sample is deformed by a creeping process. Figure 7-22 shows hardness measurements of an indentation impression in a steel

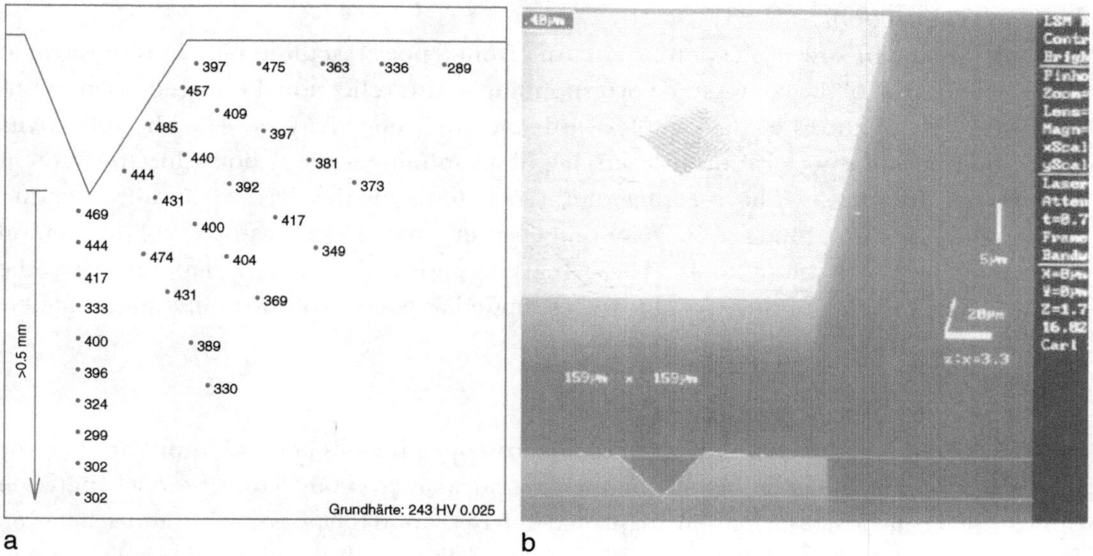

Figure 7-22

Microhardness profile (Vickers) after stamping in stainless steel (X8 Cr Ni 18 8). (a) Hardness of the nondeformed material: 243HV.

(b) A hardness indentation in low alloy steel (hardness of the nondeformed material: 108HV) measured with a laser scan microscope. Right: Depth profile.

sample. Direct detection of the deformed regions using local microhardness measurements over a larger surface area did not appear practical until several years ago, when automatic hardness instruments with fine resolution appeared on the market, thus allowing testing of larger areas [4].

C/ Relief Polishing

If the material at the bottom of an erased marking has a very different hardness from the rest of the material, the restoration of the marking appears as a relief when the sample is polished. The polish rate depends on the actual hardness. Material from different phases is removed at different rates, due to the varying hardness or wear rate of the individual phases. A surface morphology that depends on the local hardness will be produced [4].

D/ X-Rays (Transmission)

The use of radiography in efforts to recover obliterated serial numbers does not appear to be successful [2]. It seems that the radiographs are simply not sensitive enough to enable a visualization of the minute damaged areas left after the removal of a number. However, the method has been successfully used to locate serial numbers that have been hidden with paint or body filler or by welding another piece of metal on top of the original. Metallurgists commonly use this technique to inspect castings, welds, and forgings for defects.

E/ X-Rays (Reflection)

The intensity of an x-ray reflection originating from a polycrystalline sample is sensitive to the concentration of the compound corresponding to this reflection. Preforced orientations of crystals (i.e., textures of the samples) influence the reflectivity as well. The reflectivity varies from location to location in a surface of an inhomogeneous body due to different textures of this phase. The imaging of these heterogeneities has been called texture topography. Texture topography experiments were made with samples of aluminum alloys punched with blunt tools. These markings were removed by lapping away the surface layer. Imaging of these local textures showed a recovery of the obliterated numbers [16].

F/ Scanning Acoustic Microscopy

In material research, the scanning acoustic microscope provides items of information, some of which are complementary to that obtained in optical microscopy. The differences between optical and acoustical information are primarily based upon the physical difference between light and sound waves and on their different interactions with the object. Sound waves are able to reveal deformation properties of the material, because they are able to penetrate optically opaque materials. However, the results obtained in the recovery of erased numbers are still insufficient [4, 17].

G/ Electron Channeling Contrast

Electron channeling contrast imaging is a scanning electron microscopy technique that allows the imaging of near-surface crystal defects in bulk samples (Figure 7-23). The contrast arises from the variation of the backscattered electron yield in strained (deformed) regions of a crystal close to a dislocation core [18–20]. Today this method, applied in forensic science, is in an experimental research stage.

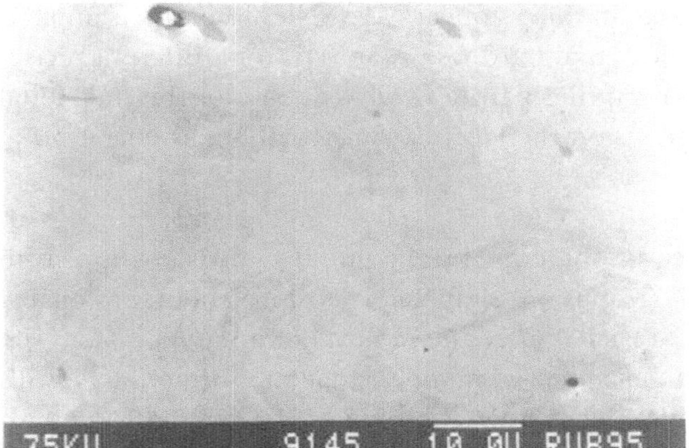

Figure 7-23

Electron channeling contrast. Deformed microstructures of a stainless steel, imaged under "channeling conditions" (40-degree tilt angle). Slip bands are shown as bright and dark lines in the crystal grains.

7.6 RESTORATION METHODS FOR PLASTICS

7.6.1 Principle

Many experiments were carried out to reveal erased numbers in technically significant polymers [1, 4, 9, 10–12]. In polymers, a memory effect is based upon frozen elastic deformation (entropy elasticity) of the long-chain molecules forming a network of physical cross-links. When the marking is erased, an area of oriented material remains at the bottom of the mark. However, the recovery methods for metals described earlier are not applicable in these instances. Indeed, there are several methods for the restoration of erased numbers in polymers. Table 7-4 gives an overview of destructive and nondestructive procedures.

Table 7-4

Restoration procedures for plastics.

Method	Remarks
Swelling	Destructive method
Heat treatment	Destructive method
Clove powder treatment	Destructive method
Relief polishing	Nondestructive method

7.6.2 Destructive Restoration Procedures for Plastics

A/ Swelling

A restoration can be made using swelling agents [1, 4, 8–11]. The thermomechanically treated regions of a polymer possess a higher swelling capacity under the influence of solvents than untreated regions. This is in accordance with the Flory-Rhener theory, which predicts that the swelling capacity increases with the application of external compression. During the swelling process, a widening of the macromolecule's cross-linked structure occurs. A variety of chemicals such as ethanol and acetone can be used to swell polymer substrates.

B/ Heat Treatment

By heating the sample to its glass temperature, in the case of amorphous polymers, or to its crystallite melting temperature, in the case of semicrystalline polymers, molecular relaxation is increased and the mark appears as a small elevation. Figure 7-24 shows the results of successful restorations in polypropylene and epoxy resin [4, 9].

C/ Clove Powder Treatment

Several years ago, while baking gingerbread men students observed that clove powder reacted with the surface of a plastic tray. The material of the tray was the polymer acrylo-

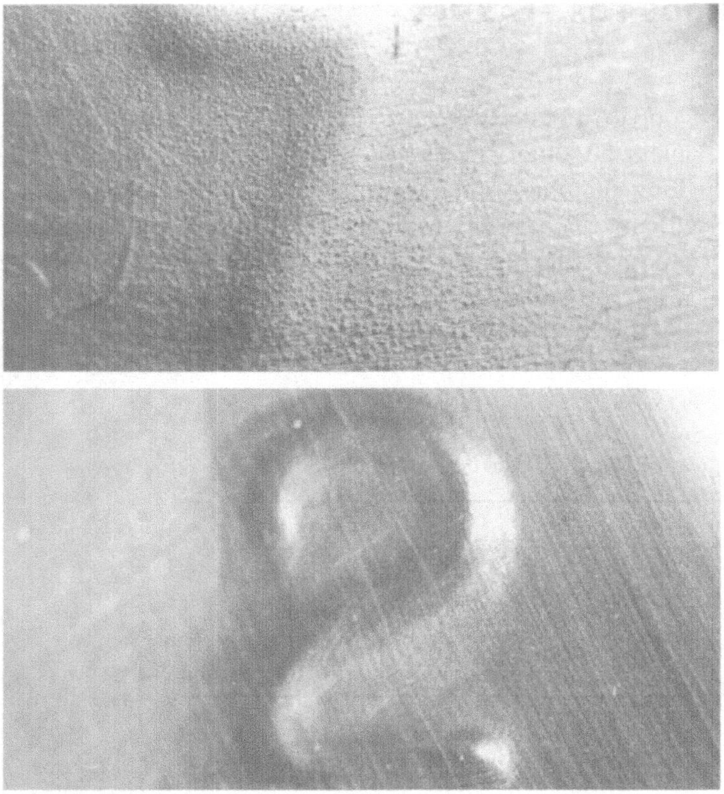

Figure 7-24
Heat treatment experiments with polypropylene (top) and epoxy resin (bottom) samples. The temperature was about 0.8 times the melting temperature of the materials in degrees Kelvin.

nitrile-butadiene-styrene (ABS). At that time a project entitled "revisualization of erased numbers in polymers" was carried out between the Forensic Science Institute of the BKA and the Polymer Physics Institute of the Technical University of Berlin. From this project, the technique of restoring erased numbers in ABS by using clove powder was developed [12, 21]. To see the differences between the original surface and the surface after the interaction with clove powder, roughness measurements with a laser profilometer were made. Figure 7-25 shows the result of a restoration experiment. The left side shows the original marking in ABS, produced by a warm punch. The right side shows the result of the restoration (after erasing of the characters) by using clove powder (on the surface for 10 hours). Because of its nontoxicity, harmlessness to the environment, and ease in handling, clove powder should be given preference compared with other swelling agents, such as a solution of toluene : n-propanol (1 : 4).

It is well known that there are so-called memory effects in some materials [7]. At first this phenomenon was observed in steel by studying the transformation process of two phases of steel. When a high-temperature form of steel named austenite (a face-centered cubic phase of steel) is cooled rapidly, a change in a body-centered structure named mar-

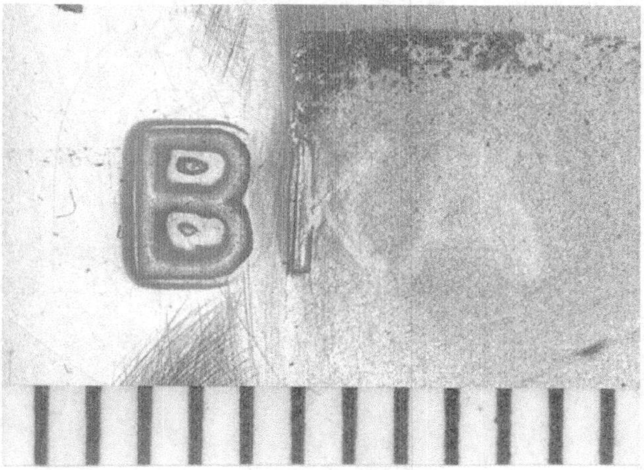

Figure 7-25
Clove powder ("Eugenol") treatment. The left side shows the original marking ("B" of BKA) in ABS, produced by a warm punch. The right side shows the result of the restoration ("KA" of BKA) by using clove powder (on the surface for 10 hours).

tensite happens by twinning and diffusionless transformations. This effect is important for the so-called shape memory effects in metals. Later shape memory effects were also observed in polymers, the first time with polyethylene by Hosemann [22].

After erasing the characters and heating the sample or when the sample interacts physically with a medium, the molecule chains are forced back into the statistically coiled conformation. The sample returns to its original shape. Figure 7-26 shows diagrammatically the memory effect by entropy elasticity: The entropy is decreased by markings (above) and is increased by swelling (below) and the erased number becomes visible. In the middle, the "erased situation" is shown. Why does clove powder interact with ABS? Butadiene-rubber-particles interact with the following component of the clove oil: Eugenol or 4-allyl-2-methoxyphenol.

7.6.3 Nondestructive Restoration Procedures for Plastics

The only nondestructive method used for the restoration of serial numbers on plastic is the technique of relief polishing and is based upon the same effect as in metal samples (see paragraph 7.5.4C). Very good results were achieved with the plastic styrene-acrylonitrile (SAN) [4, 9].

7.7 PHOTOGRAPHY

The photographic techniques used to record recovered serial numbers are the same as those used in other forensic areas, such as fingerprints and crime scene examination. When photographing the results of a chemical etching, black and white film has proved to be most suitable when using conventional photography. It allows for enhancement of the contrast between the faint image and the background. A mixture of flash photographs and available light images should be obtained where possible. Oblique lighting appears to

Figure 7-26

Memory effect by entropy elasticity. The macromolecules in the region of the punched and then erased marking are orientated. The entropy is decreased by markings (top and middle). After restoration (below), the entropy in the material is increased by swelling. The earlier punched marking is visible: The macromolecules are disorientated again.

produce the best images. It is often necessary to test the best angle for the incident light, depending on the extent of the recovery and the nature and location of the surface. A film rated at 400 ASA is suitable to produce images from faint recoveries.

Digital cameras are becoming increasingly common and useful for many purposes. Digital photography is very suitable for this purpose, and the contrast enhancement can easily be made with digital imaging software. The digital chip in the camera determines the resolution in replacement of the film. Following the latest developments of digital photography and imaging enhancement software, these new photographic techniques make it easier to achieve good results in the immortilization of restored erased numbers.

The nature of the recoveries obtained by chemical etching does not always lend themselves to photography. It is sometimes possible to observe a faint recovery, which is decipherable with the naked eye but will not produce a satisfactory image on a photographic film. This reinforces the principle that the etching process should be constantly monitored and careful detailed notes should always be maintained throughout an etching process.

7.8 EVALUATION OF METHODS

Table 7-5 presents a list of different restoration techniques along with their suitability with different metals [1]. Table 7-6 presents a list of different restoration techniques along with their suitability with different polymeric materials [1]. These tables should help the investigator in making the proper choice of a technique for a given material.

Table 7-5

Matrix for the evaluation for the restoration of erased numbers in metals and methods suitable for recovery.

Material method	Steel (ST 37)	Steel (C 10)	Stainless steel (X5 Cr Ni 18 9)	Aluminum alloy (Al Cu Mg 2)	Aluminum alloy (Al Mg 22)	Gold (Au 333)
Chemical etching	++	++	++	+	+	+
Electrolytic etching	++	++	++	+	+	–
Heat treatment	+	+	+++	++	++	–
Ultrasonic cavitation	+	+	++	++	+++	–
Magnetic particle	+++	+++	–	–	–	–
Hardness profile	+	+	+	+	+	–
Relief polishing	+	+	+	+	+	+

(–) not suitable; (+) suitable; (++) more suitable; (+++) most suitable.

Table 7-6

Summary of experiments in the restoration of erased markings in polymers of a selected group of materials and methods.

Method material	Swelling	Clove powder	Heat treatment	Relief polishing
Polyethylene	+ (ethanol)	np	++	+
Styrene-acrylonitrile (SAN)	+ (petroleum ether)	np	++	+
Acrylonitrile-butadiene-styrene (ABS)	+ (petroleum ether)	+++	++	+
Polyamide (6-6)	+ (petroleum ether)	np	++	+
Polyoxymethylene	+ (water at 80°)	np	++	+
Polybutylene terephthalate	–	np	++	+
Polycarbonate	+ (ethanol)	np	+	+
Epoxy resin	+ (acetone)	np	–	–

(np), not proved; (–) not suitable; (+) suitable; (++) more suitable; (+++) most suitable.

7.9 PRACTICAL SUGGESTIONS FOR SERIAL NUMBER RESTORATION

7.9.1 Inspection and Preparation

- Perform an initial inspection of the marked area for coatings, trace material, or any remainders of the characters.
- When possible, determine the method of obliteration.
- Record the "as received" condition of the obliterated serial number area by notation and/or photographic depiction.
- Clean the serial number area from any coatings using solvents.
- If possible, document the area and all steps taken to determine the method of obliteration. Inspect for any remains of the characters.
- Note any toolmarks of value by obliteration tool(s); if toolmarks are present, casts should be taken for possible future comparison.
- Polish the serial number area to a mirror-like finish with any variety of abrasive or abrading tool(s).
- Again, inspect the serial number area for any remains of the characters.
- Document accordingly the remnants present.

7.9.2 Application of a Processing Method

- Determine the physical properties of the serial number medium (i.e., magnetic or nonmagnetic).
- Select the best method for accomplishing a restoration. This requires knowledge of the particular material of the specimen. Fortunately, in its current state of refinement, serial number restoration work requires only identification of a general classification of the materials.
- Begin processing.
- A constant monitoring of the entire process is necessary. Note any remnants of the characters being recovered and document them accordingly.
- A secondary processing method should be used if the initial method is only partially or not fully effective. It is not restricted to the utilization of only one additional processing method.

7.9.3 Recording of Results and Conclusions

- Record all recovered characters and/or character fragments by notation and, if possible, by photographs. Even with the best techniques, a restoration attempt may recover only a part of an obliterated numeral. The investigator should have knowledge of the styles of numerals used for serial numbering to obtain a maximum of information from a partial recovery.
- After finishing the restoration work, the preservation of the specimen surface may be a concern. A surface that has been polished is subject to corrosion, particularly if it has been treated with

etching chemicals. The specimen should be thoroughly rinsed with water followed by acetone to avoid corrosion. It is advisable to apply a protective layer of oil, Vaseline (petroleum jelly), or clear lacquer.

- The restored character fragments form the basis of any conclusions about possible character types or combinations.
- In case of partially restored numbers, research the serial number make-up in an attempt to exclude certain characters based upon their location in the number. This is accomplished with various sources of information such as from the manufacturer (see Chapter 6).
- Research all appropriate processing methods, and render final conclusions. Have conclusions corroborated by another examiner and document/record it in the examiner's notes.
- Issue laboratory reports reflecting the examiner's restoration results and/or conclusions.

7.9.4 Restoration from the Reverse Side of the Specimen

If a specimen is made from thin metal, deformation from stamping can extend completely through the piece and reach its backside (see Figures 6-34 and 6-35). A recovery may be possible by the application of a conventional restoration procedure to the backside of the specimen.

ACKNOWLEDGMENTS

The author would like to thank the colleagues of the Material Technology Section of the Forensic Science Institute of the BKA Wiesbaden for very valuable discussions and excellent technical assistance in various phases of the "restoration work": Dr. J. Balzer, M. Barten, M. Braune, L. Gabriel, D. Herrmann, A. Koch, A. Körschgen, S. Lubjuhn, K.-H. Pohl, B. Radke, M. Ströbele, and B. Weimar. I express my gratitude to Prof. Dr. M. Pohl (Ruhr-Universität Bochum, Institute of Engineering Materials), Prof. Dr. G. Hinrichsen (Technische Universität Berlin, Institute of Non-Metallic Materials), Prof. Dr. E. Born (Technische Universität München, Institute of Mineralogy), and Prof. Dr. B. Schiewe (University of Applied Sciences Berlin, Institute of Medical Physics) for the very good results in common research projects (works) concerning the restoration of erased numbers. Many thanks to Prof. Dr. G. Lange (Technische Universität Braunschweig, Institute for Engineering Materials), Dr.-Ing. M. Feyer (Germanischer Lloyd, Prüflabor Mülheim), and (posthum) Dr. D. Horstmann (Max-Planck-Institut für Eisenforschung Düsseldorf) for their useful ideas and suggestions.

BIBLIOGRAPHY

[1] Katterwe H. (1996) Modern approaches for the examination of toolmarks and other surface marks, *Forensic Science Review,* **8,** pp 45–72.

[2] Treptow RS. (1978) *Handbook of methods for the restoration of obliterated serial numbers,* National Aeronautics and Space Administration, Cleveland, OH.

[3] Polk DE and Giessen BC. (1989) Metallurgical aspects of serial number recovery, *AFTE Journal,* **21,** pp 174–181.

[4] Katterwe H. (2003) *Wiedersichtbarmachung entfernter Markierungen in Werkstoffen: Basiswissen, Methoden, Anwendungen,* Bundeskriminalamt, Wiesbaden, Germany.

[5] Pohl M, Katterwe H, Feyer M, and Illenseer O. (1995) Metallurgical procedures for detection of deformations by forensic science methods/Metallkundliche Untersuchungen zum forensischen Nachweis von Verformungen, *Praktische Metallographie/Practical Metallography,* **26,** pp 405–413.

[6] Schumann H. (1974) *Metallographie,* VEB Deutscher Verlag für Grundstoffindustrie, Leipzig, Germany.

[7] Bergmann W. (1984) *Werkstofftechnik,* Hanser Verlag, München, Germany.

[8] Katterwe H. (1989) Forensic-physical investigations of stretching and swelling behaviour of epoxy resin, *Beiträge zu Elektronenmikroskopischen Direktabbildungen von Oberflächen (BEDO),* **22,** pp 301–310.

[9] Katterwe H. (1994) The recovery of erased numbers in pölymers, *Journal of the Forensic Science Society,* **34,** pp 11–16.

[10] Katterwe H. (1987) Kunststoffe merken sich ihre Prägung/Plastics und memory effects, *Kriminalistik,* **7,** pp 365–366.

[11] Katterwe H. (1995) Polymerphysical aspects of serial number recovery in plastics. In: *Advances in Forensic Sciences,* ed Jacob B and Bonte W, Verlag Dr. Köster, Berlin, **4**, pp 273–278.

[12] Katterwe H. (2003) Interactions of spices with plastics and the restoration of erased numbers in polymers, *Forensic Science International,* **136**(supplement 1), p 347.

[13] Beckert M and Klemm H. (1985) *Handbuch der metallographischen Ätzverfahren,* VEB Deutscher Verlag für Grundstoffindustrie, Leipzig, Germany.

[14] Petzow G. (1994) *Metallographisches, keramographisches, plastographisches Ätzen,* Bornträger Verlag, Stuttgart, Germany.

[15] Feyer M, Pohl M, and Katterwe H. (2001) Restoration of erased numbers. In: *Proceedings of the European meeting for shoeprint/toolmark examiners (SPTM 2001),* Berlin, Germany, pp 23–31.

[16] Born E, Schwarzbauer H, Semioshkina N, Willibald E, and Zorn G. (1986) Fundamentals and applications of imaging inhomogeneous X-rays reflectivity from polycrystalline samples (including results of restoration of erased numbers), *Zeitschrift für Metallkunde,* **77,** pp 49–53.

[17] Quate CF. (1979) The acoustic microscope, *Scientific American,* **241**(9), pp 58–66.

[18] Ng BC, Simkin BA, and Crimp MA. (1997) Electron channeling contrast imaging of dislocation structures in deformed stoichiometric NiAl, *Materials Science and Engineering A,* **239–240,** pp 150–156.

[19] Sandström R, Spence IF, and Humphreys CJ. (1974) A theoretical model for the energy dependence of electron channelling patterns in scanning electron microscopy, *Journal of Physics D: Applied Physics,* **7**(7), pp 1030–1046.

[20] Reimer L and Pfefferkorn G. (1977) Application of channelling diagrams. In: *Scanning electron microscopy/Rasterelektronenmikroskopie,* Springer Verlag Berlin, Heidelberg, Germany, pp 136–140.

[21] Katterwe H. (2004) Revisualisation of erased numbers using clove powder, *Praktische Metallographie/Practical Metallography,* **41,** pp 286–295.

[22] Hosemann R. (1972) Rückstellungseffekte an Polyethylen, *Zeitschrift für Naturforschung,* **27a,** pp 478–484.

CHAPTER 8

ANTI-THEFT SYSTEMS

Robert F. Mangine

8.1 INTRODUCTION

Theft-deterrent systems are installed in new vehicles by manufacturers to provide electronic protection in addition to the physical security, which includes the ignition lock, column lock, and gear selector lock. Original equipment manufacturer (OEM) systems include engine immobilizers and anti-theft/content alarm systems. Both systems are an integrated part of the vehicle electronics and cannot be removed. The engine immobilizer systems are passive and require no action on the part of the driver. The anti-theft/content alarms require arming, usually by locking the doors.

The electronic protection is separate from the physical protection and constitutes a more sophisticated second layer of security. Both the physical and electronic systems must be defeated or compromised to successfully start and operate a vehicle without the proper key. This is not accomplished by inexperienced thieves or "joy riders." Most require knowledge, skill, very expensive electronic tools, and professional procedures. More vehicles are equipped with OEM systems; however, there are still millions of vehicles around the world that do not integrate any anti-theft or alarm systems.

There have been three main OEM engine immobilizer systems generally available since 1986:

I The General Motors (GM) PassKey I and II systems (1986–2004);
II The GM PassLock I and II systems (1996 to present);
III The transponder system, which is the most prevalent system worldwide since 1995. GM's version of the transponder is called the PassKey III, which started in 1997.

The GM PassKey and the PassLock systems are exclusive to GM-manufactured vehicles.

The OEM anti-theft/content alarm systems are less sophisticated than the immobilizer systems and did not become widely available until 1985. The anti-theft/content alarms consist of an audible alarm, visual blinking lights, fuel/ignition interruption, or any combination of the three. These systems are active and armed by the act of locking the doors.

Aftermarket anti-theft systems are installed on the vehicle after its manufacture. Dealer-installed systems are not OEM and are also considered aftermarket. The most significant difference between OEM and aftermarket systems lies in the fact the latter's components are attached to existing vehicle electronics and are not an integral part of the on-board vehicle systems. Although they do provide varying levels of protection, any system that is installed on existing components by someone can be uninstalled or disabled by someone else. Examination of a stolen-recovered vehicle with an aftermarket system requires verification that the relays, sensors, wiring, and control modules are present, intact with no evidence of tampering, compromising, or removal.

Additionally, most aftermarket systems require action on the part of the driver to arm the system (active system) as opposed to OEM immobilizer systems, which require no action on the part of the driver (passive system). There are also several mechanical devices available that provide additional physical protection for the vehicle, as shown in Figure 8-1. Some of the most notable and popular devices are the locking bar and/or the metal steering wheel cover preventing steering wheel rotation, the locking brake pedal to floor bar that stops the brake pedal from being depressed, the gear shift lock, the steel upper steering column cover or collar that encases the upper column to protect the internal column components, and the service brake locking system installed on the hydraulic brake lines that locks the brakes using a key operated device. These systems can be effective but are

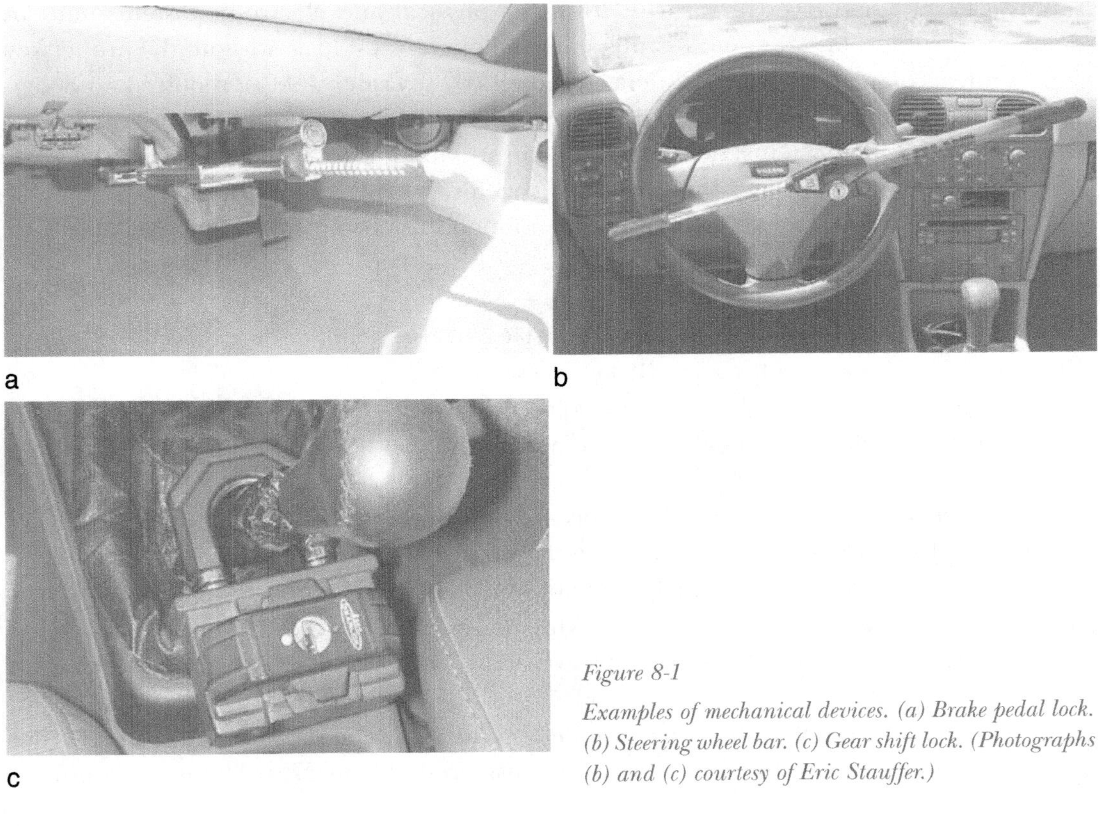

a b c

Figure 8-1

Examples of mechanical devices. (a) Brake pedal lock. (b) Steering wheel bar. (c) Gear shift lock. (Photographs (b) and (c) courtesy of Eric Stauffer.)

Table 8-1

The different OEM and aftermarket anti-theft systems and their characteristics.

	OEM	After market	Electronic	Mechanical	Passive	Active	Immobilizer	Anti-theft/ alarm
GM PassKey I and II	X		X		X		X	
GM PassLock I and II	X		X		X		X	
Transponder (and GM PassKey III)	X		X		X		X	
OEM content alarms	X		X			X		X
Aftermarket alarm/anti-theft		X	X			X		X
Steering wheel cover or locking bar		X		X		X		
Brake pedal bar		X		X		X		
Brake lock		X		X		X		
Gear shift lock		X		X		X		
Steel column collar		X		X	X*			

* This device is installed once and left in place. Thus, it is passive in the sense that no operation is required from the user after its first and only installation.

active and require the driver to physically install or activate them each time the vehicle is parked. Table 8-1 presents a summary of the different existing anti-theft and alarm systems along with their characteristics.

8.2 VEHICLE ANTI-THEFT SYSTEMS

8.2.1 Development of OEM Immobilizer and Alarm Systems

Until 1985, most OEM anti-theft systems were installed on new cars more as an afterthought rather than dedicated protection for vehicles. Most of these systems were audible alarms and most installed only as an option. In model year 1985, for example, a survey of 38 manufacturers and 261 models revealed that 12 vehicles had an anti-theft system standard and 48 as an option, with 201 vehicles having no system availability [1]. In 1986, GM introduced the vehicle anti-theft system (VATS) or PassKey I system on the Corvette [2]. The significance of this system is that it was the first system to be an integrated part of the vehicle electronics and ushered in the engine immobilizer concept. The PassKey system was a completely passive system, and when the owner left the vehicle, a vital component of the system, which is embedded in the ignition key, left with that person. The system availability expanded through the various GM product lines since 1986, and by 1994 over 66% of GM domestically produced vehicles were PassKey equipped [3]. With the advent of a sophisticated electronic system such as PassKey, a separate layer of protection was added to vehicle security. In addition to the physical protection of a vehicle such as door locks, locking column, locking gear selector, and protected starter switch, there was now an electronic barrier that required compromise. No longer could a steering column be forced with common tools; a certain level of skill, knowledge, and expertise had to be used by thieves to defeat both the physical and the new electronic barrier.

8.2.2 General Motors PassKey I and II Systems

A/ Description

The PassKey I and II systems consist of a dedicated VATS control module hard-wired to two contacts in the outer ignition lock keyway. The vehicle computer (electronic control module [ECM]) is also used as a part of the VATS system. When the key is inserted into the ignition lock, a resistor pellet embedded in the key blade, as shown in Figure 8-2, touches the contacts located on either side of the outer keyway. With the key rotated and the system energized, a signal is sent from the VATS control module through the resistor pellet now seated in the keyway. The resistance to the signal caused by the key blade pellet is measured by the control module. If the resistance is correct, a pulse width modulated signal is sent to the ECM to release engine functions. If there is incorrect resistance to the signal, engine functions are disabled and there is an approximate three-minute wait before the system allows another attempt at starting. The most noteworthy feature of the PassKey system is that when the driver leaves the vehicle, he or she takes not only the properly cut mechanical key but also the resistor pellet, because it is embedded in the key. Figure 8-3 illustrates the PassKey I and II system components and principle of operation. PassKey I and II systems both act on the starter and fuel system but present very slight differences in their operation, such as the lockout time.

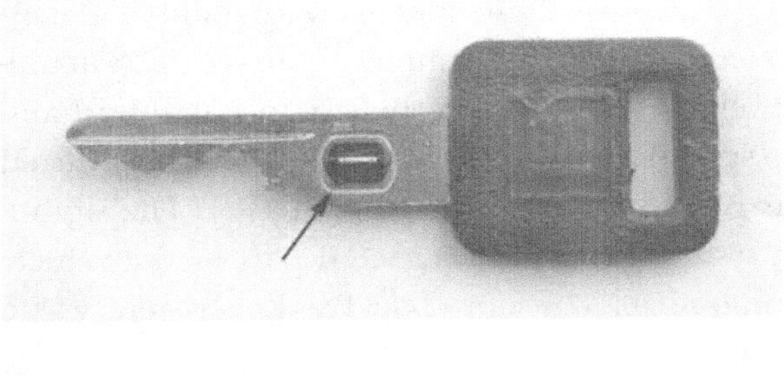

Figure 8-2

Example of an ignition key GM PassKey I or II. Note the resistor pellet embedded in the upper key blade as shown by the arrow.

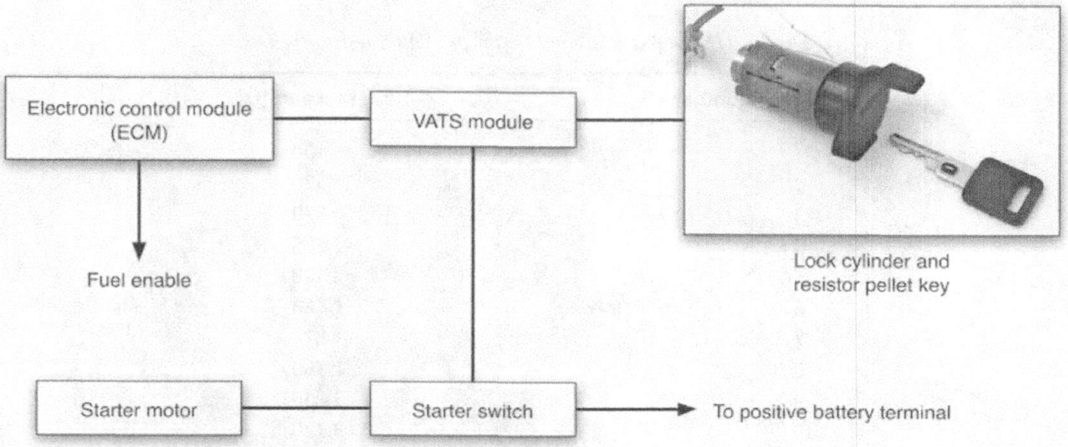

Figure 8-3
Diagram showing the principle of operation of the PassKey I and II systems. (Diagram created by Eric Stauffer.)

In 1986, the PassKey system was highly technical and state-of-the-art in vehicle protection. Not only did the driver need a properly cut mechanical key to unlock the steering column and energize the vehicle, but also the proper resistor pellet embedded in the key blade to allow the engine to start. The effectiveness of the PassKey I and II systems began to diminish as the resistor pellet key blanks became increasingly available through local locksmith stores rather than exclusively from dealerships. Another problem with PassKey was the limited number of possible resistance values of the resistor pellets. As shown in Table 8-2, there are 15 possible resistance values for the resistor pellets. With a ring of the 15 possible pellet-embedded keys and the approximate three-minute delay between start attempts, the longest a thief would need to start the engine would be 45 minutes after physically defeating the column. In the experience of the author, the average time is approximately 22 minutes.

The PassKey I and II (pellet) systems began to be replaced with the PassLock I system in 1996, the PassLock II system in 1997, and the PassKey III (transponder) system in 1997. Only the 2004 Corvette (C-5) retained the PassKey II (pellet) system, but that was replaced with the keyless (no mechanical key) electronic ignition system on the 2005 Corvette (C-6).

B/ Basic Operation Modes

Both the PassKey II (pellet system) and PassLock systems have three basic modes: the normal mode, tamper mode, and a fail-enable mode. The fail-enable mode is initiated when the system detects that a component failure has occurred only after a valid start. A system

Table 8-2
The 15 possible resistance values for the VATS resistor pellets.

Pellet number	Resistance in [Ω]
1	400
2	520
3	679
4	885
5	1,128
6	1,468
7	1,871
8	2,369
9	3,010
10	3,728
11	4,750
12	6,038
13	7,485
14	9,531
15	11,796

component failure could be a detached or separated wire, sensor failure, electronic component malfunction, or electrical shorting that would result in an interruption in the serial data. The security light in the instrument cluster turns on and remains illuminated to alert the driver that a failure is present. The power control module (PCM) becomes fail-enabled and allows the vehicle to be started without the proper codes or signals until system repairs are performed. Again, this mode is only enabled after the vehicle has been properly started and is intended to prevent a motorist with the proper key from being stranded by a system failure. It has the negative effect of neutralizing the electronic anti-theft system and leaves the vehicle vulnerable to theft until the faulty components or failure is repaired.

8.2.3 GM PassLock System

In 1996, GM introduced the PassLock I system on the Buick Skylark, Chevrolet Cavalier, Pontiac Sunfire, Pontiac Grand Am, and Oldsmobile Achieva. It should be noted that GM labels their anti-theft systems based on the components of the system. The PassKey I and II systems have a vital system component embedded in the key: the resistor pellet. The PassKey III has a microtransponder embedded in the key head. On the PassLock system, the components for the anti-theft system are located within the ignition lock cylinder, with no components found in or on the mechanically cut key. The drawback to the PassLock system is obvious in that no vital components are located in the ignition key. Besides the physical location of the anti-theft components, the PassLock I and II systems are similar to the PassKey I and II systems.

The PassLock system works with a magnet embedded in the ignition lock core and an electronic Hall effect sensor secured to the outer ignition lock housing. When the ignition lock core is rotated within the lock housing using the properly cut mechanical key, the magnet passes over the housing-mounted sensor, which creates a signal that is sent to the PassLock data circuit of the body control module (BCM) or, in certain models, the instrument panel cluster (IPC). The resulting voltage is measured by the BCM or IPC. This voltage value is compared with the value stored in the memory (a learned value). If the PassLock data voltage corresponds with the learned value, the BCM sends a pass code to the PCM. If the pass code matches, then engine functions are enabled (fuel).

If a person attempts to forcibly remove the ignition lock core, this action disables the PassLock sensor by removal of the magnet embedded in the lock core. Figure 8-4 shows a PassLock sensor secured to the outer ignition lock assembly housing.

The PassLock system uses two tamper modes. If the resistance value is not read in a start attempt, the vehicle goes into short tamper mode during which the vehicle will not operate for four seconds. After three consecutive failed starts without the proper resistance value, the vehicle goes into long tamper mode and disables the fuel injector for 10 minutes. Figure 8-5 presents a diagram of the PassLock system components.

Rotating the ignition lock core, by any means, on a PassLock-equipped vehicle disengages the steering column locking mechanism (if equipped with a locking column), unlocks the gear selector, activates the starter (ignition) switch, and creates the proper PassLock signal, thus enabling the engine functions. This is accomplished by using the OEM properly cut mechanical key, an aftermarket properly cut mechanical key, a properly cut mechanical key created by any other means, or by forced rotation of the lock core. Because the PassLock system is combined with the GM half-wafer sidebar ignition lock assembly, the sidebar can be removed (by force) from the lock core or the center lock

Figure 8-4

Ignition lock assembly with the PassLock sensor secured on the lock housing as shown by the arrow. This configuration is used for the dashboard-mounted assembly.

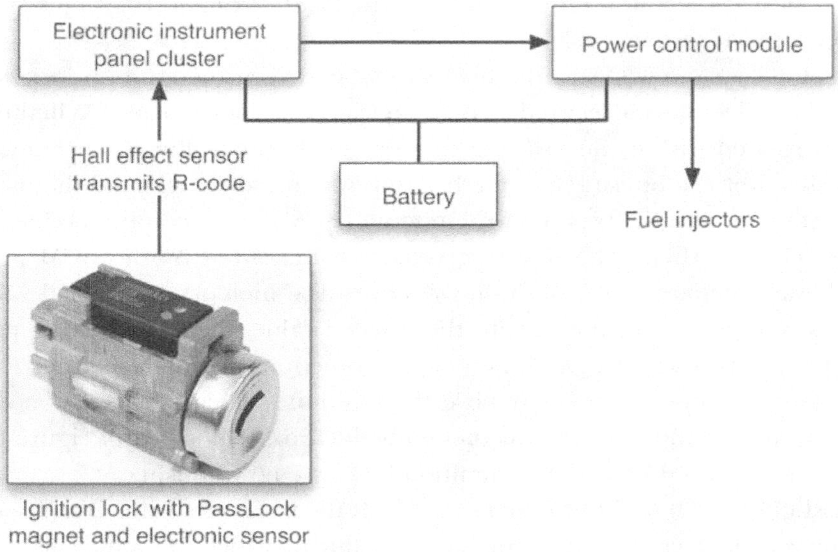

Figure 8-5
Diagram showing the principle of operation of the PassLock system. (Diagram created by Eric Stauffer.)

(keyway) can be drilled to weaken the lock wafers sufficiently to accomplish core rotation without the proper key and without core removal, although these methods are identified by significant physical damage to the ignition lock components. Figure 8-6 reveals the damage to two defeated column-mounted ignition lock, starter switch, and PassLock assemblies (cast column design).

8.2.4 Transponder or Radiofrequency Identification Systems

The transponder vehicle anti-theft system, also referred to as radio frequency identification system (RFID), is currently the most prevalent electronic protection installed on new vehicles worldwide. Vehicle applications of transponder technology emerged soon after the fall of the Berlin wall as the Russian and former Soviet Union Bloc black market demands for European cars increased dramatically [4]. The automotive transponder anti-theft system was initially developed to meet the stringent 1995 requirements of German insurance companies. BMW began phasing the new transponder system into series production during January and February of 1995 [5]. In the United States, domestically manufactured vehicles began using limited transponder technology in 1996, with selected line expansion from 1997. In 2005, many manufacturers still offer a transponder system only as an option or package upgrade. Japanese manufacturers started equipping upper end vehicles in the United States in 1997 [6]. The Korean producer Daewoo did not provide a transponder-

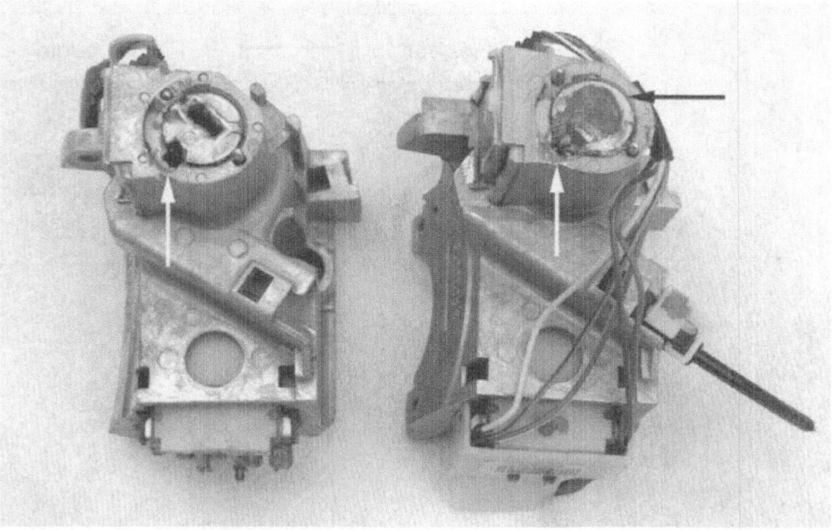

Figure 8-6

These two GM cast column assemblies were removed from recovered stolen vehicles. Note that the sidebar has been removed from the left assembly lock core and the lock wafers and sidebar forced from the right assembly lock core. Each method resulted in significant damage to the lock but still permitted lock core rotation without the proper key and allowed the vehicle to be started and driven away.

equipped vehicle until 2001 in the United States [6]. The system availability can be predicated by the requirements of various countries. Hyundai offered three transponder-equipped vehicles in Canada in 2002, but the same vehicle models in the United States are not so endowed [6]. On the GM Chevrolet Tahoe and Trailblazer, domestic models are provided with the PassLock immobilizer system, whereas export vehicles of the same models are upgraded to the superior PassKey III (transponder) system [7].

The Ford transponder system (SecuriLock), as an example, has proved to be so effective that the 2006 Ford Thunderbird was granted a full exemption from the parts-marking requirements of the US Federal Motor Vehicle Theft Prevention Standard required by the National Highway Traffic Safety Administration (NHTSA) [8]. Ford stated in its petition for exemption that there are currently 18 quintillion possible codes for the system, and in 2003 the SecuriLock transponder system was upgraded from read-only transponder technology to encrypted technology. In 1997, the Ford transponder system was installed on the entire Mustang vehicle line as standard equipment and resulted in a 70% reduction in theft compared with the 1995 Mustang. The NHTSA's theft data report on thefts of 2002 model year passenger vehicles revealed the transponder-equipped Thunderbird as having only 14 thefts out of a production of 28,639 vehicles, a theft rate of 0.488 per 1,000 vehicles produced [9].

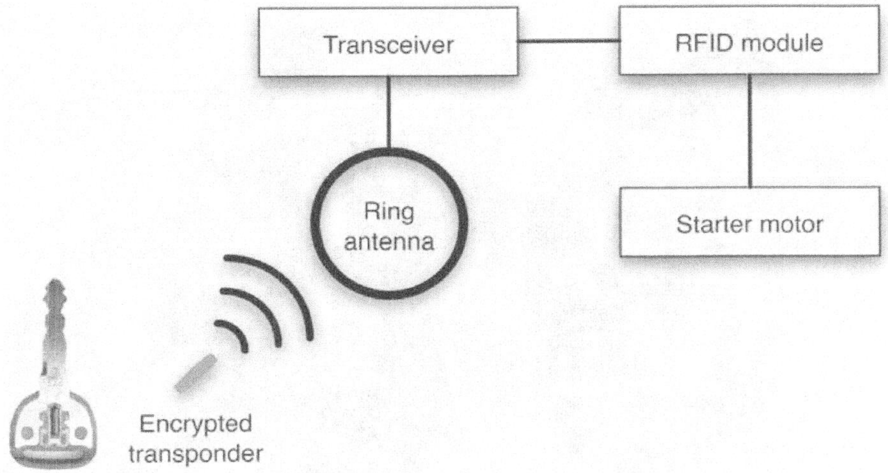

Figure 8-7
Diagram showing the principle of operation of a transponder system. Diagram created by Eric Stauffer.

Figure 8-7 details typical transponder system components. The transponder system consists of a microtransponder (transmitter/responder) embedded in the head (or bow) of the ignition key, a ring antenna (also known as an induction coil), encased in plastic and surrounding the outer ignition lock cylinder housing, a transceiver (transmitter/receiver) usually found on the steering column assembly adjacent to the ring antenna, and an ECM.

When the correctly cut mechanical key is inserted into the ignition lock and rotated, the vehicle electronics are energized. The transponder ring antenna (induction coil) sends an electromagnetic field of energy to the key. A microcapacitor in the transponder wedge or capsule embedded in the key absorbs this energy and powers the transponder to emit a unique signal. The ring antenna receives the transmitted signal and sends it to the transceiver and the control module. If the signal is recognized as being correct and accepted, the engine functions are enabled.

The concept of the transponder system remains the same in all applications, although various manufacturers may differ slightly in configuration. There are three distinct types of transponder operating systems: the single identification (or fixed) code, the challenging response code (encrypted), and the rolling code. The single identification (or fixed) code uses an alphanumeric set of digits assigned to a particular vehicle. The challenging response code (encrypted) is the most complex of the three currently in use. Communication between the key and the ECM is encrypted in both directions (bidirectional encryption of data), and the code changes during every use. The rolling code allows the vehicle control module to generate a new variable code and to send it to the transponder after each engine start. European trends indicate that of 73 million vehicles equipped with transponders, 45 million are single identification codes, 24 million are encrypted codes, and 4 million are rolling codes [10].

Figure 8-8 shows the components of a column-mounted transponder. Figure 8-9 is a photograph of a Ford column with transponder components secured around the ignition lock. There are two types of transponders: the glass capsule and the plastic wedge

Figure 8-8

Components of a column-mounted transponder system. The ignition lock cylinder, which is mounted inside the transponder ring antenna, is shown on the left (L), and the ring antenna (C) and transceiver assembly (R) are shown on the right.

Figure 8-9

Standard Ford steering column with transponder components. Note the ignition lock (L), the transponder ring antenna (C), and the transponder transceiver (R).

transponder. Figure 8-10 presents a detailed view of both types of the transponder chips. Figure 8-11 reveals the transponder chips as found embedded in the key heads. Because the key embedded transponder does not use any battery and is only energized during each cycle by the energy surge from the ring antenna (induction coil), there is a very limited range of communication, generally up to 15 cm [11].

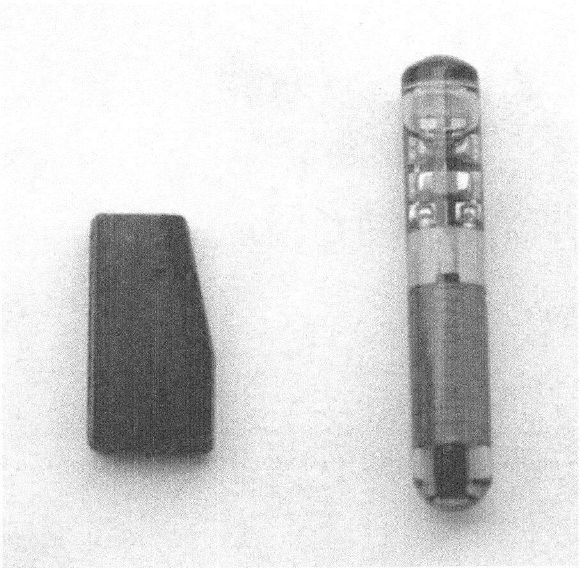

Figure 8-10

Two different types of transponder. The plastic wedge type transponder is shown on the left. The glass capsule type transponder is on the right. Magnification approximately 10×.

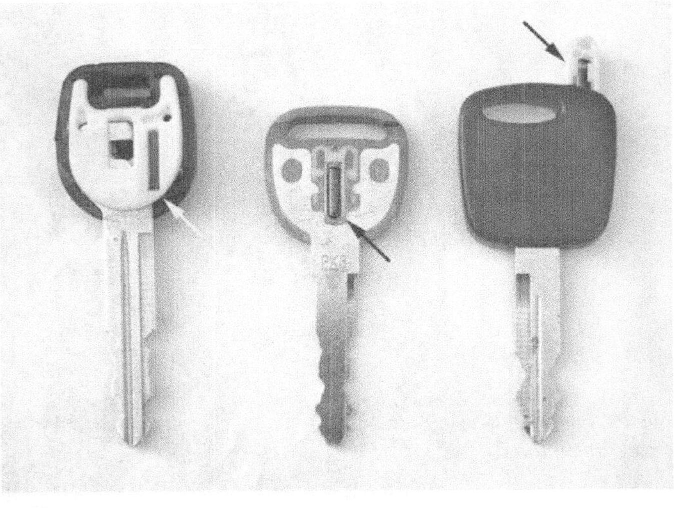

Figure 8-11

The key on the left is a Mitsubishi plastic wedge transponder in the opened key head. The center key is a GM capsule transponder. The key on the right is an early Ford design. Note the key ring hole is offset in the Ford key to accommodate the larger transponder.

8.3 FUNCTION TESTING OEM IMMOBILIZER SYSTEMS

There are methods that a field examiner can use to function test the OEM immobilizer systems on a still-operational vehicle.

8.3.1 PassKey I and II

- Using the proper key with the correct resistor pellet embedded in the blade, the vehicle is started as normal.
- The vehicle is then turned off and the key is removed from the ignition.
- The exposed resistor pellet embedded in the upper key blade is covered with a small piece of plastic electrical tape, as shown in Figure 8-12. The tape prevents the PassKey signal from passing through the resistor pellet resulting in the PassKey system defaulting to tamper mode, thus preventing engine start.
- Using the tape-covered key, attempt to start the vehicle again. If the system is intact and in good working order, the vehicle should not start.
- The key should now be removed from the ignition and the tape removed from the resistor pellet. After waiting about four minutes, the key is used to start the engine. The engine should start and run normally.

Figure 8-12
Plastic electrical tape is used to cover and insulate the upper blade embedded PassKey resistor pellet when function testing the system.

8.3.2 PassLock

This test can only be used on column-mounted lock assemblies.

- The vehicle is started using the key.
- The engine is turned off and the key removed from the lock.
- After waiting several minutes, the upper column shroud is removed and the PassLock wiring connector at the top of the ignition lock housing is located. Figure 8-13 shows the column mounted ignition lock, starter switch, and PassLock assembly.
- The PassLock wiring is separated from the housing at the connector. Figure 8-14 shows how to disconnect the PassLock components.
- The key is used to attempt to start the vehicle. The engine should start for a couple of seconds and then shut off due to fuel deprivation, indicating the system is functioning properly.
- Reconnect the PassLock components and start the vehicle again. The engine should start and run normally.

Figure 8-13

Example of a GM column-mounted ignition lock, starter switch, and PassLock assembly. Note the PassLock sensor position (L) and PassLock wiring connector (R).

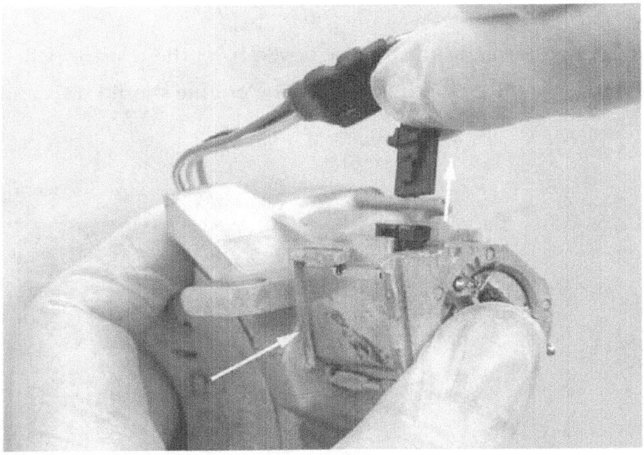

Figure 8-14

Disconnect the PassLock wiring to function test the system. Note the PassLock sensor position on the left housing.

8.3.3 Transponder

The transponder key can be field tested by completely covering the key head (containing the transponder) with aluminum foil. The metal foil blocks the radiofrequency signal and prevents transmission between the steering column-mounted components and the key-mounted transponder. This procedure confirms that the transponder system is functional.

- The key is covered with a piece of aluminum foil, as shown in Figure 8-15.
- The key is inserted in the ignition lock, as shown in Figure 8-16, and used to start the vehicle. With the key head shielded, the vehicle should not start.
- The key is removed from the ignition lock and the aluminum foil is removed from the key.
- The key is used again to start the vehicle. The vehicle should start normally.

Figure 8-15

The key blade is pushed through a square of aluminum foil before function testing the transponder system.

Figure 8-16

The foil is securely wrapped around the key head to block transmissions and inserted into the lock for testing. Note the theft light in dash that is illuminated as shown with the arrow. See Color Plate.

8.4 TRANSPONDERS

8.4.1 Aftermarket Programming and Servicing Tools

When investigating the theft of a transponder-equipped vehicle, there are several procedures used to clone or program new transponder keys that must be considered. Aftermarket programming and servicing tools started to become widely available in the late 1990s through locksmith supply companies and tool manufacturers. The tools were designed to assist independent locksmiths to service transponder-equipped vehicles and to add or delete transponder keys to vehicles in the field. These tools are connected to the vehicle through the diagnostics port. There are certain vehicles that cannot be programmed using these types of tools, and appropriate reference charts should be checked to identify current applications. A good source of information is an aftermarket transponder key manufacturer key application chart. These tools can also provide information on how many transponder

Table 8-3

Different aftermarket programming and servicing tools used to clone or program new transponder keys.

Name of tool	Tool manufacturer	Vehicle manufacturers that use the tool
Code Seeker	STRATTEC	Acura, Audi, Cadillac, Chrysler, Dodge, Ford, Honda, Infiniti, Jaguar, Jeep, Lincoln, Mazda, Mercury, Mitsubishi, Nissan, Plymouth, Volkswagen
NGS (New Generation Star)	HICKOK	Ford, Lincoln, Mazda, Mercury
SDD (Silca Diagnostic Device)	Silca Group	Acura, Audi, Cadillac, Chrysler, Dodge, Ford, Honda, Infiniti, Jaguar, Jeep, Lincoln, Mazda, Mercury, Mitsubishi, Nissan, Plymouth, Volkswagen
ETD-1 (Electronic Transponder Duplicator)	JET	Acura, Buick, Cadillac, Chevrolet, Honda, Infiniti, Mazda, Nissan, Oldsmobile, Pontiac, Porsche, Saab, Jaguar
T-Code	ASP	Acura, Audi, Cadillac, Chrysler, Dodge, Ford, Honda, Infiniti, Jaguar, Jeep, Lincoln, Mazda, Mercury, Mitsubishi, Nissan, Plymouth, Volkswagen
Quick-Code	STRATTEC	Acura, Buick, Cadillac, Chevrolet, Honda, Infiniti, Mazda, Nissan, Oldsmobile, Pontiac, Porsche, Saab
DART (Diagnostics and Reprogramming Tool)	STRATTEC	Chrysler, Dodge, Jeep, Plymouth
RW2	Silca Group	Acura, Buick, Chevrolet, Honda, Infiniti, Cadillac, Jaguar, Mazda, Nissan, Oldsmobile, Pontiac, Porsche, Saab

keys have been programmed to a specific vehicle, in particular recently programmed keys.

Table 8-3 shows 8 aftermarket programming and servicing tools currently available and the respective vehicle makes for which they are designed. Each tool is used for adding and deleting transponder keys [12].

When programming a new transponder key to a vehicle with a servicing tool, an initial on-screen prompt asks the user to turn the ignition on. During a vehicle theft, the steering column/ignition lock must be physically defeated to access the protected electrical starter (ignition) switch to energize the systems. The transponder system ring antenna must remain properly connected and intact to allow a new transponder key to be programmed to that vehicle and a new nonprogrammed key for that specific vehicle must be present and in close proximity to the ring antenna. Not all vehicles can have a key programmed using commercial servicing tools, and many require a personal identification number or vehicle code for programming.

8.4.2 Transponder Key Cloning

JET Hardware Manufacturing (Brooklyn, New York, USA) and Silca Group (Vittorio Veneto, Italy) are two popular manufacturers of transponder cloning equipment. To clone a new transponder key, an existing properly programmed key for that specific vehicle must be present as well as a new (blank) transponder key appropriate for that particular vehicle. The vehicle does not need to be present. When the properly programmed existing key is inserted in the cloning equipment, the equipment "reads" the key and retains the information. When the new transponder key is inserted, the cloning equipment sends the stored information and programs the new transponder in the key. Encrypted and rolling code transponder keys cannot be cloned, and only single identification codes can be duplicated. Additionally, any key with a Texas instrument transponder cannot be cloned.

8.4.3 Emergency Start Procedures

Certain vehicles use an emergency start procedure that permits the vehicle to be started without the transponder being present. To accomplish this, the vehicle must be energized using a correctly cut mechanical temporary key or by physically defeating the column. Vehicle-specific codes, which must be obtained from the dealer using the vehicle identification number (VIN), engine control unit (ECU) serial number, and/or immobilizer serial number (ISN) can be programmed into the vehicle by depressing the brake pedal, using the odometer counter reset button, or pressing the trip odometer reset button, depending on the vehicle. Some vehicles with emergency start procedures are shown in Table 8-4.

Table 8-4

Emergency start procedures for some vehicles.

Manufacturer	Number of digits in code	Necessary information to obtain code	Method
Acura	5	VIN	Use brake
Audi	4	ISN and VIN	Use clock and trip odometer reset
Honda	5	VIN	Use brake
Mitsubishi	5	ECU serial number	Use brake
Porsche	n/a	n/a	n/a
Volkswagen	n/a	ECU serial number	Use odometer reset

(n/a) not available.

8.4.4 Transponders for Fraud Prevention

The number of vehicles stolen in Europe declined by 50% between 1993 and 2000, with transponder anti-theft systems given the bulk of the credit for the decrease in the number of thefts [13]. Insurers in Europe found that criminals are less inclined to attempt the theft of a vehicle when they can conspire with the owner instead. A criminal can pay the owner for a copy of the key and simply drive the car away. The owner then informs the insurance company of a theft and submits a fraudulent claim (see Chapter 19). All keys are submitted to the insurance company as requested, and no record of the copied key may exist.

To prevent this type of fraud, Texas Instruments is adding an enhancement to its transponder called digital signature transponder plus (DST+) [13]. This system stores data on individual keys, with each key having a unique serial number. If a new copy is made from an original key and used in the vehicle, a unit in the steering column records the existence of the new key. The next time the original keys are used, the steering column unit provides the data to the keys so that all the keys have information on the existence of all other keys. The system also has the capability to store the date on which a key was used last. BMW has also introduced a key that stores the vehicle mileage, as well as other information, that can be downloaded at an authorized BMW dealer.

Aftermarket temporary bypass systems for transponder, PassKey I and II, and PassLock are available from various manufacturers worldwide. These components are designed for vehicle remote start functions only and are not capable of defeating the OEM immobilizer systems. When installing a bypass system for a transponder-equipped vehicle, an existing properly programmed key is required. Either the key must be cloned to the bypass module or a programmed spare key for that specific vehicle must be used in the aftermarket bypass module. A PassKey I and II kit requires the key pellet resistance value to be obtained before programming the remote start module. The PassLock system requires the vehicle to be started and running to obtain the proper code. Additionally, most remote start systems are protected by a variety of security functions, such as a brake pedal switch, door switch or gear selector switch to disable engine functions, or an audible alarm sound if the proper key is not used during an attempt to drive the vehicle.

8.5 ELECTRONIC KEY AND KEYLESS IGNITION SYSTEMS

Vehicle manufacturers are trending toward ignition systems that eliminate the need for a mechanically cut ignition key and ignition lock set. Mercedes Benz took the technology lead in 1998 with the introduction of a dedicated electronic key (no mechanical key) that used an infrared signal to communicate with the vehicle. When the electronic key is plugged into the dashboard receptacle, the ignition switch is activated by an infrared data exchange between the electronic key and the ignition module mounted in the dashboard. If the proper identification data is received, the engine starts and the steering column lock is electrically disengaged.

The keyless (no mechanical or electronic key) system for access control, driver authorization, and vehicle immobilization was jointly developed by Siemens Automotive and Mercedes Benz, who introduced the first keyless entry and ignition technology in Germany on the S-Class Mercedes in 1999. System availability continued to be expanded in newer models as an option [14].

The entire process of unlocking the doors, deactivating the immobilizer, and enabling the ignition is completely passive and automatic. The only active steps taken by the driver is to carry the electronic card (in a wallet or purse) and to approach the vehicle. When the driver touches a door handle, the electronic card is scanned by the on-board vehicle control center. There is an encrypted data exchange, and if the control unit recognizes the card information, the data is decoded and the doors are unlocked. The engine is started by pressing a start-stop button. The immobilizer is only deactivated by a bidirectional encryption process. The electronic gear selector is unlocked and the ignition enabled once the control unit has determined, by means of a second check, that the electronic card and driver are both in the passenger compartment. When leaving the vehicle, the start-stop button is pressed to turn off the engine, to activate the immobilizer, and to lock the gear selector.

The 2005 Chevrolet Corvette, 2004 Cadillac XLR, and 2005 Cadillac STS are equipped with similar keyless systems. The American version of the Siemens system differs from the European smart card system in that it uses a key fob rather than a smart card. Most other functions remain essentially the same.

The BMW 7 Series also uses the keyless electronic card system. The Japanese started using the keyless electronic type system in the Toyota Prius in 2003 and Lexus (selected models, optional) in 2004. Expansion of the keyless electronic systems will continue to increase in more expensive models over the next several years, eventually progressing to most models over the next half decade.

8.6 ALARM SYSTEMS

Before 1985, OEM alarm systems were not widely available. In 1985, for example, of 261 vehicles reviewed, only 12 models had OEM alarm systems as standard equipment [1]. OEM anti-theft/alarm systems (also known as content alarms), much like the aftermarket systems, use an audible alarm, visual alarm, fuel/electrical interrupt, or a combination of these components. These systems are armed by the locking of the doors by the driver. Alarms are activated by the opening of the doors, trunk, or hood. OEM content alarms can also be installed by the manufacturer on selected immobilizer-equipped vehicles. Vehicles equipped with interior motion or shock detectors, usually found on aftermarket systems, also activate with motion or vibration such as the ones created by glass breaking.

Most OEM and aftermarket alarm systems can be compromised without special equipment or extensive knowledge. For instance, certain OEM systems can be defeated by entering the interior through a broken door window (not opening the door) and force rotating, forcibly extracting, or otherwise defeating the ignition lock to disarm the system and start the engine. Less sophisticated aftermarket systems can be neutralized by removing or bypassing interior components, accessing the engine compartment to cut the siren or horn wiring, or to remove or replace relays. These procedures would be ineffective against an immobilizer-equipped vehicle (PassLock, PassKey, or transponder).

The aftermarket and OEM alarm/anti-theft systems provide a moderate measure of protection and deterrence against auto thieves when compared with the highly effective immobilizer systems that continues to advance in sophistication with technology. However, the dedicated, knowledgeable, and professional thief will find a way to compromise most of these systems.

ACKNOWLEDGMENTS

The author would like to thank Staci L. Rosenberger for her help in producing the photographs and diagrams and Sharon L. Mangine for her assistance in the research of information necessary to produce this chapter.

BIBLIOGRAPHY

[1] Highway Loss Data Institute (1994) Anti-theft device availability. In: *NICB 1994 Passenger vehicle identification manual,* ed National Insurance Crime Bureau, 65th edition, pp 192–201.

[2] Johnson G. and Levine J. (2004) A History of locksmithing 1939–2004, *Locksmith Ledger,* November, special 65th anniversary supplement, p 8.

[3] Levine J. (1994) *Service Manual American Car Locks,* volume 3, The Locksmith Publishing Corporation, Park Ridge, IL.

[4] O'Leary T. (2004) The changing field of automotive locksmithing, *Locksmith Ledger,* May, p 30.

[5] BMW (year unknown) *Anti-theft and Alarm Systems,* BWM Technical Service Bulletin # 61 01 95, BMW, Woodcliff Lake, NJ.

[6] Kaba Ilco (2001) *The basic transponder guide: Making transponder technology crystal clear,* 4th edition, Kaba, Rümlang, Switzerland, pp 6–8.

[7] General Motors (2002) Theft deterrent application and programming chart, *General Motors Tech Link,* **4**(4), p 4.

[8] Brewer HK. (2005) Department of Transportation—National Highway Traffic Safety Administration—Petition for exemption from the vehicle theft prevention standard; Ford, *Federal Register of March 15, 2005,* **70**(49), pp 12780–17282.

[9] Kratzke SR. (2004) Department of Transportation—National Highway Traffic Safety Administration—Final theft data; motor vehicle theft prevention standard, *Federal Register of September 1, 2004,* **69**(169), pp 53354–53359.

[10] Silca (2005) Transponder technology, available at http://www.silca.it/eng/html/body_transponder.html, last access performed on May 27, 2005.

[11] Hyde M. (2004) *Automotive transponder systems,* Auto-Security Products Inc, Redmond, WA.

[12] Street Keys (2003) *Transponder key programming tool chart,* Street Keys, LLC, Cupertino, CA.

[13] Unknown author (2002) Stealing cars will get tougher, *RFID Journal,* available at http://www.rfidjournal.com/article/articleview/33/1/1, last access performed on May 27, 2005.

[14] Siemens VDO Automotive (2005) Available at http://www.usa.siemensvdo.com, last access performed on May 27, 2005.

CHAPTER 9

EXAMINATION OF STEERING COLUMNS AND IGNITION LOCKS

Robert F. Mangine

9.1 INTRODUCTION

In 1915, Dodge introduced a 12-volt electrical system and a starter/generator connected to the motor with a chain drive on their touring car [1]. It featured a single-ignition coil with cam-actuated breaker points and a distributor. The ignition switch was mounted to the dashboard and used a stamped steel key with limited lock combinations. Around 1920, Henry Ford retired the model T hand crank used for starting the engine and replaced it with a 6-volt starter, four induction coils, and an ignition switch mounted on the dashboard [1]. The switch was protected by a key-operated lock using about two dozen different lock/key combinations. In the late 1920s, Studebaker had limited success with a steering column lock that locked both the ignition and column [1]. In 1969, General Motors (GM) made column-mounted ignition locks, locking steering columns, and locking gear selector's standard on most models. All manufacturers started to equip domestic vehicles as such by the early 1970s. This began the era of the modern steering columns as they are known today.

The early lock manufacturers used pin or disk (wafer) tumblers. In 1935, GM began using the Briggs and Stratton sidebar ignition lock with six wafers and a single-sided key [1]. In 1994 and 1995, GM began to change the six-wafer lock to a nine-wafer lock using a double-sided key. They also began to change the composition of the lock sidebar from metal alloy to steel. The increased lock wafers, from six to nine, and the steel sidebar significantly strengthened the lock. It also allowed a single-key system to be used as opposed to the older two-key system (the first for the ignition lock and the second for the doors, trunk, and glove box). In 1988, Chrysler began the process of converting from pin tumbler locks to wafer locks [2]. Chrysler completed the process in 1991 using full-wafer locks. Ford began the transition from pin tumbler locks to wafer sidebar locks in 1984. The 1993 Ford Mustang was the last domestically built vehicle equipped with pin tumbler locks, as they were being replaced with Ford's six half-wafer sidebar locks [2]. There are no pin tumbler locks currently in use. Through the mid-1990s vehicle manufacturers increased lock wafers in the locks from 6 to 8 or even 10. The increase in wafers and lock strength discouraged picking, impressioning, and forced rotation by vehicle thieves.

From the 1920s forward, ignition locks and eventually locking steering columns have continued to evolve in strength, complexity, and configuration. Still, they have always remained a physical protection provided by security hardware for over three decades. That all began to change in 1997 when the Chevrolet Malibu was introduced with a nonlocking steering column and the ignition lock was moved to the dashboard, although in 2006 GM returned the ignition lock to a nonlocking steering column in the Impala and Monte Carlo.

The locking steering column has begun to become obsolete with advances in vehicle electronic anti-theft technology. Figure 9-1 shows a typical GM dash-mounted ignition lock assembly. Ford and Chrysler components are similar. Figure 9-2 presents the GM single-component dash-mounted plastic housing and lock assembly.

Figure 9-1

GM dashboard-mounted ignition lock and starter switch assembly. There is no connection to the steering column. The gear selector locking cable and PassLock anti-theft components are also contained in the assembly.

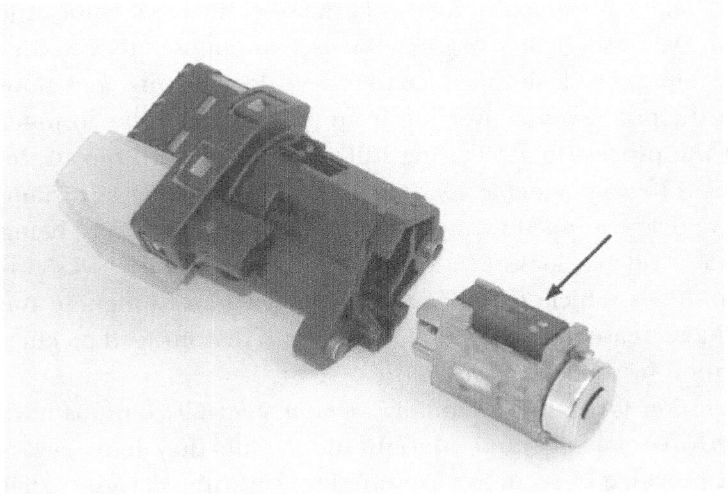

Figure 9-2

The GM single-component dash-mounted assembly removed from the dashboard with the ignition lock cylinder removed from the assembly housing. Note the PassLock sensor on the lock cylinder.

The resources spent by manufacturers in the past on redesigning ignition locks and heavier column hardware never succeeded in deterring criminals from stealing cars. They just used heavier tools to compensate for the increased security. So, rather than spending money on even more complex or heavier physical protection, it was a logical step to invest in the more effective electronic anti-theft protection systems. The dashboard-mounted assemblies are secured to the dashboard, between the column and stereo, with two mounting bolts. The assembly is not connected to the steering column mast. The compact assembly is a single component containing the ignition lock, starter (ignition) switch, gear selector locking cable connection, and anti-theft system components (PassLock or transponder).

This chapter is intended to provide the investigator with a basic working knowledge of ignition locks and steering columns. There are numerous types of columns and locks. For example, GM alone used 22 different ignition locks and 12 different steering columns during the 1990s [2]. Most columns and locks, European, Asian, or American, are usually variations of existing designs, and modifications are an extension of these design principles.

To investigate a reportedly stolen vehicle, burned or unburned, it is important to remember the basic principles. To start and operate a vehicle, without the properly cut mechanical key, three physical barriers must be overcome. Additionally, if the vehicle is electronically protected with an (original equipment manufacturer [OEM]) anti-theft system, a fourth barrier must be compromised. A locked vehicle must be forcibly entered, the steering column and/or gear selector must be unlocked, the electronic anti-theft system neutralized, and the engine started. Failure to overcome any one of these four possible barriers makes it impossible to start and operate the vehicle without the proper key.

When the ignition lock and/or steering column is physically defeated, the vital parts of the lock or column will be damaged, distorted, marked, or scared. Forensic examination can determine, with a high degree of scientific certainty, the method used to defeat the lock and column or if the properly cut mechanical key was used, even on a vehicle that has been burned to completion.

9.2 VEHICLE STEERING COLUMNS

9.2.1 Steering Column Design

The steering column was originally designed simply as a wheel mounted on a shaft used to steer the earliest vehicles. This has evolved into a highly complex component with an ever-growing array of controls and convenience features. It also currently houses physical and electronic security components in many vehicle models.

There are three types of columns in use today: the separate-component column, the single-component column, and the nonlocking column. All locking columns, regardless of the manufacturer, have a column housing assembly (which is secured to the column mast), an ignition lock, a column-locking device (GM uses mostly steel locking pins or bolts and all other manufacturers use steel locking lugs), and an electronic starter (ignition) switch.

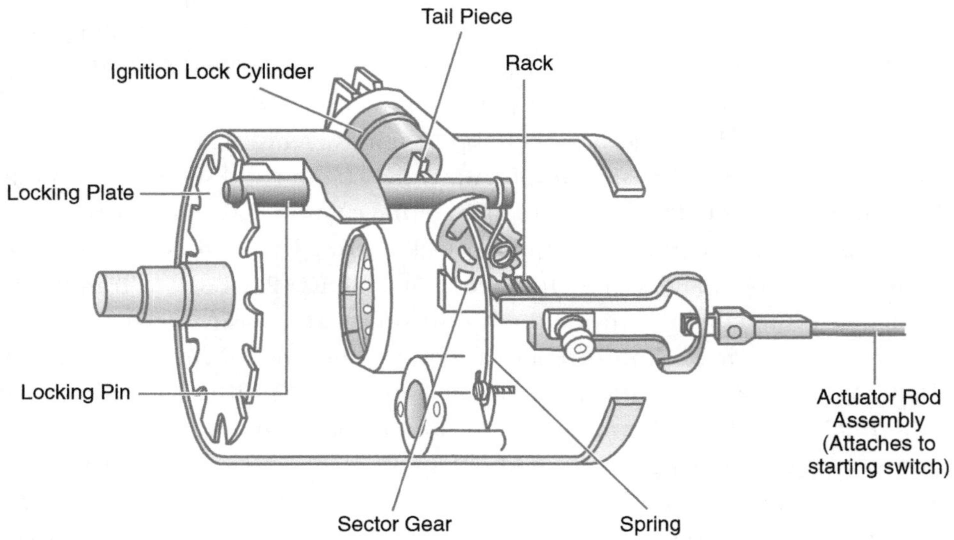

Figure 9-3

Diagram of a GM tilt wheel steering column assembly. The position and presence of the different components vary depending on the types of column, but this diagram provides a good representation of the different components.

The components are secured in or on the column housing assembly, with only the GM Saginaw column having the starter (ignition) switch mounted externally on the lower column mast.

Figure 9-3 is a diagram of a GM tilt wheel steering column assembly. Although not all steering column assemblies are designed in this fashion, this diagram allows for a better understanding of the different components usually found in the assembly and their functions. The lock cylinder (top) acts on the sector gear and rack through its tailpiece. This action pulls the locking pin back, liberating the locking plate and allowing the steering wheel to revolve. Finally, the actuator rod assembly (bottom left) connects to the starter switch.

9.2.2 Separate-Component Column

The first modern columns were constructed with separate components located in various areas. The GM standard round tilt (or Saginaw) type column was first introduced in the late 1960s with many design variations over the years. Figure 9-4 is a top view of the GM standard round tilt column. Note the breach in the left side of the column. Figure 9-5 is a close-up of the internal components of the column's left side. This column can be defeated though the left side without tampering with or damaging the right-side mounted ignition lock.

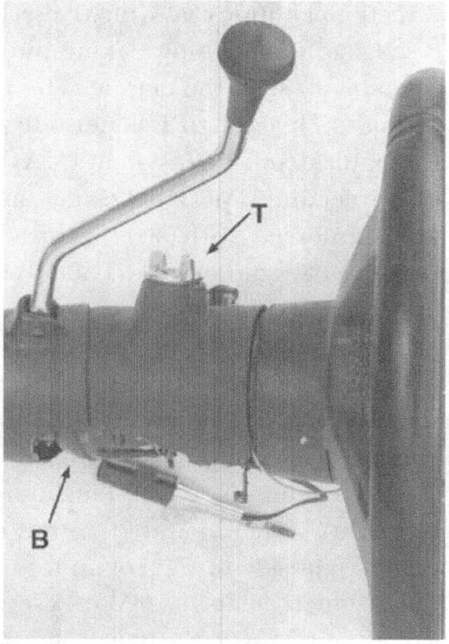

Figure 9-4

Top view of the GM standard round tilt column. Note the ignition lock (T) and the breach in the column shroud (B) used to access the internal column components.

Figure 9-5

From top to bottom: Column locking pin, sector gear connected to the tail of the lock by a shaft, locking pin return spring, and rack assembly connected to the starter switch.

The ignition lock is located on the right side of the column. The column-locking pin (or bolt) is found on the upper left side of the column, the sector gear and rack assembly in the center left side, and the starter (ignition) switch on the lower column below the dash. These components are connected by a series of drive shafts, rods, gears, and springs. The GM standard round tilt wheel (Saginaw) column could be defeated by accessing the left side column components. This is accomplished by breaking the left column shroud at the directional hump, removing the sector gear, and manipulating the components to manually disengage the column locking pin to unlock the column and pull rearward on the rack assembly to start the engine. The time required for this operation is usually less than 30 seconds. This design was periodically improved during the life of the column, with the mechanical anti-theft system (MATS) column, which is found primarily on the 1993–1994 Chevrolet and GMC trucks, vans, and sport utility vehicles (SUVs) [3]. The MATS column was reinforced with steel guards protecting vital components and used a "security" (improved) ignition lock. The MATS column was discontinued in cars at the end of 1993 but continued in trucks.

In the mid-1990s, GM introduced the much improved cast metal column, also known as the component set strategy (CSS) column [4] and modular column [3]. Figure 9-6 shows the two types of cast column assemblies, a PassLock-equipped assembly and non-PassLock equipped assembly.

This column used a cast metal housing assembly with both the ignition lock and starter (ignition) switch mounted on the right side. The column locking pin was located in the upper right side of the column, to the rear of the ignition lock assembly, and encased in the cast metal housing. GM uses several variations of the cast-style column. Chrysler started using the GM style Saginaw column in the late 1970s to early 1980s. Chrysler designed its own component column, called Acustar, which was introduced in 1989 and began to replace the GM columns, although GM-designed columns were still found on the Jeep Cherokee until 1995, the Dodge Viper until 1997, and the Dodge Ram van until 2000.

The Chrysler Acustar column used a plastic column-mounted housing assembly on the right of the column that included the ignition lock and starter (ignition) switch. Figure 9-7 is a photograph of the Chrysler Acustar mounted on the column. Figure 9-8 shows the lock retainer in the plastic housing.

The column locking mechanism and gear selector lock were enclosed in the metal upper column housing and located behind the plastic assembly. The ignition lock was secured in the plastic housing with one active retaining pin. Defeating this early Acustar column only required the metal ignition lock cylinder to be pried from the plastic housing. This could

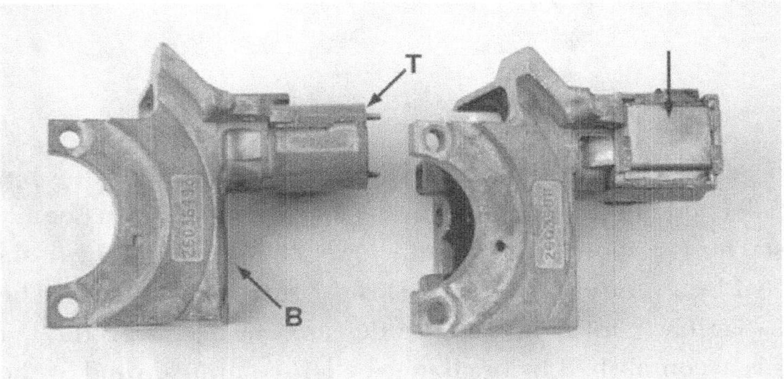

Figure 9-6

Left: This GM cast assembly is a non-PassLock type. Note the ignition lock position (T) and starter switch position (B). Right: This assembly has PassLock sensor secured to the housing and protected by the metal cover plate.

Figure 9-7

Chrysler Acustar ignition lock (T) and starter switch (B) assembly mounted on the right side of the column.

Figure 9-8

Detailed view of the ignition lock retainer in the plastic housing of the Chrysler Acustar column.

be accomplished with a flathead screwdriver in under 15 seconds. After driving and parking the vehicle, the lock could be replaced in the housing and the retaining pin reinserted in the plastic housing slot with no damage visible to the casual observer. From 1994 to 1995, Chrysler reinforced the assembly and added a security bracket to the lock assembly section of the housing. Within two years, Chrysler further reinforced the Acustar column by mounting a steel plate to the column, behind the ignition lock and starter switch assembly housing, and used a tailpiece on the rear of the lock core that would insert through the steel plate and lock behind the plate. If the ignition lock was forced out of the column during an attempted theft, the lock tailpiece would break and separate from the rear of the lock and block the slot in the actuating gear behind the steel plate. This prevented a screwdriver, or similar tool, from being inserted into the blocked slot to rotate the internal assembly linkage. This last version of the Acustar column proved to be an excellent design.

From 1991 to 2002 Ford used two types of columns, the standard Ford column and the European/Asian style single-component assembly column (only on the later Ford Escort, Mercury Tracer models, and vehicles made in partnership with Mazda, such as the Ford Probe). The standard Ford column assembly is found on all types of Ford vehicles, including cars, trucks, and SUVs. There are several different configurations of the standard column, but the basic design has survived. The Ford column uses a well-designed heavy casting. The original design included the ignition lock, column lock, and starter (ignition) switch mounted either on or in the casting. The later column moved the column-locking lug into the column mast housing, protecting it further. All the standard Ford columns use a sector gear and rack to actuate internal column functions. The sector gear is protected by an armored plate at the base of the ignition lock well.

When the ignition lock is inserted into the lock well in the housing, a tailpiece on the back of the lock core fits through a hole in the armored plate and into the sector gear slot. In the "lock" position the tailpiece locks behind the armored plate. When a proper key is used to rotate the lock core, the lock tailpiece rotates the sector gear, which allows the column to function. If the ignition lock cylinder is forced from the column housing during an attempted theft, the lock tailpiece breaks and separates from the rear of the lock and blocks the sector gear slot from tampering. The sector gear remains protected by the armored plate. A determined thief can eventually force the armored plate and sector gear from the base of the empty lock well to directly access the rack, but it takes work and is time consuming. Figure 9-9 shows a standard Ford column housing with the ignition lock cylinder forcibly removed.

Figure 9-9

Standard Ford column (nontransponder equipped) with the ignition lock cylinder forcibly removed from the right column housing assembly.

In 2002, Ford began to use a third column design by equipping the Explorer SUV with their version of the European/Asian style single-component ignition lock—column lock—starter switch assembly. This column configuration was extended to the Excursion in 2003, the redesigned F-150 in 2004, and should be seen expanding in future vehicle applications. Ford has also begun to install nonlocking columns, as seen in the Lincoln LS and the Thunderbird, with dashboard-mounted ignition locks and starter (ignition) switch assemblies protected by transponder anti-theft systems.

9.2.3 Single-Component Column (Ignition Lock-Column Lock-Starter Switch Assembly)

European and Asian vehicles have been using a single-component ignition lock-column lock-starter switch assembly for decades. This is a one-piece unit that has the ignition lock cylinder and column locking lug secured in a metal housing with a plastic rotary starter (ignition) switch fastened to the opposite side of the assembly housing from the ignition lock. The assembly is secured to the column mast with a mounting strap and two shearheads or security bolts. When the ignition lock core is rotated with the correctly cut key, the internal assembly combined cam and drive shaft also rotates. The cam disengages the column locking lug and the drive shaft continues through the assembly and turns the rotary starter switch to start the engine.

There are two assembly configurations. Figure 9-10 shows a disassembled standard single-component ignition lock-column lock-starter switch assembly. Figure 9-11 shows the standard single-component assembly secured to a column mast. The basic assembly housing attaches directly to the column with the ignition lock on the right side of the column, the column lock in the center, and the starter (ignition) switch on the left side.

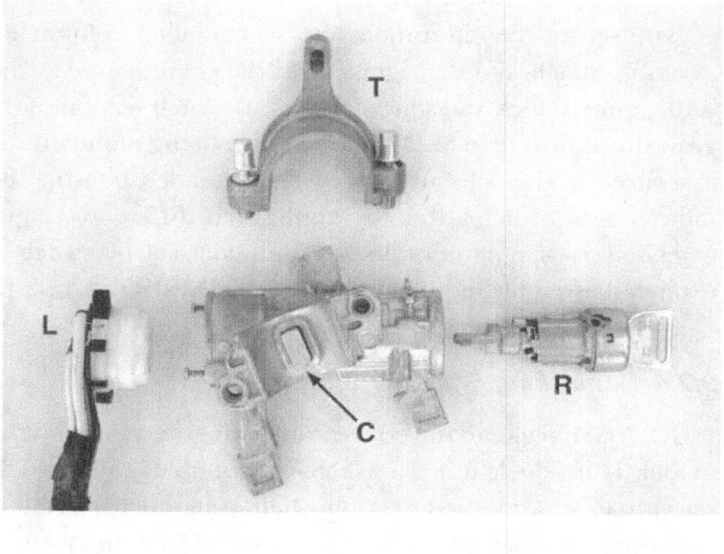

Figure 9-10

Disassembled standard single-component ignition lock—column lock—starter switch assembly. Note the assembly mounting strap with security bolts (T), column locking lug in assembly housing (C), ignition lock cylinder (R), and starter (ignition) switch (L).

Figure 9-11

Standard single-component ignition lock-column lock-starter switch assembly mounted to a column mast. The ignition lock is on the right.

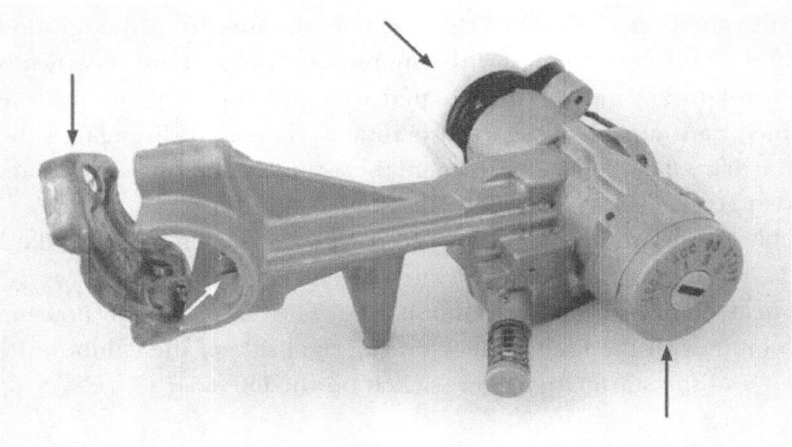

Figure 9-12

Extended housing ignition lock, column lock, and starter switch assembly. Left to right: assembly mounting strap, column locking lug, starter switch, and ignition lock.

The second configuration has an extended column locking lug housing where the assembly attaches to the right side of the column, and at the end of the extended housing is the ignition lock and starter (ignition) switch section of the assembly. This configuration gives the appearance of the ignition lock being mounted in the dashboard, when in fact it is secured to the column mast by the extended housing. Both assemblies operate on the same design principle but are configured differently. Figure 9-12 is a photograph of an extended housing ignition lock-column lock-starter switch assembly. Figure 9-13 shows the extended assembly mounted on the vehicle with the dash trim removed.

9.2.4 Nonlocking Column

In 1997, GM began to introduce nonlocking steering columns on the new Chevrolet Malibu model. This vehicle used a dashboard-mounted ignition lock, starter (ignition) switch, gear selector lock, and PassLock anti-theft components combined in a single assembly. This configuration does not include any steering column locking devices. GM has expanded the

Figure 9-13

The extended housing ignition lock, column lock, and starter switch assembly mounted on the vehicle (dash trim removed). The ignition lock appears to be in the dash, although it is actually secured to the column mast.

dashboard-mounted assembly and nonlocking columns into other selected vehicles from the 2000 model year forward. Ford uses a similar dashboard-mounted assembly with a nonlocking column on the Lincoln LS (since 1998) and the Thunderbird (since 2002). Chrysler began using this configuration on the 2003 Chrysler Pacifica, 2005 Chrysler 300C, the 2005 Dodge Magnum, and the 2006 Dodge Charger [5].

The nonlocking columns on passenger vehicles are easily identified by the ignition lock being mounted directly in the dashboard between the steering column and stereo/center dash stack and, evidently, the absence of a locking column. Nonlocking columns are not found on manual transmission vehicles, only on automatic. The nonlocking columns are electronically protected by transponder immobilizer systems or with the PassLock system (exclusive to most GM vehicles).

GM is currently the only US manufacturer installing nonlocking columns on consumer trucks. The trucks do not use the dashboard-mounted assembly but a standard cast (CSS) type column with the column locking pin and column locking plate deleted. The ignition lock is found mounted in the right part of the column. The first trucks and SUVs with the nonlocking column began to appear in 2001. Manual transmission trucks continue to retain the column locking pin assembly. For 2006, GM has relocated the dash-mounted ignition lock back to the steering column on selected vehicles and has begun to equip newly introduced vehicles with the European/Asian style single component assemblies.

9.3 VEHICLE IGNITION LOCKS

9.3.1 Principle

The typical ignition lock cylinder is a simple mechanical device designed to prevent unauthorized use of a vehicle by protecting the internal column assembly actuator, racks, cams, and switches. The only exception to this definition is the GM PassLock anti-theft ignition lock assembly. Until 2005, the PassLock system electronic components were embedded only in the half-wafer sidebar-type locks standard on most GM vehicles. Since 2005, GM began to introduce the full-wafer sidebar-locks on selected vehicles, such as the Chevrolet Malibu and Pontic G6 models, which are protected by either the PassLock system or a transponder. There is also a non-PassLock equipped half-wafer sidebar-type ignition lock. These locks should not be confused. Because this chapter deals with the physical protection of vehicles using steering columns and ignition locks, the examination of electronic protection systems is not covered.

9.3.2 Ignition Lock Components and Their Operation

It is important for the forensic examiner to have a basic understanding of the ignition lock, its components, and its functioning. Literature on locksmithing is readily available and could be useful [6–8]. Figure 9-14 is a diagram of a generic lock cylinder (pin tumbler type) and allows for a basic understanding of the functioning of a lock. When the key is inserted, it pushes the pins up and compresses the springs. If the proper key is inserted, as shown in Figure 9-14, the separation between the bottom and top pins is aligned with the shearline, allowing the cylinder core or plug to rotate freely. If no key is present or the wrong key is inserted, the separation between the top and bottom pins is not aligned with the shearline, preventing any rotation of the cylinder core. As stated previously, there are no pin tumblers in use currently in the automotive industry. Modern automotive locks use wafers, but the basic concept remains the same, with wafers in lieu of pins.

There are four basic configurations of ignition locks as well as various combinations of these configurations. These would include the full-wafer lock, the split-wafer lock, the sidebar lock (full and half wafers), and the high-security side-cut ignition lock and key sets (includes split wafers). The full-wafer and sidebar locks use a standard key, with the cuts on the top and bottom edges of the key blade that contact the wafer riding surfaces (also called lands). The side-cut lock sets use keys that have cuts (channels) in the left and right side of the key blade that correspond to protrusions on the inside of the wafer key opening. These keys are also called sidewinder or centerwinder keys (see Chapter 10). Side-cut keys

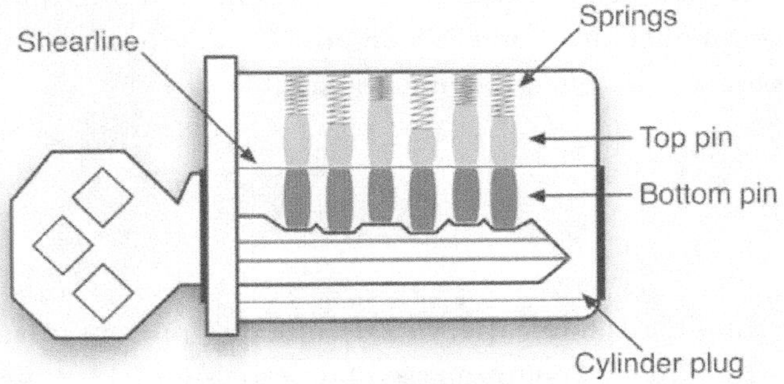

Figure 9-14

Diagram of a generic lock cylinder. Diagram courtesy of Eric Stauffer.

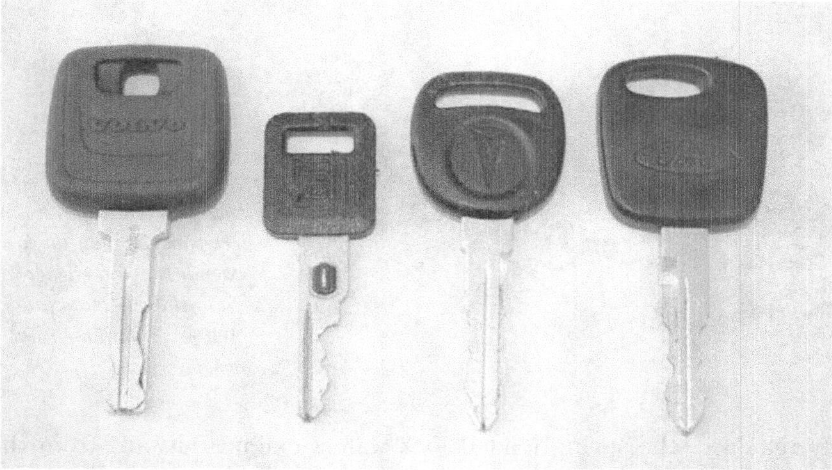

Figure 9-15

Left to right: side-cut/high security key, GM 6-cut PassKey with resistor pellet, GM 10-cut key, and Ford standard cut transponder key.

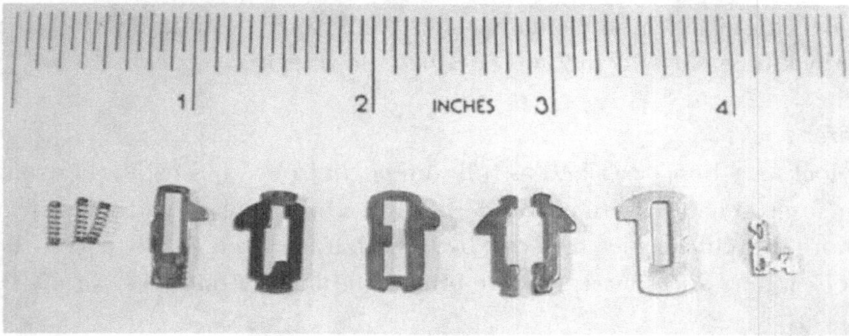

Figure 9-16

This photograph shows six wafers commonly in use along with three wafer springs. Left to right: wafer springs, full wafer, full wafer (note difference in design), high security wafer (note the protrusion in keyway), split wafer (note the two halves), full wafer for sidebar lock (note the sidebar notch on right side), and half wafer for sidebar lock (note sidebar notch on left side).

use both the full wafers and the split wafers. Figure 9-15 shows the different types of keys. Figure 9-16 illustrates the different types of lock wafers currently in use.

A/ Full-Wafer Locks

The major components of a full-wafer lock are the lock wafers, the lock core, and the outer lock cylinder housing. The lock wafers are found in separate chambers within the core and are spring-loaded. With the core seated in the lock cylinder housing and the mechanical

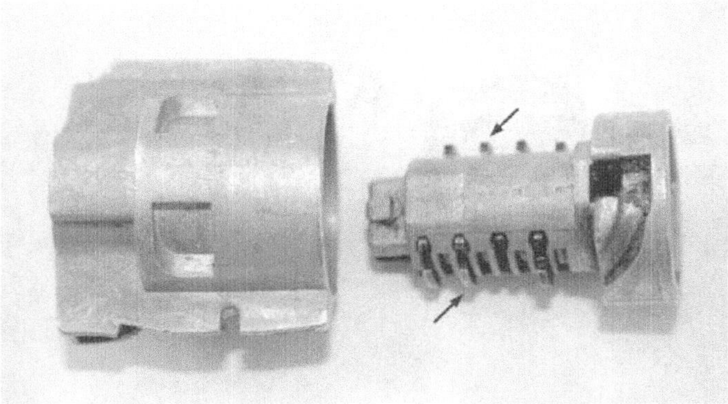

Figure 9-17

Note the lock wafers extending from the lock core (R) with the key removed. The extended wafers engage a corresponding channel in the lock cylinder housing and prevent rotation when the key is removed.

ignition key removed, the spring loaded lock wafers extend outward from the core and engage in a channel in the lock cylinder housing, which prevents core rotation. Figure 9-17 is an example of a full-wafer lock core and housing.

When the proper key is inserted into the keyway of the lock core, the wafers are withdrawn into the core and the core is allowed to rotate. The tail of the core is attached to a drive shaft or inserted into a slot in the internal column assembly. As the lock core rotates, the internal column assembly actuators and cams rotate simultaneously to unlock the column, unlock the gear selector, and energize the vehicle electrical systems.

B/ Split-Wafer Locks

Split-wafer locks are best described as full wafers cut in two halves. Rather than one solid wafer, these are two halves, with each side riding on half the key blade. This feature makes locks far more difficult to pick and can be best characterized as having two locks side by side. Full-lock wafers and split wafers are often combined in one lock core.

C/ Sidebar Locks

A sidebar lock uses the same principle as a full-wafer lock but is configured differently and generally uses half wafers. A sidebar lock wafer has a notch in the side of the wafer. With the key removed from the keyway, the notches in all the wafers in the lock do not properly align. This pushes the sidebar outward and the sidebar, rather than the wafers, engages in the channel of the lock cylinder housing. When the properly cut key is inserted into the lock core keyway, all the lock wafer notches align and the spring loaded sidebar edge is pushed into the notches. This action disengages the lock sidebar from the lock cylinder housing channel and allows core rotation. There are several manufacturers who use a combination of full-wafer locks with a sidebar to strengthen their components. Until 2004, GM used half-wafer sidebar locks exclusively on domestically built vehicles. In 2005, full-wafer sidebar locks were also being installed. Ford uses both full-wafer and half-wafer sidebar locks. Chrysler uses a full-wafer sidebar lock. Virtually all European and Asian vehicles use full-wafer or split-wafer locks.

DI High Security Side-Cut Ignition Lock and Key Sets

The side-cut high security ignition keys and locks have been a staple on European manufactured vehicles for decades, although they are becoming more widely used on Japanese models (see Chapter 10). They have yet to be seen on US-manufactured vehicles. The benefits of the side-cut keys (sidewinder and centerwinder) are that they cannot be impressioned, they are difficult to pick, and they must be cut on specialized equipment. Rather than having the blade cuts on the top and bottom of the blade, side-cut keys have the cuts along the sides (or flanks) of the key blade. When a standard key is inserted into the ignition lock keyway, the wafer riding surfaces directly contact the cuts on the top or bottom of the key blade and disengage the corresponding lock wafers. When a sidewinder or centerwinder key is inserted into the lock, protrusions on the inside of the full-wafer key opening ride the cuts (or channel) in the side of the key blade. The side-cut keys can also use the split-wafer configuration with half of a wafer riding each side of the cut key blade.

9.4 DEFEATING THE IGNITION LOCK

9.4.1 Principle

The solution to physically defeating a steering column is to access the protected internal components of the column. The most common method consists of compromising the ignition lock cylinder. This may be accomplished using several possible procedures, depending on the design of the ignition lock. Forced removal simply means the ignition lock is removed from the column housing using forced extraction techniques. Compromise can also be accomplished by forced removal of individual lock components, such as the lock core, lock wafers, or sidebar. Forced rotation indicates significant leverage is applied to the lock housing or core to facilitate rotation without the proper key. Lock picking is a professional locksmith procedure. It is an art as well as a science and takes training, skill, and, most of all, time. Lock impressioning, a technique where a key blade is actually hand-filled to create a cut key, is also time consuming and is considered a professional locksmith procedure.

Forced removal and forced rotation of the lock core are the most commonly observed methods used to defeat locks. Lock picking and lock impressioning are the least desirable methods (from a time perspective) used to compromise a lock, and indications of these procedures are rarely found on stolen-recovered vehicles.

9.4.2 Forced Removal of the Ignition Lock

The lock can be forced from the housing using a slide hammer (dent puller) to overcome the lock retainer and remove the lock cylinder. Figure 9-18 is an illustration of such a slide hammer. It is composed of a screw at its end, used to secure the slide hammer into the lock keyway, and a weight, which is rapidly pulled backward to provide the inertia necessary to

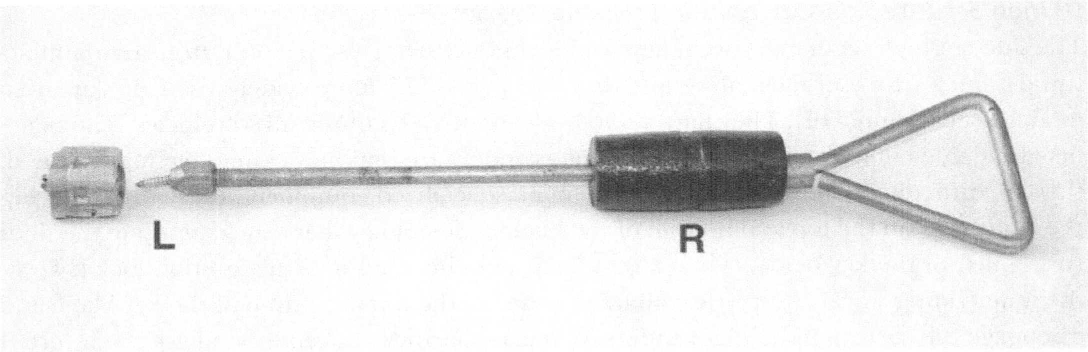

Figure 9-18

Slide hammer, also called dent puller, used to force a lock cylinder or lock core out of the assembly housing. Note sheet metal screw (L), which is twisted into the lock keyway and the weighted slide (R) to overcome resistance.

break the lock core or cylinder free from its housing. A large screwdriver, or similar tool, can also be used to pry the lock from the column housing by breaking and separating sections of the housing to free the lock cylinder. These methods are easily identified by the missing lock core or cylinder, damage to the column housing, or thread marks in the center wafers (see Figure 4-16). Once the ignition lock cylinder or core is removed from the column housing, a flathead screwdriver, or other appropriate tool can be inserted into the empty lock well to access the internal column linkage/actuators.

9.4.3 Forced Rotation

This technique implies that the ignition lock core is forcibly rotated. The locks most susceptible to this method are the ones whose core and face extend past the metal column housing assembly. When the core is sufficiently exposed, clamping-type pliers can be secured onto the unprotected outer lock. Leverage can then be applied to forcibly rotate the core within the cylinder housing. Locks that are flush with the column housing are generally immune to this type of tampering. Figure 9-19 is a good example of a lock that had been force rotated using a clamping tool.

Full- or half-wafer sidebar locks that use a metal alloy sidebar, rather than a steel sidebar, have an inherent weakness to forced rotation. When forced rotation occurs in an alloy sidebar lock, the damage can be identified by a partially sheared sidebar and damage to the ignition lock cylinder housing sidebar channel. There will also be marks, scars, and/or metal compression and distortion to the lock outer keyway or clamp marks on the outer lock or lock cap. Full-wafer locks collapse in the center from forced rotation, and this severe damage is easily identified. There should also be corresponding damage to the lock outer keyway from the use of a tool inserted in the keyway for leverage or clamp marks on the outer lock or lock cap (see Figure 4-15).

Figure 9-19

This is a GM non-PassLock half-wafer sidebar lock that has been force rotated using a vice grip or similar clamping tool. Note the clamp marks around the outer lock and the keyway in the on position.

9.4.4 Lock Picking

Lock picking is both an art and a science. It is also one of the least desirable methods to defeat the ignition lock as it entails training, skill, and time. Picking requires a tension tool to be inserted in the outer lock core keyway to maintain pressure on the spring-loaded lock wafers. Too much pressure and the lock wafers cannot be disengaged from the ignition lock housing channel. Insufficient pressure and the wafer springs reengage the picked wafers back into the housing channel. Then, a lock pick must be used to disengage the lock wafers either individually or by raking the wafers by rapidly moving the pick in and out along the wafer riding surfaces with the hope of simultaneously disengaging all the wafers to allow lock rotation, while not knowing the depths that the individual different sized wafers are engaged in the housing channel. This procedure is complicated by the amount of wafers the lock contains (typically 6 to 10), split-wafer locks, combination of split- and full-wafer locks, and sidebar locks. Figure 9-20 shows a standard lock picking set, and Figure 9-21 shows the lock picking procedure.

Picked wafers can be detected by microscopic examination of the lock wafer riding surfaces. Wafers are made from brass, and lock picks are constructed of steel. Examination of the lock wafers will reveal pick marks left by the lock pick manipulation along the wafer riding surfaces (lands). Such an example is shown in Figure 9-22. The wafer examined is shown on the left, and a magnification of the riding surface is shown on the right. Notice the toolmarks indented in the surface, which are the result of lock picking manipulation.

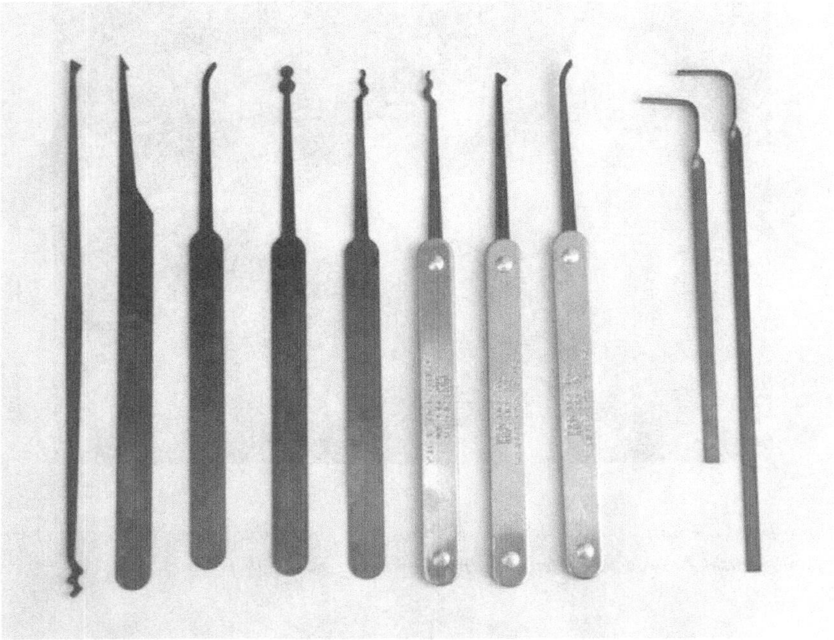

Figure 9-20
Standard lock picking set.

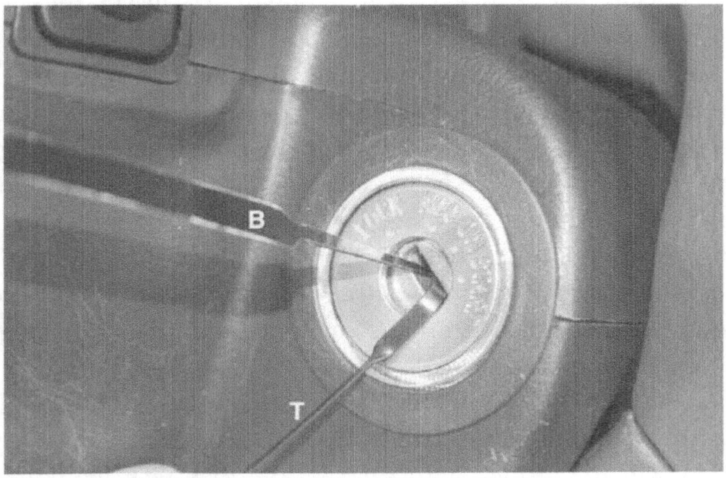

Figure 9-21
The tension bar (B) and lock pick (T) inserted in the lock keyway.

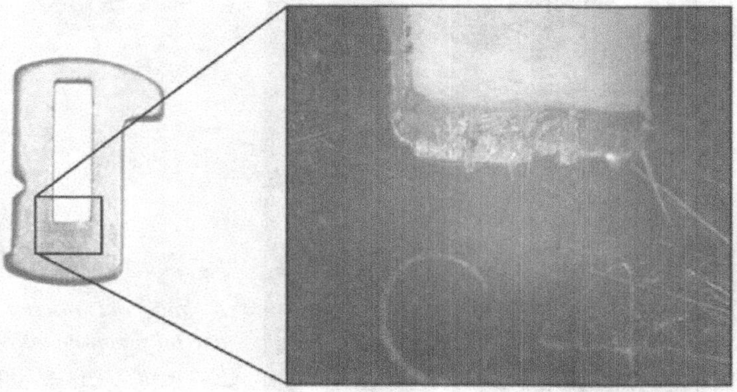

Figure 9-22
Photograph of a wafer (left) and the toolmarks (right) left on its riding surface due lock picking operation.

9.4.5 Key Impressioning

Impressioning a key is the art of creating a correctly cut mechanical key at the vehicle when no original key is available for duplication and key codes cannot be obtained. It is also one of the least desirable procedures for a thief to use. It is a specialized skill as well as time consuming. A correct uncut key blank is required that is clamped to a pair of locking pliers. The key blade is inserted into the lock core and forcefully manipulated using the locking pliers. Figure 9-23 shows a blank key inserted in the lock keyway with a clamping tool ready for manipulation.

This manipulation results in light imprints being left on the blank key blade from contact with the lock wafers. The marks are then filed, the key is reinserted into the lock, and the procedure continuously repeated until a properly cut hand-filed key is obtained. Figure 9-24 shows the wafer imprints on the key blade being filed to create a properly cut key. This repeated forced manipulation of the freshly filed key in the lock leaves numerous marks and scars on the wafer riding surfaces that are easily detected with magnified examination.

9.4.6 Key Picks

Key picks can also be described as "shim," "shaved," or "jiggle" keys. They have limited success on severely worn ignition locks, and newer locks are generally immune to their use. These keys have the cuts on the blade filed down or "shaved" to simulate wear. Figure 9-25 shows good examples of shaved keys. These keys are then forcefully manipulated, or jiggled, in a worn lock in an attempt to rotate the lock core. Marks and scars remaining on the lock

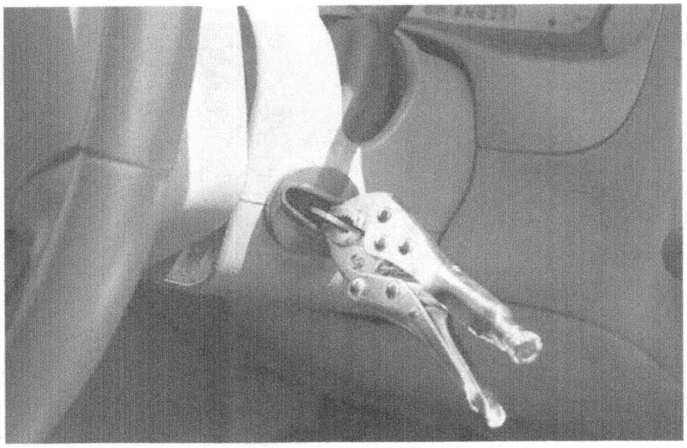

Figure 9-23

A blank key is inserted into the keyway and forcefully manipulated (turned) to obtain lock wafer imprints on the key blade for filing.

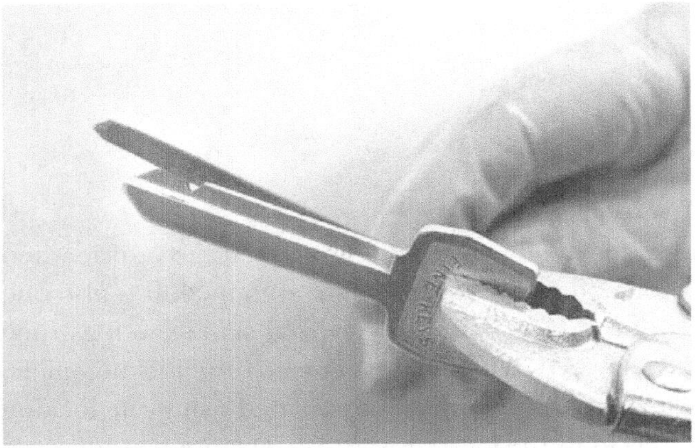

Figure 9-24

A file is used to create cuts in the blank key blade based on the wafer imprints.

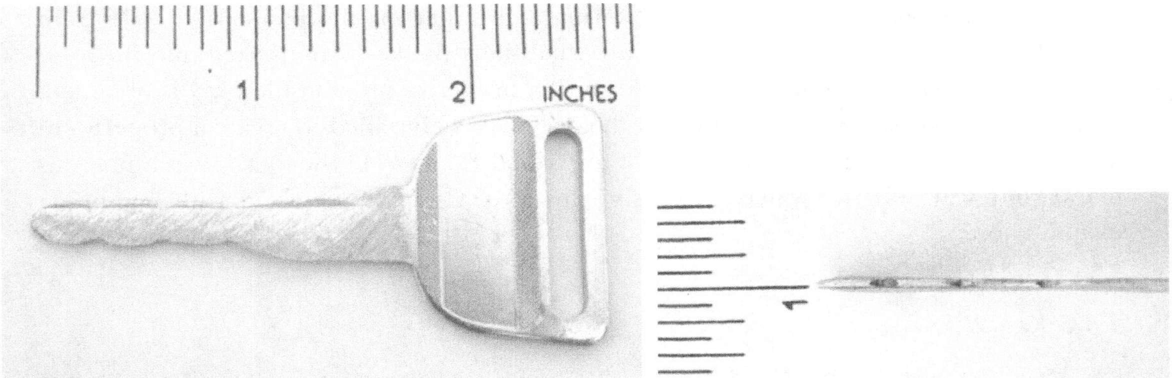

a b

Figure 9-25

(a) This is a shaved key used to manipulate a worn lock. Note the grind marks on the side to lessen the width of the key for easier manipulation. The key cuts have also been "shaved" to simulate wear.
(b) Note the diminished width of the key blade as a result of grinding or "shaving."

wafer riding surfaces from this forceful manipulation are readily detectable under magnification.

9.4.7 Master Keys

Master keys for automotive purposes are vehicle-specific keys. A master key will open the doors, trunk, and storage compartments and rotate the ignition lock, as opposed to a valet key, which will only open the doors and start the engine (see Subsection 10.5.3). Master keys that are fitted for an entire vehicle line or that properly operate all locks of a particular manufacturer do not exist. There are fleet operations (police departments, taxi companies, etc.) who request that their vehicle locks to be all keyed alike. These locks and keys are limited to specific vehicles requested by the operator. There are aftermarket keys available that are advertised as master keys but are best described as key picks and require manipulation with force to rotate even a well-worn lock. This procedure will result in damage, often severe, to the wafers.

One important fact should be remembered when investigating a reported vehicle theft: If a vehicle is to be started and operated without a properly cut mechanical key, damage will occur. Parts or components will be marked, scarred, damaged, destroyed, or missing. Identification and analysis of this evidence will assist in determining whether the damage is consistent with a theft, the method of theft, or if, in fact, an actual theft ever occurred.

9.5 EXAMINATION OF STEERING COLUMN COMPONENTS ON BURNED VEHICLES

9.5.1 Principle

Vehicles reported as stolen to commit insurance fraud or that are involved in the commission of a crime are often recovered burned, sometimes to completion (see Chapter 12). Examination, analysis, and conclusions as to the condition of any recovered physical and electronic security components prior to the fire become far more complex and specialized. When any vehicle is reported stolen and subsequently recovered burned, the resulting investigation must answer a series of questions to determine whether the theft is genuine or if the owner is involved.

Professional auto thieves, defined as those who derive their living from the theft of vehicles, do not generally burn their stolen vehicles, especially if the most valuable components have not been removed. If owner involvement is suspected, can it be proven? Does any evidence remain? The answer to both questions is yes.

On a severely burned vehicle, forced entry cannot be determined. There is simply not enough surviving evidence for examination. The glass has been destroyed, the rubber seals and weather-stripping no longer exists, paint has been consumed, the light metal or plastic outer door handle inserts have melted, and, on a convertible, the entire fabric top disappeared. Even if the door lock cylinders can be recovered from inside the bottom of the doors and examined, there are many other methods of entry remaining, whose evidence would be burned beyond recognition and forcing or tampering could not be detected.

The vehicle on-board electronic anti-theft components of PassKey, PassLock, and transponders are largely comprised of plastic housings/casings, plastic wiring connectors, and soldered connections. The plastic components are consumed by the fire and the wiring separated. The vehicle computer (power control module, electronic control module, etc.) will melt. This leaves virtually no electronic components for examination, and their prefire condition cannot be determined. The components cannot be function tested, and possible system tampering cannot be eliminated. The only physical evidence for forensic examination that can survive is the ignition lock and steering column components. Figures 9-26 and 9-27 demonstrate the damage to the electronic anti-theft components as a result of a fire.

What generally does survive in a burned vehicle are the physical protection components such as the vital ignition lock parts, steering column locking components (if so equipped),

Figure 9-26

Left: a column-mounted GM PassLock sensor metal cover plate. Right: the recovered remains of a PassLock sensor. Note the severe thermal damage to the sensor.

Figure 9-27

Left: The remains of a transponder transceiver. Right: Remains of the wire wrap from the transponder ring antenna. Both components sustained severe thermal damage.

gear selector locking hardware, metal ignition (starter) switch parts, other pieces of column, and ignition lock hardware. Using prescribed methods of recovery and examination of surviving physical evidence, conclusions can be made with a high degree of scientific certainty as to the condition of these components prior to the fire and indications of forcing or tampering activities identified.

When a reportedly stolen vehicle is recovered with severe and extensive fire damage to the interior, particularly at the steering column area, examination and processing of the driver's side floor debris area becomes necessary to recover any ignition lock, column lock, or starter (ignition) switch components that are no longer secured to the column mast as a result of the fire. Logic would dictate that if these components were secured to the column at the time of the fire, they would be located in the driver's side floor debris area below the column after the fire. Figure 9-28 shows the driver's side floor debris area as found in a 2002 Suzuki Aerio, recovered burned. Figure 9-29 shows debris recovery with embedded column and lock parts. The only exception to this rule is if the debris area has been significantly disturbed before the examination. The debris area could be disturbed during firefighting/suppression activities, tampered by a person not familiar with debris field processing techniques, or stirred up during vehicle recovery. Recovered steering column and ignition lock parts must be examined and analyzed by qualified technicians to ascertain whether or not the vehicle was operated with a properly cut mechanical key or whether there is sufficient damage to the parts or missing components that could indicate the ignition lock and/or column had been physically defeated. Figure 9-30 shows the recovered ignition lock, column lock, and starter switch components from the burned 2002 Suzuki Aerio.

Figure 9-28

Driver's side floor debris area of a 2002 Suzuki Aerio, which was a stolen vehicle recovered burned. Note the steering column remains (T) and the debris area directly below. See Color Plate.

Figure 9-29

The clump of metal being recovered is the melted ignition lock—column lock—starter switch assembly housing. The 2002 Suzuki Aerio is equipped with a standard single-component assembly. See Color Plate.

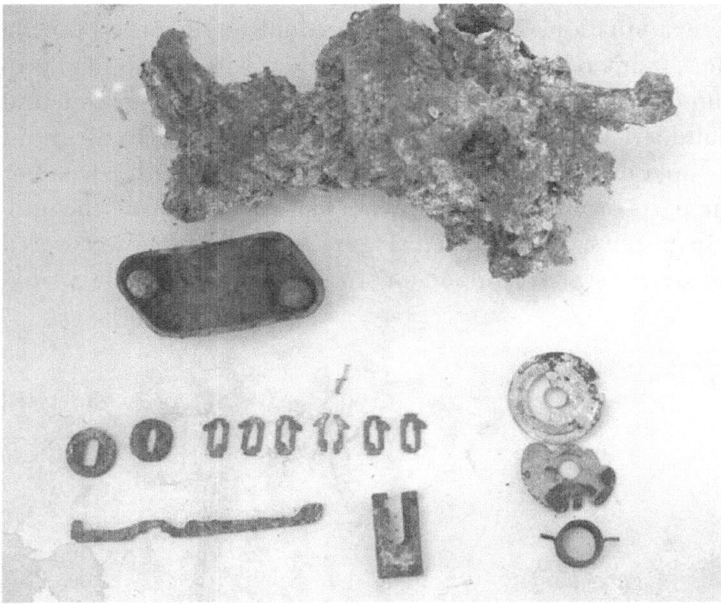

Figure 9-30

After processing the driver's side floor debris area on the 2002 Suzuki Aerio, all the relevant ignition lock, column lock, starter switch, and assembly mounting hardware were recovered. Examination of the recovered parts revealed no evidence or indication of forcing or tampering. These parts were recovered in the upper layers of the debris field, and most of the parts were found embedded in the melted housing.

9.5.2 Creation of the Debris Area

As fire consumes different interior materials, a distinct layering of debris occurs on the floor. Papers, light and heavy fabrics, light and heavy plastics, glass, and light and heavy metals, all settle on the debris area based on several parameters, including their location, amount, and melting or ignition temperatures. This is important because the exact location where the column and lock components are found in the debris layers indicates when they arrived there. The layering effectively creates a timeline based upon evidence. For instance, the ignition lock cylinder is encased in an alloy metal housing, which is secured to the steering column and covered in a plastic column shroud. If the ignition lock is properly secured in the column at the time of the fire, one could expect to recover the ignition lock and column component remains in the upper layers of the driver side floor debris area. If the lock was forced from the column during an actual theft and dropped on the floor, the lock would be recovered under the debris because it was discarded on the floor prior to the fire. Figure 9-31 illustrates a lock cylinder that was exposed to the fire and a recovered lock cylinder forced from the column prior to the fire.

It is important to carefully note the condition of the lock cylinder, lock remains, or loose internal lock components. This will serve to support the conclusion as to whether or not the lock was secured in the column or dash at the time of the fire. A relatively intact lock cylinder with minimal thermal damage or distortion recovered on the floor and under the debris would indicate the lock had been forced from the column prior to the fire. A lock cylinder that sustained severe thermal damage or loose internal components found on the upper layers of the debris area would indicate that the lock was properly secured to the column at the time of the fire and was directly exposed to the damaging effects of the fire.

Column and ignition lock components, when all are secured to the column at the time of the fire, generally maintain close proximity (within 6 to 8 inches) to the debris area. This is especially true with the European and Asian type single-component ignition lock—column lock—starter switch assemblies. There are, of course, exceptions to this rule. The most common is that drafting caused by the fire can result in individual components being

Figure 9-31

The ignition lock on the left displays severe thermal damage to the housing from direct exposure to the fire, which indicates that the lock was secured to the column at the time of the fire. Note the undamaged lock wafers embedded in the melted housing. The lock on the right was placed on the floor below the steering column of a vehicle before a test burn. After the vehicle burned to completion, the lock cylinder was recovered. The lock housing had some discoloration, however there was no thermal damage or distortion. The evidence clearly demonstrates that the lock cylinder had been removed from the column and placed on the driver floor before the fire. See Color Plate.

carried astray as they fall from the column. But, they still generally remain within 12 inches of the primary impact area in the debris area. Second, and more disturbing in terms of scattered components, falling parts may deflect off the metal knee bolster (tray) immediately below the column or other metal dash supports. This causes critical parts to be diverted to other areas of the driver's side floor debris area such as the front floor area near the pedal, under the driver seat, or even becoming trapped on the dash metal bolster/tray and failing to reach the floor debris area.

9.5.3 Recovery of Evidence

The success of the evidence recovery process in relationship to the forensic technician cannot be understated. The debris area must be carefully sifted and sorted layer by layer. The debris area should be photographed before and after processing. Individual components must be identified, recovered, and preserved with their locations noted. Parts embedded in melted metal housings must be retained for further processing and component recovery. The clumps of melted material may require remelting to recover any embedded parts.

A/ Remelting

The typical lock assembly housing is made of a pot metal (alloy) with a melting temperature of approximately 300 to 400°C [9]. The lock wafers are brass and will melt starting at about 900°C, depending on alloy content [9]. The steel parts will typically not melt below 1,400°C [9]. If processing is required, heating the melted housing remains from 430 to 440°C will allow the recovery of the embedded parts without damaging them. Any other method of embedded parts retrieval could possibly result in significant damage to the parts. Also, the investigator must be aware that an increasing number of manufacturers use a magnesium alloy in column housing construction. Although magnesium is very strong and lightweight, it is also a combustible metal and can burn well in excess of 1,100°C. These temperatures are sufficient to melt or distort the lock wafers and sometimes result in preventing a conclusive analysis of the parts. Extreme care must also be observed in remelting recovered debris due to the danger of igniting any remaining magnesium.

B/ X-Ray

It is also possible to perform an x-ray analysis of the melted debris. In some instances, x-ray allows a much clearer visualization of different materials present among a debris sample. X-rays are absorbed when going through material, and this absorption is dependent on the density of the material. Thus, it is possible distinguish different metallic parts among burned and charred debris. Figure 9-32 shows a melted single-component ignition lock assembly. It is possible to distinguish some parts such as the starter switch, the locking lug, and the lock caps. However, components such as the wafers cannot be seen. Figure 9-33 is the same assembly viewed under x-rays. This time, the wafers are clearly distinguished, as

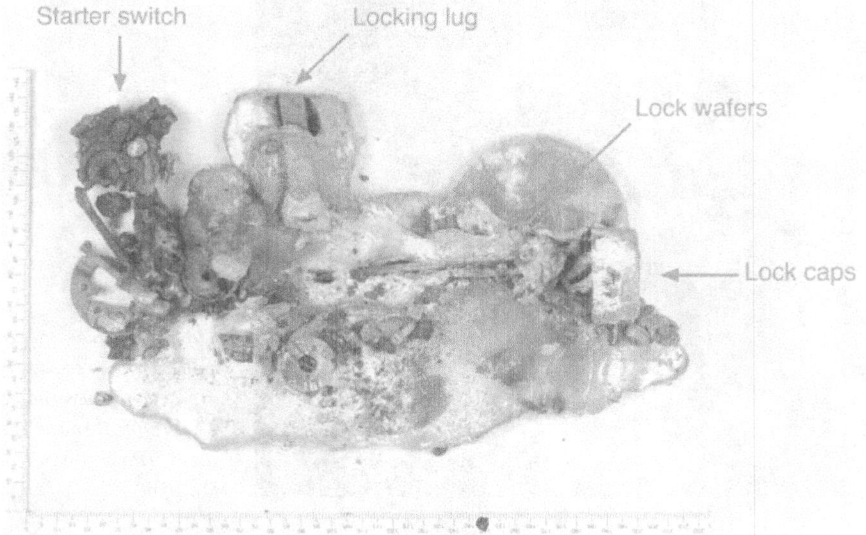

Figure 9-32

View before x-ray of the melted remains of a single component ignition lock—column lock—starter switch assembly as it would be recovered from a burned vehicle driver's floor area. It is possible to distinguish some parts such as the starter switch, the locking lug, and the lock caps. However, it is not possible to distinguish the lock wafers. (Photograph courtesy of Eric Stauffer.) See Color Plate.

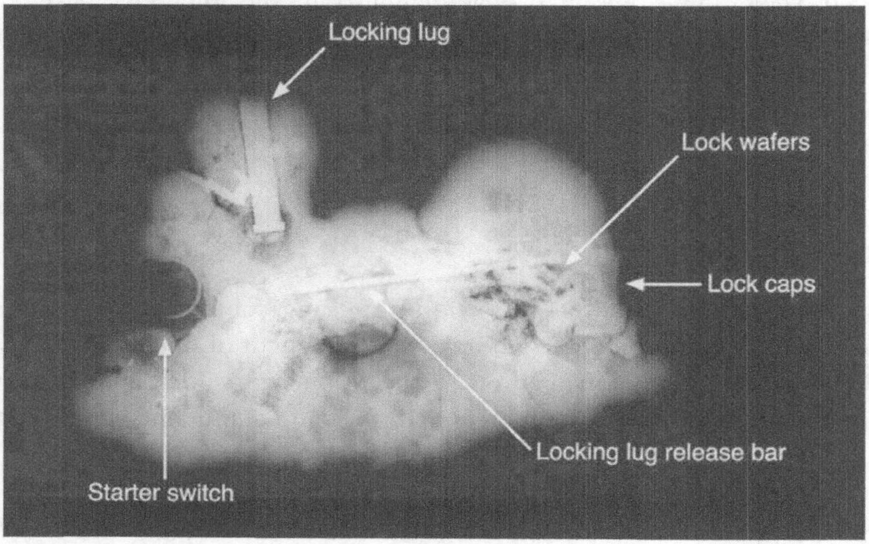

Figure 9-33

Same assembly viewed under x-rays. The x-ray clearly reveals the ignition lock components and wafers, the locking lug release bar, and the column locking lug embedded in the melted housing. The starter (ignition) switch is more difficult to see due to its lower density. (Photograph courtesy of Eric Stauffer.) See Color Plate.

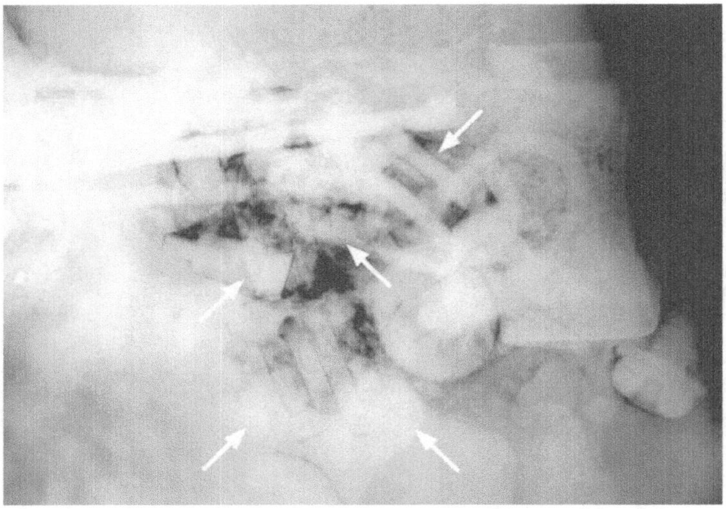

Figure 9-34

Close-up view of the area where the wafers are present. White arrows indicate the position of the wafer. Their typical shape is clearly recognized. (Photograph courtesy of Eric Stauffer.)

well as the locking lug and locking lug release bar. A close-up view of the lock wafer area is shown in Figure 9-34, where their typical shape can be easily recognized.

The use of x-ray technology is not part of the routine work because it is not readily available at the laboratory, and even less so in the field. However, it may be a very valuable tool, not only to select a portion of the debris sample, which may be remelted and more carefully examined, but also to directly identify components when enough contrast and resolution are obtained. The example presented above demonstrates the potential value of x-ray analysis. However, in many instances, it might not bring interesting results or it might not be feasible. In any case, an advantage of x-ray analysis is the fact that it is a nondestructive technique.

9.6 EXAMINATION OF EVIDENCE

9.6.1 Principle

There are two common locations to examine steering column and ignition lock evidence: on site in the field or at the laboratory. The benefit of on-site field examination is that the components remain secured to the unburned vehicle and if an opposing examiner inspects the vehicle, that person will find all components in the same condition as first observed by the initial examiner. Based on the circumstances, it might not be desirable to collect components from the unburned vehicle during the initial examination. Nevertheless, a combined vehicle examination and/or component removal, disassembly, and examination can be arranged for a later time, if required. If there is damage to the ignition lock or other components, it may be appropriate to recover the components for disassembly and further examination to perform a complete and thorough analysis. Thermally damaged (partially melted) components that remain on the column or in the dashboard of a partially fire damaged vehicle may also require recovery for laboratory processing.

The second method is the processing and examination of components or debris in a dedicated facility or laboratory. The benefit of laboratory processing or examination is that the facility is usually better equipped with examination material such as microscopes, tools, photographic equipment, and workstations. All examinations, regardless of the location, must be well documented and include detailed photographs.

9.6.2 On-Site Field Examination

Circumstances may make an on-site examination of the evidence the most desirable option in some cases. This procedure is only acceptable when examining unburned vehicles or burned vehicles with minimal fire damage with the column or dash components still secured and intact. A thorough examination can be accomplished by removal of the column shroud or dashboard trim pieces to access the required assemblies. A detailed examination is required to identify and document any indication of tampering, damage, or compromising. All parts should be present, operational, and described. Evidence of ignition lock forcing or forced rotation should be identified. The wafers must be examined for marks, scars, or distortions consistent with picking, forced manipulation, impressioning, key forced removal, or any other indication of tampering. Function testing of the components should be performed whenever possible. The ignition lock must be examined internally using a magnified microborescope with light capability and probe sufficient for locksmithing or forensic applications and conform to industry standards.

9.6.3 Laboratory Examination

Laboratory examination involves the disassembly of components and recovery of relevant parts from unburned components or the recovery of embedded parts from melted debris obtained from burned vehicles. Once the parts are removed and photographed they should be cleaned, preferably in a solvent and/or in an ultrasonic bath, and examined under magnification.

A/ Wafers

The wafers should be examined for evidence of forcing or forced rotation. The wafer riding surfaces should be examined for marks, scars, or distortions that would indicate picking, impressioning, imprecise key use, newly cut key insertion (tracks), forced removal of a key while the lock was in the on position (distortions or scars from outward forcing), or any other evidence or indication of tampering or manipulation. The owner's key, when possible, should also be examined for corresponding damage to the lock wafers when forced key removal is suspected.

As a general rule, at least half of the lock wafers should be recovered from a burned vehicle for forensic examination. It is possible in some instances to perform the examination with fewer wafers depending on condition and other components recovered. It is important to keep in mind that all the lock wafers must be damaged to defeat the lock. All must reveal damage, marks, or scars consistent with forcing, tampering, or manipulation. So, if half the wafers from a particular lock are recovered in a relatively good condition, a definitive conclusion can be made based on the examination of those wafers combined with the examination of other recovered components.

B/ Sidebar Locks

On sidebar locks, the sidebar notches on the wafers must also be examined for evidence of damage. Only GM uses a steel sidebar in their locks, which will survive a fire and can be examined, although GM has been using alloy sidebars in their later sidebar locks. Other manufacturer's lock sidebars are made of a metal alloy, which will melt and usually not survive for examination.

C/ Cap or Key Guide

The lock armored cap/key guide protects the internal lock parts and is located immediately behind the outer decorative or trim cap. These armored caps are made of steel and will survive the fire. To force or otherwise tamper with the internal lock components, the armored cap must be forcibly removed or damaged. Recovery and examination of the lock armored cap can provide essential information. Many of the outer decorative or trim caps are made of stamped steel. They will also survive a fire and should be recovered and examined for damage such as bending, metal distortion, or clamp marks. Additionally, lock actuators (such as chime), springs, lock cylinder retaining pins, roll pins, and other lock parts are also constructed of steel and should be recovered.

D/ Locking Pins/Bolts and Locking Lugs

The steering column locking pins/bolts and locking lugs (if so equipped), being made from steel, will survive. Their recovery is essential. The electrical starter (ignition) switch components consist of rotary contacts, starter switch wiring contacts, return springs, and heavy gauge wiring. These parts will most often survive the fire, can be recovered for examination, and help determine whether all components were secured to the column, column assembly housing, or dash assembly at the time of the event.

E/ Ignition Lock Component

The single-component ignition lock-column lock-starter switch assembly housings are secured to the column with a mounting strap and shearhead or security bolts. The steel mounting bolts will survive a fire. The mounting strap, which can be made from steel or an alloy material (consistent with the assembly housing), may or may not survive the fire, depending on the manufacturer's choice of materials.

F/ Nonlocking Columns

The nonlocking steering column, where the ignition lock and starter switch assembly is mounted directly to the dashboard, has been found to have different characteristics from the column-mounted assemblies during fires. The dash-mounted assemblies have a tendency to fall from the dash mount to the floor at an earlier point during the fire than the column-mounted assemblies. There are several reasons for this. The GM housings are plastic, which will obviously melt and fall from position at lower temperatures. Other manufacturers use light metal alloy housing. All of these assemblies are not as protected from the direct effects of the fire as column-mounted assemblies and are not as well secured. This results in the internal dashboard mounted components, such as the ignition lock cylinder, sustaining a lesser degree of thermal damage than the column-mounted components. Experience has shown that the dash-mounted components can most often be recovered just forward of the accelerator pedal, at the base of the center hump, in the lower to middle layers of the debris area.

ACKNOWLEDGMENTS

The author would like to thank Staci L. Rosenberger for her help in producing the photographs and diagrams and Sharon L. Mangine for her assistance in the research of information necessary to produce this chapter.

BIBLIOGRAPHY

[1] Perkins M. (2004) A brief (and interesting) history of automotive keys, *Keynotes,* January, pp 16–19.

[2] Levine J. (1994) *Service manual American car locks,* Volume 3, The Locksmith Publishing Corporation, Park Ridge, IL.

[3] Arthur R. (1999) *GM steering column repair: A guide for the locksmith,* The Locksmith Publishing Corporation, Park Ridge, IL.

[4] Trepanier J. (2005) Fitting keys to GM 10-cut CSS columns using an A-1 pick set, *Locksmith Ledger,* April, pp 92–96.

[5] Goodwin C. (2005) The 2006 Dodge Charger: Modern muscle, *Dodge: The Magazine,* **4**(1), pp 26–29.

[6] Rathjen JE. (1995) *Locksmithing—From apprentice to master,* TAB Books, New York, NY.

[7] Phillips B. (2001) *The complete book of locks and locksmithing,* McGraw-Hill, 5th edition, New York, NY.

[8] Phillips B. (2000) *Locksmithing,* McGraw-Hill, New York, NY.

[9] National Fire Protection Association (2004) NFA 921 guide for fire and explosion investigations, 2004 edition, Quincy, MA.

CHAPTER 10

EXAMINATION OF VEHICLE KEYS

Emmanuel Fivaz
Monica S. Bonfanti

10.1 INTRODUCTION

The study of keys, and more particularly their production, functioning, duplication, and security, is of great interest from a forensic perspective. In some instances, it is important to determine how a key and its corresponding lock work and whether the key matches the lock. In the scope of auto theft investigation, another question often requires an answer: whether a key has been duplicated or not.

When a complaint for a stolen vehicle is filed with the insurance company, all available keys are usually requested from the owner of the vehicle (see Chapters 2 and 19). This is done as a security measure from the insurance company. Unless the vehicle was stolen by carjacking or homejacking, the owner should still have all the keys in his or her possession. It is possible that the theft is not legitimate and that the owner attempted to fraud the insurance company. Often, the owner sells (or gives away) the vehicle with a copy of the original keys, keeping the original keys to surrender to the insurance company. In such instances, the insurance can request an examination of the keys in possession of the vehicle's owner to determine whether these are original keys, if they are all present, if they correspond to the vehicle reported stolen, and/or if copies were made. To answer the latter, the expert observes the keys and searches for possible traces left by the duplicating machine during the reading of the original key. These traces are not always present because they depend on the duplication process used and the history of the original key.

In the scope of this chapter, the different types of car keys have been classified in a definite number of categories, for which the principle of reproduction is explained. Then, the different traces left by the duplicating machine on the original key are presented along with their examination.

10.2 CAR KEY TYPES

10.2.1 Evolution and Generalities

The first cars did not include a key as a security feature. As passenger cars became more popular, criminals became more interested in their value and convenience. Thus, vehicles were equipped with locks on doors and ignition. Until approximately 15 years ago, most car keys

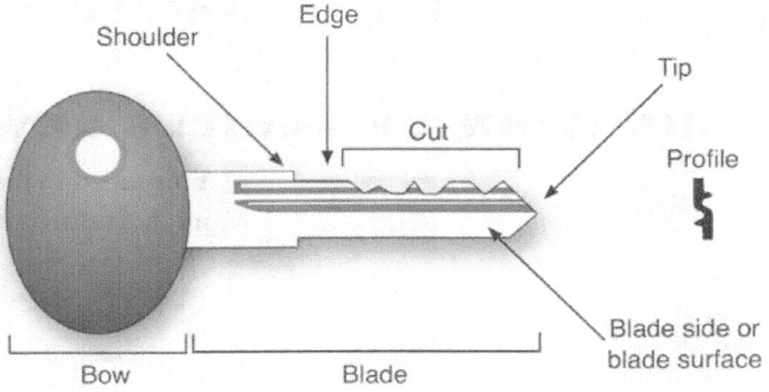

Figure 10-1

Nomenclature of a key. (Diagram courtesy of Eric Stauffer.)

were of the pin/wafer tumbler type. In the late 1980s, other types of key appeared on the market such as sidewinder keys, Ford and Jaguar Tibbe keys, Ford Chubb keys, and keys equipped with an electronic anti-theft system. More recently, keyless remote systems have appeared on the market. Figure 10-1 shows a general sketch of a key with its typical elements.

10.2.2 Pin/Wafer Tumbler Keys

Pin/wafer tumbler keys, also called cylinder keys, are used widely around the world.[1] The blade of such keys is profiled. This profile consists of lands and grooves milled on the blade. For the key to be inserted in a cylinder, the key's profile must be compatible with the profile of the cylinder keyway. This profile is the same on all blank keys of a same series. In addition, the top or the top and the bottom of the blade are cut. When the number, location, and dimension of these cuts are matched on the cylinder lock, this latter can be rotated and the lock can be opened [1–3]. Pin/wafer tumbler keys are usually classified by the number of cuts. For example, Chrysler equips its vehicles with 5-, 7-, or 8-cut keys and Ford with 8- or 10-cut keys [4]. Figure 9-15 presents more examples of different pin/wafer tumbler keys.

Car keys are often bigger than residential keys. In addition, the logo of the vehicle's manufacturer is engraved in the metal or in the plastic head. Some pin/wafer tumbler keys are shown in Figure 10-2.

Before the 1980s, car keys were only cut on one edge of the blade. Thus, the key could only be inserted in one fashion in the keyway. Most vehicles of later generation were equipped with reversible or symmetrical keys (cut on both edges). Pin/wafer tumbler keys are now being slowly replaced by other types of keys and keyless systems. However, they are still widely encountered on older vehicles.

---

[1] Pin and wafer tumbler keys can be identical because the difference lies in the lock (see Chapter 9). A pin tumbler key is used in a pin tumbler lock while a wafer tumbler key is used in a wafer tumbler lock. Nowadays, vehicles are no longer equipped with pin tumbler locks.

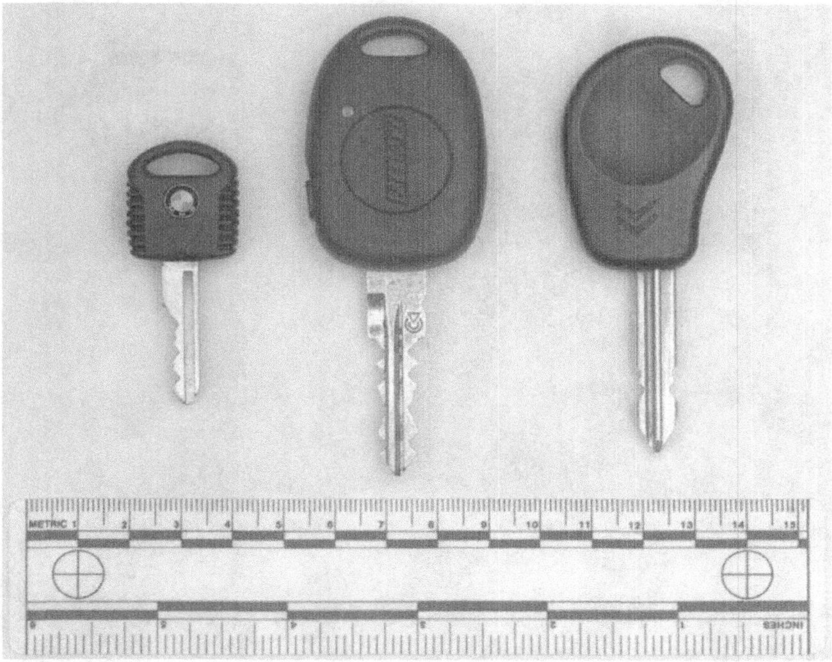

Figure 10-2

Pin/wafer tumbler keys. From left to right: A BMW motorcycle asymmetrical key, a Fiat symmetrical key, and a Citroën symmetrical key.

10.2.3 Sidewinder and Centerwinder Keys

Sidewinder keys (also called side-cut, side-milled, or laser-cut keys) are becoming more widespread in modern vehicles.[2] Sidewinder keys are reversible because they are cut on both sides of the blade. Instead of being cut on the edge as for pin/wafer tumbler keys, a channel is milled on the side of the blade. There can be one cut on each side only (2-track type), two cuts on each side (4-track type), or one cut located in the middle of each side (internal cut type, also called center-cut type or, more simply, centerwinder) [5]. Figure 10-3 presents the three types of sidewinder keys. The outer edge of the blade is never cut and remains flat. With 4-track type keys, the two cuts present on one side are not symmetrical.

Some Mercedes, BMW, and Honda are equipped with 4-track sidewinder keys. The 2-track sidewinder key is the most common of all three types and the centerwinder is the least common [5].

---

[2] Sidewinder and centerwinder keys are used in wafer tumbler locks and therefore, are a subcategory of wafer tumbler locks. Due to their particular design, they are treated separately from the regular wafer tumbler key described in the previous paragraph.

Figure 10-3

Sidewinder and centerwinder keys. From left to right: An Opel 2-track sidewinder key, a Volkswagen centerwinder key, and a Honda 4-track sidewinder key.

10.2.4 Ford/Jaguar Tibbe Keys and Ford Chubb Keys

Ford cars intended for the European market come with keys of very different shapes than pin tumbler keys and sidewinder or centerwinder keys. These keys are called Tibbe and Chubb keys (Figure 10-4) [6]. Tibbe keys and Chubb keys are used on Ford vehicles, and Tibbe keys are also used on Jaguar vehicles. Chubb keys were discontinued a few years ago.

Tibbe keys consist of a solid metal cylinder in which the tip bears the profile of the key. The cuts do not consist of hills and valleys, as for the pin tumbler key, but rather of a combination of different inclined surfaces [7]. Chubb keys consist of a similar solid metal cylinder but differ in the shape of the tip. The tip has two opposed small blades onto which the cuts are made.

10.2.5 Dimple Keys

A dimple key is a key whose cuts are drilled or milled into its blade surfaces, as shown in Figure 10-5 [8]. These are symmetrical keys because the cuts are identical on both sides.

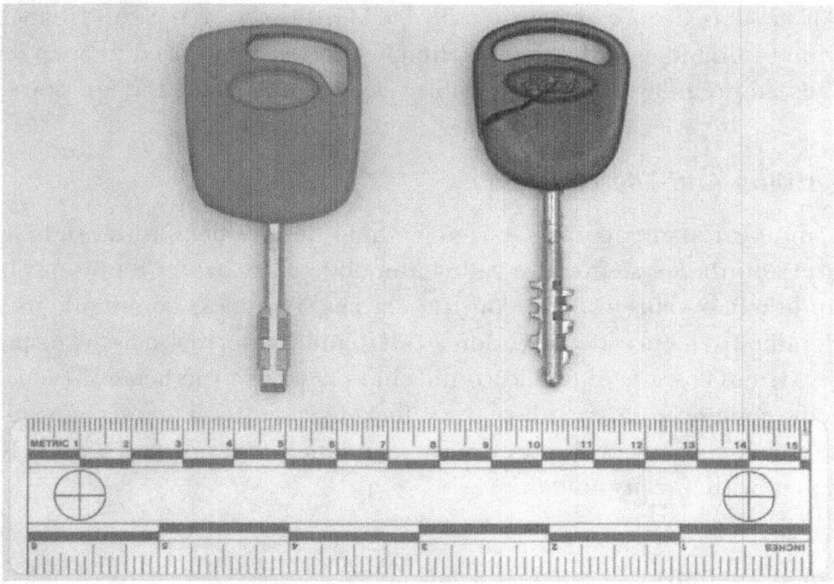

Figure 10-4

Left: Ford Tibbe key. Right: Ford Chubb key.

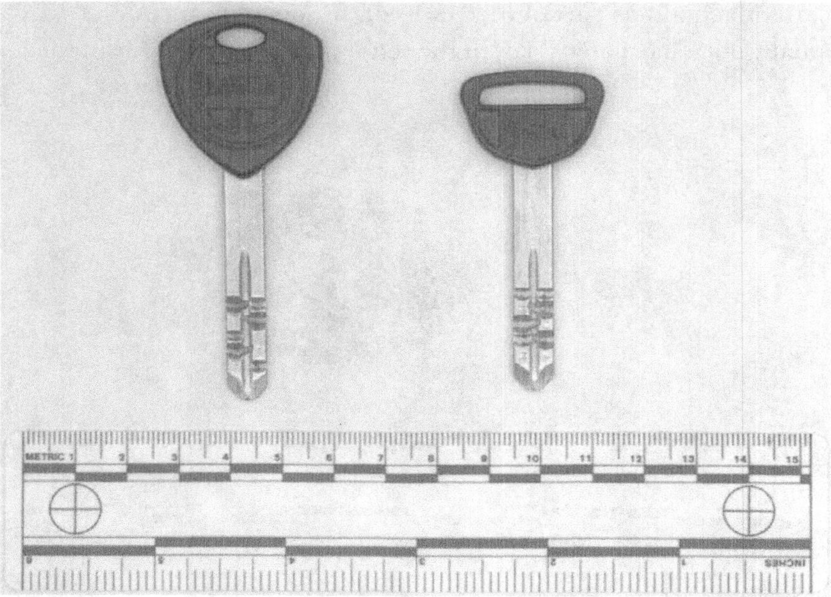

Figure 10-5

Two original Lancia dimple keys. (Keys from the personal collection of Eric Stauffer.)

This type of key has been extensively used by Alfa Romeo, BMW, and Lancia until the 1980s. It is interesting to note that Lancia dimple keys were only used to open doors, trunk, and glove box and regular pin/wafer tumbler keys were used for the ignition lock.

10.2.6 Keys With Anti-Theft System

Since 1995, most car manufacturers started to equip their vehicles with original equipment manufacturer anti-theft systems, also called immobilizers. Chapter 8 presents in detail the different anti-theft systems available on the market. The most commonly used system is based upon radio frequency identification (RFID) and is referred to as "transponder." The transponder system consists of an electronic chip present in the head of the key that contains a specific code necessary to allow the vehicle to start [9]. The mechanical part of the key (blade) consists of any of the previously mentioned types of key. Figure 10-6 shows a blank equipped with a transponder.

10.2.7 Keyless Systems

During the last few years, keyless systems have become more common. These systems allow the user to open doors, trunk, and even start the vehicle remotely up to a certain distance away from the car. Furthermore, most of these systems also include a panic button, which trips the vehicle's alarm. The latest technical advance is a system that automatically unlocks the vehicle upon approach of the person carrying the key. Keyless systems are usually found as a small card or actually as part of the key itself. In some instances, the "key" part of the system is actually not a mechanical key in the sense that although it fits in the ignition locks

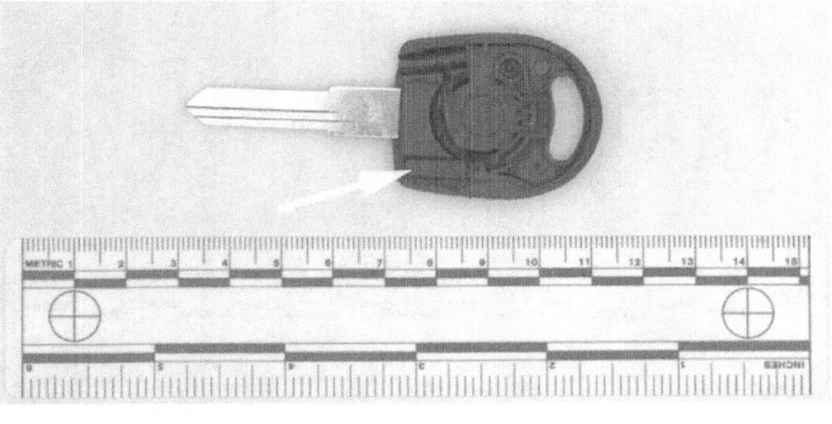

Figure 10-6

An Audi blank key equipped with a transponder. The arrow shows where the transponder is located inside the key head, whose cover has been removed.

and rotates it, the mechanical part is no longer a security. In such instances, the key is often reffered to as a key fob. An example of a keyless system is shown in Figure 10-7a. This particular system is found on a 2005 Volvo S40 and still requires the "blade" of the key to be inserted in the cylinder as for a regular cylinder lock. However, there is no mechanical security with the blade of the key. Figure 10-7b presents another keyless system used on a 2002 Renault Laguna. This system consists of a card that can be carried in the driver's pocket. Upon approach to the vehicle, the doors automatically unlock without pressing any buttons. The vehicle is started by pressing a "start/stop" button located on the dashboard

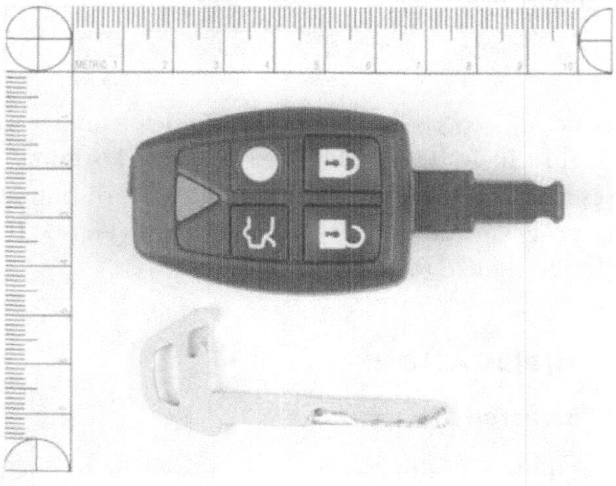

a

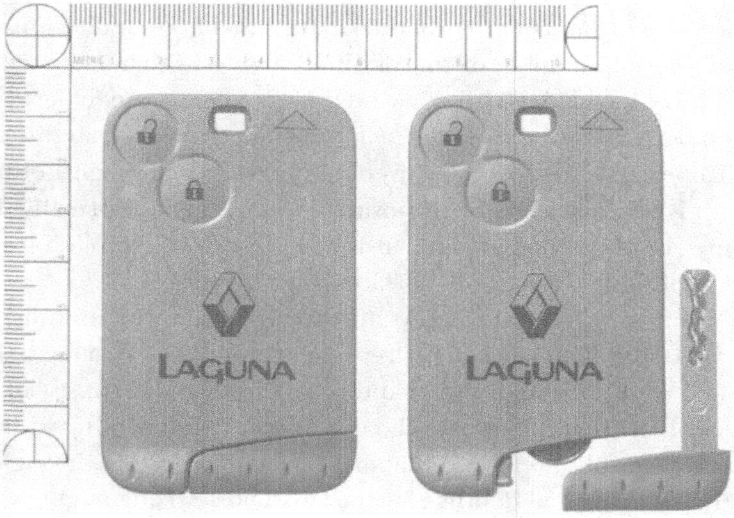

b

Figure 10-7

(a) Example of a 2005 Volvo S40 keyless system. The main key, shown on top, is a remote control operating doors, trunk, and alarm (panic button). The "mechanical" part of the key merely rotates the ignition system, without providing any security. The bottom key, a 2-track sidewinder key, is normally located inside the main key and is used to open doors if the remote no longer works. Photograph courtesy of Eric Stauffer.

(b) Example of a 2002 Renault Laguna keyless system. The card is carried in the driver's pocket and upon approach to the vehicle, the doors automatically unlock. The vehicle is started by pressing a button located on the dashboard. A spare key is located inside the card, as shown on the right part of the photograph, and permits to unlock the doors, but not to start the vehicle.

C

Figure 10-7 Continued

(c) Example of a 2003 Renault Scenic dashboard. The button "start/stop" is used to turn the vehicle's contact on and to start or stop the engine. The vehicle can be programmed so the car key must be placed in the slot located below the button to enable the vehicle to start. (Photograph courtesy of Eric Stauffer.)

without the necessity to place the card in physical contact with the vehicle. Such a button is shown in Figure 10-7c. The dash-board illustrated originates from a 2003 Renault Scenic. Note the presence of a slot just below the "start/stop" button. On this particular model, it is possible to program the vehicle so the key card (Figure 10-7b) has to be inserted in the slot to enable the vehicle to start.

10.3 KEY DUPLICATION

10.3.1 Key Characteristics

A blank is cut, milled, or dimpled to make it specific to one particular lock. The combination of cuts (number, type) and their characteristics (size, position) are often referred to as "spacing and depths" of the key. In case of pin/wafer tumbler and Chubb keys, it consists of a certain number of cuts of different depths that make the serrated edge on one or both sides of the blade. With sidewinder keys or centerwinder keys, it consists of milled portion(s) on both sides of the blade. Each milled portion also has different spacing and depths. With Tibbe keys, it consists of a given number (normally eight) of angled cuts along a "rib." Each cut is characterized by its position and its angle (normally three possible angles). Finally, in case of dimple keys, it consists of a series of dimples of different depths located at different positions on both surfaces of the blade.

Since 1970, most car keys are characterized with a numerical code, called the lock/key code [10]. The key code is representative of the spacing and depths or cut combination. The code can be direct or indirect. Direct codes consist of a series of numbers (one for each cut) that quantify each cut's depth. When dealing with indirect codes, the locksmith must consult a reference database, usually available through the manufacturer, to obtain the "spacing and depths" for the key. The key code might be present in different locations, such as [10]:

- In the original owner's manual;
- On the original bill of sale;
- On a tag that came with the original keys;
- On a vehicle identification card;
- On a plate located inside the vehicle;
- On the outside rim of the lock keyway;
- Stamped or engraved on the original keys;
- Stamped on lock housing or body;
- On a decal attached to or near the lock;
- Stamped on lock cylinder.

The access to some of these locations may require disassembly. In addition, for some makes and models, the key code is available from the manufacturer's database upon production of the vehicle identification number (VIN).

10.3.2 General Duplication Process

The copy of a key is a process performed in four main steps, described below [11].

A/ Step 1: Identifying the Blank

First, it is necessary to choose the blank corresponding to the original key (also called the pattern key). This is performed by observing the general type, shape, and profile of the key and referring to existing key blank directories and databases. In some instances, the key blank type may be engraved in the key, which can help in identifying the right blank. Often, this character sequence is raised lettering [10]. The key blank type should not be confused with the key code, which is representative of the spacing and depths. On some occasions, the key blank type precedes the key code. Also, the key blank type and the key code are occasionally found in the locations described in Subsection 10.3.1 [6]. There are many different types of blanks, and it is not unusual for a locksmith to have several hundred blanks in his or her shop.

B/ Step 2: Reading the Original Key

The key is read to determine its spacing and depths. This step can be performed using four different methods. First, it can be read using the key-cutting machine as described in Subsection 10.3.3 hereafter. Several types of instruments can be used for this reading step, such as a mechanical machine or electronic machine with mechanical or optical detection. Second, it can be read using a key-decoding gauge, which is a very simple instrument with several marked cut depths that are matched to the different cuts of the key. Also, it can be read with the naked eye, which could be a straightforward process in some instances such as with a Kaba 8 key, and an impossible process in some other cases. It depends on the key type and the locksmith's experience. It has been shown that Ford/Jaguar Tibbe keys are very easy to read [12]. Finally, the spacing and depths can be obtained from the key code as explained previously.

C/ Step 3: Making Corrections

Keys undergo mechanical frictions due to their regular use, which lead to a certain wear of the key in general and of its cut characteristics in particular. Occasionally, this wear diminishes the precision of the key to operate. Hence, if the copy is created by a machine that uses the pattern key as a direct template, the wear will be reproduced on the copy. This situation is not desirable, and thus it is important to correct these small defects on the copy whenever possible.

This step is not always feasible because the correction can only be performed by knowing the original spacing and depths of the key. This can be achieved by obtaining the key code using one of the methods previously presented. Once the key code is known, a machine capable of cutting keys by code input is used or the cut is adjusted with shims on the original key or by using a combination of space key and depth cam. These corrections are usually in the order of micrometers.

D/ Step 4: Cutting the Blank

Once the corrections are made, if needed, the new key can be cut. This step is identical on all key-cutting machines independently on how the original key was read. It consists of cutting, milling, or dimpling the proper characteristics on the key blank. To achieve this step, different types of cutters can be used depending on the key type.

10.3.3 Key-Cutting Machines or Duplicators

There are two main types of key-cutting machines: mechanical and electronic. For both types, some duplicators could be operated by code input (i.e., when configured with the key code), and the machine will directly cut a key to the right specifications. Each type is described in more detail in the following paragraphs.

A/ Mechanical Duplicator: Manual Lever Operated (Pantograph)

This machine is among the oldest used to duplicate keys. Even though more modern electronic machines have been developed, the manual lever operated machines are still used on a daily basis. The original manually operated machine has also undergone improvements over the years, with models that are now semiautomatic or completely automatic, rendering the duplication process faster and more reliable.

This machine reads the pattern key simultaneously as it cuts the blank key. The machine used will be slightly different depending on the types of key to be copied; however, the principle of the pantograph remains the same [7, 13–15]. An example of such a machine is shown in Figure 10-8.

Figure 10-8

Mechanical duplicator (manual lever operated) Börkey Rexa used in a Swiss locksmithing shop to duplicate pin/wafer tumbler keys.

The pattern key and the blank key are clamped with vise jaws in two separate vices located on a yoke, as shown in Figure 10-9. Using the lever, the yoke is moved and the pattern key is brought in contact with the depth guide, as shown in Figure 10-10. The cutter is located at the same level as the depth guide. The guide then follows the pattern key, while the cutter cuts the blank key with the same pattern.

This type of machine can cut pin/wafer tumbler, sidewinder, centerwinder, and dimple keys. When dealing with symmetrical keys, the original key is almost never turned around to cut the second edge/side of the blank key. The user usually follows the same pattern on the original key to cut both edges/surfaces of the blank key. This means that one edge/side of the original key will not contact the depth guide and will be left free of any marks.

An inconvenience of this machine is that it is very difficult to make any corrections on the new key (see Subsection 10.3.2C). However, it is possible to place small shims under the original key to slightly modify its position in the vise jaws and proceed to corrections. This technique will only work with a uniform wear of the key. Also, if the key code of the original key is known, it is possible to use space keys and depth cams corresponding to the key code as the pattern key from which the blank is cut [16].

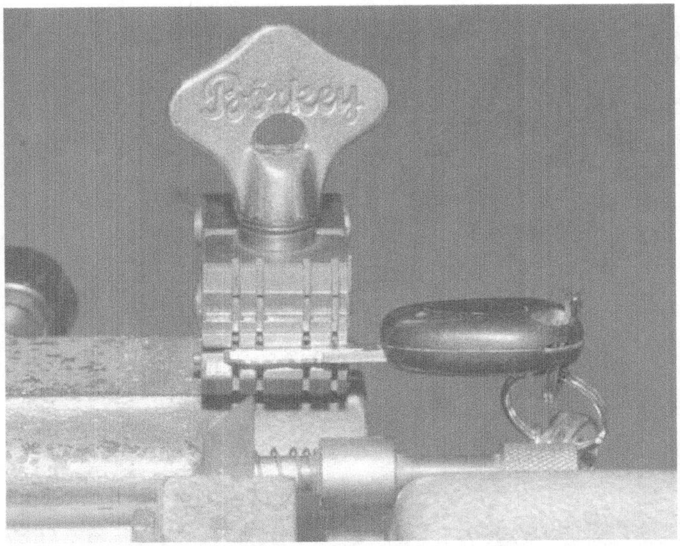

Figure 10-9

Detailed view of the vise jaws used to clamp the key in place on the Börkey Rexa mechanical duplicator.

Figure 10-10

Demonstration of when the depth guide comes in contact with the pattern key during the reading process on a Börkey Rexa mechanical duplicator with an original Renault key.

B/ Mechanical Code Duplicator

This machine operates in the same manner as the previous machine regarding the cutting process; however, the pattern of the original key does not need to be read by a depth guide. The key code is entered in the machine, which cuts a key accordingly [7, 13].

With this type of machine, the original key is not needed, only the key code is needed. Furthermore, step 3 (corrections) is always carried out with this duplicator, because the wear is not reproduced since the original pattern is not used as a direct template. Tibbe and Chubb keys cannot be read with a mechanical duplicator and thus must be produced by a code machine, equipped with particular vise jaws and cutting wheels. Mechanical code duplicators are becoming less popular because of the new electronic code duplicators.

C/ Electronic Duplicator with Optical Detection

This duplicator uses a light beam (laser) to read the spacing and depths of the pattern key. Thus, it differs from the manual lever operated and mechanical code machines in step 2 of the duplicating process. The remaining steps are identical.

The key is simply placed in vice jaws located between the light source (a laser) and the detector. The detector is connected to an electronic controller, which interprets the information and returns the spacing and depths of the key. Then, the machine cuts the blank key [17]. Figure 10-11 shows such a machine.

D/ Electronic Code Duplicator

This machine is the electronic version of the mechanical code duplicator. This type of machine is often used with vehicle and commercial keys. Connected to a computer, a large database of reference key codes is directly available to perform proper cuts [7, 13, 18]. Databases of different spacing and depths of car keys are updated every year. Some manufacturers transmit the data as soon as new models are available. Others, particularly luxury brands such as BMW, Mercedes, and Porsche, wait a few years. This duplicator also automatically integrates the correction step. Figure 10-12 shows such a machine.

E/ Combined Duplicator

There are duplicators that combine electronic code, optical detection, and mechanical guide. With such a machine, it is possible to read the key with the laser and then perform

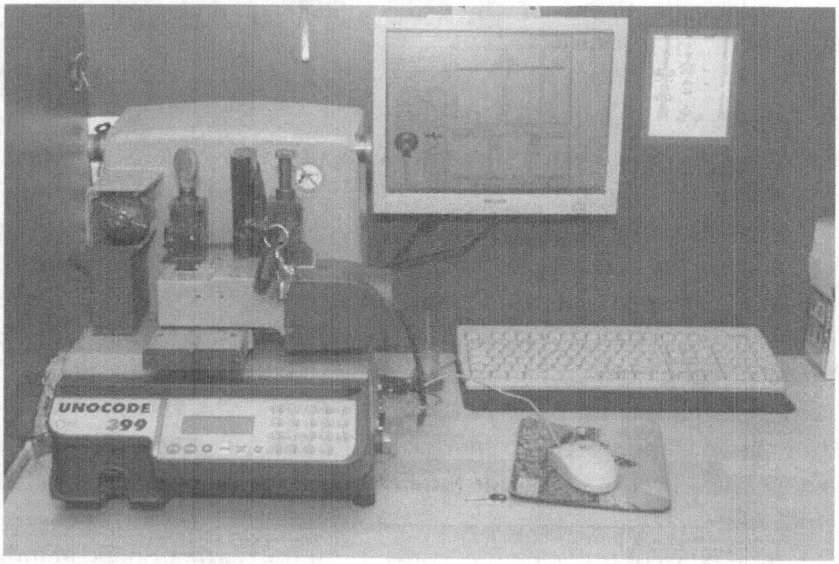

Figure 10-11

A Silca Unocode electronic duplicator with optical detection used in a Swiss locksmithing shop to duplicate pin/wafer tumbler keys.

Figure 10-12

A Silca Quattrocode electronic duplicator with pattern key reading performed either by mechanical reading (depth guide) or code entry. This machine is used in a Swiss locksmithing shop to duplicate sidewinder and center-winder keys. It is interesting to note here that the pattern key is placed in the same vise jaws as the blank key: The reading of the key is performed first and memorized and then the pattern key is replaced by the blank key to proceed to the cut.

corrections by referring to the database. In other instances, such a machine can help in retrieving the spacing and depths of a broken key. Such equipment is highly technical and offers many applications [18, 19].

F/ Electronic Duplicator for Anti-Theft Systems

With the recent advances in anti-theft systems integrated in car keys, new machines were developed to respond to the latest technology in the matter of key reproduction. This machine allows for the duplication of the transponder [20]; it reads the transponder or the original key and formats a blank key to include the proper electronic information. An example of such a machine is shown in Figure 10-13.

The original key is inserted in the machine and the make, model, and year of the vehicle are provided to the duplicator. The duplicator then reads the code emitted by the transponder, memorizes it, and transmits it to the blank key. The transponder code can be memorized by the machine to create more copies without the need for the original key. Then, the mechanical part of the key is duplicated using one of the previously described key-cutting machines.

There are three types of transponders available on the market: single identification (or fixed) code, challenging response code (encrypted), and rolling code. Only the fixed code transponders can be copied by machines available on the market [21]. Some keys from luxury car makers such as BMW or Mercedes are not allowed to be copied and must be

Figure 10-13

A Silca RW2 electronic duplicator for anti-theft system used in a Swiss locksmithing shop to duplicate keys equipped with transponder.

ordered from the manufacturer. For instance, in Switzerland, copies made with a duplicator for anti-theft systems do not need to be recorded. However, in Germany, the copy of transponder-equipped keys is much more strict and controlled. A person willing to make such a copy must present an identity card, and all copies are kept on record.

Finally, there exists an apparatus that can extract data from transponder keys when connected to the vehicle. When such an apparatus is placed in the vehicle, it can format a blank transponder key, such as the encrypted type [20]. Some diagnostic instruments used by mechanic shops or car dealers can also format transponder keys (see Chapter 8). One advantage over this technology is the possibility of rendering inactive an original key stolen from the owner.

Electronic duplicators for anti-theft systems are also attractive to professional thieves. Many Eastern European organized crime groups target automobile dealers to steal such equipment. Once the equipment is in their hands, they note VINs of vehicles to be stolen, generate a key, and steal the vehicle with minimum damage.

10.4 COPY TRACES ON THE ORIGINAL KEY

10.4.1 Principle

The duplication process of a key can leave two types of trace on the original key: vise jaw marks and depth guide marks [22]. Each type of mark is explained in detail in the following paragraphs. It is important to keep in mind that if the original key is not placed in physical contact with the duplicator, no traces will be present on it. This would be the case if the key code is extracted from the original key and a code machine is used. Nevertheless, if a machine equipped with optical detection is used, no depth guide marks will be present.

The examination of original keys for evidence of duplication process can only be carried out a specific number of key types and can only be successful with an even smaller number.

However, it is important to bear in mind that this examination has proven itself successful in the past, and even if it will become less and less possible in the future, there will still be a small number of cases where it will be useful.

10.4.2 Vise Jaw Marks

The vise jaws are used to squeeze the original key in place in the yoke so it does not move during the reading process. Although designed not to damage the original key, these jaws can leave traces on the pattern key. The high pressure used in squeezing the key in place may result in small indentations impressed on the key. These traces do not indicate that the key was read or copied but merely that the key was held in the vice. However, studies have failed to demonstrate the presence of this type of trace during the examination of copied keys [22]. As for guides, there are several different types of vise jaw that depend on the type of key duplicated. Original factory keys do not present such traces.

10.4.3 Depth Guide Marks

The mechanical machine with guide is the only duplicator for which the reading of the spacing and depths implies a physical contact between the pattern key and the depth guide. As a matter of fact, the guide contacts the edge or the pattern of the original key to follow its line, so the cutter can reproduce the same shape on the blank key (Figure 10-10). The guide is pressed with more or less force on the original key and then dragged from end to end. Thus, this contact creates toolmarks, usually in the form of striations parallel to the blade of the key as seen in Figures 10-14 and 10-15.

If the guide is only used to measure the depth of each cut by jumping from cut to cut without being dragged along the cut line, impressed toolmarks can result from the contact

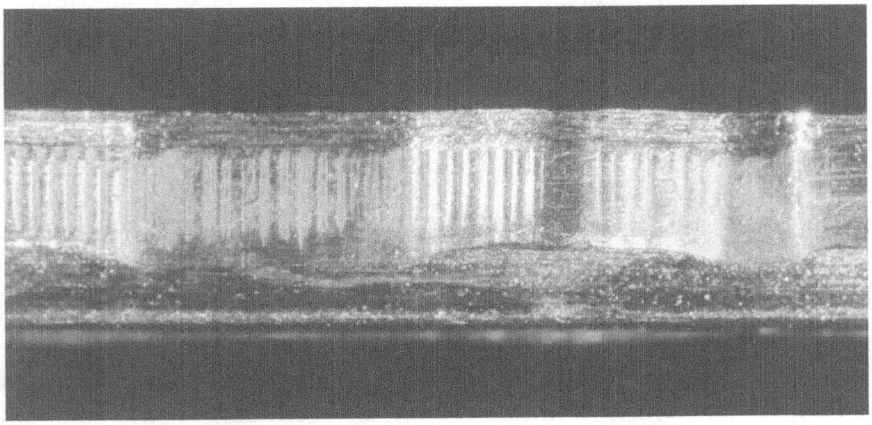

Figure 10-14

Striated marks (top of edge) from the depth guide present on the edge of a pin/wafer tumbler key.

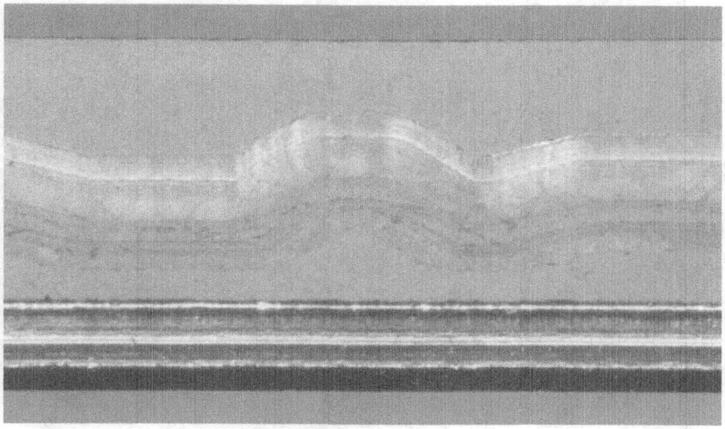

Figure 10-15

Striated marks left by the depth guide at the bottom of the cut of a sidewinder key (in the middle of the channel). Source: Zanetta S. (2001) Duplication des clés à fraisage horizontal II. Etude d'éléments d'interprétation (persistance, traces d'usure) et de datation. (Photograph courtesy of the Ecole des Sciences Criminelles, University of Lausanne, Lausanne, Switzerland.)

with the guide. These toolmarks can be observed under magnification and are indicative of a mechanical reading of the key (Figures 10-16a and b). When dealing with symmetrical keys, these traces will not necessary be present on both sides as only one edge is usually read for the key-cutting process. There are a certain number of different guides used for different types of keys.

When dealing with striations along the cut of the key, these are usually continuous. However, in some cases, it is possible that they will be interrupted or that multiple striations are superimposed (due to the back and forth movement of the guide on the key induced by the user) [23]. The striations are usually shorter than the width of the edge. With pin/wafer tumbler keys, these striations are perpendicular to the cut marks (as seen vertical in Figure 10-14). With sidewinder and centerwinder keys, the striated marks are normally present at the bottom of the milled channel and are parallel to the milled line [6]. In other instances, these marks might be seen only on one edge and not at the bottom of the milled channel. The presence of such traces is characteristic of the mechanical reading of a key.

When using an automatic machine, the guide does not continuously follow the edge of the original key as with a manual machine: the guide just lightly touches the edge at regular intervals (as a matter of fact for each spacing), with very little pressure. Thus, there will not be any continuous striated marks on the edge but instead localized marks at different intervals. Furthermore, because the machine uses very little pressure, these marks are usually very superficial and thus, very faint.

Centerwinder keys cut using an automatic machine will present traces at the bottom of the cut exclusively. These are small impressed circles, not necessarily complete [6]. These

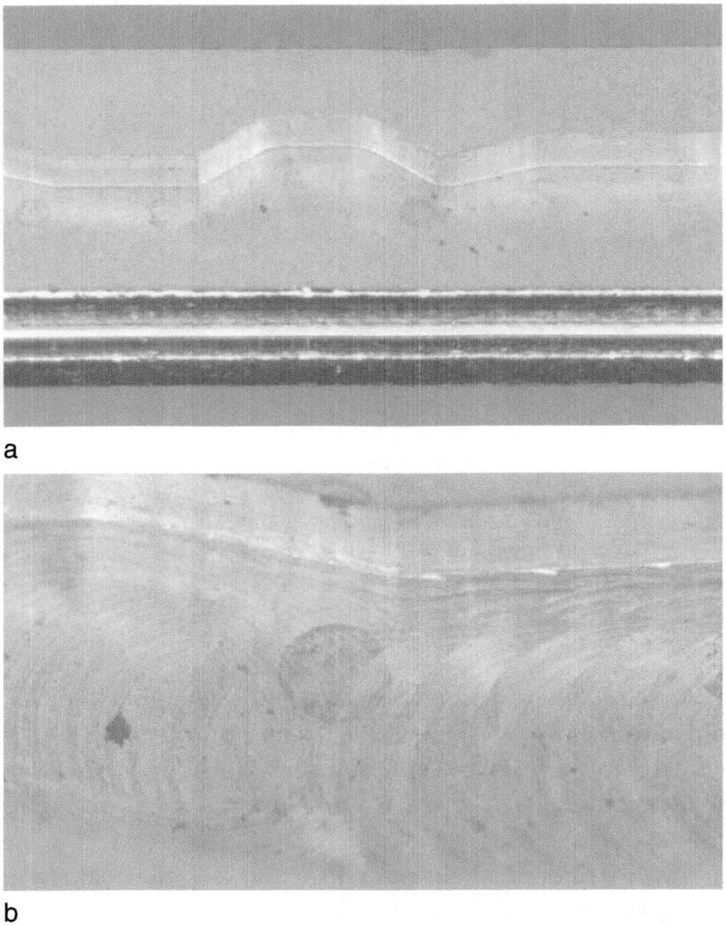

Figure 10-16

(a) Marks left on a pattern (centerwinder) key by the depth guide of an automatic machine.
(b) Detailed view of one of the round mark shown in a. Note that these marks are clearly distinguished from the marks left by the milling of the key as seen in the photograph. Source: Zanetta S. (2001) Duplication des clés à fraisage horizontal II. Etude d'éléments d'interprétation (persistance, traces d'usure) et de datation. (Photographs courtesy of the Ecole des Sciences Criminelles, University of Lausanne, Lausanne, Switzerland.)

circled traces are very characteristics and cannot be attributed to any other operation but the reading of the key during a duplication process [6]. Figure 10-16 shows an example of such marks.

Sidewinder keys cut with an automatic machine present traces on the edge of the original cut at regular intervals. However, these traces are very faint due to the extreme delicateness with which the guide is touching the edge; thus, they are often absent [6].

10.4.4 Examination Outcome

A/ No Marks Found

When vise jaw marks and depth guide marks are absent from an original key, four possible situations can be considered:

I No copies of the original key were made.

II One or more copies were made using a duplicating process that does not involve any physical contact between the original key and the duplicating machine. With (mechanical and electronic) code duplicators, the original key is not needed to make the copy because the key code suffices. Thus, in this case no traces will be present on the original key indicating if it had been copied or not.

III When in presence of a sidewinder key, the copy was made with an automatic machine, which leaves faint or nonexistent traces.

IV One or more copies have been made with a mechanical machine with guide or an electronic machine with optical detection but the resulting traces have fainted away or wore off due to regular use of the key. In this case, the important parameters to consider are the time span between the moment the copy was made and the observation as well as the use frequency of the key during this period. Also, it is important to remember that vise jaw traces are rarely, if ever, visible.

Consequently, an absence of copy traces does not indicate that the key has not been copied.

B/ Marks Found

When an electronic machine with optical detection has been used, there could only be vise jaw traces, on a very rare basis. On the other hand, when a mechanical duplicator with guide has been used, depth guide marks (and theoretically vise jaw marks) should be present. Thus, the presence of such traces is an indication that the key has been mechanically read.

10.4.5 Other Indicators

Some locksmiths keep in memory the copies made from a key code in a database, which includes the contact information of the person who ordered the key. When a request for duplicate keys is made through the manufacturer, records are kept accordingly. Thus, the investigator should always consider requesting information through locksmiths, dealers, and manufacturers regarding possible copies made.

10.5 FORENSIC EXAMINATION

10.5.1 Questions

Forensic laboratories are sometimes required to examine car keys to determine whether they have been copied. In general, the expert is requested to answer the following questions:

- How many keys are part of the original set?
- What is the exact function of each key (master key, valet key, etc.)?
- Do the present keys correspond to the cylinders (doors, trunk, ignition, and glove box) of the vehicle?
- Are the keys original or copies?
- Do the keys present regular wear and tear linked to their normal use?
- Have any of the keys been copied?

These questions can be answered by performing a proper forensic investigation and examination of the keys using the steps described hereafter.

10.5.2 Original Set

The original set of keys is the set that is provided by the factory with the vehicle. In this context, it is important to distinguish the number of keys provided to the new owner of the vehicle at the time of its first purchase from the number of keys provided by the manufacturer to the car dealership. It is also possible that the vehicle was sold used one or more times and that the original set of keys underwent some modifications (loss of key, replacement key, etc.). Information regarding the original set of keys provided with a new vehicle should be gathered from the manufacturer and/or local dealership. Another great source of information is the database EuVID (see paragraph 2.4.2D).

10.5.3 Key Functions

It is important to accurately identify the function of each key. The primary or master key opens all the locks (doors, trunk, glove box, and when applicable the gas trap) as well as the ignition lock. The secondary key (also called valet key) opens doors and ignition lock but does not allow for the opening of the trunk and glove box. In some instances, a wallet key is also provided with the vehicle. The wallet key presents all the same functions as the master key but does not have the same bow, so it can fit in a wallet. Also, the wallet key might not contain any electronic anti-theft system such as the transponder. In such instances, in addition to the wallet key, the vehicle's owner is provided with a separate card containing the anti-theft electronic device. The card must be placed in the vicinity of the receiving antenna at the time of the vehicle's start.

10.5.4 Key and Lock Matching

This step is performed by comparing the spacing and depths of the keys to ensure they are designed for the same lock. It is important to keep in mind that secondary keys will slightly differ from primary and wallet keys because they are restricted to some locks in the vehicle. However, all primary keys and wallet keys should present the same profile and spacing and depths. Questions regarding the cylinders can be asked to the dealership or manufacturer. Usually, this information can be obtained based upon the VIN.

10.5.5 Key Originality

This examination consists of determining whether the blank is an original or if it is generic. Original blanks are found with original keys and duplicates made through the manufacturers, dealerships, and some locksmiths. Original blanks are available to locksmiths, but often a generic blank is used for cost-saving purpose. If the blank is an original, observations of the cut must be made very carefully. In case of pin/wafer tumbler keys, the serrated edge will present cut marks perpendicular to the blade of the key and their morphology will differ depending on whether the key was cut at the factory or if it is a copy. In general, a key cut by the manufacturer presents metal coating (such as chrome) on top of the cut. This is demonstrated in Figure 10-17. Similar differences will be observed on other types of keys (dimple, sidewinder, etc.).

10.5.6 Key Wear and Tear

Cut marks faint and key edges round due to the wear and tear engendered by the regular use of the key. Fivaz and Zanetta demonstrated that copy traces of good quality can still be seen on the edge after 3,000 introductions in a cylinder [23, 24]. This would approximately correspond to the regular use of a key for a period of one year [25]. As these traces are not altered at this stage, it is possible to imagine that they would persist for a much longer period of time. On the other hand, traces of weak quality disappear after several

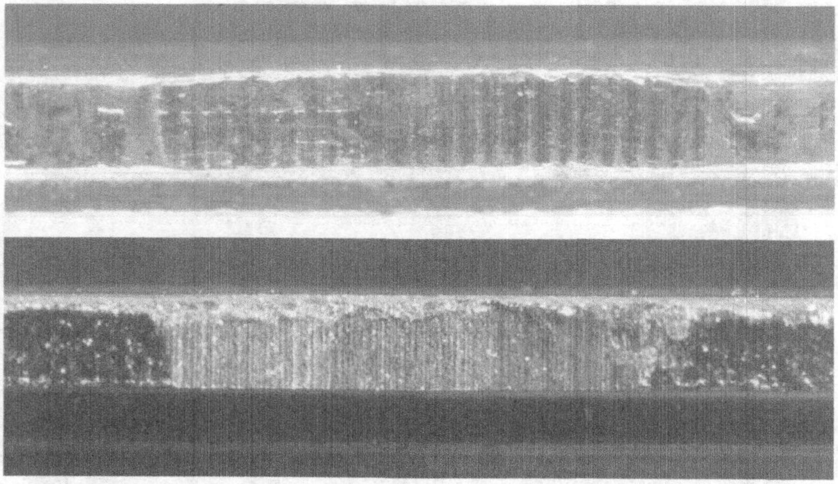

Figure 10-17

Example of cut marks. Top: Cuts typical of an original key from the manufacturer. Bottom: Cuts typical of a duplicate key.

introductions of the key in a cylinder. In this case, it will be difficult to see any evidence of the reading of the key if it has been used several times after the duplication process [24]. It is not possible to quantify the use of a key based upon its wear. However, it is possible to determine which one of two keys is most commonly used.

10.5.7 Copy Traces

Determining whether a key has been copied is based upon the presence or absence of copying traces, as explained in Section 10.4. Zanetta demonstrated that original cutting marks, oxidation marks, and wear marks cannot be confused with copy marks (Figure 10-18) [24].

Based upon the guide marks, it is not possible to determine with which machine the original key was read, nor is it possible to determine when the copy was made [24]. Braune

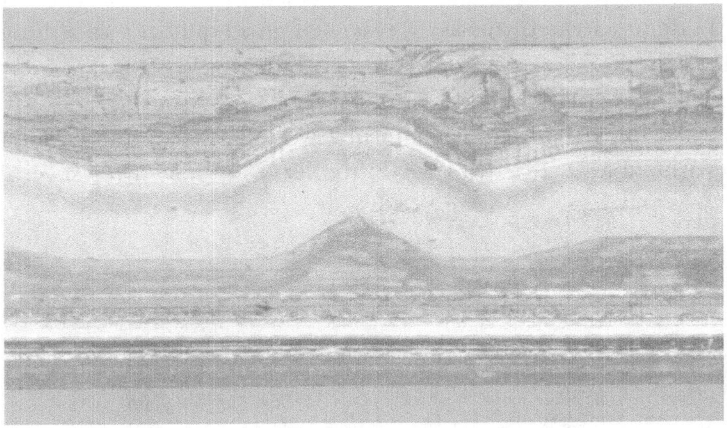

a

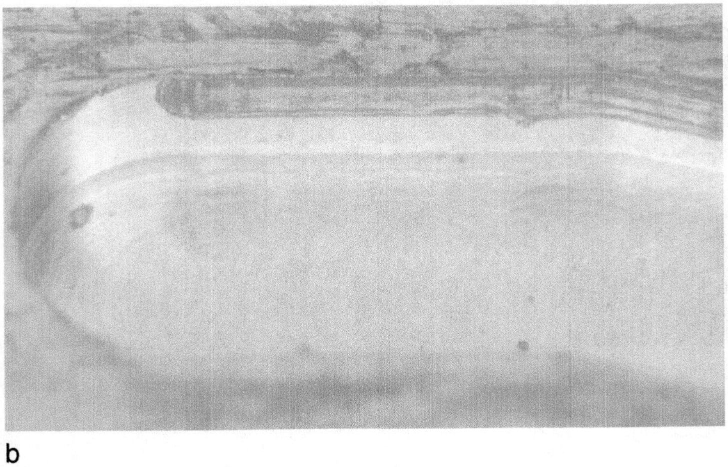

b

Figure 10-18

(a) Deposit of grease on a center-winder key, indicative of key usage. (b) Detailed view of the grease deposit. (Photographs courtesy of the Ecole des Sciences Criminelles, University of Lausanne, Lausanne, Switzerland.)

and Göbel demonstrated that it was not possible to determine the number of insertions of the key in a cylinder after it had been copied based upon copy marks [26]. These results were confirmed by Fivaz and Zanetta [23, 24].

ACKNOWLEDGMENTS

The authors would like to thank Mr. Alexandre Anthonioz, scientific collaborator at the Ecole des Sciences Criminelles, University of Lausanne, and Laurence Barby, Barby Clés (locksmithing shop), in Lausanne, Switzerland for their help. Also, the authors would like to thank Eric Stauffer for translating this chapter into English.

BIBLIOGRAPHY

[1] Bauer Kaba AG (1996) *Manuel pour systèmes de fermeture*, personal communication, Wetzikon, Switzerland.

[2] Keso (1994) *KESO . . . la solution la meilleure pour votre sécurité*, Richterswil, Switzerland.

[3] Zeiss Ikon (1984) *Profil à nervures de blocage avec sûreté à labyrinthe*, Berlin, Germany.

[4] Wurth USA (2003) *Key machine quick reference guide*, available at http://www.wurthusa.com, last access performed on October 22, 2005.

[5] Framon Manufacturing Company (year unknown) *Framon sidewinder instruction manual*, available at http://www.framon.com, last access performed on October 22, 2005.

[6] Zanetta S, Anthonioz A, and Neumann C. (2006) Etude des traces de duplication sur les clés à fraisage horizontal, *Revue Internationale de Criminologie et de Police Technique et Scientifique*, **59**(1), pp 95–119.

[7] Silca S.p.A. (1996) *Machines à reproduire les clés*, personal communication, Vittorio Veneto, Italy.

[8] Citiloc Systems LTD (2005) *Glossary*, available at http://www.citiloc.com/glossary.html, last access performed on October 30, 2005.

[9] Silca S.p.A. (1996) *RW 100*, personal communication, Vittorio Veneto, Italy.

[10] AP Workshop (2004) *Frequently asked questions: code-cut keys*, available at http://www.keys4classics.com, last access performed on October 22, 2005.

[11] Kummer S, Bonfanti M, and Gallusser A. (1997) Le processus de reproduction des clés et son intérêt en sciences forensiques (partie 1), *Revue Internationale de Criminologie et de Police Technique*, **50**(4), pp 479–492.

[12] AP Workshop (2003) *Code reading Jaguar "Tibbe" high security keys*, available at http://www.keys4classics.com, last access performed on October 22, 2005.

[13] August Börkey Nachf. GmbH (1993) *Catalogue pour des clés à cylindre, à croix et pour automobiles*, Gevelsberg, Germany.

[14] Silca S.p.A. (1996) *Matrix*, personal communication, Vittorio Veneto, Italy.

[15] Silca S.p.A. (1996) *Bravo, personal communication*, Vittorio Veneto, Italy.

[16] Framon Manufacturing Company (year unknown) *DC-300 Duplicating code machine instruction manual & parts book*, Framon Manufacturing Company, Alpena, MI.

[17] Silca S.p.A. (1996) *Prima · Laser*, personal communication, Vittorio Veneto, Italy.

[18] Silca S.p.A. (1989) *UNOCODE*, Vittorio Veneto, Italy.

[19] Silca S.p.A. (1996) *Quattrocode*, personal communication, Vittorio Veneto, Italy.

[20] Silca S.p.A. (year unknown) *RW3*, available at http://www.silca.it, last access performed on October 22, 2005.

[21] Silca S.p.A. (2005) *Transponder types*, available at http://www.silca.it, last access performed on October 22, 2005.

[22] Kummer S, Bonfanti M, and Gallusser A. (1998) Le processus de reproduction des clés et son intérêt en sciences forensiques (partie 2), *Revue Internationale de Criminologie et de Police Technique et Scientifique*, **51**(2), pp 229–237.

[23] Fivaz E. (1997) *Etude des traces laissées par le processus de duplication des clés et leur persistance*, Institut de police scientifique et de criminologie, Université de Lausanne, Lausanne, Switzerland.

[24] Zanetta S. (2001) *Duplication des clés à fraisage horizontal II. Etude d'éléments d'interprétation (persistance, traces d'usure) et de datation*, Institut de police scientifique et de criminologie, Université de Lausanne, Lausanne, Switzerland.

[25] Hitzemann M and Kleinhaus T. (1995) *Untersuchung von Fahrzeugschlüsseln. Aussagefähigkeit von Schlüsselbefunden für die Fallaufklärung Fahrzeugschlüsseln*, Internal publication, Dekra, Stuttgart, Germany, pp 119–130.

[26] Braune M and Göbel E. (1997) *Investigation on the frequency of use of mechanically copied car keys*, Information Bulletin for Shoeprints/Toolmarks Examiners, 3, p 14.

CHAPTER 11

ANALYSIS OF VEHICLE FLUIDS

Eric Stauffer

11.1 INTRODUCTION

Modern motor vehicles require many different types of fluids to operate. There are fluids used as fuel (gasoline, diesel), to lubricate (engine oil, gear lubricant), to transfer power (brake fluid, automatic transmission fluid), to control temperature (coolant), and for other miscellaneous purposes (washer fluid). Some of these fluids may fulfill two or more purposes simultaneously. When in use, some of these fluids are in direct contact with crucial internal mechanical parts of the vehicle. This contact results in the fluid carrying small particles in suspension, usually at very small concentrations. If the malfunction of a given mechanical component occurs (excessive wear of a gear, leaking gasket), different particles or unusual amounts of some of these particles will be present in the fluid. Thus, the content of the fluid provides an indication of the operating state of the vehicle.

One reason a vehicle is allegedly stolen or allegedly catches fire is the commission of insurance fraud (see Chapter 19). There could be many different motives for the owner of the vehicle to proceed with such a criminal act. One reason could be that the vehicle no longer functions properly, due to severe mechanical damage (common wear or accidentally induced), for which the cost of repair is too much for the owner. Therefore, it is very pertinent for the forensic investigator to evaluate the conditions of different mechanical components of the vehicle, most particularly the engine and (automatic) transmission, to corroborate or rebut the potential motive of the owner. Normally, a thorough investigation can only be performed when a competent mechanic takes these components apart and examines the internal parts. This represents a long, tedious, and expensive endeavor. Fortunately, it is possible to simply analyze the vehicle's fluids to provide a clear indication of the operating state of the vehicle's components. Naturally, this can be followed by further more expensive examination of the vehicle by a mechanical expert.

Analysis of vehicle fluids is inexpensive and easily accessible to anyone, because many commercial laboratories offer this service. Some private forensic laboratories also offer this service, with a more extensive experience in court testifying. First, a representative sample of the fluid must be collected from the vehicle. This is usually a simple task requiring very little effort, but it can become more complex under specific circumstances, such as with burned vehicles. The sample is then submitted to a competent laboratory, which performs the desired analyses. Finally, the interpretation of the results, which may be partially provided by the laboratory, is the most delicate and complicated part of the process. It is important for the investigator to understand that this science is not exact and that the interpretation can be vague.

The intent of this chapter is to present the concept of vehicle fluid analysis to the forensic investigator. A brief review of the different fluids found in a vehicle, followed by the proper collection procedures, are presented. Then, a section on the analyses and interpretation of the results is offered. Finally, a few practical cases conclude this chapter.

11.2 VEHICLE FLUIDS

11.2.1 Different Types of Fluids

Most fluids are common to many vehicles; however, some vehicles require particular fluids based upon their design and function. This chapter covers engine oil and, to a lesser extent, automatic transmission fluid (ATF) and gear lubricant. These three fluids are the most important witnesses of the conditions of the engine and transmission, two vital organs of a vehicle that can generate high repair cost. A vehicle owner most likely will not commit insurance fraud for a broken master cylinder; however, he or she might do so when the automatic transmission fails and needs to be replaced.[1] Engine oil, ATF, and gear lubricants are all lubricants. The most important function of a lubricant is to leave a thin film of oil between two metal parts that are moving close to each other. This lubrication reduces or prevents wear and tear by decreasing the friction between these two parts.

A/ Engine Oil

The engine oil is used to lubricate the internal components of the engine. It is found in the engine, where it flows in a closed circuit. In some vehicles, the engine oil is routed outside the engine through rubber lines to a cooling radiator, providing extra cooling capability. Most vehicles contain between 2 and 8 liters of engine oil. Trucks and heavy equipment can contain up to a few dozen liters of engine oil. When new, engine oil is usually a clear-brown to dark-brown liquid. It quickly darkens as it is used in the engine. With diesel-powered engines, a new oil can become pitch black during the first few minutes after an oil change. This is due to the heavy presence of soot in the oil. In a properly working gasoline-powered engine, the oil remains brown and translucent for a certain period of use after its change.

---

[1] By way of illustration, in October 2002, a man rented a backhoe and buried his 1997 BMW in his father's property in the rural northeastern part of the State of Ohio in the United States. He then filed a claim with his insurance company reporting his car as stolen. As a matter of fact, he decided to cheat the insurance company after his engine blew up and he realized that it would cost him too much to repair. After the insurance company discovered the fraud, thanks to an anonymous tip provided to the police, he was sentenced in early 2006 to one year of imprisonement.

B/ Automatic Transmission Fluid (ATF)

Automatic Transmission Fluid is used to move the propeller and to lubricate the transmission. It functions in dual capacity as both lubricant and power transmitter. ATF is not found in vehicles equipped with manual transmission. ATF is located in the transmission case and is usually routed outside the transmission case through a combination of hard steel lines and flexible rubber lines to a separate circuit of the (water) cooling radiator or sometimes to a separate radiator entirely dedicated to the ATF. This provides temperature control for the transmission fluid. ATFs are all red but present different shades of red. The fluid is dyed red to clearly identify it when leaking [1]. ATF is also translucent and can become much darker when used, particularly when its changing cycle is grossly overdue. However, even with a very dark sample of used ATF, its red tint is distinguishable when held to the light.

C/ Gear Lubricant

Gear lubricants are used to provide lubrication in gearboxes (manual transmissions), differentials, and in some transfer cases (ATF is also used in some transfer cases). Not all vehicles use gear lubricants in their gearboxes. Gearboxes that operate under light load use either engine oil or ATF. Gear lubricants are found in different shades of brown to dark brown, and when new, they are slightly translucent. They are very thick fluids and become much darker after extensive use.

11.2.2 Chemical and Physical Properties of Fluids

Fluids have literally dozens of different properties. It is possible to characterize fluids using several different analytical techniques, looking at many different attributes. In the perspective of oil or fluid analysis, there are several pertinent characteristics. In the scope of this chapter, only viscosity and elemental composition are presented. The reason is that the interpretation of some of these attributes is really complex, should only be performed by trained professionals, and does not necessarily add much relevant information to the investigation.

A/ Viscosity

The kinematic viscosity of a fluid represents its resistance to flow [2]. The higher its viscosity, the less easily oil flows. Viscosity also represents the thickness of an oil. A thick oil has a high viscosity and does not flow well, whereas a thin oil has a low viscosity and flows relatively easily. Viscosity decreases with increasing temperature. Thus, the viscosity of an oil is always expressed for a given temperature. From a mechanical perspective, the viscosity of an oil is an extremely important characteristic, because it directly relates to its lubricating capability. The thicker an oil, the more it sticks to metal surfaces and leaves a film between two metal parts. However, the thicker an oil, the more difficult it is for the parts to move. Thus, a good balance between the efficiency of the motion of metal parts and their lubrication must be reached. In oil analysis, the kinematic viscosity is expressed in centistokes ($1\,cSt = 1\,mm^2/s$). The viscosity is usually measured at 40 and 100°C.

The grade of an oil, a measure that is known by most everyone, is a different scale developed by the Society of Automotive Engineers (SAE) used to express the viscosity of an oil. The higher the number, the higher the viscosity. Multigrade oils have been developed to exhibit a greater viscosity stability over a wider range of temperatures than monograde oils [3]. This means that the viscosity of a multigrade oil is not as readily influenced by temperature as that of a monograde oil. Table 11-1 shows different grades of engine oils and gear lubricants with their respective viscosities [4, 5]. ATFs present very similar kinematic viscosities, which range from approximately 33 cSt at 40°C to 5–8.5 cSt at 100°C [6].

B/ Elemental Composition

The elemental composition of an oil is a crucial characteristic when the operating state of an engine and, to a lesser extent, the condition of the oil itself must be determined. Oils contain different levels of elements when new. Table 11-2 shows approximate levels of elements found in brand new engine oil (a) and in ATF (b). The first column contains the elements representative of contamination of the fluid; these elements are almost absent from a new fluid. The second column contains the elements from the additives present in the fluids. Thus, these elements are indicative of the condition of the fluid itself. These concentrations must be considered cautiously, because serious variations may occur in the additive package between different brands of oils or fluids. For example, one brand of ATF presents 13 ppm of calcium (Ca), whereas another brand presents 2,849 ppm of Ca. The last column contains the elements representative of the operating state of the engine or transmission; these elements are generally absent from a new fluid.

Table 11-1

Kinematic viscosities and SAE grades of engine oils and gear lubricants.

	SAE grade	Prefix	Kinematic viscosity in [cSt] at 100°C
Engine oil	20	0W, 5W, 10W	5.6–9.3
	30	0W, 5W, 10W	9.3–12.5
	40	5W, 10W, 15W, 20W	12.5–16.3
	50	5W, 10W, 15W, 20W	12.5–16.3
	60	10W, 15W, 25W	21.9–26.1
Gear lubricant	70W	—	>4.1
	75W	—	>4.1
	80W	—	>7.0
	85W	—	>11.0
	90	—	13.5–24.0
	140	—	24.0–41.0
	250	—	>41.0

Table 11-2

Elemental compositions of (a) new unused engine oil and of (b) new unused ATF.

(a) Engine oil

Oil contamination		Oil condition		Engine wear	
Element	**Concentration in [ppm]**	**Element**	**Concentration in [ppm]**	**Element**	**Concentration in [ppm]**
Silicon	<10	Boron	up to 250	Iron	<3
Potassium	<10	Barium	<10	Chromium	<1
Sodium	<5	Calcium	>1,800	Nickel	<1
		Magnesium	10–60	Aluminum	<3
		Molybdenum	up to 150	Lead	<2
		Phosphorus	>1,000	Copper	<1
		Zinc	>1,000	Titanium	<1
				Tin	<1
				Silver	<1

(b) Automatic Transmission Fluid

Oil contamination		Oil condition		Transmission wear	
Element	**Concentration in [ppm]**	**Element**	**Concentration in [ppm]**	**Element**	**Concentration in [ppm]**
Silicon	<10	Boron	1–200	Iron	<3
Potassium	<10	Barium	up to 2,000	Chromium	<1
Sodium	<5	Calcium	up to 3,000	Nickel	<1
		Magnesium	<10	Aluminum	<3
		Molybdenum	<5	Lead	<2
		Phosphorus	up to 1,000	Copper	<1
		Zinc	up to 1,000	Titanium	<1
				Tin	<1
				Silver	<1

11.3 FLUID SAMPLING

11.3.1 Preliminary Observations and Level Check

A/ Observation of Fluid Circuits

Before collecting any fluids from a vehicle, it is important to observe the circuit of that particular fluid to check for any possible leaks or breaches. When vehicles are involved in road accidents, it is possible that a hydraulic line breaks or the bottom of a pan cracks, allowing for liquid to escape. Also, if the vehicle underwent fire, it is extremely important to determine whether the fluid circuit was affected or not. In some instances, it is possible that a hose burned through and that the liquid completely drained from the circuit.

Furthermore, it is also possible that the lids (or caps) to these circuits burned away and contamination, such as water from the fire suppression efforts, could have entered the circuit. Thus, the investigator must observe the entirety of the circuit (exterior, underside, and engine compartment) to ensure of its integrity. If the circuit has been compromised, this must be recorded because contamination could have penetrated the circuit. These data are important when interpreting the results of the analyses. Then, the investigator must check the levels of the fluids to be collected.

B/ Engine Oil Level Check

The engine oil dipstick is usually easily located under the hood. In general, the engine oil dipstick is marked with minimum and maximum level values. The oil level should be between these two values. If the engine oil circuit has been compromised, no oil may be present on the dipstick. If fire suppression activities led to the addition of water to the engine oil circuit, it is possible that the oil reached a level much higher than the maximum value. The investigator should first pull the dipstick and note the position of the oil level. Then, the dipstick should be cleaned with a towel, reinserted in the tube, and pulled out again for a confirmatory reading. In case of burned vehicles, this second step might not be possible: If the tube is burned and filled with debris, the reinsertion of the dipstick might be prevented. A photograph of the dipstick tip can be taken if necessary. To obtain optimum contrast, it is a good practice to use a light blue background when taking such a photograph. When the oil is very dark, it is relatively easy to distinguish it on the dipstick tip. However, if the oil is very fresh and clear, it might be difficult to see the level.

C/ ATF Level Check

It is not possible to properly check the level of ATF when the engine is turned off. It is designed to be checked while the engine is running. Often, the dipstick for this fluid is located close to the firewall, because the transmission is located behind the engine in propulsion vehicles. The transmission is located on the side of the engine in traction vehicles. The tip of the dipstick usually shows two positions, one for cold engine (transmission) and one for hot. Some vehicles are equipped with a sealed transmission. In this instance there is no ATF dipstick, and it is not possible to check the level without removing the filling plug. Only under a very particular set of circumstances should the investigator turn the engine on to check the proper level of ATF. This should be done only when it has been ascertained that such an operation is possible and will not create any further alterations. With burned vehicles, it is often impossible to start the engine. In this instance, even if the level of the transmission fluid cannot be properly checked, it is good practice to check the level with the engine off. However, a difference between the level and the marks on the dipstick is not necessarily alarming. The procedure used is the same as with engine oil, and the results are carefully recorded in notes and/or by photography. The automatic transmission circuit typically does not exhibit a separate opening with a lid for filling the circuit. This is done by pouring liquid through the mouth of the dipstick tube.

D/ Gear Lubricant Level Check

If one desires to collect gear lubricant, either from the gearbox of a manual transmission vehicle, from the transfer case of some four-wheel drive vehicles, or from differentials, it is good practice to first check the level. Unfortunately, these components do not usually exhibit a dipstick. The only way to check the level is to remove the filling plug and to ensure that the liquid is leveled with the opening.

11.3.2 Sampling

Proper sampling procedures are very important. The sample of the fluid must accurately represent the contaminants present in the system, the wear conditions of the components, the additives, the oxidation and degradation by-products present in the oil, and the particulates present in the system [7]. If the sample is not properly collected, one or more of these five conditions might be seriously misrepresented. A good sampling must be performed when the oil is at normal operating temperature and is well mixed [2]. Using an on-line system, where a valve placed somewhere in the circuit delivers the proper sampling quantity while the engine/transmission is running, provides the best oil sample. Unfortunately, this is not an option in the scope of the examination of stolen-recovered or burned vehicles. In these instances, it is possible to rely on two methods of sampling a fluid from its container: drainage and extraction. Before extracting the oil, if circumstances permit it (which is extremely rare), the engine could be turned on and run for a couple of minutes to warm the oil and to agitate it. This step should only be performed if no further damage to the vehicle and/or evidence alteration would occur. Otherwise, samples should be taken cold. Although this is not an optimal situation, it is not meaningless either.

A/ Drainage

When possible, drainage is the optimum method of collecting a fluid sample. This is performed by removing the drain plug, usually located under the crankcase or transmission, and letting the fluid drain into a clean recipient. In some instances, it might be necessary to lift the vehicle to access the oil pan or transmission case. It is good practice to use new, clean, lined metal (paint) cans of either one-quart or one-gallon capacity. It is usually undesirable to collect the first few deciliters that are draining out, because they might not be representative of the overall fluid. It is also possible that an important amount of water would drain out of the engine or transmission first, particularly with burned vehicles. This is because the water from fire extinguishing activities or rain may flow inside openings created by fire in hoses or caps. Because water is denser than oil, it eventually settles near the bottom of the pan and drains first. Thus, two cans should be used: the first to collect the first fraction and the second to collect the sample to be analyzed. A can should never be filled more than 80% of its volume. This allows for agitation of the fluid before sending a representative portion to the laboratory for analysis.

B/ Extraction

Extraction is performed using a special vacuum pump (manually operated) connected to a sampling bottle on one side and to plastic tubing on the other side. The plastic tubing should be disposable to minimize contamination. The tubing is inserted through the dipstick tube of the engine or transmission. The tube should hit the bottom of the container and then be pulled back a couple of inches, which provides the most representative sample for this sampling technique. The pump is activated, which pulls the fluid out to the sampling bottle. This technique is simple, fast, and clean. However, if the engine or transmission are cold and have not run for a long period of time, the sample taken in this fashion will not be fully representative due to settling of particles in the fluid. In such instances, it could be useful to collect more than just the regular sampling bottle and fill up a one-quart or one-gallon can to obtain a bigger sample. In any event, the first portion of the sample can be discarded for reasons previously explained; however, the investigator must always be aware of spoliation of evidence issues.

C/ Quantity

It is crucial to consult with the analyzing laboratory before taking any sample to ensure of the proper quantity of fluid that must be submitted. The quantity of fluid needed depends on the nature and number of tests desired. In general, for a basic run of the tests described in the next two sections, 100ml of fluid is enough. If other analyses are required, more sample volume will be needed. It is also important to note that all these analyses are destructive, and the sample sent to the laboratory will no longer exist after analysis. Thus, if the investigator wants to keep a portion of the sample as evidence, more volume must be taken. It is good practice to sample a one-quart volume and to send only the necessary portion to the laboratory.

11.3.3 Information Accompanying the Samples

The two main factors that lead to a poor oil analysis are improper sampling and lack of information about the sample [2]. To properly interpret the results, it is crucial to gather as much pertinent information as possible about the vehicle from which the sample is taken and about the fluid itself. Table 11-3 presents the information that should accompany the oil sample. This information is not always available, but the data in bold are required for a minimal interpretation.

If the vehicle is burned, it is pertinent to determine the extent of the damage, such as if the engine compartment has undergone heavy heat damage. This is used to help in the interpretation of the results, as explained in the next section. In general, elements representative of the engine or transmission wear are not influenced by fire [8]. However, it is possible that the fire impacts other conditions, such as the viscosity.

Table 11-3

Data that should be collected at the time of the oil sampling and necessary to perform proper interpretation of the results.

Vehicle	
Characteristics	**Make** **Model** **Year**
Engine type	**Gasoline** **Diesel** **Other**
Mileage	Be aware of engine and/or transmission change
Vehicle burned?	**Yes/no**
Comments	Known mechanical damage, problem, condition of fluid circuit
Fluid	
Type	**Engine oil** **Automatic Transmission Fluid** **Gear lubricant**
Characteristics	Make Model Quantity
Level	Insufficient/normal/excessive
Time since last fluid change	
Mileage since last fluid change	
Additives	Special additives used with the fluid
Comments	

Data in bold are required for a minimal interpretation.

11.4 ANALYSIS OF ENGINE OIL

11.4.1 Oil Contamination

A/ Fuel Dilution

Fuel dilution is the relative amount of unburned fuel present in the oil. Fuel (both gasoline and diesel) is completely miscible in engine oil, and its dilution in the oil has two main effects: It decreases the viscosity of the oil and decreases the additive levels in the oil [2]. The viscosities of fuel are much lower than those of engine oils. The presence of about 14% of fuel in an oil would decrease its viscosity by half. This would lead to a poor lubricating capability of the oil. The presence of fuel is typically determined by infrared spectroscopic technique or by gas chromatographic analysis and is expressed in percentage of fuel in oil (volume/volume). A high dilution of fuel is usually indicative of a malfunctioning fuel system assembly such as a pump delivering too much fuel or a faulty injector [9]. Fuel dilutions of less than 2% are typically not worrisome. A fuel dilution of 4% or more would be indicative of a problem.

B/ Water Concentration

When water is present in the oil, different harmful effects may occur, such as corrosion (rust) of the internal metal parts, decreases in oil viscosity, reduction in lubricating capability, and steam in the engine. Water can come from either a leak in the cooling circuit, from an external source of water, or when condensation of the air occurs as the system cools down [11]. The concentration of water can be determined using many different techniques, usually depending on the amount of water present [2]. For routine work and very small concentrations of water, infrared spectroscopic techniques or chemical reactions with calcium carbide or hydride (and the measurement of the volume of gas produced) are used. It is expressed as percentage of water in oil (volume/volume). Normally, no water should be present in the oil. However, a very small amount (up to approximately 0.10%) is tolerable. Water might be present at this very low level if the oil analyzed was sampled cold. Because the engine oil temperature is much higher during its normal operation, this small amount of water might evaporate. Any concentration of 1% and above is indicative of a serious problem. It is important to interpret the presence of water in conjunction with the presence of glycols to properly diagnose whether a leak of the cooling circuit has occurred or not. Also, it is necessary to know if the vehicle is burned or not (and the extent of the fire damage) when evaluating the presence of water. If the engine compartment is burned, it is very likely that water will be present in the oil due to suppression efforts and the fact that the fire created openings in the oil circuit.

C/ Presence of Glycols

The presence of glycols in engine oil is due to a leak from the cooling system. This could be due to a faulty head gasket, intake manifold, or cracked engine block. The presence of glycol is detected using the Schiff's reagent glycol test, a colorimetric reaction test. In some instances, the laboratory reports a percentage of glycols by volume in the oil. The important point is to determine whether glycols are present or not. Normally, no glycol should be present in the oil at all.

D/ Soot

The presence of soot in engine oil is due to the combustion of the fuel in the cylinders. A high concentration of soot indicates that the air/fuel mixture is too rich, thus indicating an insufficient combustion in the engine. Soot is a useful measure of the air-to-fuel ratio as well as combustion and exhaust efficiency. An excessive amount of soot is typically due to a blowby. A blowby is a leakage of combustion gases between the piston and the cylinder wall into the engine crankcase due to a significant wear of the piston rings. Soot is undesirable in an engine because soot particles are extremely hard and can contribute to excessive wear of the internal components by grinding. Diesel engines present much higher levels of soot than gasoline-powered engines. Thus, oils designed for diesel engines usually contain an additive (detergent) to keep the soot in suspension. Soot is usually measured by infrared spectroscopic technique and is expressed in percentage by mass [12].

E/ Silicon

Silicon is the main indicator of natural soils and dusts, because they are mostly made of silica and silicates [13]. In this regard, silicon is an excellent indicator of the contamination of an oil with external dusts. These very small particles of dust are very hard and highly abrasive [2]. They usually accumulate in the engine oil via the air intake. Once in the engine oil, they stay in suspension and lead to excessive wear by interfering with the lubricating films by grinding metal surfaces. The measurement of the amount of silicon in a fluid is performed by spectrochemical analysis. The amount is expressed in parts per million (ppm). In general, with engine oil a level of silicon of 25 ppm or lower is not alarming [11]. A level of 30 ppm and above is abnormal, and a level of 75 ppm or above is severe. Although in most instances silicon is representative of dirt and dust contamination of the oil, it could also come from other sources such as silicone gasket sealant or coolant leak. Thus, it is important to determine whether glycols are present. As a rule of thumb, if silicon is only due to dirt entry, the ratio of aluminum to silicon should stand between 1:10 and 1:2 [14]. This is because dirt is mostly composed of silicon oxide and aluminum oxide.

11.4.2 Oil Condition

A/ Viscosity

The viscosity should not deviate too much from the values shown in Table 11-1 and Paragraph 11.2.2A. If deviation of more than 30% from these values occurs, it should be considered abnormal [11]. The most common reason for an increase in viscosity is overheating: The light fraction of the oil evaporates and the remaining fraction becomes thicker with time, even with regular top off. Common causes of decrease in viscosity are fuel dilution and degradation of the oil with extended use.

B/ Elements

Elements representative of the oil conditions come from the additive package of the oil. These are boron (B), barium (Ba), calcium (Ca), magnesium (Mg), molybdenum (Mo), phosphorus (P), and zinc (Zn). It is not possible to provide some level limits, because additive packages are highly variable among oils. If the oil brand and type are known, it is always possible to analyze a new sample to obtain baseline numbers. In general, if the engine is overheating, an increase of these elements will occur as a result of the evaporation of the oil and regular top-off, and because the additives do not evaporate. If the oil is leaking and regular top-off is performed, the levels should remain constant. A decrease of these additives would be consistent with a degradation of the oil leading to their evaporation.

11.4.3 Engine Wear

Different elements are analyzed to determine the engine wear. Table 11-4 shows these elements with their limit values. Again, these values must be interpreted with caution, as some

Table 11-4

Recommended values for abnormal and severe levels of elements representative of engine wear.

Element	Abnormal value starts at . . . [ppm]	Severe value starts at . . . [ppm]
Iron	150	400
Nickel	5	10
Chromium	20	40
Copper	155	250
Aluminum	40	90
Tin	10	25
Lead	50	150
Silver	2	5

deviations might occur depending on the oil, conditions, and vehicle. However, they provide a good indication in most instances.

Upon wear or failure, the different components of the engine contribute separately to the levels of elements. Thus, the interpretation of the results regarding the attribution of the levels of elements to the failure of a specific part can be extremely complex and is outside the scope of this chapter. But it is important for the reader to understand that in-depth interpretation is possible and regularly practiced by oil analysts [10, 11, 13–19].

11.5 ANALYSIS OF ATF

The interpretation of ATF is slightly simpler. There are fewer components involved and, thus, less possible contamination. Water level should not exceed 0.5% in ATF [11]. No glycol should be detected as the cooling system is not routed through the transmission. No fuel should be present either, because the fuel lines do not share any common location with the transmission. The viscosity should also be within the 30% deviation rule as explained for the engine oil. See Paragraph 11.2.2A for the viscosity values of ATF and gear lubricant.

When silicon is present in the transmission, it is usually not due to dirt contamination but rather to an ingress of seal material, and it will be present in a high amount. The transmission does not have openings that would let dirt or dust come in contact with ATF, as the air intake of the engine does for the oil. The level of the additive's elements is to be taken with great caution as with engine oil. The additive package of different fluids varies tremendously. Table 11-5 presents some level limits for elements indicative of wear. These limits must be taken with caution, as some deviation may occur depending on the fluid, conditions, and vehicle.

Table 11-5

Recommended values for abnormal and severe levels of elements representative of automatic transmission wear.

Element	Abnormal value starts at . . . [ppm]	Severe value starts at . . . [ppm]
Iron	200	400
Chromium	10	25
Copper	200	400
Aluminum	25	75
Tin	10	25
Lead	50	150

11.6 PRACTICAL CASES

11.6.1 2000 Renault Espace

This vehicle burned while being driven on the highway. It has a gasoline engine, and a sample of the engine oil was collected and sent to the laboratory for analysis. At the time of the fire, the vehicle had logged more than 221,000 km (137,323 miles). The engine oil circuit had not been compromised by the fire. Other data were not available. Table 11-6 shows the results of the laboratory analysis.

The silicon and potassium levels do not show any dirt contamination. The level of sodium is elevated, which is consistent either with the presence of an additive or with contamination from either seawater or a coolant leak. The level of fuel is not alarming. The presence of glycol, along with some water, is indicative of a leak of coolant in the engine oil. This is probably due to a defective head gasket and/or intake manifold. The presence of coolant likely contributed to the presence of sodium, thus explaining its high level. The absence of soot indicates that the injection, combustion, and exhaust systems work properly. The elemental composition presented in the second set of columns (oil condition) does not present any abnormal levels. The viscosity at 100°C (12.43 cSt) is consistent with an oil of grade SAE 30 or 40, which is recommended for this type of vehicle. The levels of the elements representing the engine wear are extremely low for an engine with such a mileage, indicative of a very good operating state. Thus, the oil analysis revealed that the engine suffers from a small leak of coolant but is in good operating state.

11.6.2 1997 Chevrolet Venture

A sample of engine oil was collected from a 1997 Chevrolet Venture, equipped with a gasoline engine. No further information was provided, thus limiting the interpretation. Table 11-7 presents the results from the laboratory.

Table 11-6

Results of the analysis of the 2000 Renault Espace engine oil.

Oil contamination		Oil condition		Engine wear	
Attribute	**Result**	**Attribute**	**Result**	**Attribute**	**Result**
Silicon	7.7 ppm	Boron	48 ppm	Iron	5.4 ppm
Potassium	0.0 ppm	Barium	0.2 ppm	Nickel	0.0 ppm
Sodium	56 ppm	Calcium	1,803 ppm	Chromium	1.5 ppm
Fuel	<2.0 %	Magnesium	38 ppm	Titanium	0.2 ppm
Glycol	0.10 %	Molybdenum	27 ppm	Copper	3.3 ppm
Water	0.160 %	Sodium	56 ppm	Aluminum	2.2 ppm
Soot	0 %	Phosphorus	683 ppm	Tin	0.0 ppm
		Sulfur	5,520 ppm	Lead	5.3 ppm
		Zinc	802 ppm	Silver	0.1 ppm
		Viscosity at 100°C	12.43 cSt		

Table 11-7

Results of the analysis of the 1997 Chevrolet Venture engine oil.

Oil contamination		Oil condition		Engine wear	
Attribute	**Result**	**Attribute**	**Result**	**Attribute**	**Result**
Silicon	70 ppm	Boron	15 ppm	Iron	441 ppm
Potassium	>999 ppm	Barium	<10 ppm	Nickel	<1 ppm
Sodium	433 ppm	Calcium	1,628 ppm	Chromium	4 ppm
Fuel	3.0 %	Magnesium	33 ppm	Titanium	<1 ppm
Glycol	0 %	Molybdenum	42 ppm	Copper	42 ppm
Water	0.70 %	Sodium	433 ppm	Aluminum	188 ppm
		Phosphorus	1,603 ppm	Tin	41 ppm
		Zinc	1,427 ppm	Lead	126 ppm
		Viscosity at 100°C	89.4 cSt	Silver	<0.1 ppm

The oil presents elevated levels of silicon, potassium, and sodium, indicative of serious contamination with dirt and dust. Some fuel contamination is present, which could originate from an air/fuel mixture that is too rich or from a blowby. Unfortunately, data regarding the soot level, which could have corroborated or discounted this hypothesis are not available. The presence of water should also be investigated. It would be interesting at this point to determine whether the cooling system is properly filled with coolant. If this is the case, then the water would come from a source other than the cooling system. If the cooling system is filled with pure water, the engine might present a leak between the cooling system and the engine oil circuit. The levels of the elements in the second column are normal. However, the viscosity is extremely elevated, which is abnormal for this vehicle. The elements representative of the engine wear also present abnormally and severely elevated levels of iron, aluminum, lead, and tin. This is typical of an extensive wear of the engine. The high viscosity of the oil is consistent with an overheating engine accompanied by a regular topping-off of the oil level. This engine exhibits severe internal wear. This example also demonstrates how important it is to obtain comprehensive circumstantial information about the case in order to proceed to a more proper and complete interpretation of the results.

11.6.3 2002 Ford Explorer

A sample of ATF was collected from a 2002 Ford Explorer (gasoline engine) with 157,711 km (97,997 miles). The ATF had just been changed and had 744 km (462 miles) in the transmission when the owner realized that the transmission was slipping. Table 11-8 presents the results from the analysis.

The values of silicon and water are normal. The additives values do not present any abnormalities, and the viscosity at 40°C for that fluid is right on target. However, the values of some elements from the transmission wear column are alarming for a fluid with only 744 km. Although these values are not above the values found in Table 11-5, it is important to remember that with such a low mileage on the new fluid, these values should not be that high. Elevated levels of iron (184 ppm), copper (148 ppm), aluminum (41 ppm), and tin (11 ppm) are consistent with severe gear wear of the transmission, which will probably fail shortly.

Table 11-8

Results of the analysis of the 2002 Ford Explorer ATF.

Oil contamination		Oil condition		Transmission wear	
Attribute	Result	Attribute	Result	Attribute	Result
Silicon	20 ppm	Boron	36 ppm	Iron	184 ppm
Water	<0.1%	Barium	1.4 ppm	Nickel	0.9 ppm
		Calcium	92 ppm	Chromium	0.9 ppm
		Magnesium	15 ppm	Titanium	0.3 ppm
		Molybdenum	2.6 ppm	Copper	148 ppm
		Sodium	18 ppm	Aluminum	41 ppm
		Phosphorus	257 ppm	Tin	11 ppm
		Zinc	83 ppm	Lead	17 ppm
		Viscosity at 40°C	33.97 cSt	Silver	0.0 ppm

ACKNOWLEDGMENTS

The author would like to thank Doug Byron, Forensic & Scientific Testing, for his invaluable technical assistance and Sarah Brown, Georgia State University, for her help in the manuscript review.

BIBLIOGRAPHY

[1] Crouse WH and Anglin DL. (1984) *The Auto Book*, 3rd edition, McGraw-Hill, New York, NY.

[2] Evans JS and Hunt TM. (2004) *The oil analysis handbook*, 1st edition, Coxmoor Publishing Company, Oxford, United Kingdom.

[3] Evans JS. (1997) How to read a can of oil (Part 1), *WearCheck FYI*, available at http://www.wearcheck.com, last access performed on September 28, 2005.

[4] Society of Automotive Engineers (1997) Engine oil viscosity classification—SAE J300 DEC95, *SAE Handbook, Volume 1 Materials, Fuels, Emissions, and Noise*, SAE, Warrendale, PA, pp 12.26–12.29.

[5] Society of Automotive Engineers (1997) Axle and manual transmission lubricant viscosity classification—SAE J306 OCT91, *SAE Handbook, Volume 1 Materials, Fuels, Emissions, and Noise*, SAE, Warrendale, PA, pp 12.49–12.50.

[6] Society of Automotive Engineers (1997) Fluid for passenger car type automatic transmission—SAE J311 FEB94, *SAE Handbook, Volume 1 Materials, Fuels, Emissions, and Noise*, SAE, Warrendale, PA, pp 12.56–12.58.

[7] Lubrigard Ltd (2005) *A guide to taking oil samples*, WearCheck, Burlington, Ontario, Canada.

[8] Byron D. (2005) Personal communication, Forensic & Scientific Testing, Inc., September 10, 2005.

[9] Smith M. (2000) *Basics of oil analysis*, Analysts, Inc., Torrance, CA.

[10] Geach A. (year unknown) Detecting particles in oil (Part 2), *WearCheck Technical Bulletin*, **25**, available at http://www.wearcheck.com, last access performed on September 28, 2005.

[11] Mayer A. (year unknown) Greek for beginners (Part 1) or the tests and what they tell us, *WearCheck Technical Bulletin*, **19**, available at http://www.wearcheck.com, last access performed on September 28, 2005.

[12] Evans JS and Robinson N. (2003) We are ready for more soot, *WearCheck Technical Bulletin*, **26**, available at http://www.wearcheck.com, last access performed on September 28, 2005.

[13] Blevins B. (year unknown) Silicon. enemy number one, *WearCheck FYI*, available at http://www.wearcheck.com, last access performed on September 28, 2005.

[14] Evans JS. (1999) Wear limits versus trends, *WearCheck Technical Bulletin*, **15**, available at http://www.wearcheck.com, last access performed on September 28, 2005.

[15] Duncanson M. (2005) Detecting and controlling water in oil, *Practicing Oil Analysis*, September–October, pp 22–28.

[16] Fitch J. (2005) A much closer look at particle contamination, *Practicing Oil Analysis*, September–October, pp 2–5.

[17] Geach A. (year unknown) Detecting particles in oil (Part 1), *WearCheck Technical Bulletin*, **24**, available at http://www.wearcheck.com, last access performed on September 28, 2005.

[18] Mayer A. (year unknown) Greek for beginners (Part 2) or the tests and what they tell us, *WearCheck Technical Bulletin*, **20**, available at http://www.wearcheck.com, last access performed on September 28, 2005.

[19] Robinson N. (year unknown) Monitoring oil degradation with infrared spectroscopy, *WearCheck Technical Bulletin*, **18**, available at http://www.wearcheck.com, last access performed on September 28, 2005.

CHAPTER 12

EXAMINATION OF BURNED VEHICLES

Eric Stauffer

12.1 INTRODUCTION

In 2003, there was an estimated 312,150 vehicle fires in United States [1]. In 2002, the 329,500 vehicle fires led, in addition to the dollar amount, to 565 deaths and 1,825 injured people [2]. It is important, however, to note that the total number of vehicle fires greatly diminished in the United States during the last 20 years (from 471,500 in 1980 to 312,150 in 2003) [2]. Indeed, it is estimated that in the United Kingdom the probability of accidental fires decreased from 67 fires per billion kilometers in 1993 to about 38 fires per billion kilometers in 2003 [3].

For the year 2003, it is estimated that approximately 10% of the vehicle fires in the United States were intentionally set, causing USD 132 million in loss [1]. An earlier study from the US Fire Administration reported that about 18% of vehicle fires were intentionally set [4]. They also cautioned that arson fires reported to the National Fire Incident Reporting System may be undercounted. Du Pasquier reported that the proportion of arsons among vehicle fires is about 15% in the United States and about 50% in the United Kingdom [5]. He partially explained this difference by the fact that causes suspicious in nature but not proven as arson are classified as undetermined in the United States, whereas they are classified as arson in the United Kingdom. Furthermore, the active functions of terrorism in the United Kingdom, in which vehicles are often involved (as discussed in Chapter 17) add to the statistics. Official statistics from the United Kingdom reflect 92,800 vehicle fires for the year 2003, which led to 79 fatalities and 651 people injured [3]. Although these figures represent a decrease of the total number of vehicle fires, the proportion of deliberate fires is estimated to be 81% [3]. It is important to note that deliberate fires in the UK report are considered as "those where deliberate ignition is merely suspected, and those recorded by the brigade as 'doubtful'" [3].

In the United States, the National Fire Protection Association reports that the dollar amount for intentionally set vehicle fires is disproportionate compared with that for accidental vehicle fires [2]. During the period of 1994 to 1998, intentional vehicle fires were in the proportion of 16 to 17% of total fires, whereas the dollar amount represented 23 to 25% of the total losses. The proportion of stolen vehicles that are recovered burned is quite small. In 1996, 8.3% of stolen-recovered vehicles were retrieved burned in the United Kingdom [6]. The same year, Australia recorded only 2.7% of stolen-recovered vehicles that were burned; however, this figure jumped to 4.3% for the year 2000, for a total amount of almost USD three million [6]. For the year 1994 in the United States, there were approximately 4% of stolen-recovered vehicles that were recovered either burned of flooded [7]. From this author's experience, about 25% of vehicle fires are caused by arson and only a few percent are stolen-recovered.

Fire investigation is a specialized discipline of forensic sciences. However, in many countries, and particularly in the United States, most fire investigators are often not forensic scientists and did not undergo any formal general forensic training. Thus, their approach is purely from a fire investigation perspective, which consists of determining the origin and the cause of the fire. They typically do not and are not trained to approach a stolen-recovered vehicle from a criminalistics perspective and to answer other pertinent questions, such as "Was the vehicle stolen?", "How was it stolen?", and "Who stole the vehicle?"

Examining a vehicle to determine the origin and the cause of the fire that destroyed it requires the removal of different parts and debris from it. Pieces of evidence that present a forensic interest to the auto theft investigator might be seriously disturbed by the fire investigator and would prevent the auto theft investigator to conduct his or her work properly. The opposite is also true. For this reason, the fire investigation portion of the forensic examination should be conducted simultaneously with the full forensic (criminalistics) examination of the vehicle, with both scientists working together as a team. The ideal situation is to have one forensic expert, proficient in both auto theft investigation and fire investigation, conducting the examination.

In any instance, it is very important for the auto theft investigator to fully comprehend how a complete vehicle fire investigation is performed. This allows for better integration of the fire investigation in the overall forensic examination of the vehicle. It also helps the auto theft investigator to work in a team setting with the fire investigator and for each to respect the other's needs. Finally, it shows the auto theft investigator the important elements to conducting a proper fire investigation.

This chapter does not portend to cover the topic of vehicle fire investigation in a comprehensive manner. To do so, a whole book or even a series of volumes could be written. On the contrary, this chapter concentrates on the examination of burned stolen-recovered vehicles. The accidental causes of vehicle fire, which cover more than half of the vehicle fires in the United States, are mostly ignored. The reason is that this book focuses on stolen-recovered and crime-related vehicles, and most, if not all, vehicles recovered burned after being stolen result from arson.

Although the odds for the auto theft investigator to encounter a burned stolen-recovered vehicle are quite small, it still does happen around the world on a daily basis. The investigation of the fire that burned the vehicle is a separate science to the regular auto theft investigation, which requires different knowledge, skills, and experience. The author invites the reader who is interested in learning more about fire investigation to read important publications in the field and to gain the necessary hands-on experience by examining burned vehicles [8–10]. Unfortunately, there is no good, accurate, up-to-date, and comprehensive publication dealing with the investigation of vehicle fires available on the market.

12.2 BASIC PRINCIPLES OF FIRE

12.2.1 Fire Triangle

Fire is an exothermic reaction involving a combustible and an oxidizer that occurs at a rate rapid enough to produce both heat and light. For a fire to occur, three conditions must be present and interact together:

- Combustible (or fuel);
- Oxidizer (or oxygen);
- Thermal energy (or source of ignition).

The theory of the fire triangle describes this phenomenon [11]. Each condition is represented by one side of the triangle, as shown in Figure 12-1. When the three conditions are brought together (assembly of the three sides to form a triangle), fire occurs.

This theory is simple and works well. The fact that each side is connected to the other symbolizes the continuous interaction between all three conditions. The fire is stopped by breaking away one of the sides forming the triangle, thus inhibiting the interaction of this condition with the two others [12]. This is the principle of fire extinguishment [13]. If the fuel (combustible) is removed, the triangle is not complete, and without fuel, no fire is possible. If the oxidizer is removed, again, the triangle is not complete, and therefore no fire occurs. If combustible and oxidizer are present together without sufficient thermal energy, the triangle is not complete and no fire results. In the everyday world, many combustibles are in constant contact with oxidizers; however, fire typically does not spontane-

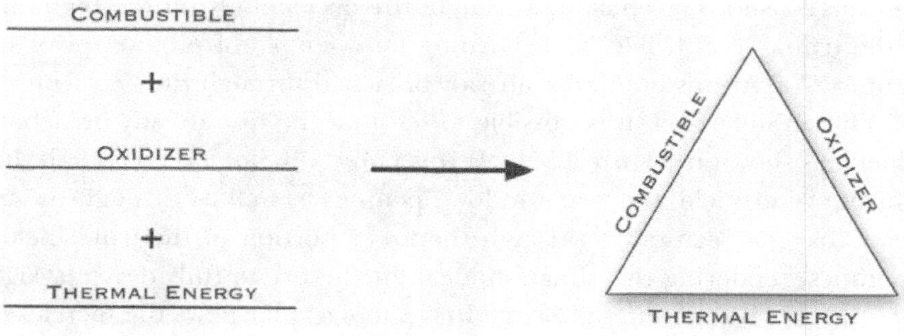

Figure 12-1

Illustration of the principle of the fire triangle. The three sides, representing the three necessary conditions for a fire to occur, are shown on the left as combustible, oxidizer, and thermal energy. When assembled together in a triangle shape, as shown on the right, the continuous interaction between these conditions is established and fire occurs.

ously occur. Only with the addition of sufficient thermal energy (an ignition source) does combustion occur.

12.2.2 Combustible

Combustibles are comprised of any materials that exist in a chemical state in which they can be oxidized in the presence of a suitable ignition source [11]. Combustibles, or fuels, are present almost everywhere. Some common fuels such as gasoline or wood are well known. However, there are many combustible materials that are not necessarily thought of as fuel, such as some metals and other inorganic compounds [14]. Fuels can be organic or inorganic materials. They can be simple elements or complex molecules. Fuels can be found in the solid, liquid, or gaseous phase [12]. In vehicles, fuels can be present in all three phases. Table 12-1 presents a list of some common liquid, gaseous, and solid fuels with their major chemical composition [5, 15].

The list in Table 12-1 is not exhaustive. There could be many other polymers and materials found in different proportions within a vehicle. Also, one object might be made of many different materials, and one material may be used for many different objects. Plastics and polymers typically account for one third of the total mass of a modern vehicle. Some vehicles can present more parts made of plastics and thus may be more combustible than the average vehicle. For example, the car manufacturer Saturn has installed plastic fenders on all its vehicles until 2005, when metal fenders were used on some of its vehicles for the first time, and Chevrolet Corvettes are equipped with fiberglass body parts. Also, some wheels are made of magnesium, which burns violently and to completion once ignited.

Some of these solid fuels contain fire retardant materials to satisfy different regulations. In the United States, the *Code of Federal Regulations* contains a section dedicated to the requirements needed for a certain combustible material to be present in a vehicle [16]. However, even in the presence of fire retardant materials and even if federal regulations are exceeded, these materials burn very well once ignited and provide a tremendous heat release. Figure 12-2 shows a vehicle that caught fire accidentally on the highway. The fire started in the trunk area. It had been burning for a couple of minutes prior to the first view in Figure 12-2a. At this point, fire already breached through the rear window and the moon roof. One minute later, it is possible to see that the fire already breached through the windshield, as shown in Figure 12-2b. At this point, the hot gases can exit through the top parts of the windshield and rear window openings as well as through the moon roof and fresh air (oxygen) can enter through the lower portion of the windshield and rear window openings, rendering this situation ideal for the fire to fully develop. Figure 12-2c, taken approximately two minutes after the first photograph, shows the increased intensity of the fire. The rear side windows failed under the heat, adding more dynamics to the fire. Then, melted plastic falls to the ground on fire, spreading the fire to the lower parts of the vehicle. The rear bumper also starts to burn. Figure 12-2d, taken approximately 6.5 minutes after the first photograph, shows the fire consuming the rear lights and spreading under

Table 12-1

List of fuels typically encountered in a vehicle.

	Object/material	Major components
Liquids	Gasoline (also called petrol)	Hydrocarbons
	Diesel fuel	Hydrocarbons
	Engine oil	Hydrocarbons
	Brake fluid	Glycols or silicone compounds
	Power steering fluid	Hydrocarbons
	Automatic transmission fluid	Hydrocarbons
	Gear lubricant	Hydrocarbons
	Hydraulic fluid (usually on heavy equipment)	Hydrocarbons
	Coolant	Glycols
Gaseous	Liquefied petroleum gas (LPG)	Hydrocarbons
	Camping fuel (in residential vehicles)	Propane/butane
	Air-conditioning refrigerant	Tetrafluoroethane (R134a), polyalkylene glycol
Solids	Exterior body parts and interior components (bumper, hood, fender, quarter panel, trim, door panel, dashboard, molding, etc.)	Acrylonitrile-butadiene-styrene (ABS)
		Polypropylene
		Polyethylene
		Polycarbonate
		Polyvinyl chloride
		Polyurethane
		Polyamide
		Polystyrene
		Fiberglass
	Interior carpet and floor mats	Nylon
	Upholstery	Polyurethane
		Leather
		Polyester
		Polyvinyl chloride
		Nylon
		Cotton
	Tires	Styrene-butadiene copolymers
		Polyisoprene
		Polyisobutylene
	Electrical insulation	Polyethylene
		Polyvinyl chloride
	Headlight and rear lights	Polycarbonate
	Hoses	Polybutadiene
	Rubber parts	Polyisoprene

This list provides a good indication of the types of material available in a vehicle but is not a comprehensive list.

Figure 12-2

(a) Photograph of a vehicle that accidentally caught fire on the highway. The fire started in the rear part of the passenger compartment and had been burning for a couple of minutes before this view. T = 0:00 [min:sec]. (b) T = 0:54. (c) T = 1:56. (d) T = 6:23. See Color Plate.

the vehicle. It is easily imaginable that in the next few minutes, the fire will reach the engine compartment and continue to destroy every single combustible in the vehicle.

A vehicle that caught fire accidentally or that was intentionally set on fire can burn to quasi-completion in as little as 15 minutes. A recent full-scale fire test of a parked van, which was set on fire in the cargo compartment with an open flame and some paper, showed that flashover inside the van was reached about six minutes after ignition [17]. It is important to note that in this particular test, a window was mechanically broken after about four minutes.

When all windows and doors of the vehicle are closed, and particularly if ignitable liquids are used, it is not uncommon to see the fire dying due to lack of oxygen. However, if proper ventilation is provided, the fire will grow to flashover and burn the vehicle to completion unless extinguishment procedures are conducted. Figures 12-3 and 12-4 show two examples of vehicles that burned to completion. At this stage, although a proper fire investigation should be performed, the chances of finding the origin and cause of the fire are highly reduced. Some original burn patterns might still have survived the long fire and

Figure 12-3
Example of a stolen-recovered Chevrolet Cavalier that burned to completion (post-flashover). This vehicle must have burned for several dozen minutes and every combustible present in the vehicle was consumed.

Figure 12-4
Example of a Chevrolet Tahoe that burned to completion (post-flashover). This vehicle must have burned for several dozen minutes and every combustible present in the vehicle was consumed.

may indicate a compartment of origin. However, every combustible in that compartment will be burned to completion and eventual physical evidence will be greatly damaged, if not completely destroyed, particularly if the fire was intentional.

12.2.3 Oxidizer

The oxidizer reacting with combustible materials in almost all fires is oxygen (O_2). Oxygen is readily available in earth's atmosphere, as air contains approximately 21% of O_2. For a flaming combustion to be sustained, a minimum proportion of 15% O_2 is required [18]. Below that, the burning rate is highly diminished [8]. Eventually, flaming combustion ceases and smoldering combustion (or glowing combustion) usually occurs and persists even to very low levels of oxygen. The presence of an oxidizer is normally not an issue of concern when investigating a fire, because it is readily available in the atmosphere. However, under certain circumstances, the lack of oxidizer can prevent a fire from occurring or spreading. For example, Figure 12-5 shows a Ford Ranger that was deliberately set on fire and abandoned.

The front seat was lit with an accelerant (ignitable liquid) and the doors were closed with all the windows up. The ignitable liquid burned and quickly used the available oxygen in the passenger compartment. At this point, the combustion was incomplete, as evidenced by the heavy soot deposit on the upper surfaces of the passenger compartment, and more

Figure 12-5

Example of a Ford Ranger that was deliberately set on fire using an ignitable liquid as an accelerant. The doors were closed immediately after ignition, with the windows rolled completely up. The fire extinguished itself by starvation of oxygen. Heavy soot deposits on the surface of the interior, such as on the windshield, are typical of an incomplete combustion, resulting from a low availability of oxygen.

particularly on the windshield. Once the oxygen level falls below 15%, flaming fire would stop and smoldering fire would occur. Because this vehicle was recovered several hours after it was set on fire, the fire had completely extinguished by itself. A strong smell of ignitable liquid was present when opening the doors, because not all the accelerant burned. When fire cannot sustain because there is not enough oxygen available but still plenty of fuel, it is said that the fire dies by starvation or lack of oxygen.

12.2.4 Thermal Energy or Source of Ignition

There are several forms under which a source of ignition can be found. These different thermal energies all originate from the basic thermal sources, which are electrical, mechanical, chemical, and biological. Also, there are many different manners to classify these sources. More details concerning the different sources of ignition found in a vehicle are provided in Section 12.4. Table 12-2 presents the different sources of ignition with some examples [19].

Table 12-2
Different types of source of ignition with some examples.

Sources of ignition	Examples
Direct flame	Match, cigarette lighter
Heating	Electrical (high resistance), mechanical (friction), radiant (sunlight), chemical (spontaneous ignition)
Arcs/sparks	Electrical or mechanical

12.2.5 Heat Transfer

Heat can be transferred from a zone of higher temperature to a zone of lower temperature via three different modes [10].

A/ Convection

Convection is the transfer of heat from a moving fluid to a solid surface. This fluid can be a gas or a liquid. When a fire is burning, flames, smoke, and hot gases are traveling upward. In a vehicle, when the engine is turned on, hot exhaust gases are produced. They are confined to the space defined by the exhaust system. The exhaust manifold and exhaust pipes become hot very quickly, because they have been heated by convection from the hot exhaust gases.

B/ Conduction

Conduction is the transfer of heat energy by direct contact between two solids or within a solid. The molecular activity, increased by heat, is transferred to neighboring molecules.

For example, in a vehicle, the bolts securing the wheel to the axle are usually very hot after extensive driving and braking. These bolts have been heated by conduction of the heat from the brake disk or drum to the bolts.

C/ Radiation

Radiation is the transfer of heat energy by electromagnetic wavelengths. There is no physical contact between the heat source and the surface being heated. Often, this transfer of heat is invisible but can be felt at a great distance. An example of such heat transfer is the heating of the passenger compartment of a vehicle by sunlight.

D/ Final Considerations

It is very important for the fire investigator to be familiar with these concepts of heat transfer, because they play a crucial role in the determination of the cause of a fire. The presence of a heat source is one element to establish; however, the feasibility of the ignition of a given material from that heat source is another event that needs to be evaluated with great accuracy. Some heat sources are not suitable for some materials. For example, it is almost impossible to ignite gasoline on a hot exhaust pipe, even if the pipe's temperature is above the ignition point of gasoline. Also, gasoline will not be ignited by a smoldering cigarette [20]. By contrast, although it is very difficult to ignite automatic transmission fluid with the open flame from a lighter, it will readily ignite on a hot exhaust manifold. Each material and ignition source have their own characteristics, and one must ensure that heat transfer between them can take place in a manner suitable for ignition to occur.

12.3 GENERAL PRINCIPLES OF FIRE INVESTIGATION

12.3.1 Purpose

The role of fire investigation is to answer the following two questions: "Where did the fire start?" and "Why did the fire start?"

A/ Where Did the Fire Start?

This question refers to the determination of the origin of the fire. The goal is to identify the first combustible ignited and its location. This location is called either the point of origin or the seat of the fire. It is crucial to identify this location, because it is impossible, except under some particular circumstances, to answer the second question without answering the first one. Narrowing the point of origin to a 1- to 10-cubic decimeter volume is typically sufficient to begin looking for clues to the second question. In some instances, it is not possible to determine a very precise area of origin, which complicates the subsequent process.

B/ Why Did the Fire Start?

This question refers to the determination of the cause of the fire. The goal is to identify the first combustible that caught fire and the ignition source that started the fire. This ignition occurred at the point of origin. To identify the cause of the fire, the investigator needs to be aware of the different sources of thermal energy and different combustibles available in the area of origin. As previously explained, for a fire to occur the three conditions of the fire triangle must be present and interact together. Although it is good to identify an ignition source and a combustible, the investigator must demonstrate that the ignition source is suitable for that particular combustible and that the proper heat transfer was completed.

C/ Root Cause Analysis

Although determining the first combustible ignited and the source of ignition is good, it is often not sufficient for a comprehensive fire investigation. It is also important to determine the exact event(s) that led to the contact between the particular source of ignition and the fuel. This process, which is not always performed by the fire investigator, is commonly called "root cause analysis." This is an important step in fire investigation, particularly with accidental fires, because it allows for the determination of who (or what) is ultimately responsible for the fire. For example, a fire starts under the driver's seat right where an electrical wire is located. The wire arcs on the bottom rail stop of the front seat, which sets the carpet on fire. This would be the cause of the fire. However, it is important in this case to determine why the cable arced on the bottom rail. Who installed the cable? Was the cable loose? Was the cable located at the location where it is supposed to be? Did the insulation of the cable degrade beyond a certain safety point? Was the cable overloaded? Several parties could be responsible for this fire, such as the manufacturer of the vehicle, the owner, or the repair shop who performed some electrical repairs. After a detailed investigation, it was determined that the owner of the vehicle installed the cable by himself, running it under the seat. The seat was moved back and forth on a regular basis because different drivers were operating the vehicle. The movement of the seat combined with the fact that the cable was placed loose under the seat with no particular attention degraded the cable's insulation to the point that it arced with the seat's rail stop. Thus, the responsible party is the owner who improperly installed this cable.

12.3.2 Fire Causes

A/ In General

Fire causes can be classified into three categories: accidental, natural, and incendiary [21]. A fourth category, undetermined, is used when the investigation fails to reach a conclusion as to the cause of the fire. The accidental category comprises fires that have been caused by direct or indirect human intervention, without the intent of creating and spreading such a fire. This includes but is not limited to mechanical and electrical failures, negligence, carelessness, or ignorance. Natural causes encompass all fires that are not caused by a direct or indirect human intervention. Examples of such causes are volcano lava spills, lightning strikes, and some cases of spontaneous combustion. The incendiary category represents all fires that have been caused by the deliberate intervention of a human being with the intent of setting and spreading fire where it should not be. The term "arson" is a legal definition that can vary from jurisdiction to jurisdiction. The fire investigator often uses the undetermined category because only a certain proportion of fire causes can be clearly determined. If the fire cannot be classified in any of the first three categories, then the cause should be listed as undetermined.

It is very difficult to substantiate that a fire was incendiary. Fire, by nature, destroys evidence, and if not extinguished at an early stage, the destruction might be carried out beyond any possible recognition. Also, there are many mechanisms for igniting a fire that would leave no physical evidence. Finally, although it is often possible to demonstrate a human intervention as the cause of a fire, it is impossible in many instances to substantiate the intent. The modus operandi might be identical in both fires, which differ only by the state of mind of the perpetrator.

The demonstration of deliberate fires can be achieved in two different fashions: positive and negative corpora. The *corpus delicti* means literally "body of offense" and is a Latin term used to describe the concrete evidence of a crime. If the investigation reveals direct evidence of the malicious act, such as the presence of gasoline poured on the front seat of a vehicle with a burned match lying there, the demonstration of arson is made by positive corpus. For example, Figures 12-6 through 12-8 show a 1995 Lexus SC300 that was deliberately set on fire. Figure 12-6 is a view of the front seat, where a plastic jug filled with gasoline was found only partially burned. Gasoline was splashed over the interior of the vehicle and then ignited. The perpetrator closed the doors and the windows of the vehicle, and the fire, after burning for a short period of time, eventually died by starvation of oxygen. The damage in the vehicle is limited to the upper part of the passenger compartment, as

Figure 12-6

Example of a deliberately set fire of a 1995 Lexus SC300. A plastic jug filled with gasoline was found on the front seat, partially burned. The vehicle did not undergo flashover, as the perpetrator closed the doors and windows. The fire died by starvation of oxygen. (Photograph courtesy of Mark D. Culver, Key Fire Investigations, Inc.) See Color Plate.

Figure 12-7
View of the interior of the 1995 Lexus SC300 showing fire damage on the upper part of the passenger's compartment. Note the heavy soot deposit throughout the vehicle, due to an incomplete combustion reaction. (Photograph courtesy of Mark D. Culver, Key Fire Investigations, Inc.) See Color Plate.

Figure 12-8
View of the trunk of the 1995 Lexus SC300, which was splashed with gasoline but not ignited. Notice the unburned match lying on the carpet of the trunk. (Photograph courtesy of Mark D. Culver, Key Fire Investigations, Inc.) See Color Plate.

shown in Figure 12-7. The trunk was also splashed with gasoline but was not ignited. An unburned match was found lying in the trunk, as shown in Figure 12-8.

Another example of deliberate fire where the positive corpus is easily demonstrated is shown in Figure 12-9. This is a 2003 Ford Escort that was stolen and then set on fire to cover the theft. Fortunately, the fire went out very quickly and the car was well preserved. It is possible to see the two canisters of cigarette lighter fluid on the floorboard. A large burn pattern is located on the left part of the passenger seat, with no source of ignition in

Figure 12-9

View of a 2003 Ford Escort that was stolen and recovered burned. This is a clear example of deliberate fire by positive corpus as it is possible to see the very clear origin with no accidental sources of ignition and the two canisters of cigarette lighter fluids still sitting on the floorboard. Several matches (burned and unburned) were also recovered throughout the vehicle. (Photograph courtesy of Mark D. Culver, Key Fire Investigations, Inc.)

the area (besides the matches). Other smaller burn patterns can be seen on the right side of the seat, apparently not connected to the large burn pattern, which could suggest multiple points of origin.

If the demonstration of deliberate fire is done by logically and scientifically eliminating all other possible causes of fire (natural and accidental), it is referred to as a negative proof of corpus delicti, or more simply negative corpus [22]. Although in some instances it is possible to eliminate all natural and accidental sources of ignition, hence demonstrating human intervention, it is rarely possible to demonstrate a malicious intent versus an unintentional act. Therefore, the demonstration of incendiary fires is a difficult task, even more than the demonstration of accidental fires, for which physical evidence of the failure or event that brought the ignition source in contact with the combustible often survived.

BI With Vehicles

As any other premises or objects, vehicles can catch fire naturally, accidentally, or as a result of a deliberate human intervention. Table 12-3 shows statistics of vehicle fires in United States for the year 1999 [2].

Natural occurrences are extremely rarely encountered and typically represent less than one percent of the total number of fires, as seen in Table 12-3, where only 540 vehicles caught fire naturally in 1999. Accidental causes include both the category "failure of equipment or heat source" and the category "unintentional," found in Table 12-3. Thus, more than 80% of fires are accidental in nature. A vehicle carries many potential ignition sources, because it is a very complex machine, with many different mechanical and electrical components that move or rotate in a very harsh environment. A deliberate human intervention is the final category from which vehicles catch fire. As seen in Table 12-3, about 16% of the fires were intentionally set. Interestingly, only 3% of the fires are reported as undetermined,

Table 12-3

Statistics of fire causes for vehicles in the United States for the year 1999.

	Fires		Direct property damage		Injuries		Deaths	
	Number	%	USD amount in M	%	Number	%	Number	%
Failure of equipment or heat source	192,800	66.9	496.1	54.9	478	41.4	35	9.9
Intentional	45,900	15.9	204.5	22.6	85	7.3	28	8.0
Unintentional	40,370	14.0	174.2	19.3	557	48.2	264	75.3
Unclassified	8,610	3.0	28.2	3.1	35	3.0	24	6.8
Act of nature	540	0.2	1.2	0.1	0	0.0	0	0.0
Total	288,220	100.0	904.1	100.0	1,154	100.0	350	100.0

Source: Ahrens M. (2004) US vehicle fire trends and patterns. *National Fire Protection Association, Quincy, Massachusetts.*

which is an unusually low number. This is probably because these figures are extracted from the National Fire Incident Reporting System and data input is carried out by the responding fire departments. In many instances, no formal investigation was performed or no competent fire investigator examined the fire and the fire is merely reported as "witnessed" by the responders. Among fire investigation professionals, the proportion of undetermined fires is much higher.

C/ With Stolen-Recovered Vehicles

Realistically, any vehicles reported stolen can be placed in two categories: vehicles actually stolen by a third party and vehicles subject of a fraud scheme involving their owner. Vehicles actually stolen by a third party (criminal) are sometimes set on fire after their use. The vehicle could have been stolen as a joyride or to commit another crime (see Chapter 1). The vehicle is set on fire to eliminate traces and evidence that would lead to the identification of the perpetrators, to delay the identification of the vehicle, or simply to vandalize, for the pleasure of the perpetrators. Vehicles reported stolen by their owners and set on fire as a part of insurance fraud schemes could involve different motives, which are ultimately all financial. Such motives include mechanical problems with the vehicle, high mileage on leased vehicles, or simply the need for immediate cash (see Chapter 19).

Vehicles are also set on fire without being stolen. There are a number of intentionally burned vehicles that are part of an overall scheme of insurance fraud that are not reported as stolen by their owners. In many instances, the malicious owner will set the vehicle on fire, claim a possible mechanical defect as the cause of the fire, and try to obtain reparation for the loss.

Also, vandalism is a cause of vehicle arson. For example, in France, every year it has become a sort of tradition for young hooligans to set hundreds of vehicles on fire during the New Year celebration [23, 24]. This phenomenon affects particularly the cities of Paris and Strasbourg. These crimes are not committed for any particular reason and are thus stupid and gratuitous acts of vandalism. Other similar isolated acts of vandalism are found all around the world.

12.3.3 General Approach to the Fire Investigation

The auto theft investigator periodically investigates a stolen-recovered vehicle that has been burned. In such instances, the vehicle has almost always been set on fire intentionally. The situation of a thief stealing a vehicle that accidentally catches fire during (or after) the perpetration of the crime has yet to be reported in the literature. Even if this situation is plausible (but very unlikely), the thief would have a very hard time convincing any investigator of such facts unless strong evidence of accidental fire is present on the vehicle.

In any case, the investigator approaching the burned stolen-recovered vehicle should always proceed with a full fire investigation as part of the forensic examination of the vehicle. A fair and scientific investigation of a vehicle fire must remain unbiased. The investigator must at first consider all possible scenarios and all possible causes. It is not an acceptable practice to start the examination of a vehicle believing it is an accidental or deliberate fire. Open-mindedness and integrity are important qualities in this line of work. This is the reason why a fire investigator must be trained and experienced in all causes of fire and not only in arson cases. As the investigation progresses, the investigator can narrow the possible causes of fire by eliminating different scenarios and eventually finding evidence of one and only one scenario. At this point, it is possible to express an opinion as to the origin and cause of the fire. Because this chapter does not cover natural and accidental causes, it is important to note that it does not relate how a complete investigation is performed. This chapter is only written to inform the auto theft investigator of what a fire investigation is and to illustrate concepts with examples of burned stolen-recovered vehicles, which represent only a small proportion of vehicle fires.

Also, the fire renders the full forensic examination much more difficult. The auto theft investigator must be aware of the limited observations that can be made on a burned vehicle compared with an unburned vehicle. For instance, the identification of the vehicle can become very cumbersome, if impossible. Fire destroys many elements of identification on the vehicle. Vehicle identification number (VIN) plates may survive some fires or may be completely destroyed. VIN plates made of aluminum will likely melt completely as the fire progresses. If the fire is long and intense enough, all safety certification labels and anti-theft labels will completely disappear in the fire. Figure 12-10 shows the conditions of some VIN plates after a fire. The identification of a completely burned vehicle might be very difficult. In such instances, the investigator will need to rely on stamped VINs or partial VINs and confidential VINs as explained in Chapter 6.

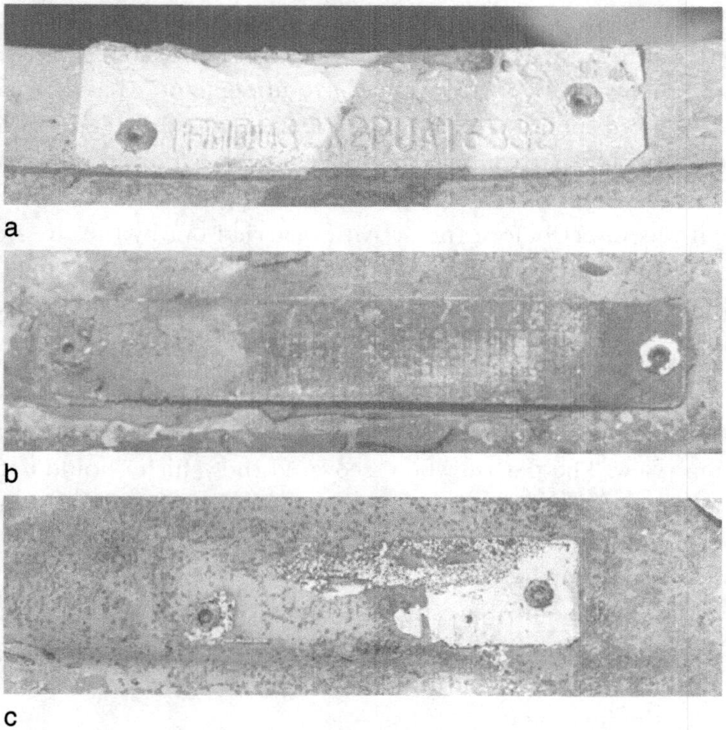

Figure 12-10

View of different degradations of VIN plates due to fire. Some plates might survive (a), some might be partially degraded (b), and some might be completely destroyed (c). This renders the identification of the vehicle more difficult and the investigator will need to rely solely on stamped (usually partial) VINs and confidential VINs.

12.3.4 Interviews and Information Collection

A/ In General

Interviewing any witnesses or suspects to the fire is a mandatory step in the investigation. This step should be performed as early as possible and prior to the examination of the fire scene. Witness and/or suspect statements should be carefully recorded. It is also important for the investigator to stay unbiased and not to develop a prejudice toward the fire scene based upon the statements. Nevertheless, these statements should be evaluated and used to attribute different degrees of significance to different locations, items, examinations, and observations at the fire scene.

The driver of the vehicle should be extensively interviewed. This can reveal the conditions under which the vehicle was operated and any possible malfunctions that could have occurred prior to the fire. This step should be performed even if the vehicle was parked at the time of the fire. It is always possible to gather some information regarding the previous immediate use of the vehicle and its history.

The fire department personnel who responded should also be interviewed. Although these workers are not investigators, they might have very important observations to relate. Furthermore, it is necessary for the investigator to understand what activities were carried out at the scene by the responding personnel. For example, if firefighters forced open the driver's door (that was intact) from a burning vehicle, it is important for the investigator to know this. Thus, he or she will not spend time trying to understand how the door was forced, because it was intact before the activities carried out by the fire department. Fire reports should be obtained and carefully studied. Some fire reports might not offer any more information than the date and time of the emergency call, whereas some other reports might offer a plethora of very pertinent information.

B/ With Stolen-Recovered Vehicles

The problem with stolen vehicles that are recovered burned is that there is usually no (reliable) driver to interview. The person who discovered the vehicle should definitely be interviewed, as well as the fire personnel who responded to the incident. Many times, such a vehicle is found several hours or days after it burned and no fire department responded to the scene. In such instances, there are virtually no interviews to be conducted for the fire investigation. In any case, the fire investigator should obtain and review all interviews conducted in conjunction with the theft investigation of the vehicle. This could offer pertinent information regarding the history of the vehicle (particularly mechanical history), which could be helpful to the investigation.

If the driver or owner is suspected of the theft and fire of the vehicle, a crime scene investigator or forensic scientist should examine the suspect, his or her clothing, and his or her shoes for any potential physical evidence. For example, if an ignitable liquid was used to set the vehicle on fire, it is possible that residues will be present on the suspect's hands or clothing [25, 26]. In such instances, it would become pertinent to collect these items as evidence.

12.3.5 Personnel Protection and Equipment

Fire investigation is a very dirty job. It involves crawling, digging, and sifting through burned debris. When doing his or her job properly, the fire investigator cannot stay clean. The auto theft investigator who desires to participate in or to conduct a fire investigation must wear proper attire. Not only should this attire be comfortable enough for the investigator to perform all necessary movement and to get into any necessary positions, but it should also be strong enough to protect against broken glass, sharp objects, charred debris, hot surfaces, and some chemicals. The basic personal protective equipment includes a strong jumpsuit, gloves, waterproof steel-toed boots, goggles, hard hat, and facemask. It is strongly recommended to use disposable gloves, such as the nitrile type, which are much more resistant to tear and chemicals than regular latex gloves. The gloves should be changed regularly to avoid contamination and to guarantee their integrity and thus their level of protection. A good practice involves wearing two or three pairs of such gloves simultaneously, which increases the safety and facilitates the change of gloves. Also, a pair of working gloves should be available when moving debris in situations where sharp and dangerous items are handled. It is always important to remember that such gloves can carry contamination throughout a fire scene because they are typically not disposable. Thus, they must be thoroughly cleaned after each fire scene. Facemasks are extremely helpful when investigating vehicles that burned to completion, particularly under hot and dry weather. Under such conditions, remnants of fiberglass and other particulates quickly become airborne and the investigator breathes them unless wearing such as mask.

When working inside a vehicle, one must be able to get into the vehicle and to move around the passenger compartment and trunk area. These areas are not necessarily very large spaces and one might need to bend, get on his or her knees, and move in a crouched position. When the outside temperature approaches 37°C (100°F) and there is no shade, the investigator may also greatly suffer from heat and dehydration, particularly if the investigation requires several hours under these conditions. Thus, refreshments and sun protection should be available throughout the investigation to preserve the well-being of the investigator. Similarly, when the outside temperature is subzero (°C), the investigator must wear proper winter attire to prevent any hypothermia. In addition, the investigator must be very careful with his or her hands as to prevent any possible frostbite.

Several tools are required when performing an investigation. Table 12-4 is a nonexhaustive list of the equipment and tools that should be available to conduct the investigation of a vehicle fire.

Table 12-4

Recommended equipment and material used to investigate vehicle fires.

Evidence documentation	Fire debris removal tools	Mechanic tools
Professional camera equipment (SLR system with lens 24–80 mm and a macro lens) including flash	Small shovel	Complete set of wrench and sockets for automotive work
Photographic ruler	Small rake	Pliers
Measuring tape	Small broom	Bolt cutters
Evidence marker	Small bucket	Hammer
Pens and pencils		Screwdrivers
Notepads		Pry bar
		Chisel
		Saws
		Set of Allen wrenches
		Sheet metal scissors
Other items	**Evidence collection items**	
Flashlight or other lighting system	Tweezers	
Compass	Brushes	
Thermometer	Precision knife	
Multimeter (ohmmeter, voltmeter)	Bags and container of all sorts for evidence	
	Clean metal paint cans (or other fire debris containers)	
	Evidence labels	
	Empty jars	
	Vehicle fluid collection kit	

12.4 DETERMINATION OF THE ORIGIN

12.4.1 General Principles

To determine the origin of a fire, one relies on fire, smoke, and heat patterns [8, 27]. Fire evolves following physicochemical principles and leaves patterns as evidence of this evolution on different surfaces. The task of determining the origin of a fire can be quite complicated and sometimes impossible. The earlier the fire is extinguished, the easier it is to determine its origin. If the fire grows strong enough and the circumstances permit, flashover occurs. Flashover is the sudden event in a compartment fire that leads to the full involvement of the compartment [10]. Thus, after undergoing flashover, every combustible present in the compartment is burning, even at floor level. If such an event occurs, original fire patterns are typically destroyed and replaced by new patterns, which might not allow for a clear determination of the origin of the fire. For example, Figure 12-6 shows the interior of a vehicle that did not undergo flashover. This is evidenced by the bottom part of the passenger compartment not being burned. Figures 12-3 and 12-4 show vehicles that underwent flashover. Figure 12-2a shows a vehicle that has not yet undergone flashover. However, a few minutes later, as shown in Figure 12-2d, the passenger compartment underwent flashover. Figure 12-11 shows the interior of a vehicle that underwent flashover and burned to completion. In this case, it is impossible to distinguish any burn patterns inside the passenger compartment that would indicate, for example, that the fire started on the driver seat rather than the passenger seat.

Figure 12-11

View of the interior of a vehicle that underwent flashover and burned to completion. All burn patterns that would have been indicative of the origin of the fire have been destroyed.

In other instances, the vehicle bears fire and smoke patterns that indicate the direction of travel and the intensity of the fire. By analyzing these patterns, it is possible in many instances to trace the fire back to its point of origin. The fire investigator first observes the vehicle from the outside and then from the inside to analyze and record these patterns. During this examination, photographs should be taken as a record of the examination. It is also a good practice, although not always necessary, to make a diagram of the vehicle and to indicate the direction of fire travel. In any instance, the investigator must take photographs of each area before removing any debris and/or other materials.

12.4.2 Compartments

When determining the origin of the fire, it is common practice to consider different compartments in the vehicle. These four compartments are illustrated in Figure 12-12.

Compartment 1 is the area defined by the engine compartment. It extends from the back part of the front bumper to the front part of the dashboard support panel or bulkhead (formerly known as firewall), between the two fenders and under the hood. Compartment 2 is known as the passenger compartment, also often referred to as the interior. This comprises the space delimited by the side doors and quarter panels of the vehicle, the roof, and the floorboard, located between the back part of the dashboard support panel and the rear seat backrests or trunk separation panel.

Compartment 3 is defined as the trunk or cargo space and is delimited by the trunk lid, the two rear quarter panels, and the floorboard. It extends from the rear of the rear seat backrests or trunk separation panel to the front part of the rear bumper. Note that in many vehicles (sport utility vehicles, station wagons, etc.) compartments 2 and 3 are not actually separated. In such instances, only one compartment is considered, because there is no physical separation between the trunk area and the interior. Compartment 4 is actually not a compartment per se but is defined as the area outside the vehicle. This area extends below and above the vehicle.

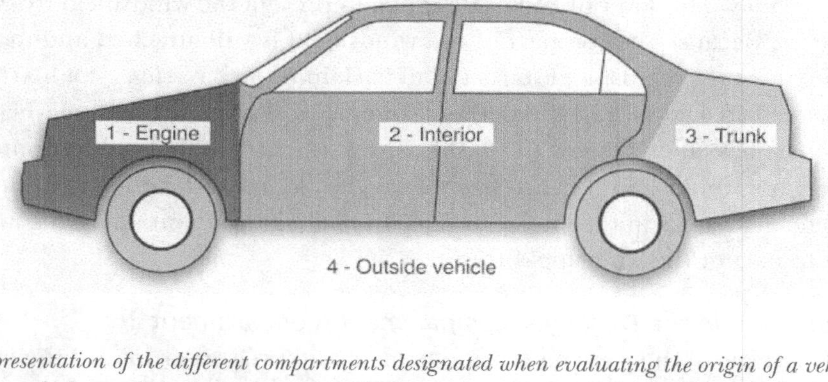

Figure 12-12

Schematic representation of the different compartments designated when evaluating the origin of a vehicle fire. Note that with some vehicles, such as station wagon or SUV, compartments 2 and 3 are one unique space.

12.4.3 Patterns on the Vehicle Body

A/ Limitations

Although it is not always possible to determine from which area the fire started, this operation is relatively easy when the proper patterns are present. The outside body of the vehicle usually bears patterns characteristic of the progression of the fire. The following describes the patterns of progression of fire from the engine compartment to the interior and vice versa. It is important to understand that although this progression often occurs as described, there are many exceptions where different routes can be taken by the fire. When the patterns described are encountered, they are usually characteristics of the fire travel associated with them. However, there are many instances when different patterns, or no patterns, might be present. The investigator is urged not to consider the following text as a rule but rather as a guide that fits the greatest number of cases.

The determination of the origin of the fire can help to corroborate or to contradict the testimony of the witnesses, if any. When an owner, for example, describes that he or she was driving the vehicle down the road and that smoke came out from under the hood, when clearly the fire started in the back part of the passenger compartment, there is a problem with the testimony.

B/ Fire Originating in Engine Compartment (Compartment 1)

When a fire starts in an engine compartment, it evolves throughout that area and eventually breaches into the passenger compartment. The fire travels from the engine to the passenger compartment through the openings of the dashboard support panel. The fire leaves a radiant heat pattern on the hood and fenders, as shown in Figure 12-13. These heat patterns are developed as waves would be by a stone thrown in water. The epicenter of these radiant patterns is usually located above the origin of the fire. The patterns seen on top of the hood are also seen under the hood in most instances.

Because fire tends to rise due to the heat, it starts to damage the bottom part of the windshield. The windshield glass will start to soften and break. The windshield is made of laminated glass and thus, contains a layer of polyvinylbutyral (PVB) sandwiched between the two glass panes. This layer of PVB is designed to prevent the windshield from scattering in small pieces. Because the top part of the windshield is still attached and mostly intact, the bottom part of the windshield starts to fall back into the passenger compartment. This can be observed in Figure 12-12. Another example is shown in Figure 12-14, where the windshield glass has fallen beyond the dashboard. If the fire would have continued to burn, most of the glass would have been retrieved on the floorboard rather than on the dashboard. When a fire starts in the engine compartment, it is common not to see any glass left at the bottom part of the windshield frame.

C/ Fire Originating in the Passenger Compartment (Compartment 2)

When the fire starts in the passenger compartment and progresses toward the engine compartment, the radiant patterns exhibit waves originating from the base of the windshield

Figure 12-13

Example of a burn pattern evolving from the engine compartment (no. 1) toward the interior (no. 2) on a Dodge Intrepid. Notice that the heat pattern originated in the middle of the hood close to the windshield and spread around. Notice how the windshield is first attacked by the fire at the bottom, whereas the top part is still intact.

Figure 12-14

Example of an engine compartment (no. 1) fire that progressed toward the passenger compartment (no. 2) on a Ford Explorer Sport Trac. Notice how the glass started to fall back. Most of the fallen glass will be retrieved on the floorboard. The vehicle was equipped with a composite hood, which disappeared with the fire, leaving only the frame of the hood.

toward the front bumper or from the doors (or rear part of the fenders) toward the front bumper. An example of radiant heat patterns originating from the passenger compartment on the hood is shown in Figure 12-15. This pattern appears as half-circles growing from the base of the windshield directed to the front bumper.

In some instances, the patterns found on the vehicle are very faint or nonexistent. An example of very faint and small patterns is shown in Figure 12-16. It is possible to observe very small wave patterns just ahead of the passenger door, on the top left part of the fender. These patterns show a progression of the fire from the passenger compartment to the engine compartment.

Figure 12-15

Example of radiant heat patterns present on a hood originating from the passenger compartment (no. 2) and progressing toward the front of the vehicle. This demonstrates a fire originating from the interior of the vehicle.

Figure 12-16

In some instances, the heat patterns might be very faint or not at all present. The only heat pattern that can be seen on this Ford F-150 is located at the upper right corner of the door, just under the base of the windshield and consists of a few faint wave patterns, which indicates an interior (no. 2) fire that progressed toward the front. See Color Plate.

Figure 12-17

View of the base of the windshield on a Chevrolet Blazer where the fire started in the passenger compartment (no. 2). It is possible to notice not only the fire pattern (on the hood) traveling from the interior to the engine compartment (no. 1), but also the glass from the windshield lying at the base of the windshield. The presence of glass in this fashion at this location is usually characteristic of an interior fire. See Color Plate.

Also, when the fire starts in the passenger compartment, the top part of the windshield is heated first. This makes the top part of the windshield collapse on the floorboard and seats before the bottom part of the windshield gets affected. The bottom part usually fails last and just lies down on the dashboard. In such instances, the base of the windshield frame usually bears some glass, as shown in Figure 12-17. Fires originating from compartment 3 (trunk) develop very similar patterns to the ones present with fire from the passenger compartment that evolve toward the front and the rear of the vehicle.

D/ Fire Originating from Outside of the Vehicle (Compartment 4)

Fires originating from outside the vehicle do not necessarily leave very significant patterns. It depends on many factors, the most important factor being the size of the external fire. If the fire starts on a taillight, some wave patterns originating from the taillight could be seen on the body parts. If the fire originates from a house fire or from a garage in which the vehicle is stored, there could be a complete absence of patterns, as the fire reaches all parts of the vehicle simultaneously. Figure 12-18 shows a Ford F-150 in which the front grill was set on fire. Fortunately, the fire was quickly noticed and extinguished. It is interesting to note the very clear wave pattern emanating from the front grill of the vehicle. If that vehicle would have burned to completion or at least reached the passenger's compartment, it is very likely that patterns showing the progression of the fire from the engine compartment (no. 1) to the passenger compartment (no. 2) would still have been present. However, the epicenter of these patterns, located at the edge of the vehicle, would strongly suggest an external fire.

Figure 12-18

Example of a fire that was deliberately set on the front grill of this Ford F-150. It is possible to observe the radiant pattern that originated from the front end of the hood and traveled toward the passenger compartment. The fire was extinguished at a very early stage, preventing further spread, and conserving the original fire patterns. No physical evidence of a source of ignition was present, which suggests that the grill was ignited with an open flame.

12.4.4 Narrowing the Point of Origin

Once the compartment of origin is determined, the investigator observes the area in more detail to narrow the origin to a smaller volume. The smaller the determined area of origin, the less volume the investigator will have to search for a source of ignition. This operation is performed by observing heat, smoke, and fire patterns throughout the compartment. In some instances, it is not possible to narrow the origin further than the whole size of the compartment, which renders the determination of the cause much more difficult.

12.4.5 Multiple Points of Origin

When multiple points of origin are present, it is extremely important for the investigator to determine whether there are any incidental connections between them or whether these are independent events. There have been many fires that created two points of origin, such as some electrical overloads of circuit or traveling hot gases that heat two separate specific spots on the vehicle. If there is no logical explanation for the presence of two points of origin, the fire was deliberately set. Figure 12-19 shows a Chrysler 300 that was deliberately set on fire at the front left and rear right corners. The perpetrator either set the plastic components of the bumper, molding, or lights on fire or used an ignitable liquid as an accelerant. In any instance, this fire is clearly deliberate, because there are two points of origin with no heat source and logical connection, except deliberate human intervention.

a b

Figure 12-19

(a) View of the front left of a Chrysler 300, clearly showing a point of origin from outside the vehicle. (b) View of the rear right of the same Chrysler 300 showing a second point of origin, with no link to the first point of origin. Nevertheless, the fact that there are no ignition sources present in these areas of origin it clearly establishes a human intervention. (Photographs courtesy of Mike Carlson, Probe Inc.) See Color Plate.

12.4.6 Other Patterns

Fire patterns can also be used to determine the position or presence of some objects at the time of the fire. This could be extremely useful when comparing these observations with statements from witnesses and, more importantly, from victims and suspects. For example, the study of fire patterns can reveal whether a door was closed or open at the time of the fire. Also, it can reveal whether a window was rolled up or down. An example is shown in Figure 12-20, with a Chevrolet van that has been set on fire as a result of vandalism. The perpetrator(s) opened all doors and set the front seats on fire. The remnants of glass at the base of the windshield can be seen on Figure 12-20a, which are characteristic of an interior fire. Interesting patterns are found on the driver's side door. The burn pattern on the door does not match the pattern present on its counterpart on the side of the van, indicating that the door was open at time of the fire. The passenger side door and the right side doors were also open during the fire, as shown in Figure 12-20b. The fire patterns on these doors do not match the patterns on the body where the doors would have touched had they been closed.

12.5 DETERMINATION OF THE CAUSE

12.5.1 General Principles

Once the investigator determines an area of origin, he or she examines this area very carefully to identify all the different sources of heat present. It is very possible, particularly in cases of arson, that no sources of ignition will be detected in the area of origin because these may have been removed by the perpetrator after the ignition. As an example, a lighter

a

b

Figure 12-20

(a) View of the front driver's side door of a Chevrolet van that was set on fire by vandals through its front seats. Note the remnants of glass at the bottom of the windshield, whereas there is a total absence of glass at the top, which is typical of an interior fire. Also, note the fire pattern on the door, which does not match those on the body of the vehicle, indicating that the door was open at the time of the fire.

(b) View of the passenger's side door and side doors of the van, exhibiting patterns that indicate that they were open at the time of the fire. (Photographs courtesy of Mike Carlson, Probe Inc.)

will rarely be left at the area of origin once flames sustain on a bumper or seat that has been lit.

It is very important to also understand the difference between a source of heat and a source of ignition. Not all sources of heat are suitable sources of ignition for a given fuel. Thus, once the different sources of heat are identified, the investigator needs to eliminate the ones that are not suitable sources of ignition for the surrounding combustibles. Once this process is completed, if the investigator is left with more than one possible source of ignition, each one must be carefully examined to determine whether there is any evidence of the ignition itself.

12.5.2 Sources of Ignition in a Vehicle

A vehicle is a complex machine using electrical, mechanical, and chemical systems to operate. All these systems provide heat at some point and can be potential sources of ignition. In addition, when one of these systems fails, it is much more likely that it could become a source of ignition, because it does not operate in the fashion it is supposed to or in the environment it was designed for. Thus, there are many different sources of heat and/or ignition that exist in a vehicle [5, 28].

When considering the potential sources of heat in a vehicle, it is important to determine which of the three situations pertains to the vehicle at the time of the fire:

I Vehicle is moving, engine is running.
II Vehicle is stopped, engine is running.
III Vehicle is stopped, engine is stopped:
- Ignition contact is on.
- Ignition contact is off.
- Vehicle just stopped.
- Vehicle has been stopped for an extended period of time.

This helps to determine which systems were or were not operating (or energized) at the time of the ignition. Some sources of heat can be eliminated under certain circumstances. For example, if the vehicle was parked for several hours, it is possible to eliminate an ignition of grass due to contact with a hot catalytic converter. If the contact was off, all electrical circuits that are energized only when the contact is in the on position can be eliminated. If the engine was not running, a backfire from the exhaust system could not have been created and thus could not have been the source of ignition. If the vehicle was moving with the engine running, all possibilities are usually considered. A vehicle provides four possible types of sources of ignition: hot surfaces, open flames, electrical arcs and sparks, and mechanical sparks.

A/ Hot Surfaces

There are several different hot surfaces found throughout the vehicle. These include the engine block, cooling system components (hoses, radiator, etc.), exhaust system (manifold, catalytic converter, mufflers, exhaust pipes), brakes, electrical components (heater, ignition system, lighting systems, etc.), and electrical failures (high-resistance heating or overload). There could also be the development of a hot surface by mechanical friction of a failed part.

B/ Open Flames

The only situation where an open flame emanates from a vehicle is due to a backfire/ misfire or engine malfunctions. Open flame briefly appears, usually at the exit of the exhaust system.

C/ Electrical Arcs and Sparks

There are many electrical sparks that occur in a vehicle under normal operation. Such sparks are created by components including spark plugs, relays, and the starter. When electrical failures occur, many arcs or sparks can be created.

D/ Mechanical Sparks

Mechanical sparks are the results of friction between two surfaces. This friction occurs either between an object from the vehicle and the road or between two objects within the vehicle. During the normal operation of a vehicle, no mechanical sparks should result.

12.5.3 Causes with Stolen-Recovered Vehicles

As stated before, a stolen vehicle that is recovered burned could have caught fire accidentally, as could any other vehicle driving down the road. Because the odds of such an event are extremely low, it is not usually, if ever, encountered. However, all the sources of ignition described in Subsection 12.5.2 can be found in stolen-recovered vehicles. This is the reason burned vehicles must be fully investigated from a fire perspective using an open-minded approach before reaching any conclusions.

In most instances, when a vehicle is stolen and then burned, the thief (and arsonist) sets the vehicle on fire in the passenger compartment. The most convenient way to set the vehicle on fire is typically chosen. If the perpetrator has a lighter or match, he or she will use it to set the upholstery or some paper or plastic on fire and let the vehicle burn. In such cases, there is an almost nonexistent chance of retrieving any physical evidence. If the criminal has an ignitable liquid handy, he or she will use it as an accelerant. Most often, however, too much liquid is used and no doors or windows are left open to provide enough oxygen to sustain the fire. This type of situation is ideal, because the evidence will incur very little damage and can be easily recovered by the investigator, as shown in Figures 12-5 to 12-9. If an ignitable liquid was used to burn a vehicle, it might be possible to sample debris for further laboratory analysis [29]. This is very briefly described in the next section.

12.5.4 Causes with Insurance Fraud Scheme Vehicles

If the vehicle is the subject of an insurance fraud scheme, the criminal usually has access to many different methods of igniting the vehicle, because the malicious act is planned over time. There are still some owners who merely set the interior of the vehicle on fire with a lighter or match and let it burn. In such instances, it is very important to cross-reference their statements with the evidence at the scene. If the owner stated that fire came out of the hood as he or she was driving down the road and the fire patterns demonstrate the fire started in the trunk, serious suspicion would arise. Conversely, some criminals think for a long time about how the vehicle will be set on fire and develop very complicated modus operandi. An example of such an operation is in the modification of the wiring of the cigarette lighter and its placement close to a diesel fuel line in the engine compartment. This melted the fuel line, ignited the fuel, and the vehicle caught fire as it was driven down the road [30]. Figure 12-21 shows a vehicle that was parked on the side of the highway after one of the rear tires exploded. The owner reported that he left the vehicle as it was and

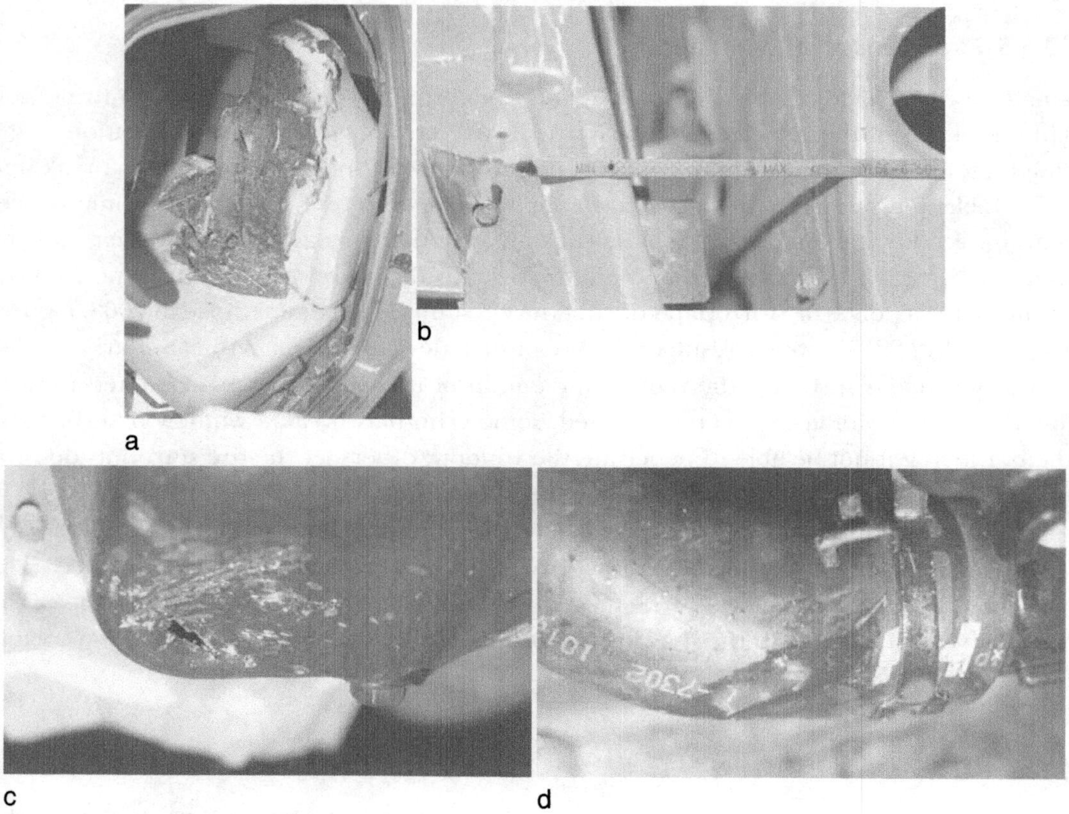

Figure 12-21

(a) View of the driver's seat of a Ford Ranger that was set on fire by spreading an ignitable liquid on the seat. The fire died by starvation of oxygen, which preserved all the necessary evidence to establish its incendiary nature.

(b) View of the engine oil dipstick, showing less than minimum level.

(c) View from the undercarriage, showing sabotage action on the oil pan.

(d) View from the undercarriage, showing sabotage action on the lower radiator hose. See Color Plate.

walked away to call for help. He stated that he returned a few hours later to discover the vehicle burned. Figure 12.21a shows the origin of the fire, which is the driver's front seat. An ignitable liquid, lighter fluid in this instance, was poured on the seat and set on fire. The presence and nature of the liquid was demonstrated by laboratory analysis.

Fortunately, the perpetrator closed the doors and windows, and the fire died from starvation of oxygen. More interestingly, when the oil level was checked (see Chapter 11 for fluid analysis), it was noticed that it was below the minimum level, as shown in Figure 12-21b. The undercarriage of the vehicle was then checked, only to discover sabotage. A hole was made in the oil pan, as shown in Figure 12-21c, and several holes were made in the lower radiator hose, allowing the coolant to drip, as shown in Figure 12-21d. After examining the vehicle and cross-referencing the evidence with the owner's statement, it was clear that this was a case of insurance fraud.

12.5.5 Fires Caused by Vandalism

Finally, if the vehicle is subjected to an act of vandalism, any deliberate cause of fire is possible. Typically, if the vehicle is in a busy area where criminals may attract attention or get caught, it will be ignited from the outside, because they will not take the time to break into the vehicle. Outside parts of the vehicle contain many different plastics and combustibles that are readily ignitable [5]. Molotov cocktails or other incendiary devices may also be used. Figure 12-22a shows different Molotov cocktails that were seized from demonstrators by the Geneva police in Switzerland during the G8 summit in Evian, France in 2003. Figures 12-22b and 12-22c shows the result of the use of such devices on a Chevrolet Suburban 1500. In this particular instance, the rapid intervention of the fire department limited the fire damage. When Molotov cocktails are used, some criminals break a window first, because the bottle might not be able to penetrate the window; others let the fire start outside (on) the vehicle.

If the vehicle is found in a location where the criminals have plenty of time to "play" with it first, they typically break into it through the windows and other body parts. The interior will be ransacked, and then the vehicle will be set on fire. In such instances, the fire may very well be set in the interior compartment, because most of the fuel is available there.

12.6 PHYSICAL EVIDENCE

Physical evidence, as explained in detail in Chapters 4 and 5, might still be present in the vehicle, even if partially burned. This is the reason why the fire investigator must work in conjunction with the auto theft investigator and/or crime scene officer to make sure that these items of evidence are not overlooked, destroyed, or contaminated. Thus, it is important to always use a criminalistics approach to the examination of the vehicle, even if burned, because it might contain several different types of evidence. In addition to these

Figure 12-22

(a) Examples of Molotov cocktails used by demonstrators during the G8 summit in Evian, France. These were seized by the Geneva police in Switzerland.
(b) View of a Chevrolet Suburban 1500 that was set on fire by demonstrators during the same G8 summit.
(c) View of a Chevrolet Suburban 1500 that was set on fire by demonstrators during the same G8 summit. The fire was extinguished very rapidly, preventing its spread throughout the vehicle. (Photographs courtesy of the Brigade de Police Technique et Scientifique, police of Geneva, Switzerland.)

typical traces found in cases of auto theft, there is one fire-specific type of evidence that can possibly be recovered in vehicles: ignitable liquids residues.

When it is suspected that an ignitable liquid might have been used to start and accelerate the fire, it is possible to sample fire debris from the area where the liquid has been poured and send these debris to the laboratory for analyses [31, 32]. The analysis of fire debris samples has been performed for many years and is a reliable scientific examination. There are several standards devoted to the analysis of fire debris samples, and most crime laboratories offer that service [33]. Such analysis can reveal the presence of different ignitable liquids, including gasoline, diesel fuel, and mineral spirits. It is always important for the investigator to communicate with the laboratory analyst to provide information regarding the circumstances surrounding the fire and the location of the collection of the fire debris samples. This greatly aids the interpretation of the results, because many modern polymers contain background levels of chemical components similar, if not identical, to the ones present in ignitable liquids [34].

ACKNOWLEDGMENTS

The author would like to thank Sarah Brown, Georgia State University, for her help in the manuscript review. Additionally, the author thanks Mark D. Culver, Key Fire Investigation, Inc., Mike Carlson, Probe Inc., as well as Monica S. Bonfanti, Geneva Police Department, for providing photographs of very interesting vehicle fire investigations.

BIBLIOGRAPHY

[1] United States Fire Administration (2005) *USFA arson fire statistics,* available at http://www.usfa.fema.gov, last access performed on August 1, 2005.

[2] Ahrens M. (2004) US vehicle fire trends and patterns, National Fire Protection Association, Quincy, MA.

[3] Office of the Deputy Prime Minister (2005) *Fire Statistics, United Kingdom, 2003,* ODPM Publications, London, England.

[4] United States Fire Administration (2002) Highway vehicle fires, *Topical Fire Research Series,* **2**(4).

[5] Du Pasquier E. (2003) *Investigation des incendies de vêhicules automobiles,* Presses polytechniques et universitaires romandes, Lausanne, Swizerland.

[6] Potter R. (2000) Arson on the increase: motor vehicle theft and arson in South Australia, *Information Bulletin on Motor Vehicle Theft Issues,* Comprehensive Auto-Theft Research System, Adelaide, Australia.

[7] State Farm Insurance (1998) Auto theft has new faces by old crime still big trouble, *Insurance Backgrounder,* 1/98, available at http://www.statefarm.com, last access performed on August 1, 2005.

[8] DeHaan JD. (2002) *Kirk's Fire Investigation,* 5th edition, Prentice Hall, Upper Saddle River, NJ.

[9] Nic Daéid N. (2004) *Fire investigation,* CRC Press, Boca Raton, FL.

[10] Quintiere JG. (1997) *Principles of Fire Behavior,* Delmar Publishers, Albany, NY.

[11] Thatcher PJ. (2000) Fire Investigation: Chemistry of Fire. In: *Encyclopedia of Forensic Sciences,* ed Siegel J, Knupfer G, and Saukko P, Academic Press, London, pp 900–905.

[12] Friedman R. (1989) *Principles of Fire Protection Chemistry,* 2nd edition, National Fire Protection Association, Quincy, MA.

[13] Haessler WM. (1974) *The Extinguishment of Fire,* National Fire Protection Association, Quincy, MA.

[14] Davletshina TA and Cheremisinoff NP. (1998) *Fire and Explosion Hazards Handbook of Industrial Chemicals,* Noyes Publications, Westwood, NJ.

[15] University of Southern Mississippi (2002) *The macrogalleria—A cyberwonderland of polymer fun,* available at http://www.pslc.ws/macrog/, last access performed on August 1, 2005.

[16] National Highway Traffic Safety Administration (2004) Standard No. 302; Flammability of interior materials, *Code of Federal Regulations,* Title 49, Volume 5, Chapter V, Part 571, pp 789–791.

[17] DeHaan J and Fisher FL. (2003) Reconstruction of a fatal fire in a parked motor vehicle, *Fire and Arson Investigator,* **53**(2), pp 42–46.

[18] Mahoney E. (1992) *Fire Suppression Practices and Procedures,* Brady, Englewood Cliffs, NJ.

[19] Martin JHC and Pepler RS. (2000) Fire Investigation: Physics/Thermodynamics. In: *Encyclopedia of Forensic Sciences,* ed Siegel J, Knupfer G, and Saukko P, Academic Press, London, England, pp 928–933.

[20] Holleyhead R. (1996) Ignition of flammable gases and liquids by cigarettes: a review, *Science & Justice,* **36**(4), pp 257–66.

[21] National Fire Protection Association (2004) NFPA 921 Guide for Fire and Explosion Investigations, 2004 edition, National Fire Protection Association, Quincy, MA.

[22] O'Hara CE and O'Hara GL. (1981) "Fundamentals of criminal investigation", 5th, Charles C. Thomas, Springfield, IL.

[23] TUNeZINE (2005) *Violences urbaines en France: plus de 300 véhicules incendiés,* available at http://www.tunezine.com, last access performed on August 1, 2005.

[24] LCI (2003) *Cotillons, champagne . . . et voitures brûlées,* available at http://www.lci.fr, last access performed on August 1, 2005.

[25] Coulson SA and Morgan-Smith RK. (2000) The transfer of petrol on to clothing and shoes while pouring petrol around a room, *Forensic Science International,* **112**(2–3), pp 135–41.

[26] Folkman TE et al. (1990) Evaporation rate of gasoline from shoes, clothing, wood and carpet materials and kerosene from shoes and clothing, *Canadian Society for Forensic Science Journal,* **23**(2 & 3), pp 49–59.

[27] Ide RH. (2000) Fire Investigation: Fire-scene Patterns. In: *Encyclopedia of Forensic Sciences,* ed Siegel J, Knupfer G, and Saukko P, Academic Press, London, England, pp 918–922.

[28] Stauffer E. (2004) Sources of ignition in vehicles, *Metro Atlanta Fire Investigators' Association monthly meeting,* Atlanta, GA.

[29] Service AG and Lewis RJ. (2001) The forensic examination of a fire-damaged vehicle, *Journal of Forensic Sciences,* **46**(4), pp 950–953.

[30] Maor S. (2005) IAAI annual photography contest winners—First place—arson, *Fire and Arson Investigator,* **56**(1), pp 32–33.

[31] Newman R. (2004) Modern laboratory techniques involved in the analysis of fire debris samples. *In: Fire investigation,* ed Nic Daéid N, CRC Press, Boca Raton, FL.

[32] Newman R. (2004) Interpretation of laboratory data. In: *Fire investigation,* ed Nic Daéid N, CRC Press, Boca Raton, FL.

[33] Stauffer E and Lentini JJ. (2003) ASTM standards for fire debris analysis: a review, *Forensic Science International,* **132**(1), pp 63–67.

[34] Stauffer E. (2004) Sources of interference in fire debris analysis. In: *Fire Investigation,* ed Nic Daéid N, CRC Press, Boca Raton, FL.

CHAPTER 13

EXAMINATION OF VEHICLES RECOVERED UNDERWATER

Jean-François Voillot

13.1 INTRODUCTION

It is often believed that vehicles retrieved from underwater need to be extracted to dry ground before they can be processed as a crime scene. This is a misconception that is commonly made by the forensic community around the world. It is paramount to understand that many steps of the crime scene investigation can and should be carried out underwater before any movements or extraction of the vehicle. Such processes have to be conducted by police officers or forensic scientists certified to scuba dive and trained in underwater forensic investigation.

There are many myths that lead to the misconception that no investigation can be performed underwater [1]. The first myth is that police divers are only able to search and bring back pieces of evidence to the surface. On the contrary, police divers are like the other members of the team in charge of the inquiry. They are not simply divers but underwater crime scene investigators. They are trained and experienced in such processes, as they are fully involved with other law enforcement personnel, scientists, and judges. The second myth is that pieces of evidence brought back from underwater are deprived of forensic value. Scientific studies and case reports confirm the opposite [2]. Many items of evidence remain usable underwater and, in some instances, even longer than at the surface. The third myth is that vehicles found underwater are always simply stolen, without a link to any other offenses. As a matter of fact, vehicles are pieces of evidence themselves and also comprise other evidence. Fingerprints, documents, firearms, and sometimes bodies are often lost from vehicles when they are extracted from water. An inspection of the car in situ is needed before any movement is induced. Finally, the last myth is that localization of items underwater with geographical precision is impossible. Using various methods, accurate localization of objects present underwater, even in the black water of rivers, is possible, sometimes without a great effort [3].

When dealing with the recovery of the vehicle from underwater, the challenge is to first find the vehicle, then manage the crime scene, and finally collect the different items of evidence. Contrary to land-based crime scene investigation, vehicles underwater are rarely visible from the surface, even at a shallow depth. In many cases, a first search step, sometimes very time consuming, is required. Because the crime scene is underwater, certain limitations dictate the parameters necessary to properly search, examine, record, collect, and preserve evidence from the scene for future analysis.

13.2 LIMITATIONS IN UNDERWATER CRIME SCENE INVESTIGATION

13.2.1 Principle

Obstacles hindering the search for an underwater vehicle or crime scene can have various origins. These origins depend on the limited accuracy of witness testimonies, physical limitations of divers, and environmental conditions of the aquatic environment.

13.2.2 Limitations of Witness Testimonies

As with witness statements related to crimes that occurred on land, witness testimonies of underwater events are often imprecise. Lack of caution and poor questioning from the inquirer contribute to great losses of time. Also, in many cases a lack of witnesses leaves the investigators without a starting point, and relying on suspicion is often imprecise. Given these difficulties, police diving teams practice a larger search than the sector delimited by witnesses or police officers' statements. It has also been useful to retain witnesses until the team arrives, so more appropriate information can be asked by the team members.

13.2.3 Physical Constraints

Police divers are also limited by many physical constraints created by the underwater environment, such as pressure due to depth, temperature, decrease of intellectual performance, and stress. Diving underwater leads to biophysical change. One of these changes is the dilution of nitrogen, oxygen, and carbon dioxide in the diver's body. The intensity of the dilution depends on the depth. When breathing air, the safe diving limit is 60 meters (200 feet). Such depths are often encountered in the sea or in mountain lakes, but more rarely in rivers or low country lakes. This depth leads to a shorter diving time and a diminution of the police diver's attention. Nevertheless, vehicles lying at such a depth are very rarely encountered.

Most police diver teams use compressed air for breathing rather than other mixtures. It is important to note that when mixed gases are used instead of compressed air, divers could dive much beyond that limit, but technical constraints are so important that many police dive teams refuse to dive with such mixtures. In fact, police, technical, commercial, and military divers making dives that require enriched air and/or pure oxygen for decompression are more likely to reach oxygen exposure limits than recreational divers.

The water environment exposes human physiology to a wide temperature range. The normal core body temperature (approximately 37°C or 99°F) is critical to normal chemical processes in the body. Hence, deviation from the normal core temperature of just a few degrees for more than a short period of time may be life threatening. Although the body responds protectively to different heat and cold conditions within a specific temperature range, diving can expose the body beyond its capability to successfully overcome the effects. Water conducts heat away from the human body 20 times faster than air [4]. As an example, the temperature of rivers and lakes are often under 10°C (50°F). This induces difficulties principally because searching operations are often very long. Diving in dry suits has greatly improved some of the cold temperature problems because the dry suit uses a layer of air between the suit and the body to provide better insulation.

As soon as the diver enters the water, he or she loses part of his or her intellectual performance (decreased speed, commission of errors, etc.). This phenomenon is well documented by underwater archeological research that checked the accuracy of measurements made by divers in an underwater site: four percent of these measurements were incorrect [3]. This fact obviously affects the accuracy of crime scene investigation.

Similarly to land-based situations, stress accrues according to the circumstances of the crime scene, to the presence or absence of dead bodies, their stages of decomposition, and many other parameters. Nevertheless, working in an underwater environment is more stressful because of low visibility, which requires working in very close proximity to the evidence without the ability to leave the crime scene easily.

13.2.4 Physical Limitations of the Aquatic Environment

The water affects light, which can lead to a very poor visibility. Criminals do not throw vehicles in water by chance; the lack of visibility is a good ally. In fact, water affects light through turbidity, diffusion, absorption, and refraction, each of which create different effects [4]. Although only an average of 20% of sunlight reaches an approximate depth of 10 meters (30 feet) in clear water, there is enough penetration to sustain photosynthesis to depths approaching 100 meters (300 feet) in the open sea. By contrast, high concentrations of suspended particles in the water can keep light from reaching even three meters (10 feet). The relative concentration of suspended particles is referred to as turbidity. Freshwater scenes, often called "black water," are full of mud and sediments of various origins, which result in a certain opacity and poor visibility, as shown in Figures 13-1 and 13-2.

Figure 13-1
Divers of the Gendarmerie Nationale (French military police) ready to dive in "black water" for stolen vehicles search (July 1999, Oise river, France). See Color Plate.

Figure 13-2

Example of poor visibility conditions during the examination of a vehicle in which a body was discovered. See Color Plate.

Suspended particles may be organic, as with plankton, or inorganic, as with stirred-up sediments (silt). Diluted in water, they constitute a permanent filter between the eyes of the diver and the different items to recover. The presence of these particles has to be managed in the recording process of the vehicle crime scene: Photographing and filming the scene requires powerful lights and flashes. Police divers have to use caution during photographic activities in the presence of particles in suspension between the camera and the item, because these particles reflect the light and spoil the picture. Also, it is important to know that video charged coupled device (CCD) captors have a better resolution in such circumstances than the human eye.

Even extremely clear water scatters and deflects light, due to the phenomenon called diffusion. Diffusion reduces the amount of light reaching a certain depth, and tends to disperse the available light, making it more even. This reduces or even eliminates shadows underwater.

The concept of absorption first requires an understanding of the nature of light energy and how the eye perceives it. Electromagnetic energy (of which visible light is one form) travels in waves; the length of these waves is determined by their energy. The human eye sees only a narrow portion of the electromagnetic spectrum: from about 400 nanometers (nm) to about 760 nm [5]. When light strikes an object, the object absorbs some wavelengths and reflects some others. Because natural light is a mix of lights from various wavelengths, it is easy to predict water's tendency to absorb light. As light penetrates water, the absorption process begins by filtering out wavelengths with the least energy. This is illustrated in Figure 13-3. Red color disappears at 4.5 meters (15 feet), yellow at 30 meters (100 feet), and green around 75 meters (250 feet). Hence, at depth, objects that are red,

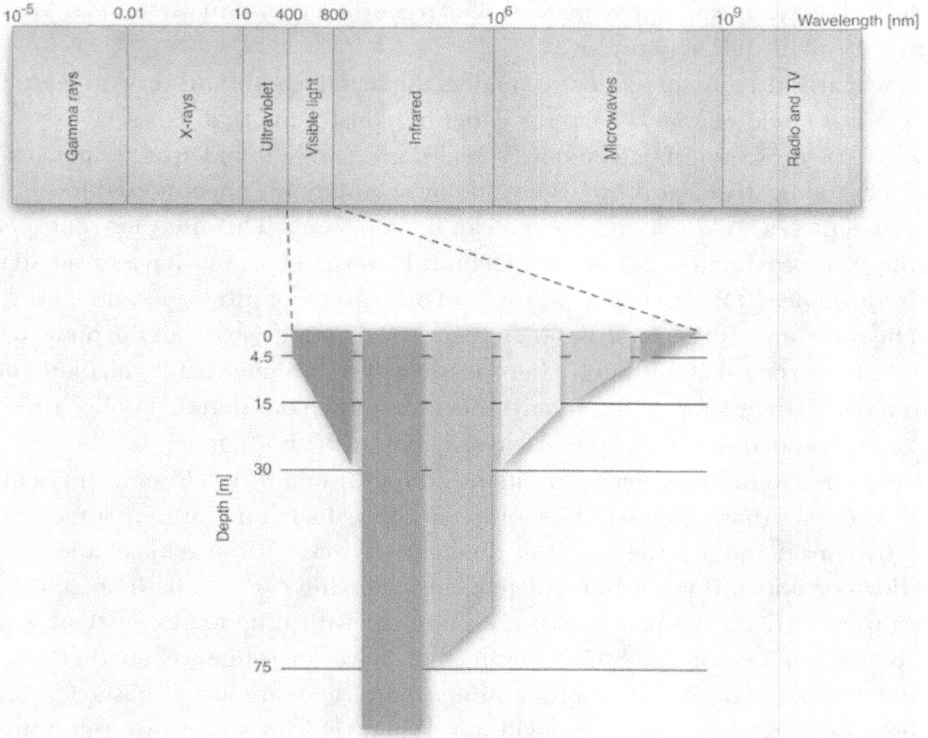

Figure 13-3

Illustration of the electromagnetic spectrum and visible light. The different wavelengths (colors) of the visible spectrum are absorbed by the water at different intensities. Notice how all the colors, except for blue, are completely absorbed after approximately 75 meters. (Diagram courtesy of Eric Stauffer.) See Color Plate.

orange, or yellow appear black or gray because the wavelengths responsible for these colors are absorbed by the water column above these objects. Clear water provides maximum transparency to wavelengths of approximately 480 nm (corresponding to a blue light). However, in turbid water, maximum transparency shifts to wavelengths of approximately 530 nm (yellow-green). This phenomenon explains why blue dominates clear water and why yellow-green dominates turbid water.

Experiments have shown that turbidity, depth, salinity, particle size, and pollution all affect light absorption by water, and contrast as well. This is because anything that affects color filtration properties affects which colors contrast with each other.

The last property of light that concerns divers is refraction. This is the tendency of light to bend as it passes from medium to medium, such as from air to water. Refraction takes place when light travels at different speeds through various substances due to different densities. To reach the diver's eyes, light must travel through water, glass (or plastic), and air. At each interface, the light wave refracts, because each medium has a different density.

As a result, objects appear approximately 33% larger than they actually are [4]. Also, objects appear closer than they actually are.

The mechanical action of water also renders the search conditions very difficult. Because of the natural circle of water, currents generate constraints that can seriously handicap underwater crime scene management. Vehicles are rarely found in seawater, except for harbor operations. Rivers and canals are the most commonly encountered locations. Currents in such places are often present and can be important. The safety current speed limit for diving is approximately 2.5 meters/second. Patterns of search have to be adapted to these circumstances. Of course, the accuracy of measurement suffers because of the current but can be overcome: If the angle between the vertical and the string of the buoy that marks the vehicle from the surface is known, it is possible to trigonometrically calculate the actual measurement. Dangers for diving in currents have to be considered; a police diver can be carried away trapped in debris (trees, wires) lying on the bottom.

Current has a significant action on items falling in water. In addition, there are many different variables that influence the movement of a vehicle underwater. Some of the variables to be considered are the speed of the current, mass of the vehicle, and opening of the windows or doors. If water immediately enters the interior, the pathway of the car will be very different from the pathway of the same car with little or no entry of water. The nature of the bottom of the body of water can have an influence on the travel of an immersed vehicle; a rocky bottom or a bottom where there are lots of obstacles might trap a sinking vehicle, while a flat bottom will not. When vehicles are thrown into water, there are no rules to calculate the exact route between the entry point and the arrival point at the bottom. However, it is possible, in some instances, to try to estimate the position using many different variables. This science is not exact and provides approximate results that might not reflect the actual position of the vehicle due to many other parameters outside of the control of the forensic scientist.

13.3 SEARCH METHODS: FROM SIMPLE TO SOPHISTICATED

13.3.1 Search from the Surface

If the immersion is recent, the use of the helicopter is helpful, because the presence of fuel can be checked from above. It can also be of great value to look for entry points into the water. This is achieved by surveying the shores from land or from the water by following the shoreline in a small boat. Marks left by vehicles can be spotted in the grass or on the edge of a harbor side. In such cases, the crime scene begins at that point, where tire tracks are collected.

Some police dive teams use an echo sounder in inflatable boats to check the bottom of rivers, harbors, or canals (Figure 13-4). In a defined zone, they sweep-search the bottom with this device. Each time an anomaly on the profile is detected (Figure 13-5), a buoy is positioned and the bottom is checked at a later time with divers.

Figure 13-4

Search from the surface: echo sounder's screen in action.

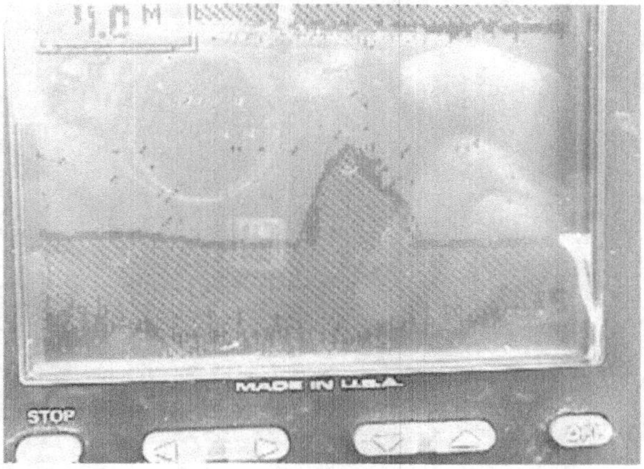

Figure 13-5

Search from the surface: a car is detected on the echo sounder's screen.

13.3.2 Search in the Water

Search in the water uses methods similar to the ones presented in Chapter 3. Most police divers use methods that include ropes to locate a vehicle in the water. The principle of these methods is similar, regardless of whether the search is started from the bank, from a point at the bottom of the body of water, or from a static boat: it consists of systematically sweeping the bottom using ropes.

The circular method consists of choosing a central point where a rope is anchored. Then, the diver swims around the anchor while keeping the rope extended at all times. After each circular sweep, the length of the rope is increased. This is illustrated in Figure 13-6.

The circular method is useful in water with weak currents. In case of stronger currents, a linear method is preferred. In this case, police divers swim in lines rather than in a circle, as shown in Figure 13-7. Several ropes are used to delineate the area of search, and the diver swims back and forth between the ropes.

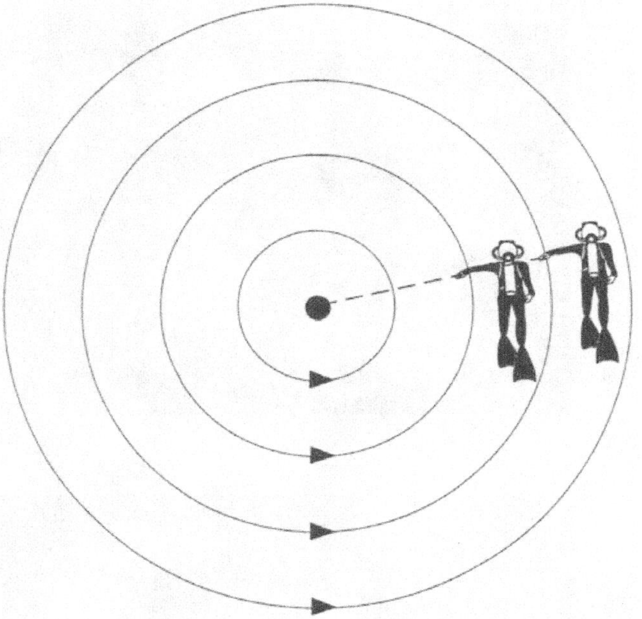

Figure 13-6

Search in the water: the circular method. (Source: Figure 4a in Becker RF. (1995) The Underwater Crime Scene: Underwater Crime Investigative Techniques, Charles C. Thomas, Springfield, IL, p. 35. Reproduced with the permission of Charles C. Thomas.)

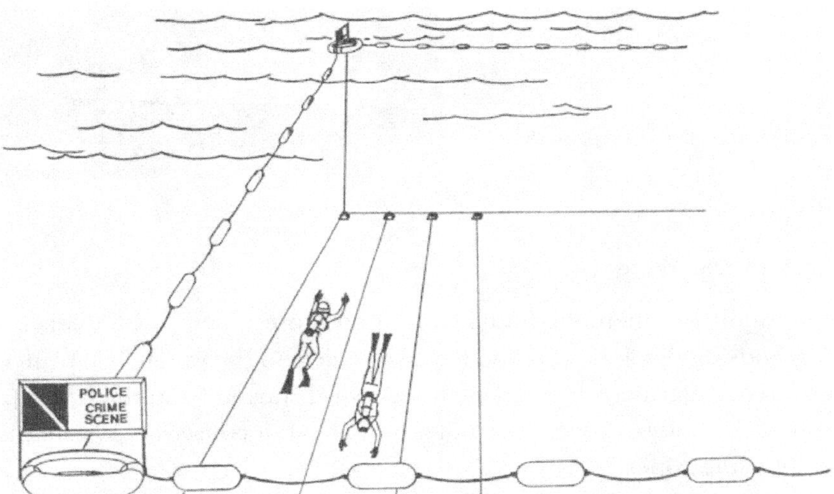

Figure 13-7

Search in the water: the linear method. (Source: Figure 3 in Becker RF. (1995) The Underwater Crime Scene: Underwater Crime Investigative Techniques, Charles C. Thomas, Springfield, IL, p. 35. Reproduced with the permission of Charles C. Thomas.)

13.3.3 Sonar Methods

Developed for mine hunting, the side-scan sonar is a little torpedo, called a fish or a towfish, that is pulled behind a boat (Figure 13-8). Antennas that transmit and receive microwaves are located on each side of the apparatus. The range and accuracy of such a system depend on the frequency of the emission. When the waves encounter an obstacle lying on the bottom, a shadow appears on the monitor, as illustrated in Figure 13-9.

The multibeam sonar works based upon the same principle but is permanently hung under the hull of boat rather than being pulled with a rope. Thus, such a system cannot be used without the proper boat. The accuracy exhibited by such a system is better than that of a side-scan sonar, because it shows an image in all directions rather than only on the sides. This system is typically used by professionals to draw maps of the bottom of canals and harbors. Currently, no European police diving team is equipped with either side-scan sonar or multibeam sonar.

In some instances, the accuracy of the system is enough to recognize the type of the vehicle. If the sonar is coupled with differential global positioning system (DGPS), the electronic image allows the diving team to come back to the car at a later time. This image is also very convenient to draw a sketch of the crime scene. Global positioning system (GPS) navigation offers accuracy down to a few meters (or several feet). This technology pinpoints location through a receiver that reads signals from a satellite and is commonly used by boaters, hikers, or travelers. Chapter 20 also presents the use of GPS to track (stolen) vehicles. The limitation of GPS in an underwater environment is that signals do not penetrate water very well. Research is currently being conducted to find a way to improve this.

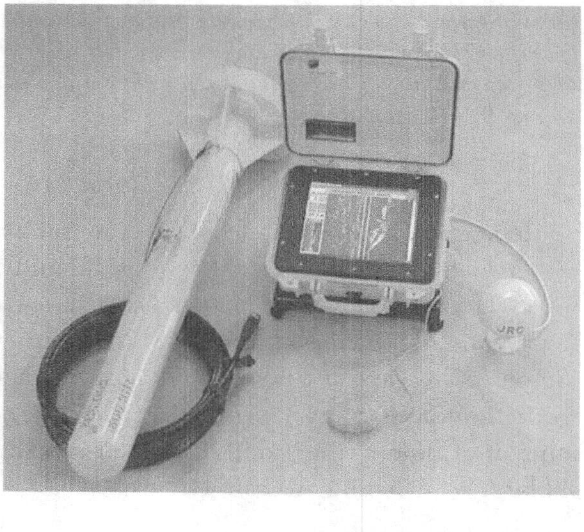

Figure 13-8

The Centurion Sea Scan PC side scan sonar system presented here with a single frequency towfish (600 kHz). This particular towfish is constructed of polyvinyl chloride, measures approximately 1.1 m long and 10.2 cm in diameter, and weighs approximately 15.9 kg. Its typical maximum range is 75 m. (Source: Centurion System image provided by Marine Sonic Technology. Reprinted with permission of Marine Sonic Technology, Inc., White Marsh, Virginia.)

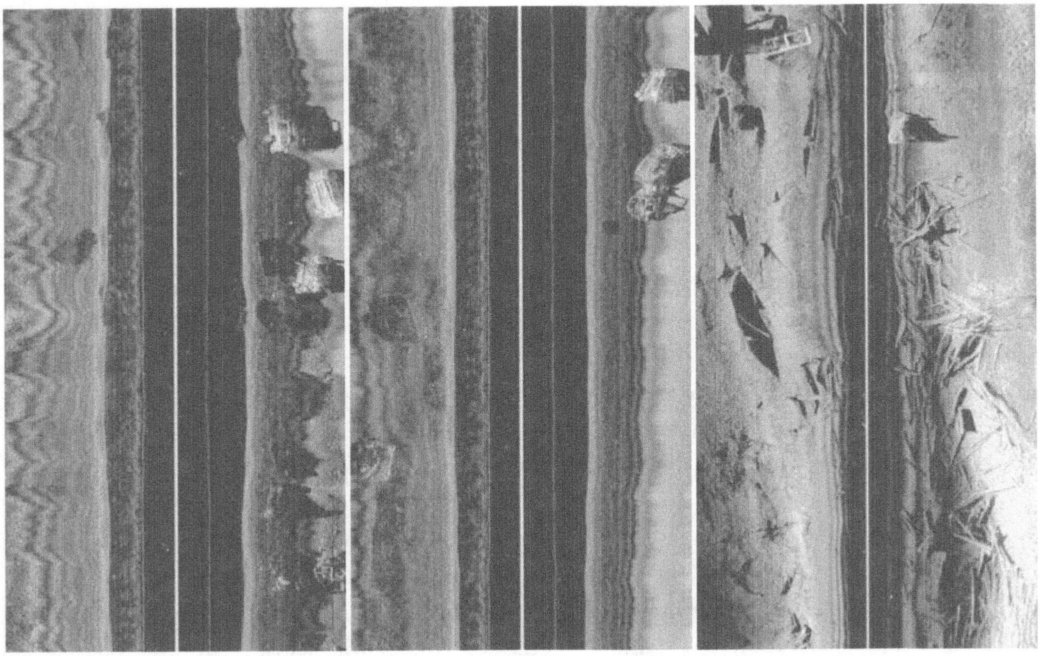

a b c

Figure 13-9

(a) and (b) Vehicles found by a Marine Sonic Technology representative while assisting the French police. The police were looking for a missing child that was allegedly placed inside a vehicle dumped in a river. Surprisingly, during the search of the river, several other vehicles were discovered as seen on the images (particularly on the right side of the images). These images were taken with a Centurion side scan sonar system.

(c) Two vehicles are present in this image taken in North Carolina: a pick-up truck on the top left and an up-side-down vehicle slightly below, on the right. The vehicle on the right is a 1996 Nissan, found after its owner placed it in the water and reported it as stolen to the police. He attempted to collect the insurance settlement, however the car was found by the police. Interestingly, while looking for this car, the Ford pick-up truck on the top left was also discovered. It was later determined that the pick-up truck was also stolen several years earlier, from the same owner of the Nissan, who collected this insurance money. The owner was charged with two counts of insurance fraud. (Source (a, b, and c): Marine Sonic Technology. Reprinted with permission of Marine Sonic Technology, Inc., White Marsh, Virginia.) See Color Plate.

Remotely operated vehicles (ROVs) are submarine robots used to perform different tasks underwater. These devices are very useful and are used by some police forces around the world. In a forensic setting, they typically have two uses: to search and recover items at great depth, and as a reconnaissance tool in conjunction with different sonar systems when the identity of the detected object is not clear enough on the screen. If a ROV can be sent down to the bottom to get a much clearer view of the object instead of a team of divers, great amount of time can be saved. ROVs were used during the recovery process of debris from flights TWA 800 and Swissair 111.

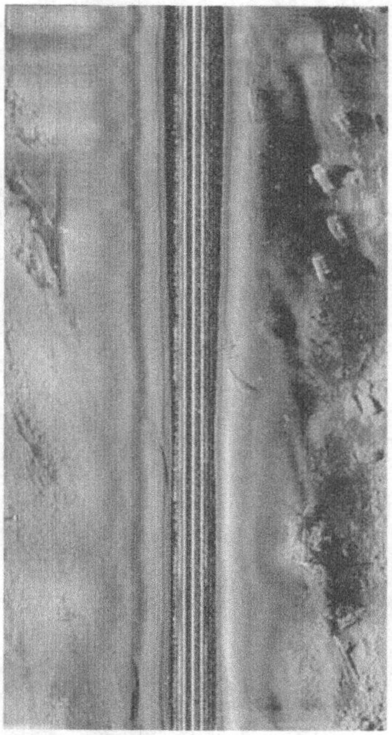

Figure 13-9 Continued.

(d) Several vehicles found in the Arkansas River (US). It is possible to distinguish at least four vehicles on the upper right part of the side scan sonar image. The second vehicle from the top is laying on its back, as it is possible to distinguish the two axles and the four wheels. (Source: Marine Sonic Technology. Reprinted with permission of Marine Sonic Technology, Inc., White Marsh, Virginia.) See Color Plate.

d

13.4 CRIME SCENE DELIMITATION AND SEARCH METHODOLOGY

13.4.1 At Discovery of the Scene

When the buoy arrives at the surface, showing the position of the car at the bottom of the body of water, the forensic examination of the scene begins. Time is of the essence. The other personnel, judges, and coroners on the shore are eager to begin their own work as fast as possible. If necessary, the chief of the police dive team explains to his or her colleagues the different steps to be taken before the car is returned to the surface.

13.4.2 Step by Step

Figure 13-10 summarizes the different steps that are taken when examining underwater crime scenes, and more particularly vehicles. Each step is performed by a different dive team as well as by personnel standing at the surface. Underwater, depending on the depth and complexity of the scene, it is possible that divers participate on more than one team.

First, a dive team performs a quick preliminary reconnaissance, followed by general observation of the vehicle. As stated in Subsection 13.3.1, the crime scene might not be confined to the vehicle and its immediate surrounding only. The path (both on land and underwater) take by the vehicle to get to its current location should also be considered as a crime scene because different items of evidence might be present. Also, it is important

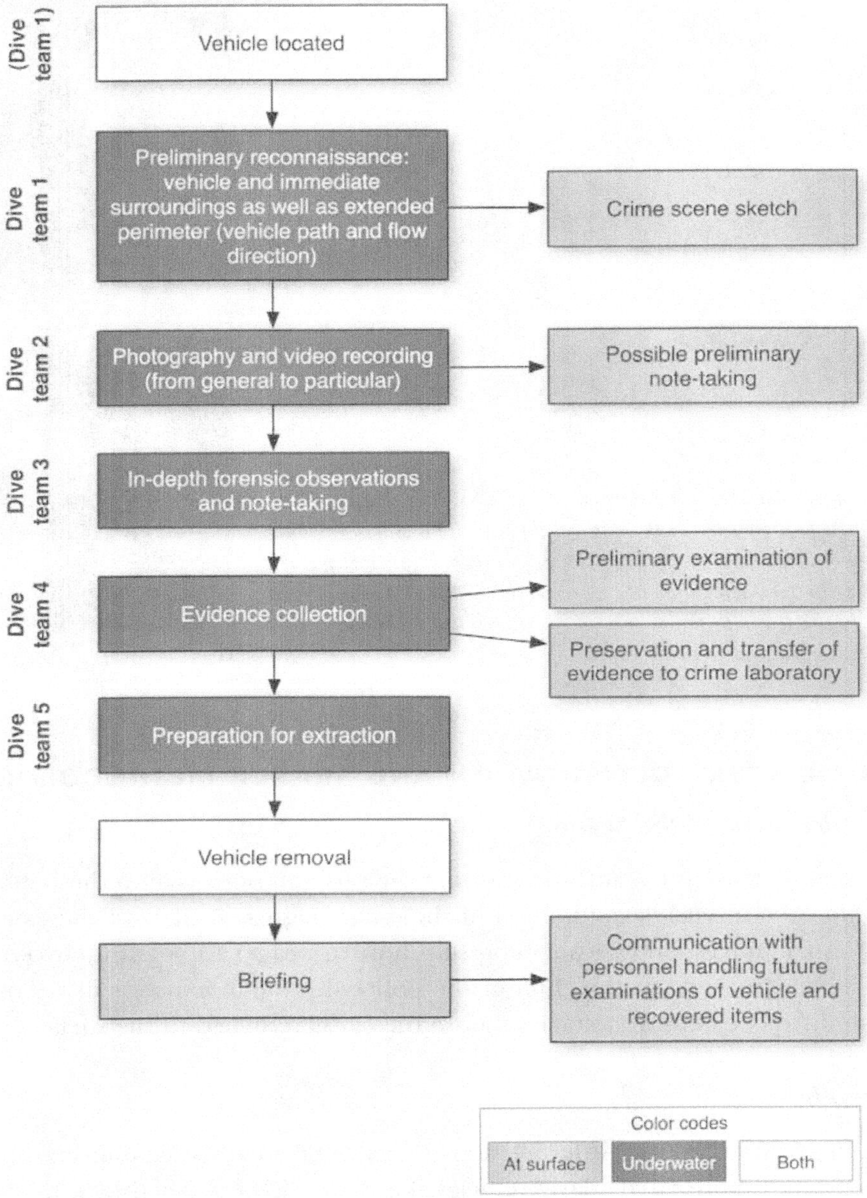

Figure 13-10

Crime scene investigation process followed by police divers after discovering a vehicle. Note that although most steps are performed underwater, some are performed at the surface. (Diagram courtesy of Eric Stauffer.)

to take into account the direction of the flow of the current because objects might have been spread out from the vehicle toward that direction. In such instances, it is crucial to examine these locations to check for the presence of additional evidence. It is possible to use underwater metal detectors when checking these extended areas. These devices can be very efficient when looking for firearms, shells, or projectiles because they can easily penetrate several centimeters inside mud, sand, and vegetation. However, as they are very sensitive, they can detect many different metallic objects that are of no interest to the forensic scientist, particularly in bodies of water that are polluted and that contain many trash items. Also, the diver's equipment contains many metallic parts, which can interfere with the use of metal detectors. It is important to use these devices properly and to check manually and visually every area where a positive response is obtained. Figure 13-11 shows a firearm that was detected away from the vehicle, in the direction of the flow. This stresses again the importance of extending the search area away from the vehicle in the direction of the water flow.

Once the crime scene is delimited, a dive team performs all the photographic and videographic recording of the scene. This is carried out at an early stage, so particles raised by other police divers do not interfere with visibility conditions vital to the recording of the scene. This is followed by an in-depth observation of the scene with extensive note-taking. The divers carefully observe the scene outside and inside the vehicle, as well as the sur-

Figure 13-11

A firearm discovered by a police diver away from the main crime scene (vehicle) is placed inside an evidence container. This shows the importance of checking the extended area in the direction of the water flow for objects that could have been carried away. (Photograph courtesy of Gendarmerie River Company of the Rhine, March 2004, Strasbourg, France.) See Color Plate.

roundings. The note-taking process is tedious and slow. Also, during this step the different items of evidence that need to be collected are identified, so the next team can collect all items. To avoid contamination of the crime scene, police divers wear gloves and work carefully without moving the sediment. Then, the vehicle is prepared for extraction, and the removal is done with the aid of a crane. Finally, a briefing is conducted with the different diving teams and forensic technicians and investigators that will examine the vehicle on land.

13.5 CRIME SCENE EXAMINATION AND RECORDING

13.5.1 Preliminary Reconnaissance

The method is similar to the practice on land (see Chapter 3). After the discovery, a police diver makes an overall preliminary examination of the surroundings of the vehicle and the outside of the vehicle. During that time, he or she can note the make and model, color (sometimes impossible), license plate, and general conditions of the vehicle (Figure 13-12). The diver also determines the orientation of the vehicle (resting on its wheels, on its back, on its side, etc.).

13.5.2 Photography and Video Recording

A large number of police diving teams are now well trained and well equipped with photographic equipment (Figure 13-13). The Gendarmerie Nationale (French military police) uses Nikonos-V cameras for underwater photography. Even if the water is black and full of

Figure 13-12

Example of preliminary reconnaissance: the police diver notes the license plate number of the vehicle.

mud, photographs can be successfully obtained. Of course, the quality of photographs in such instances greatly suffers but can still be of great forensic interest.

Better results are obtained with video cameras, like the model used by the Gendarmerie Nationale divers (Figure 13-14): A digital Sony TRV 900E in a sealed container from Extrem Vision (Toulouges, France). CCD captors of video cameras are now of very good quality and often result in better observation than the human eye in black water. The advantages of digital imaging are numerous: The camera operator knows immediately if the picture was successfully captured, image enhancements are relatively easy, and it is possible to have the camera directly connected with a monitor located at the surface, where an archivist or an expert can assist the diving team.

Figure 13-13

A police diver team taking pictures of a vehicle recovered underwater.

Figure 13-14

Digital video camera Sony TRV 900E in a sealed Extrem Vision container used by the Gendarmerie Nationale divers.

All underwater imaging systems must be protected against water and pressure. In many cases, there is also a need for an artificial light source to restore lost colors. Rapid technological advancements in cameras, portable monitors, and lighting systems have occurred in the last few years. Simultaneously, the cost of such equipment has greatly decreased. With the emergence of digital cameras for the public and the general miniaturization of electronics, technical devices and cameras are less expensive and much easier to use today.

Photographers or camera operators have to work from the general scene to particular, more detailed, views. Sometimes, it is impossible to photographically capture a general scene; in such cases, a succession of particular shots or views (if possible, overlapping each other) is recorded and should suffice. In any case, the following views or details must be filmed and/or photographed: position of the car, signs of impact and scratches, wheel positions, windows and windshield, door positions, trunk, and hood. Before or upon entrance of the vehicle, photographs must be taken through the windows or other openings. The following must be photographed extensively: body(ies) (position, wounds, etc.), driving commands (keys, steering wheel, control levers, switch positions, etc.), and other pieces of evidence (weapons, papers, etc.). When entering the vehicle, the divers must ensure that nothing is moved and that more detailed photographs, when necessary and possible, are taken. During all these operations, the buoyancy of the police divers has to be exact, not too heavy to avoid moving the sediment and not too light to stay at the level of the crime scene. The video link with the surface is a great advantage because the police diver can be guided, but more importantly, the note-taking process can begin right away with the archivist watching the monitor, before being completed underwater by the other police divers.

13.5.3 Sketch from the Surface

When there is no current, police divers can knot a buoy to each wheel of the vehicle, ensuring that the strings are tight. The archivist of the team then measures the distances between the four balloons and two fixed points on the shore with a laser telemeter.

When the current is too strong, it may be possible to attach only one buoy to the center of the car. This method is not as accurate but is sufficient for the reconstitution of the crime scene. When the divers knot the buoys or attach the sling of the crane later (for lifting), police divers have to take every possible precaution not to damage the vehicle.

13.5.4 Note-Taking

The best way to circumvent the diminution of intellectual performances in water is to use a notepad. If the police dive team is equipped with radio helmets, the archivist at the surface can write notes simultaneously as the police diver announces his or her observations.

The other solution is to dive with a notebook. There are various solutions, such as a special paper designed for underwater archeologists [6]. This special paper is prepared by mixing regular cellulosic fibers with plastic fibers. This kind of paper can be preprinted with instructions to follow during the dive and then completed underwater. A plastic shield printed with instructions to follow may also be useful (Figure 13-15).

In the same manner, bodies lying inside the vehicle have to be recorded with a similar notepad. By experience, for a vehicle, with body(ies) aboard or not, the group of divers should be kept to the minimum number of divers possible; only one police diver can do the job of observing and taking notes. He or she must be escorted by another police diver for safety reasons, but the second diver should keep distance from the crime scene to prevent disturbance.

The primary diver begins by observing the outside of the vehicle, with a long examination of the wheels and surroundings because these parts of the car can be altered during the operation of removal from underwater. Then, he or she checks the locks of the doors and trunk. Then, the inside of the car as well as the body(ies), if present, are carefully observed and properly recorded in notes. This task is long but has to be carried out in great detail, because the result is essential in writing the final report. The different pieces of evidence that have to be collected are carefully noted and described. Based upon these observations, the next group is briefed before the divers collect these items of evidence.

Figure 13-15

The note-taking process is a long and difficult process underwater, but can be realized using special supports such as plastic shields.

13.6 COLLECTION OF EVIDENCE

13.6.1 Principle

At this point of crime scene management, it is necessary to choose one of two methods. Police divers can either collect evidence from the vehicle while sitting underwater or proceed to the recovery of the vehicle and then collect the evidence when the vehicle is at the surface.

A vehicle represents an enclosed cage containing pieces of evidence. It can be brought back to the surface as a whole during the removal operation (Figure 13-16). In this case, the vehicle is lifted to the surface with the aid of a crane. The water present inside the vehicle represents a very important mass; the total mass of the vehicle to be brought back from the bottom to the surface can be enormous. There is a risk that the windshield and/or the rear window fails under the pressure, which would result in an important shifting of items inside the vehicle and, more importantly, a loss of evidence from the interior. Is the gamble worth it? Figure 13-17 shows a condom that was found floating inside a vehicle, just under the ceiling. This type of evidence will certainly be lost during the removal operation.

The collection of evidence necessitates the use of particular equipment. This equipment consists of containers of different sizes, sealed or unsealed, which can be submerged. Some of these containers must be attached to a lift buoy, which can be filled with air at a given

Figure 13-16

Extraction of a stolen vehicle (July 1999, Oise river, France).

Figure 13-17

A condom was found floating inside the vehicle, retained by the ceiling. If this precious item of evidence had not been recovered before the vehicle removal, it would have certainly been lost. (Photograph courtesy of Gendarmerie River Company of the Rhine, March 2004, Strasbourg, France.)

time. The reason is that these containers might bear a certain weight once filled, which cannot be easily brought back to the surface by the diver. Examples of such items are shown in Figure 13-18.

In some other instances, it is impossible to collect evidence before lifting the vehicle. For example, when a body is totally saponified and stuck to the ceiling of the vehicle, it can be very hard or impossible to collect underwater. At the Gendarmerie Nationale, pieces of evidence that can be removed from the vehicle are collected while the vehicle is underwater. The vehicle is then removed after this collection procedure.

13.6.2 Documents

Even if they are degradable, documents are very often found floating inside the interior compartment of a vehicle. This kind of evidence can provide valuable information such as that acquired from various tickets or receipts, restaurant receipts, credit cards, maps, driver's licenses, or vehicle registrations [7, 8].

If the vehicle is removed directly without collection, the documents are usually lost from the interior as the water escapes the vehicle during the removal operation. This is an important reason why documents must be collected from the vehicle before it is moved. The collection of documents is relatively easy: A plastic box with the correct dimension is used to package the document, with water around it, and is closed before returning it to the surface. This is illustrated in Figure 13-19. Using this method, the degradation of the document is limited and the final packaging can be delayed until the end of the recovery operations, once the forensic technicians return to the shore. At this point two options are possible: an immediate analysis of the document or the possibility of sending it to the forensic laboratory for examination.

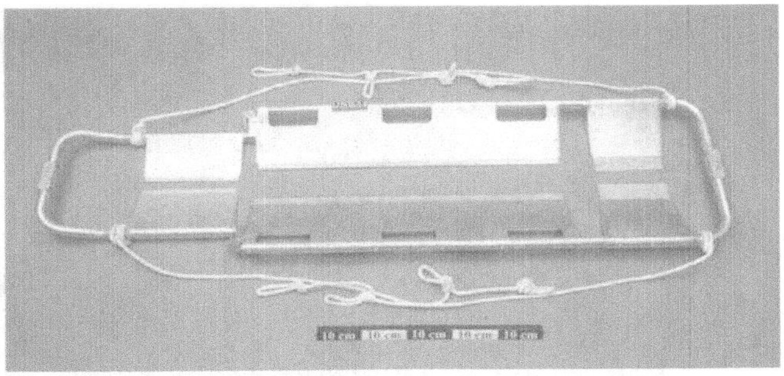

a

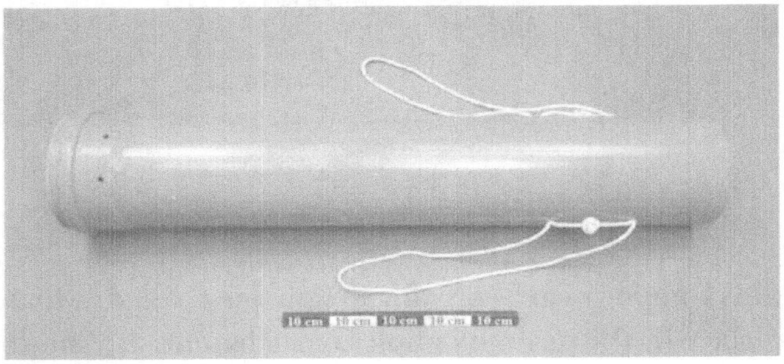

b

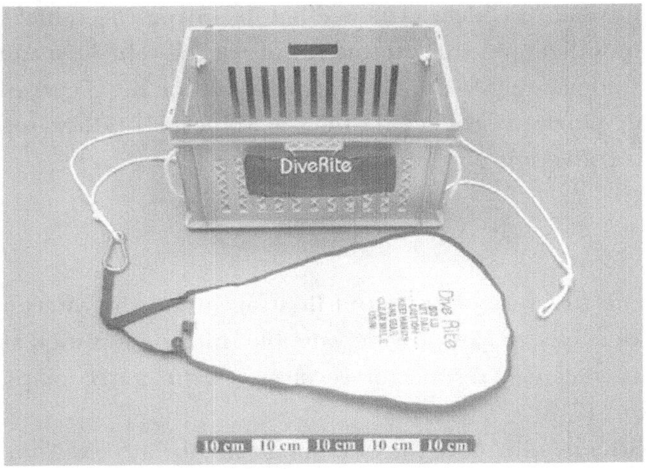

c

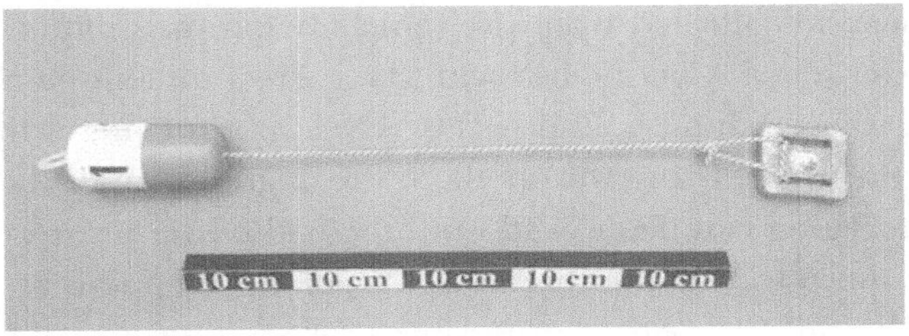

d

Figure 13-18

(a) A stretcher used by diving teams to remove and bring bodies back to the surface.
(b) A container designed to collect and recuperate long items such as shoulder weapons. Note the cords that can be attached to an inflatable buoy.
(c) A box designed to bring objects back to the surface. Note the inflatable buoy that eases the lifting of the filled container to the surface.
(d) An evidence marker, consisting of a lead weight and a small buoy. This type of evidence marker is placed by the dive team 3 to notify the exact evidence to be collected by dive team 4. (Photographs courtesy of the Brigade de Police Technique et Scientifique of the Geneva Police Department, Switzerland.)

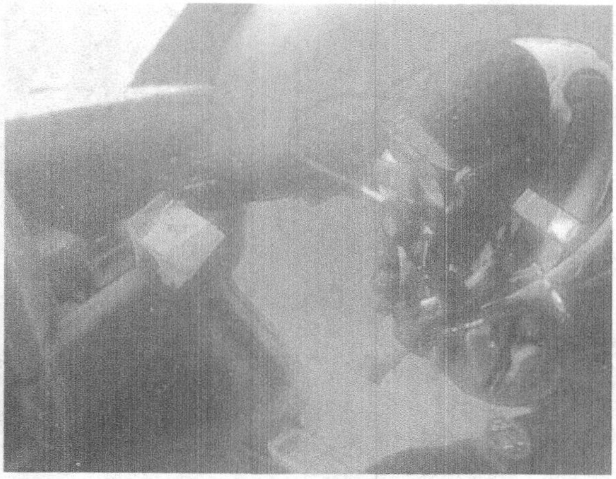

Figure 13-19

Example of collection of a document found in the glove box of a stolen vehicle. (Photograph courtesy of Gendarmerie River Company of the Rhine, March 2004, Strasbourg, France.)

A/ Immediate Analysis

For this operation, a large box (50 × 50 × 30 cm) is required. The box is filled with clear water and the document is transferred from the "collection box" and immersed in the clear water. A very smooth and patient cleaning process is carried out. The written or printed information observed on the documents is then recorded page by page by forensic technicians.

B/ Delayed Analysis

When delayed analysis is planned, the document has to be prepared to avoid the proliferation of bioorganisms. First, a smooth and patient cleaning process, as presented in the previous paragraph, is carried out. This operation must be repeated until the water surrounding the document is exempt of sediments and particles. Then, a protocol in accordance with the forensic laboratory is chosen. This could involve either drying or freezing the document. If the document is dried, it must be left in a dry room without any light (or soft light). The pages should be kept separately. If the document is too thick, such as a book, drying is problematic. A better solution is to freeze it. When performing this step, plastic sheets should be inserted at least every 20 pages to prevent the sticking of all the pages together. The forensic lab, receiving this document frozen, lyophilizes it in a special oven before examination.

13.6.3 Firearms

After noting the exact location and configuration of the firearm, the police diver collects it with a tank or a plastic box of the correct size. Of course, cartridges, shells, and projectiles are also collected from the vehicle and its surroundings (Figure 13-20). Weapons have to be handled with care, for evident safety reasons, but also because DNA traces and fingerprints can be found on them.

Figure 13-20

Example of collection of a cartridge shell from inside a vehicle. If this evidence was not collected before vehicle removal, it would probably have been lost during the vehicle's extraction. (Photograph courtesy of Gendarmerie River Company of the Rhine, March 2004, Strasbourg, France.) See Color Plate.

When the police diver reaches the surface, the water around the weapon in the box is replaced with distilled water to limit the oxidation process. This kind of evidence must never be exposed to the air; it must stay immersed in water until reaching the crime laboratory.

A rapid examination of the weapon is necessary and should be performed as early as possible. Upon arrival at the forensic laboratory, the weapon is forwarded to a different section to search for DNA traces, fingerprints, and other forensic traces, and finally it is forwarded to the firearms examiner. This sequence has to be followed to ensure the chances of successful recovery of the different potential physical evidence.

13.6.4 Electronic Devices

Cellular phones, personal digital assistants, personal computers, digital cameras, other electronic devices, and electronic supports such as CDs and DVDs are collected using the same technique as with firearms. They are put in a box directly at the bottom and are brought back in it to the surface, where they are quickly rinsed out and reimmersed in distilled water in a box of the correct size. This box is sent to the forensic laboratory for future examination. An example of the collection of a cell phone found in a vehicle underwater is shown in Figure 13-21.

13.6.5 Trace Evidence

Following Locard's principle of exchange, it is of great value to collect different trace evidence such as paint traces located on body panels of the vehicle involved in a road accident

Figure 13-21
Collection of electronic devices such as this cell phone is also very important. (Photograph courtesy of Gendarmerie River Company of the Rhine, March 2004, Strasbourg, France.)

(see Chapter 4) [9]. Usually, this collection is carried out at the surface after the vehicle has been removed from the water. Sometimes, however, it is necessary to collect the traces in situ in the underwater setting. The police diver follows the same techniques as those used at the surface: With a scalpel and a test tube, chips of paint from the impact zone are collected. Paint traces brought back to the surface in test tubes are forwarded to the forensic laboratory for future examination.

Other trace evidence includes fibers, glass, and soil. These items should be collected before any movement of the vehicle. Soil can be present on the pedals, or pieces of clothing can be found in the vehicle or at the openings of the vehicle, as shown in Figure 13-22.

13.6.6 DNA Traces

According to a recent study, DNA traces persist underwater for a certain period of time, ranging from several hours up to several days [10]. In an immersed car, it is possible to find DNA traces on various surfaces like cigarette butts, towels, condoms, and hair. These types of evidence are easy to collect when they are detected by the police diver; they are placed in a box and brought back to the surface. Figure 13-23 shows a police diver collecting a bottle of an alcoholic beverage from a vehicle.

For blood, sperm, saliva, sweat, or other biological fluid found on different items such as car seats, carpets, steering wheel, gear shift knob, or hand brake lever, traces can be collected when the car is brought back to the surface or, in exceptional circumstances, in situ; these traces are collected by cutting off the tissue or surface directly around the area or by wiping the surface with a sterilized gauze or cotton pad, as shown in Figure 13-24.

Figure 13-22

A piece of torn apart clothing was found in the latch of the vehicle's door. This piece is collected before vehicle removal, because it would probably be lost during that operation. (Photograph courtesy of Gendarmerie River Company of the Rhine, March 2004, Strasbourg, France.)

Figure 13-23

Example of collection of a bottle of an alcoholic beverage found inside the vehicle. This item is susceptible of containing DNA material. (Photograph courtesy of Gendarmerie River Company of the Rhine, March 2004, Strasbourg, France.) See Color Plate.

Placed in a small sterilized box, this item is directly dried or frozen before being sent to the forensic laboratory.

13.6.7 Bodies

Although dead bodies are psychologically hard to manage, they have to be brought back after other operations have been completed. They are rarely well positioned in the seat, but are often in other locations, where the person died either after a panic phase or by having been placed there.

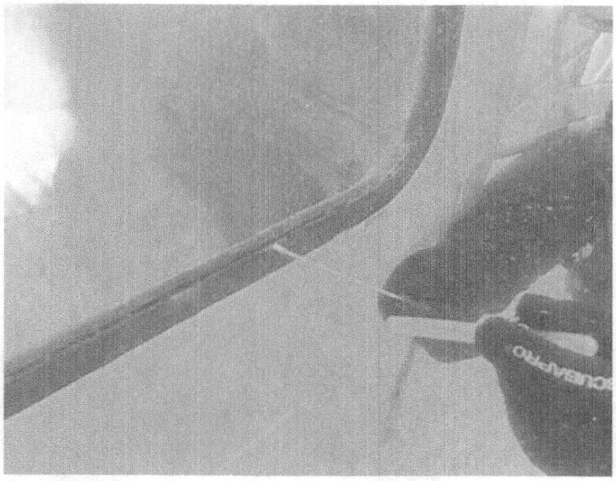

Figure 13-24

Example of collection of blood using a cotton swab. (Photograph courtesy of Gendarmerie River Company of the Rhine, March 2004, Strasbourg, France.)

At the Gendarmerie Nationale, the bagging of bodies directly performed underwater is preferred. Before bagging the body, the investigator first searches for diatoms. Diatoms are microcellular organisms living in bodies of water (Figure 13-25). If a live person is immersed in such water and if he or she breathes this water, diatoms enter the bloodstream. After the discovery of the body, if the pathologist finds diatoms in some specific parts of the organism, it means that the deceased was alive upon entrance to the water [11]. Diatom sampling

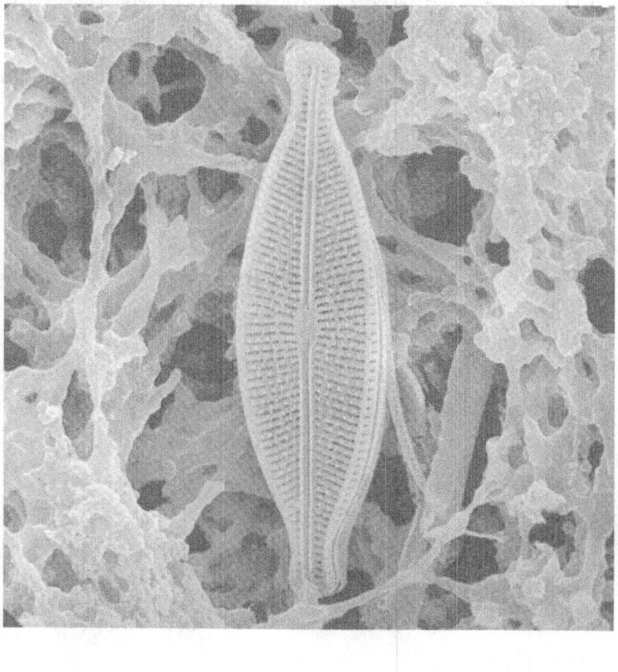

Figure 13-25

An electronic photograph of a diatom. Photograph courtesy of the Institut Universitaire de Médecine Légale, University of Lausanne, Switzerland.

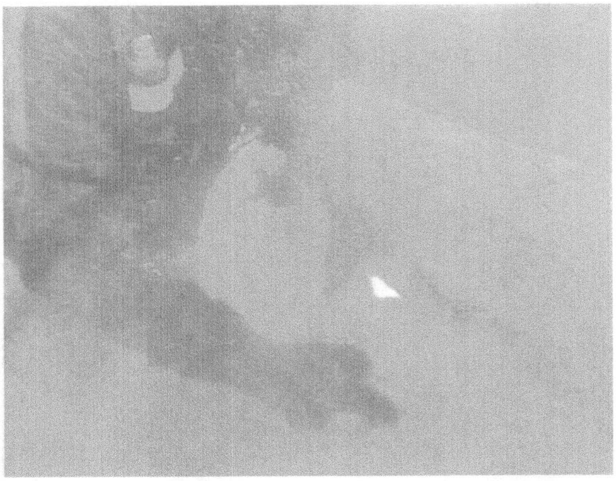

Figure 13-26

Collection of diatoms is performed by sampling the water in a jug at different locations. (Photograph courtesy of Gendarmerie River Company of the Rhine, March 2004, Strasbourg, France.) See Color Plate.

is carried out with three large bottles per body (one liter each minimum); the police diver samples the water around the body, at the surface, and at the estimated entry point (Figure 13-26). These three bottles are recorded in the notes and subsequent report and are forwarded to the laboratory with the body. The technique of the determination of diatoms is described in more details by Pollanen [11].

The next step consists of the extraction of the body; if the car has not undergone severe accident-related damage and if the body is not saponified, it should be relatively easy to extract the body [12, 13]. On the contrary, if the body is saponified or if the body is trapped in the interior due to mechanical damage, the vehicle can be brought directly to the surface with the body inside. The risk for a whole body to fall into the water during the extraction is limited. In any instance, when possible, it is better to remove the body before the vehicle's removal from the water. To avoid the loss of evidence, like teeth, hairs, or skin gloves, the body has to be prepared with bags around the head, the hands, and the feet. After that step, the body is bagged in a body bag and brought back to the surface (Figure 13-27).

13.6.8 Fingerprints and Palm Prints

Skin ridge patterns such as on fingers suffer from submersion to some extent, however often not to the point of preventing their observation [14]. Fingerprints are very good to identify a deceased person found underwater because the skin stays in place on the hands for a long period of time, as shown in Figures 13-28 and 13-29. During an autopsy, the pathologist can collect this glove of skin, and after a special treatment, fingerprints can be photographed or printed with ink.

When fingerprints are applied on a surface, like a door or a windshield of a car, and if sebaceous or grease secretions are deposited, it is still possible to recover these prints once the vehicle has been brought back to the surface. Forensic scientists generally use

Figure 13-27

A body is put in a body bag and brought back to the surface.

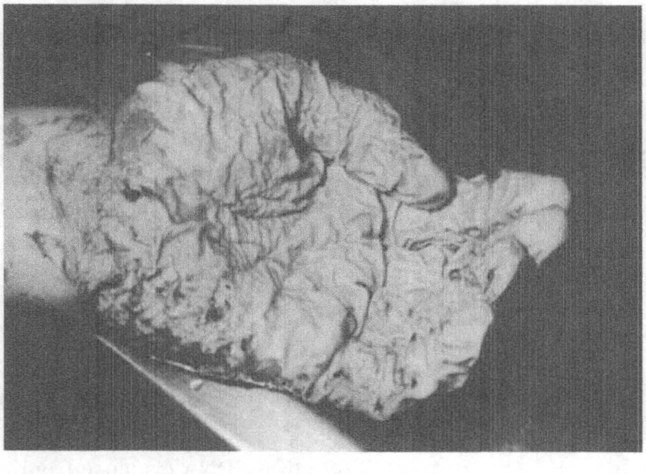

Figure 13-28

Effect of adipocere on the hand of a deceased person that was located underwater for a certain period of time. (Photograph courtesy of the Institut Universitaire de Médecine Légale, University of Lausanne, Switzerland.)

molybdenum particles in suspension in a fluid that is sprayed on the wet surface (SPR reagent as presented in Subsection 4.3.3). Molybdenum particles affix on the greasy prints; the excess of particles is rinsed by a smooth water jet [15]. If the surface is immersed in salt water, the search surface has to be soaked in fresh water before proceeding. In laboratory experiments and in case reports, these pieces of evidence are said to be usable, even after months of immersion. Thus, it is very important for the divers to be extremely careful

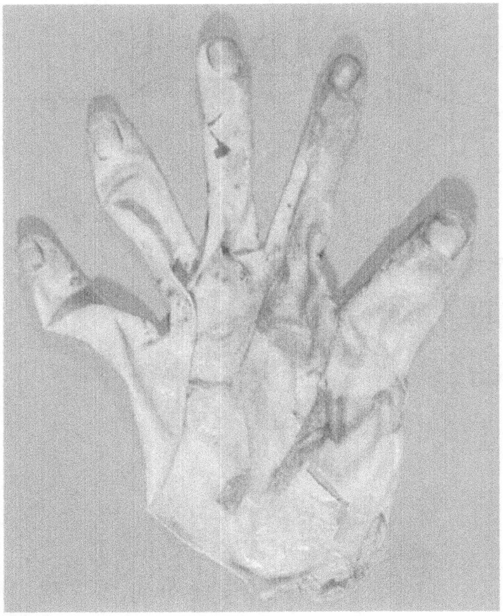

Figure 13-29

Skin glove due to adipocere retrieved from a body underwater. (Photograph courtesy of the Institut Universitaire de Médecine Légale, University of Lausanne, Switzerland.)

when examining and handling the vehicle not to wipe out possible latent prints that would be present on the different surfaces.

ACKNOWLEDGMENTS

The author would like to thank Eric Stauffer and Monica S. Bonfanti for their input in the writing of this chapter.

BIBLIOGRAPHY

[1] Becker RF. (2000) Myths of underwater recovery operations, *FBI Law Enforcement Bulletin,* **69**(9), pp 1–5.

[2] Sweet D and Shutler GG. (1999) Analysis of salivary DNA evidence from a bite mark on a body submerged in water, *Journal of Forensic Sciences,* **44**(5), pp 1069–1072.

[3] Rule N. (1995) Some techniques for cost-effective 3D mapping of underwater sites. In: *Computer Applications and Quantitative Methods in Archaeology,* ed Wilcock J and Lockyear K, British Archaeological Reports International Series, Oxford, England.

[4] International PADI (2001) *The Encyclopedia of recreational diving,* 2nd edition, International PADI, Rancho Santa Margarita, CA.

[5] Kirkbride KP. (2000) Spectroscopy: basic principles. In: *Encyclopedia of Forensic Sciences,* ed Siegel J, Knupfer G, and Saukko P, Academic Press, London, England, **1,** pp 191–194.

[6] Becker RF. (1995) *The underwater crime scene: underwater crime investigative techniques,* Charles C Thomas Publisher, Springfield, IL.

[7] Taylor LR. (1986) The restoration and identification of water-soaked documents: a case study, *Journal of Forensic Sciences,* **31**(3), pp 1113–1118.

[8] Santacroce G. (1995) The forensic examination of fire and water-damaged documents. In: *Advances in Forensic Sciences,* ed Jacob B and Bonte W, Verlag Dr. Köster, Berlin, Germany, **3,** pp 222–227.

[9] Locard E. (1934) *La police et les méthodes scientifiques,* Les Editions Rieder, Paris, France.

[10] Servettaz J and Myskowiak JB. (2004) Etat des connaissances en matière de police technique et scientifique (ADN, diatomées, anthropologie, . . .) en rapport avec le milieu subaquatique, *Conference of underwater technical and scientific polices,* Strasbourg, France.

[11] Pollanen MS. (1998) Forensic diatomology and drowning, *Elsevier Science,* Amsterdam, The Netherlands.

[12] Haglund WD. (1993) Disappearance of soft tissue and the disarticulation of human remains from aqueous environments, *Journal of Forensic Sciences,* **38**(4), pp 806–815.

[13] Kahana T, Almog J, Levy J, Shmeltzer E, Spier Y, and Hiss J. (1999) Marine taphonomy: adipocere formation in a series of bodies recovered from a single shipwreck, *Journal of Forensic Sciences,* **44**(5), pp 897–901.

[14] Cotton GE, Aufderheide AC, and Goldschmidt VG. (1987) Preservation of human tissue immersed for 5 years in fresh water of known temperature, *Journal of Forensic Sciences,* **32**(4), pp 1125–1130.

[15] Champod C, Lennard C, Margot P, and Stoilovic M. (2004) *Fingerprints and other ridge skin impressions,* CRC Press, Boca Raton, FL.

CHAPTER 14

EXAMINATION OF VEHICLE LICENSE PLATES

Didier Brossier

14.1 INTRODUCTION

Vehicles are involved in many different types of crime. Criminals often use counterfeit, forged, or stolen vehicle license plates to minimize the chances of being identified or spotted on the street. The license plate (also referred to as "tag") is an important item of the vehicle, because it can be considered as its identity card on permanent public display.

The registration of a vehicle and the issuance of license plates respond to two essential functions: administration and law enforcement. The registration is the first step in the recognition of ownership of a vehicle. In some countries, it also demonstrates the payment of an ad valorum or other types of vehicle tax. Furthermore, it allows for the government or official authority (such as the department of motor vehicle) to keep an official count of the number of vehicles and to control the vehicles authorized to drive on public roads. From a law enforcement perspective, the registration of vehicles allows the police to perform prevention and repression acts in regard to the vehicles on the road.

14.2 LICENSE PLATES REGULATIONS AND DELIVERY

14.2.1 Principle

To prevent the use and spread of counterfeit or forged license plates, some countries have limited the number of plate-manufacturing plants and have placed security features on the plate itself. The process of registration of a vehicle and the delivery of a license plate greatly varies from country to country. Even within a country, several dozens of different license plate formats may exist, such as for the registration of a new vehicle, a rental car, a car dealer vehicle, a government vehicle, a commercial truck, a postal service vehicle, or simply for a temporary registration. Hence, the information presented in this chapter cannot cover all the different plates from all countries around the world. Several volumes would be necessary to realize such a task. As a matter of fact, examples of only one very specific type of plate (regular passenger cars) for a very small number of countries are presented. Even though few examples are presented, the most important information the auto theft investigator must obtain is the basic concept behind the manufacturing, counterfeiting, forging, and examination of license plates. This is presented throughout the chapter.

As a good practice, auto theft investigators must always gather information from their department of motor vehicle (DMV) to know the laws and regulations applicable to license plates in their jurisdiction. For example, in the United States each state has a different set of license plates, and the related regulations may also vary. Furthermore, each state can have several different types of plates, which vary in design (including personalization) and function. In some countries, the numbering sequence can incorporate some information such as the jurisdiction issuing the plate, the vehicle's year, or the vehicle's type. Thus, the auto theft investigator should be familiar with the license plates issued in his or her jurisdiction and should be able to quickly read them (i.e., perform the first steps to ensure their authenticity). The problem arises when a stolen-recovered vehicle bears tags from another state or country with which the investigator is not familiar. In these instances, the establishment of a good network of professionals such as the International Association of Auto Theft Investigators (IAATI) can be very handy [1].

14.2.2 General Characteristics of Different Plates

Table 14-1 presents the general characteristics of some plates from some European and North American countries. Since 1998, all European Union (EU) member states must place a distinctive sign on the left of the license plate [3]. This sign, also called "euroband," must be on a blue retroreflective background (strip) and contain 12 yellow retroreflective stars as well as the distinctive sign of the member state of a white or yellow retroreflective color. The same EU regulation states that the blue background must be minimum 98 mm high, between 40 and 50 mm wide, and the stars must be centered on a 30-mm diameter circle, with the distance between two opposed points of a same star between 4 and 5 mm. Furthermore, the distinctive sign of the member state must be at least 20 mm high and the lines of the characters must be between 4 and 5 mm thick. With plates containing two lines of characters, the overall dimension of the sign can be reduced consequently. Table 14-2 presents the different country codes that must be used on the euroband. In practice, not all EU members have integrated this sign on their plates, yet. The euroband is not available in Denmark and on Belgian rear plates; is optional in United Kingdom, Sweden, and on Belgian front plates; and Poland still uses its national flag.

Note that in many countries there are older types of plates still circulating. These may be of different colors, dimensions, characteristics, and material than the currently issued plates.

14.2.3 Registration in Austria

The character sequence is provided by one of the 90 local offices, connected together via a computer database. The manufacturing and distribution of license plates are under the control of the Austrian government. The plates are the property of the government, and thus it is not legal to keep them once expired or no longer in use. To prevent counterfeiting, a reflecting film marked with the logo of the local office and with the Austrian Eagle is placed on the plate.

Table 14-1

Characteristics of license plates of some European and North American countries.

Country	Ownership	Dimensions[a] in [mm]	Material	Characters
Austria[b]	License plates belong to and stay with the vehicle.	520 × 120	Aluminum	Black on white background with 2 pairs of horizontal red lines
Belgium[c]	License plates belong to and stay with the owner.	Rear: 340 × 110 Front: 340 × 110 or 520 × 110 with euroband	Aluminum	Red on a white background
Canada	In most provinces, license plates are lent to and stay with the owner. If the owner no longer wishes to register the vehicle, he or she has to surrender the plates to the DMV. Also, the owner has to surrender the license plates when moving from a particular province to another and then obtain a new set at the new province's DMV or authorized stores. Prior to this system, license plates belonged and stayed with the vehicle.	303 × 152	Aluminum	Varies between provinces
Denmark	License plates belong to and stay with the vehicle.	505 × 120	Aluminum	Black on a white background with a red border
France[b,d]	License plates are issued to a given vehicle, but change with the owner's change of address (even within a same department) or with a change of owner.[e]	520 × 110	Aluminum Plastic	Front: black on a white background Rear: black on a yellow background
Germany[b]	License plates belong to and stay with the vehicle.	520 × 110 plates maybe shorter if fewer characters	Aluminum	Black on a white background Characters are 77 × 50 mm, with 14-mm thick lines
Great Britain[c]	License plates can be transferred from vehicle to vehicle.	520 × 110	Plastic	Front: black on a white background Rear: black on a yellow background

Table 14-1

Continued.

Country	Ownership	Dimensions[a] in [mm]	Material	Characters
Italy[b]	License plates belong to and stay with the vehicle.	Front: 360 × 110 Rear: 520 × 110	Aluminum	Black on a white background
Netherlands[b]	License plates belong to and stay with the vehicle.	520 × 110	Aluminum	Black on a yellow background
Norway	License plates belong to and stay with the vehicle.	520 × 110 (495 × 110 prior to December 2001)	Aluminum	Black on a silver-white background
Portugal[b]	License plates belong to and stay with the vehicle.	520 × 110	Aluminum Plastic	Black on a white background
Spain[b]	License plates belong to and stay with the vehicle.	520 × 110	Aluminum Plastic	Black on white background
Sweden[c]	License plates belong to and stay with the vehicle.	520 × 110 with euroband 478 × 114 without euroband	Aluminum[f] Plastic	Black on a white background with a black border
Switzerland	License plates are lent to and stay with the owner. If the owner no longer wishes to register the vehicle, he or she has to surrender the plates to the DMV. Also, the owner has to surrender the license plates when moving from a particular canton to another and then obtain a new set at the new canton's DMV.	Front: 300 × 80 Rear: 300 × 160 or 500 × 110	Aluminum	Black on white background
United States	Most often, license plates belong to and stay with the owner.	303 × 152 (12 × 6 inches)	Aluminum	Varies among states

[a] Unless specified, the dimensions apply to both front and rear license plates when two license plates are issued for one vehicle.
[b] A blue strip "euroband" conforming to European standard is located on the left side of the plate.
[c] The "euroband" is optional.
[d] Starting in 2008, front and rear plates will both be black on yellow [2].
[e] Starting in 2008, plates will stay with vehicle even if sold out of the department [2].
[f] All plates issued since 2002 are made of aluminum.

Table 14-2

European Union member state codes that must be used on the euroband.

Country	Code	Country	Code	Country	Code
Austria	A	Greece	GR	Poland	PL
Belgium	B	Hungary	H	Portugal	P
Cyprus	CY	Irish Republic	IRL	Slovakia	SK
Czech Republic	CZ	Italy	I	Slovenia	SLO
Denmark	DK	Latvia	LV	Spain	E
Estonia	EST	Lithuania	LT	Sweden	S
Finland	FIN	Luxembourg	L	United Kingdom	GB, SCO, ENG
France	F	Malta	M		
Germany	D	Netherlands	NL		

14.2.4 Registration in Belgium

When purchasing a vehicle, a certificate of conformity and the bill of sale are provided to the new owner. These documents are necessary to obtain insurance. Once the insurance is obtained, the documents are presented to the direction of vehicle registration (Direction des Immatriculations des Véhicules or DIV), which delivers the registration certificate and the rear license plate. The front plate is provided by a local private embosser. An example of a front plate with the euroband is shown in Figure 14-1a. Figure 14-1b shows the corresponding rear plate. To prevent counterfeiting, the rear plate bears a stamp above the dash separating the letters from the numbers. In addition, another impression with the letters "DIV" and a numbering sequence is located under the dash. The front plate does not bear any stamp.

Figure 14-1

(a) Front Belgian license plate with the euroband. (b) Rear Belgian license plate. (Photographs courtesy of Jean-François Chevalley.) See Color Plate.

14.2.5 Registration in Canada

There are two ways for licensing vehicles in the provinces of Canada. This could be performed through a government department such as the DMV, which applies to provinces that have public vehicle insurance programs, such as British Columbia, Manitoba, and

Saskatchewan. Other provinces have privatized the process, and the registration and issuance of license plates is performed through independently owned stores. These stores then send the data back to the government's DMV. These provinces only offer privatized vehicle insurance. Each province has its own sets of license plates. The provinces of Alberta, Nova Scotia, Prince Edward Island, Québec, and Northwest Territories issue one (rear) license plate [4]. All other provinces issue two license plates. There is also the possibility of having license plates personalized (i.e., to choose which sets of characters [if available] to appear on the plate). A registration sticker must be purchased every year and affixed on the plate. This system is very similar to the system adopted in the United States, and more information in this regard can be found in Subsection 14.2.17.

14.2.6 Registration in Denmark

The plates are delivered, along with the registration certificate, by the government central registry office upon presentation of the characteristics of the vehicle.

14.2.7 Registration in France

The character sequence is provided by the different prefectures, which are connected together via a computer database. The plates are prepared by any shoe repair shop, mechanical shop, or shopping center. A security feature consisting of a retroreflective film bearing the homologation number of the film is present. This number starts with the letters "TPMR." The homologation of the tag is engraved in the upper left corner. This number starts with the letters "TPPR" [5]. This number is the homologation number of the retroreflective component (white in color for the front plate and yellow in color for the rear plate), provided for the ministry of transportation (Ministère des Transports, de l'Equipement, du Tourisme et de la Mer) to each manufacturer of license plates. It is thus possible to identify the manufacturer of the plate through that number. Furthermore, on the back of some plates, a coded number containing the identity of the manufacturer is printed. This code also includes the date and time at which the plate was manufactured [5]. Starting in 2008, France will introduce a new centralized system [2].

14.2.8 Registration in Germany

The character sequence is provided by a local bureau (örtliche Zulassungsstelle) of the vehicle registration center (Kraftfahrtbundesamt or KBA). The format of the character sequence is one to three letters, one to two letters, and one to four numbers. The prefix letters denote the district and/or town. Local private companies emboss the plates. Thus, there are many different embossers in the country. With new vehicles, the license plate is brought to the local KBA, which places the stickers on them. On the rear plate, the sticker of the technical control (Technischer Überwachungsverein or TÜV), as well as the insurance and tax seal (silver with black letters and the coat of arms of the federal state) are

Figure 14-2

Front German license plate. Note the euroband on the left. (Photograph courtesy of Jean-François Chevalley.) See Color Plate.

Figure 14-3

Rear Italian license plate issued in Rome, Italy. Note the euroband on the left and the Italian-specific blue strip on the right. See Color Plate.

placed. On the front plate, the colored emission control test (Abgas-Untersuchung) sticker and the insurance and tax stickers are placed. With used vehicles, because the plates are already on the vehicle, a verification procedure is carried out by the KBA to match the plates to the registration card. Once this is done, the appropriate stickers are affixed on the plate. The control stickers are located after the first set of characters. An example of a front plate is shown in Figure 14-2.

14.2.9 Registration in Great Britain

The character sequence is issued by the Driver and Vehicle Licensing Agency (DVLA), an executive agency of the Department for Transport (DfT). British plates must conform to the standards as described by the DVLA [6]. There are four types of registration number: prefix, suffix, dateless, and current style. In the case of personalized registration, any number style can be purchased; however, the number sequence cannot postdate the registration of the car. In other words, the number cannot be used to make the car appear "younger" than it is. The license plates can be provided by a license plate supplier, which since January 1, 2003 must be registered with the DVLA [7]. Prior to that date, anyone could manufacture license plates.

14.2.10 Registration in Italy

The character sequence is issued by the ministry of infrastructure and transportation (Ministero delle Infrastrutture e dei Trasporti). The character sequence is composed of two alphabetical characters followed by three numerical and two alphabetical characters. The official symbol of the Italian Republic (a small star including the letters "RI" and circled by leaves) is located after the two first letters. Figure 14-3 shows a plate issued in Rome.

The production and distribution of plates is controlled by the government. The plates are covered with a retroreflective adhesive film. On this film are impressed the three letters PGS and the code of the authorized factory. On the right side, another blue stripe bears the year of first time in circulation (circled). The provincial code can be located just below.

14.2.11 Registration in the Netherlands

The character sequence is issued by the service of road traffic of the ministry of transportation and navigation. The plates are provided by private embossing shops. They bear a manufacturing code.

14.2.12 Registration in Norway

The character sequence is issued by the state bureau of transportation (Statens vegvesen). The plates are embossed by a few private companies authorized by the government. Car dealers as well as traffic stations can carry out the registration procedure. When someone buys a new vehicle, the car dealer does all the paperwork and delivers the car with the plates fitted. Plates for passenger car are black on white, whereas light commercial vehicles have green plates. All plates bear a security sticker. This sticker changes color annually and attests of the validity of the automotive tax and insurance. The plate number also appears on the sticker.

14.2.13 Registration in Portugal

The character sequence is issued by the bureau of registration of motor vehicle. License plates are provided by private embossing shops. On the Portuguese plates, a yellow strip is present on the right side of the plates, where the month and year of registration are displayed.

14.2.14 Registration in Spain

The character sequence is issued by the general direction of traffic (Dirección General de Tráfico or DGT). The plates are provided by local private embossers. The name of the plate manufacturer is engraved in the plate and the embossing shop engraves an identification number. There is a security feature consisting of a retroreflective film bearing an optical variable device (OVD) representing the blazon of Spain and the name of the film's manufacturer.

14.2.15 Registration in Sweden

The character sequence is provided by the office of road security and the plates are distributed by the government. License plates also bear a security sticker attesting to the payment of taxes and insurance as well as the passing of technical control. This sticker also bears the plate number and its color changes every year.

14.2.16 Registration in Switzerland

The character sequence and the plates are issued by the DMV upon presentation of the vehicle, proof of insurance, and the successful completion of a stringent technical and emission control. Plates belong to the government and must be surrendered when no longer used or when expired. It is illegal to keep plates that are no longer valid and affixed to a vehicle. On rear plates, characters are located between the national Swiss flag on the left and the canton flag on the right. Switzerland also offers a system of "interchangeable plates" where an owner can register two vehicles with one set of plates. Only one vehicle (the vehicle bearing the plates) can be driven at a time. This allows for a reduction of the insurance premium and governmental taxes.

14.2.17 Registration in the United States

There are many different regulations, which depend on the state. When the owner moves from state to state, a new set of plates must be obtained. Most often, the character sequence and the license plate(s) are issued by the DMV of the given state. Each state has different plates. The States of Alabama, Arizona, Arkansas, Delaware, Florida, Georgia, Indiana, Kansas, Kentucky, Louisiana, Michigan, Mississippi, New Mexico, North Carolina, Oklahoma, Pennsylvania, South Carolina, Tennessee, and West Virginia issue only one (rear) license plate [4]. All other 31 states require and issue two license plates. As in Canada, it is possible to get, in most states, a personalized (sometimes referred to as "prestige") tag. The design of the plates can greatly vary from a generic design to different specific designs such as the logo of a university, of an athletic team, or of a given organization. Figure 14-4 presents three generic plates from the States of Florida, Pennsylvania, and North Carolina.

Although differences exist among the license plates from the different states, there are many common elements. Figure 14-5 presents a personalized license plate with the sequence of characters "SBRINZ" from the State of Georgia.

This particular design was issued in 2004 and replaced the previous design. In this plate, it is possible to note the main elements: The sticker boxes are often found on US plates (sometimes in the upper corners), the logo (often centered or on the left), the slogan (each state has its own slogan, which is more and more often replaced by the website of the state), the location where the plate was issued (usually the county, but it is not found on every state), and the bolt holes. In Georgia, the renewal month of the registration is the month

a b c

Figure 14-4

(a) License plate from the State of Florida. (b) License plate from the State of Pennsylvania. (c) License plate from the State of North Carolina. (Photographs courtesy of Eric Stauffer.) See Color Plate.

Figure 14-5

A personalized (prestige) Georgia license plate. Note the different elements that are found in most tags in the United States. (Photograph courtesy of Eric Stauffer.) See Color Plate.

of the owner's birthday. This sticker is never changed on the plate, because the plate follows the owner. However, the right sticker is changed every year with the renewal. Depending on the state, the sticker might bear the license plate's character sequence (as seen in Figures 14-4c and 14-5) or not (as seen in Figures 14-4a and b). Also, in some states, there might be only one sticker bearing both the month and the year (as seen in Figures 14-4a and b). In some states, the sticker color changes every year.

Each state has a great number of different plate styles available. In 2005, for regular passenger cars, the State of Georgia proposed 20 styles of college plates, 10 styles of government plates, 35 styles of military plates, 12 styles of special interest plates, and 7 styles of standard plates. Figure 14-6 presents some of these different plates. These different plates can be either openly available to the regular citizen or are restricted to a certain category of citizens or vehicles.

Because plates vary from state to state, so do security features. Some states do not offer any security features, whereas some others offer some security. For example, OVDs, such as the ones in Figures 14-7a and b, can be present on the plate. Figure 14-7a is an example of a Florida tag, which exhibits a circle with the letters "FL" followed by a code "BOO2." Also, the outline of the state is located at the bottom part of the circle. This logo is repeated vertically across the plate. Figure 14-7b shows an example of a Georgia tag with the same circle repeated vertically across the plate. The circle contains the letters "GA" followed by a code "BOR1" and the outline of the state at the bottom of the circle. It is possible to distinguish these security marks by tilting the plate when observing it. At some point, these marks should appear to the eye. They are completely invisible when the plate is directly facing the eyes. Also, the State of Georgia has included a similar device in the yearly sticker, as shown in Figure 14-7c. Several circles with the letters "GA" appear on the sticker when properly observed.

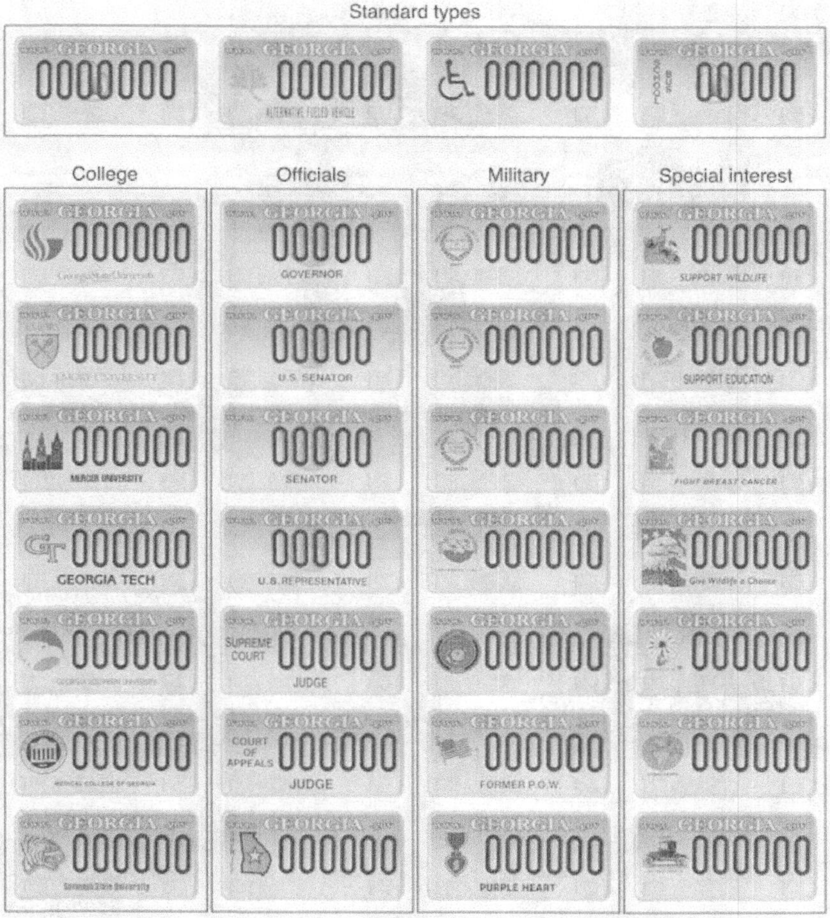

Figure 14-6

Examples of some of the different plates available in Georgia in 2005. (Source: Georgia Department of Revenue, Motor Vehicle Division, available at http://www.dmvs.ga.gov.) See Color Plate.

14.3 MANUFACTURING OF LICENSE PLATES

14.3.1 Principle

Before being taken over by officials, license plates were manufactured from any types of materials, mostly leather, rubber, or tin [8]. Then, license plates started to be made of steel, followed by aluminum, which is more expensive but also more durable. The supports of modern license plates are manufactured using two different materials: aluminum or plastic. Plastic license plates are almost exclusively used in Europe.

14.3.2 Aluminum Plates

Two methods exist for the manufacture of aluminum plates: embossing and riveting.

Figure 14-7

(a) Example of the security feature (OVD) located on a Florida tag. It consists of a circle with the letters "FL" followed by a code "BOO2" and the state outline. This circle is repeated vertically across the plate.
(b) Example of the security feature located on a Georgia tag. It consists of a circle with the letters "GA" followed by a code "BOR1" and the state outline. This circle is repeated vertically across the plate.
(c) Example of the security feature located on the yearly renewal sticker of a Georgia tag. It consists of several circles with the letters "GA" inside them. (Photographs courtesy of Eric Stauffer.) See Color Plate.

A/ Embossing

The blank plate is constituted of two or three elements. Three-element plates consist of a retroreflective film, a thin aluminum tinsel, and a thick aluminum support. The support is sturdy enough to hold the shape of the plate but soft enough to be embossed. The tinsel is cut out during the embossing process, which reveals the color of the support behind the tinsel and brings the contrast necessary to read characters. Two-element plates consist of a retroreflective film and a thick aluminum support. The characters are embossed in the support and then are colorized either by inking (painting) or thermal transfer.

To emboss the plates, a stamping machine is used. The plates are placed between dies located on each side of the plates and a mechanical or hydraulic press compresses the dies on the plate, resulting in the embossing [5]. The character sequence is unique on each plate, but the background relief (such as a logo or names common to all

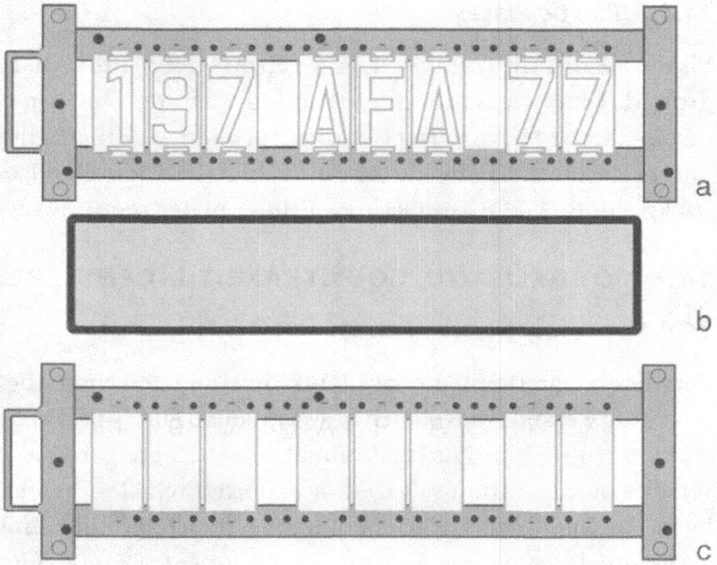

Figure 14-8
(a) Frame onto which dies are laid out.
(b) Blank plate before embossing.
(c) Frame onto which (counter)dies are laid out.

a b

Figure 14-9
(a) Riveted characters on a French license plate. (b) View of the rivet from the back side of the plate.

plates) remains the same. A sketch of the dies used in such a process is shown in Figure 14-8.

B/ Riveting

Riveted plates are composed of two elements: a retroreflective film and a thick aluminum support into which holes have been predrilled. The characters are riveted on the plates through the holes as shown in Figure 14-9.

14.3.3 Plastic Plates

Plastic plates are composed of a retroreflective film that constitutes the colorized background. Characters are then pasted, transferred, or printed on the background. In some instances, the background is directly printed with the right sequence of characters. A transparent sheet (typically about 3 mm thick) is then glued on top of the character-bearing background. The manufacture of these plates requires very little investment.

14.4 FORGED AND COUNTERFEIT LICENSE PLATES

14.4.1 Altered License Plates

Very few counterfeit plates are found in countries where the manufacturing of license plates is centralized and controlled by governmental authorities and where the plates contain security features. In such instances, criminals prefer using genuine plates, either stolen from parked vehicles or taken from salvaged or exported vehicles. In addition, these plates can be forged by modifying some characters to reduce the chance of being caught by law enforcement.

Figure 14-10 shows an example of a forged plate. Belgium criminals were using a stolen vehicle to commit an armed robbery. They modified the vehicle's original plates using red paint. The character sequence "PTT-050" was modified into the sequence "BTT-858" using red and white paint. The forgery is easily detectable upon close inspection of the plate, because the modified characters are not embossed and the paint does not present the same glossiness. However, such a forgery might not be readily noticed when the vehicle is driving down the road.

In countries where it is extremely easy to obtain a blank support, even if security features are present, criminals can easily proceed to the embossing by themselves. At this point, they can emboss any desired characters.

14.4.2 Replica License Plates

In some countries, and now also available on the Internet, there are stores that manufacture replica license plates, which in some instances are very similar to the authentic ones [9, 10]. It is possible to choose many plates from different countries and/or states and to choose the desired sequence of characters. Although a close examination of the plate might quickly reveal the farce, it is very likely that such a plate might be mistaken as genuine when affixed on a vehicle and observed from a certain distance.

14.4.3 Authentic Plates

Although in some countries such as Austria, Switzerland, or Canada (some provinces) the license plates must be surrendered to the authority when not used anymore or criminal charges may apply, some countries do not control the history of license plates once the registration has expired. Thus, it is possible to discard, give, collect, or sell these license plates. Such "traffic" of license plates mostly goes to license plate collectors; however, criminals can also obtain authentic license plates with minimal effort. In addition, with the

Figure 14-10
(a) Forged Belgian vehicle license plate.
(b) Detailed view of the modification of the characters "0" into the characters "8" by the addition of red and white paints.
(c) Detailed view of the modification of the character "P" into the character "B" by addition of red paint. See Color Plate.

constant development of the Internet, it is also possible to obtain many of these used plates on-line at little cost. For example, on the eBay auction store, there are more than 8,000 plates on sale on an almost permanent basis, including plates from Switzerland [11]. There are also other on-line stores that are specialized in such items [12]. Although some of these license plates might be expired, they are still very precious items for the criminal who desires to swap license plates on a stolen vehicle to temporarily disguise its true identity.

14.5 FORENSIC APPROACH TO THE EXAMINATION OF LICENSE PLATES

14.5.1 Principle

To determine the authenticity of a license plate, several characteristics must be verified and compared with exemplary plates from the given jurisdiction. The following steps should be followed when examining a vehicle license plate:

- Verification with the DMV that the license plate number or sequence of characters is listed and that it corresponds to the vehicle (make, model, color, year, vehicle identification number) onto which it is affixed.
- Verification of the correspondence between the national, state, or county flag and the authority of issuance of the plate.
- Verification of the compliance of the plate with laws and regulations (size, shape, color, characters, security features). It is important to note that in some countries the lifetime of the license plate can be very long. Thus, the colors can fade away and the license plate official features could have changed (change of security features, shape, logo, size, etc.). Also, it is important to keep in mind that errors in the manufacturing of plates can occur. In such instances, a plate issued by the competent authority might present deviation from the standard plates.
- Verification of the validity of the registration (date on security stickers).

- Examination of the plate by comparison with an official exemplary plate.
- Verification of the character font.
- Verification of security features. It is important to determine whether logos or stickers are genuine or if they have been added, modified, forged, or counterfeit.

As for other evidence discovered at the scene, the metallic support constituting the license plate must be immediately protected and photographed in situ at the scene [5]. When the investigator takes questioned plates off the vehicle, it is important to proceed with great precaution to prevent the loss of any other types of physical evidence. Fingerprints and DNA can be found on the plate, particularly to its reverse side, because this side is often protected and only accessible by the person installing or removing the plate. In addition, it is important to search for possible traces of glue and fibers [5].

This first step in the investigation is followed by the technical examination of the license plate, which is performed in two separate steps: First, the manufacturer of the plate is identified, and second, the search for individual characteristics from the embossing process is carried out [5].

14.5.2 Source of Information

If the investigator lacks information regarding the design and regulations of the particular plate(s) because the country or state of origin is not local, he or she should contact the proper authority. Nevertheless, it is possible to consult reference books on license plates from around the world, such as *Registration Plates of the World* [13, 14]. In addition, pertinent information is available on the Internet. For example, in the United States, a website offers links to all official DMVs from all 50 states and the District of Columbia [15].

When these steps are not feasible in a reasonable period of time or are too inconvenient, the investigator should contact a network of professionals, such as the IAATI, to find a person from that particular location who is familiar with the plates [1]. Furthermore, there are many associations not involved in auto theft investigation but that unites collectors of license plates, such as the Automobile License Plate Collectors Association (ALPCA) or the European Registration Plate Association (Europlate) [16, 17]. There are many other associations around the world [18–20]. These collectors can be an invaluable resource of information in this particular instance.

Many websites are also available that include thousands of license plate photographs from around the world [21–24]. These websites are invaluable resources of information. They offer many links to other websites, such as official DMV websites, which can be very convenient to the investigator. The investigator who is interested in license plates is strongly encouraged to visit these sites.

Finally, it is also possible to contact the manufacturer of the plate when more detailed information regarding its manufacturing process or distribution is sought. For example, the company Erich Utsch AG supplies most of the plates for European countries (and some other countries as well) [25].

14.5.3 Identification of Manufacturer and/or Embosser

In countries where license plates are not distributed by the government, local and private manufacturers or embossers must follow established regulations. For example, some regulations mandate that the name of the embosser be indicated in plain view on a particular portion of the plate or that a specific code (related to the embosser) is placed on the retroreflective film. In other countries, security marks only allow for the identification of the manufacturer of the blank plate and not of the embosser.

14.5.4 Comparison of Embossing Defects

During the embossing process, when the die hits the plate it creates a deformation of the plate of the same shape every time [5]. If the die bears a small defect, it will be reproduced on every plate that was embossed by that particular die. This defect can be either due to a manufacturing artifact or to a characteristic accidentally acquired by the die during its life [5]. This could be due to a cut or a shock that would remove a minute amount of material (indent) from the die itself. Figures 14-11a is a schematic representation of a die without defect that embosses a surface. Figure 14-11b shows a similar die bearing a small defect

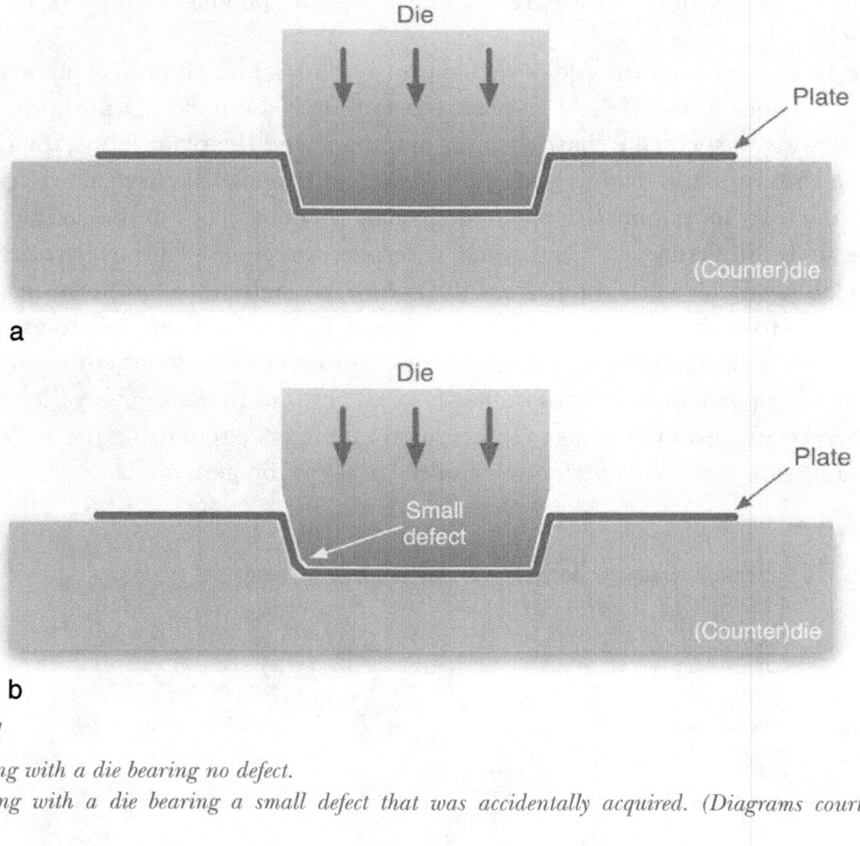

Figure 14-11

(a) Embossing with a die bearing no defect.

(b) Embossing with a die bearing a small defect that was accidentally acquired. (Diagrams courtesy of Eric Stauffer.)

accidentally acquired. The figure shows how this small defect would leave a different impression on the plate during the embossing process. This small defect is an individual characteristic of the die that is reproduced on the embossed impression every time. The proper observation of this characteristic and its comparison with the original die may allow for its identification as the source of the impression.

If an embosser is suspected of manufacturing a set of counterfeit license plates or if several counterfeit license plates are suspected to originate from the same embosser, it is possible to examine the embossed characters for defects. First, the shape, font, and size of the characters are compared. If they match, the examination can continue with the search for these small defects or individual characteristics. This is performed by observing the embossed characters under magnification. In general, the examination should be performed first on the back side of the plate, because these defects would mostly appear on the part that was in contact with the die. If the plate is debossed (opposite of unembossed), then the observation should be made on its front side. If the original dies are available, the potential corresponding defects are carefully noted. Ideally, the examiner should compare plate with plate rather than plate with die. However, when the manufacturing of a plate is not possible from the seized dies, then a cast of the die can be made and compared with the questioned plate(s). If enough unique similarities between the defects present on the embossed characters and the dies are found, the die can be identified as having made the embossment [5].

Figure 14-12 shows an example of such a process with a French license plate bearing the character sequence 8816 YR 93. The plate was found on a vehicle not properly registered. An embosser was suspected to have illegally manufactured the plate. Thus, the investigator went to the embosser and made a plate with the same sequence of characters as comparison material. Macroscopic examination revealed small characteristics in the characters "8" "9" and "3" as shown in Figures 14-13a through c, respectively. These characteristics were found on both plates, which clearly established the common origin of the impressions: The same dies were used to emboss both plates. Using this technique, it is possible to establish links between plates manufactured by the same embosser or more correctly the same stamping machine. This is particularly useful when the same embosser creates many illegal plates for an organized crime group. In this case, criminal intelligence regarding the activities of the group and its involvement in different criminal acts can be generated.

Figure 14-12

View of the questioned license plate.

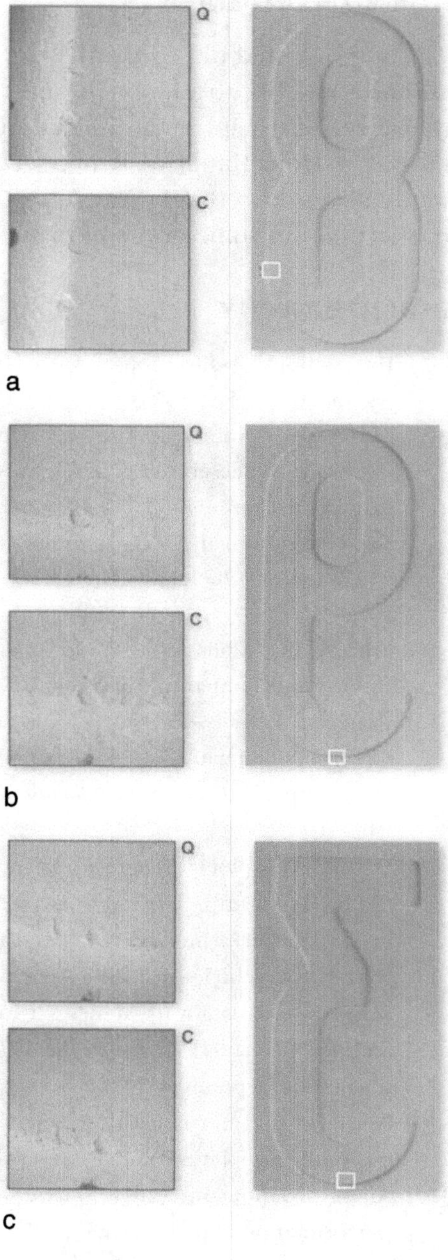

Figure 14-13

(a) View of the "8" character from the reverse side of the plate. Note the presence of the same 2 defects in both the questioned plate (Q) and the comparison plate (C).
(b) View of the "9" character from the reverse side of the plate. Note the presence of the same defects in both the questioned plate (Q) and the comparison plate (C).
(c) View of the "3" character from the reverse side of the plate. Note the presence of the same defect in both the questioned plate (Q) and the comparison plate (C).

ACKNOWLEDGMENTS

The author would like to thank Eric Stauffer for translating this chapter into English. The author would like to address his most sincere acknowledgments to the following people for their precious help in the redaction of this chapter: Monica S. Bonfanti, Jean-François Chevalley, Horst Katterwe, Rebecca Pepler, Mark Sandercock, and Eric Stauffer. The author would also like to thank Olav Arne Brekke (http://www.olavplates.com) for the review of the accuracy of some of the information presented in the chapter.

BIBLIOGRAPHY

[1] International Association of Auto Theft Investigators (2005) *What is IAATI?,* available at http://www.iaati.org, last access performed on September 23, 2005.

[2] Ministère de l'intérieur et de l'aménagement du territoire (2005) *Le nouveau système d'immatriculation des véhicules,* Secrétariat général, Service de la communication, France.

[3] European Commission (1998) Council Regulation (EC) No 2411/98 of 3 November 1998 on the recognition in intra-Community traffic of the distinguishing sign of the Member State in which motor vehicles and their trailers are registered, *Official Journal of the European Union,* **L 299,** pp 1–3.

[4] Murray TC. (2004) *The official license plate book 2004–2005,* Digest edition, Interstate Directory Publishing Company, Mill Neck, NY.

[5] Rouger P, Bons B, and Surowiec P. (1999) L'identification des traces sur une scène criminelle A. Les traces de chaussures dans la neige, B. Examen des plaques d'immatriculation réfléctorisées et estampées, *Revue internationale de criminologie et de police technique et scientifique,* **52**(2), pp 240–46.

[6] Driver and Vehicle Licensing Agency (2003) *Current requirements on the display of number plates,* available at http://www.dvla.gov.uk/vehicles/regmarks/reg_marks_current_requirements.htm, last access performed on October 13, 2005.

[7] Driver and Vehicle Licensing Agency (2003) *The new register of number plate suppliers,* available at http://www.dvla.gov.uk/vehicles/rnps.htm, last access performed on October 13, 2005.

[8] Sallmen JP. (2004) *Canadian license plates,* available at http://www.canplates.com, last access performed on September 23, 2005.

[9] Auto Tags International (2005) *"Create your own" auto tags,* available at http://www.customlicenseplates.com, last access performed on September 23, 2005.

[10] Autogeardepot.com (2005) *Licenseplates.tv,* available at http://www.licenseplates.tv, last access performed on September 23, 2005.

[11] eBay (2005) *Collectibles—Transportation—Automobilia—License Plates,* available at http://www.ebay.com, last access performed on September 23, 2005.

[12] Dream On Technology (2005) *License plates for collectors,* available at http://www.platesusa.com, last access performed on September 23, 2005.

[13] Parker N and Weeks J. (2004) Registration Plates of the World, Europlate, Bridgewater, United Kingdom.

[14] Parker N and Weeks J. (2004) The Interpol Guide to Vehicle Registration Plates, Interpol Lyon, France.

[15] eDriver (2005) *Welcome to DMV.org,* available at http://www.dmv.org, last access performed on September 23, 2005.

[16] Automobile License Plate Collectors Association (2005) *About ALPCA,* available at http://www.alpca.org, last access performed on September 23, 2005.

[17] European Registration Plate Association (2005) About Europlate, available at http://www.europlate.org, last access performed on October 13, 2005.

[18] Francoplaque (2005) *The Francoplaque license plate collectors site,* available at http://plaque.free.fr/index-english.html, last access performed on September 23, 2005.

[19] Associazione Italiana Studio Targhe Automobilistiche (2005) *Cos'è A.I.S.T.A.?,* available at http://www.aista.it, last access performed on September 23, 2005.

[20] Number Plate Collector's Club (2004) *About the Club,* available at http://www.npcc.org.au, last access performed on September 23, 2005.

[21] Nicholson D. (2005) *15q.net—License plates of North America, 1969-present,* available at http://www.15q.net, last access performed on September 23, 2005.

[22] Brekke OA. (2005) *Olav's license plate pictures,* available at http://www.olavsplates.com, last access performed on September 23, 2005.

[23] Mangiagli M. (2003) *Plateshed—the #1 forum for plate collectors,* available at http://www.plateshed.com, last access performed on September 23, 2005.

[24] Kustermann M. (2005) *License plates of the world,* available at http://www.worldlicenseplates.com, last access performed on September 23, 2005.

[25] Erich Utsch AG (2005) *Utsch at a glance,* available at http://www.utsch.com, last access performed on October 13, 2005.

CHAPTER 15

EXAMINATION OF VEHICLE REGISTRATION DOCUMENTS

Diana Ombelli

15.1 INTRODUCTION

Vehicle registration documents could be circumstantial evidence when a stolen-recovered or crime-related (registered) vehicle is examined. The documents may include relevant information for the investigation, such as a holder name, a date of first registration, or a registration number.

"Registered vehicles are those which bear a permanent number-plate issued by the competent authorities [. . .] and those bearing a provisional number-plate which, although issued to meet a particular situation, proves that tax and roadworthiness requirements have been fulfilled" [1]. This definition is used in European legislation and links the presence of license plates (or tags) on the vehicle to the process of registration. In other parts of the world there are registration documents available for vehicles registered by an official competent authority, but not necessarily tag-bearing. The competent authority issues (paper) documents to certify that the vehicle is allowed to circulate on public roads. The document issued by the authority is called vehicle registration document or certificate. This is used to identify the vehicle and to enable the owner of the vehicle (or the vendor) to sell the vehicle on the market or re-register it in another country should their place of residence change [2].

In some countries the ownership proof is combined with the registration certificate. Other countries issue separate documents or do not even link ownership to the registration. In those countries, such as the Netherlands, the person mentioned on the registration document is the keeper of the vehicle. In the United States, there are typically three separate documents assigned to a vehicle bearing a tag: the title, which declares ownership of vehicle; the registration card with the department of motor vehicles (DMV), which is an official record of the assignment of the tag to the vehicle, allowing it to be used on the road; and the proof of insurance card delivered by the insurance company.

This chapter provides information about the manufacture, personalization, issuance, and disposal of vehicle registration documents and relates those steps with the fraud risks and types. The examination methodology is supported by several references from the current literature. Finally, an examination checklist is proposed.

15.2 CONCEPT OF REGISTRATION OF VEHICLES

15.2.1 General Principle

To recognize the value of a registration document, it is important to know the processes taking place during the registration of a vehicle by the authorities. According to the European Union Council Directive 1997/37/EC, Article 2b, registration is the administrative authorization for the entry into service in road traffic of a vehicle, involving the identification of the latter and the issuance of a serial number, to be known as the registration number.

Several conditions need to be fulfilled for the registration of a vehicle [3]. First, the holder registers the vehicle in the country of his or her normal residence. Then, the vehicle is examined and inspected according to the national legal rules in relation to the following two aspects:

I Examination of the technical characteristics of the vehicle and the relevant documents (type-approval): The type-approval is mostly related to new cars and may involve car manufacturers, distributors, intermediaries, and dealers. The vehicle must satisfy technical requirements. In some cases the vehicle is not physically examined and the approval is based upon paper work.

II Inspection of the physical condition of the vehicle (roadworthiness tests): The inspection of the physical condition of the vehicle is carried out periodically to check the safety of the vehicle on roads. Also, for these aspects, the vehicle needs to meet minimum requirements during the inspection. Because of environmental issues, more countries have included an exhaust emission test as a part of the vehicle inspection procedure.

Not all countries require both conditions to be fulfilled. For example, in many states of the United States there are no physical examinations of vehicles, whereas Swiss registration authorities examine both aspects and the technical examination is stringent.

Finally, in the event the car has already been registered and needs to be re-registered, the authority examines the current registration document and checks the consistency of the information, such as license plate number, vehicle identification number (VIN), and make/model of the car, in databases or in the document itself (e.g., redundant coded information). As an example of redundant information, the Dutch registration document shows a caption named "Meldcode" which consists of the last four digits of the VIN, as seen in Figures 15-1 and 15-2.

Statistical figures regarding the registration of vehicles show a constant increase of the number of registered passenger cars and trucks or lorries in Europe, Israel, Canada, and the United States, as shown in Table 15-1 [4]. Re-registration is estimated to tens of millions solely in the European Union [5].

Citizens can find on-line information about the current procedures used in most western European and North American countries. It is important to emphasize that the decentralized registration of vehicles occurs often in large countries, particularly federal ones. This

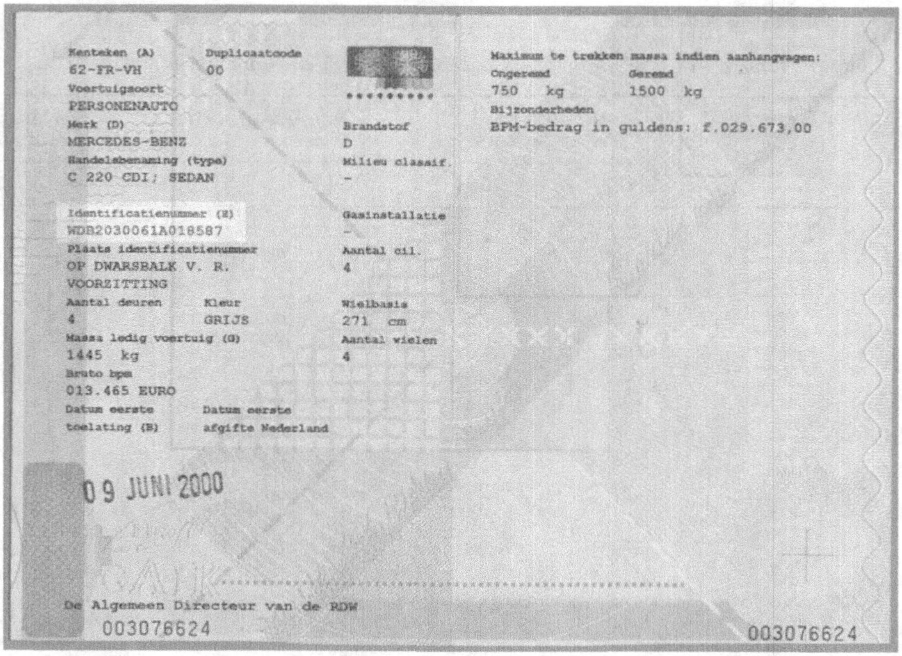

Kenteken (A) Duplicaatcode
62-FR-VH 00
Voertuigsoort
PERSONENAUTO
Merk (D) Brandstof
MERCEDES-BENZ D
Handelsbenaming (type) Milieu classif.
C 220 CDI; SEDAN -
Identificatienummer (E) Gasinstallatie
WDB2030061A018587 -
Plaats identificatienummer Aantal cil.
OP DWARSBALK V. R. 4
VOORZITTING
Aantal deuren Kleur Wielbasis
4 GRIJS 271 cm
Massa ledig voertuig (G) Aantal wielen
1445 kg 4
Bruto bpm
013.465 EURO
Datum eerste toelating (B) Datum eerste afgifte Nederland
09 JUNI 2000

Maximum te trekken massa indien aanhangwagen:
Ongeremd Geremd
750 kg 1500 kg
Bijzonderheden
BPM-bedrag in guldens: f.029.673,00

De Algemeen Directeur van de RDW
003076624 003076624

Figure 15-1

The Dutch vehicle registration document part IA (Voertuigbewijs, the vehicle document). The highlighted caption (E), called Identificatienummer, is the VIN of the vehicle. See Color Plate.

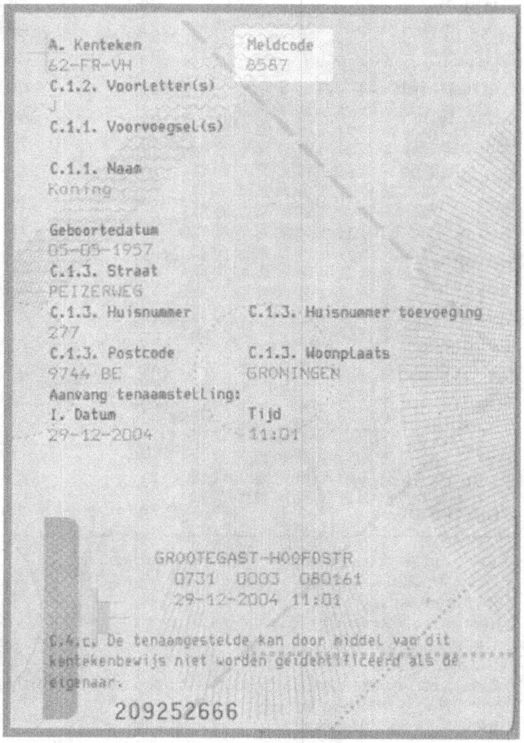

A. Kenteken Meldcode
62-FR-VH 8587
C.1.2. Voorletter(s)
J
C.1.1. Voorvoegsel(s)

C.1.1. Naam
Koning

Geboortedatum
05-05-1957
C.1.3. Straat
PEIZERWEG
C.1.3. Huisnummer C.1.3. Huisnummer toevoeging
277
C.1.3. Postcode C.1.3. Woonplaats
9744 BE GRONINGEN
Aanvang tenaamstelling:
I. Datum Tijd
29-12-2004 11:01

GROOTEGAST-HOOFDSTR
0731 0003 080161
29-12-2004 11:01

C.4.c. De tenaamgestelde kan door middel van dit kentekenbewijs niet worden geïdentificeerd als de eigenaar.
209252666

Figure 15-2

The Dutch vehicle registration document, part IB (Tenaamstellingbewijs, the keeper's document). The highlighted caption "Meldcode" shows the last four digits of the VIN (written in full in Figure 15-1). See Color Plate.

Table 15-1

Statistical figures from the UNECE regarding the number of passenger cars and trucks or lorries.

	Number of passenger cars		Number of lorries		New passenger cars registered
	1985	2001	1985	2001	2001
Albania	—	29[e]	—	11[e]	—
Austria	335	504	27	41	36
Belgium	339	462	28	51	49
Bulgaria	119	262	—	31	24
Canada	442	550	—	21	—
Croatia	146	269	9	28	24
Cyprus	225	354	81	148	31
Czech Republic	200[b]	343	13	29	19
Denmark	294	350	50[b]	71	18
Estonia	114	299	39	59[e]	19
Finland	316	416	36	60	21
France	379	483	54	89[c]	38
Germany	333	534	17	32	41[a]
Greece	127	253[d]	60	93[d]	25[e]
Hungary	136	244	14	35	12[e]
Iceland	429	561	53	68	27
Ireland	201[f]	361[f]	26	57[e]	42[f]
Israel	147	241	28	53	21
Italy	394	557[e]	31	56[e]	40[e]
Latvia	83	249	24	38	16
Liechtenstein	493	646[d]	41	73[d]	60[d]
Lithuania	96	337	21[h]	26[h]	33[a]
Luxemburg	422	635	25	49	98
Malta	234	485	54[a]	108[a,g]	33[a,d]
Netherlands	319	408	26	55	33
Norway	365	414	55	93[a,g]	28[d]
Poland	99	259	21	46	13
Portugal	170[g]	481	51	162	35[e]
Romania	—	139[a]	—	20	7
Serbia and Montenegro	—	161[e]	—	—	—
Slovakia	136	240	23	29[h]	12
Slovenia	255	433	17	26	28
Spain	241	437	40	99	38
Sweden	377	455	26	44	33
Switzerland	405	506	31[h]	39[h]	44
The Former Yugoslav Republic of Macedonia	125	148	11	10	7
Turkey	20	67	8	18	4
United Kingdom	313	422	32	50	45
United States	554	776	25	27	29

Values are expressed in per 1,000 population.
[a] Data for 2000.
[b] Includes vans.
[c] Restricted to vehicles less than 15 years of age.
[d] Data for 1998.
[e] Data for 1999.
[f] Refers to private cars under current license and vehicles and new private cards registered for the first time.
[g] Including combined vehicles and vans.
[h] Includes road tractors.

phenomenon makes the checking of registration documents difficult, because of the different procedures and the variety of documents issued. Moreover, it is extremely difficult to find public information about the security features of registration documents, which would help an occasional forensic examiner.

15.2.2 International Standardization and Cooperation

For decades governmental institutions have cooperated to facilitate the mutual understanding and acceptance of procedures and documents and to prevent vehicle-related crimes. The United Nations Economic Commission for Europe (UNECE) provides a forum for governments to cooperate on standardization policies. This international body cooperates closely with other organizations aiming toward standardization, such as the International Organization for Standardization (ISO). The UNECE drew up the Convention on Road Traffic in Vienna on November 8, 1968, which defines the principle of registration of vehicle by the contracting parties [6]. Furthermore, this Convention defines the registration certificate and the information it should contain. The following items should be present in any vehicle registration documents:

- A serial number, to be known as the registration number, composed of numerals or a combination of numerals with letters. Numerals need to be Arabic and letters are capital Latin letters.
- The date of first registration of the vehicle.
- The full name and home address of the holder of the certificate.
- The name or trademark of the maker of the vehicle.
- The serial number of the chassis (VIN).
- In case of a vehicle intended for carriage of goods, the permissible maximum mass (gross vehicle weight) and the unladen mass (base curb weight).
- The period of validity, if not unlimited.

The information mentioned in the vehicle registration document, as well as the layout of the document itself, varies from country to country and from state to state, because there is no ISO standard recommending how to organize this information. This situation complicates the genuineness check. Therefore, authorities competent in car registration have made an effort over the past years to digitize and share car information with other entities and countries.

The characteristic of the European Union (EU) is the principle of free movement of people and goods. Consequences of this single European market are the abolishment of the border-crossing controls and the increase of movements within the EU. The Benelux countries (Belgium, the Netherlands, and Luxembourg) played a pioneering role in sharing information on vehicles. Since 1994, those countries are parties of the EUCARIS treaty (EUropean CAR and driving license Information System). This legal instrument allows car registration authorities appointed by each country to exchange data related to motor vehicle and driving licenses. EUCARIS is a communication network, aiming to be an instrument for prevention, detection, and prosecution of violation of the law and to play a role in keeping national vehicle and driving license records accurate and up to date. EUCARIS allows the participating authorities to consult quickly on-line motor vehicle and driving license data kept in the national registers of countries affiliated to EUCARIS. Currently, seven countries are parties to the treaty: Germany, Great Britain, Belgium, Luxembourg, Latvia, Sweden, and the Netherlands. Other countries use the EUCARIS system for limited applications, such as bilateral and unilateral exchange of vehicle information. Practically, the checking operation is performed by entering a VIN, a license plate number, and/or a driver's license number.

Law enforcement agencies can submit questioned VINs to the Interpol ASF-SMV database (Automated Search Facility-Stolen Motor Vehicle) to determine whether they correspond to a stolen vehicle or not. In the United States, VIN Assist is an online service provided by the National Insurance Crime Bureau that supports insurance and law enforcement professional in checking VINs. Another example of cooperation can be found in the Netherlands. The National Vehicle Crime Information Centre (LIV) is a center of expertise for tackling vehicle crime and assisting investigations. The LIV is a joint initiative of the Centre for Vehicle Technology and Information (RDW), the National Police Forces (KLPD), and the Missing Vehicles Register (VAR Foundation). This is the only body in the Netherlands in which public and private parties cooperate in the fight against vehicle crime.

15.2.3 EU Legislation

In the EU, the internal market of vehicles is subjected to national and community legislation. The latter aims to harmonize the rules related to registration. Within this frame the EU issued in 1999 a Council Directive on the registration document for vehicles [7]. More recently, the Council directive 2003/127/EC amended the older directive to provide EU member countries with specifications for the issue of vehicle registration documents in microprocessor smart card format instead of paper [8]. Background of the EU directive is the harmonization of form and content of the registration certificate for vehicles in order to facilitate comprehension and thus help toward the free movement.

The harmonized registration document facilitates the inspection and the cross-check with the driving license (verifying if the holder drives solely those categories of vehicles for which he or she is authorized). Efficiency is improved by making it easier to understand the registration documents and to identify a vehicle owner or keeper when a vehicle has been registered in another member state. The Directive specifies the procedure for re-registration of a vehicle in another member state.

In both formats—paper or microprocessor smart card—the European format consists of two parts:

I It contains technical information, is carried by the driver of the vehicle, and allows for the identification of the vehicle on the road.

II It is required to re-register the vehicle. The name appearing on the document is the keeper of the vehicle.

The European legislation foresees an option to indicate the owner of the vehicle on both parts of the registration document. Moreover, EU member states are allowed to use another format for temporary registration of vehicles. The deadline for EU member states to comply with the newest directive has already expired (January 15, 2005).

A European initiative aims to improve fraud prevention by giving the registration authorities access to the Schengen Information System (SIS). It is a resolution adopted by the European Parliament in spring 2004 defining that vehicle registration authorities have the right to consult whether vehicles presented to them for registration have been stolen, misappropriated, lost, or invalidated [9].

15.2.4 World Tour

Countries such as Australia, Canada, Mexico, and the United States have delegated the responsibility of issuance of registration documents to the lower level political entities (e.g., states, provinces, territories). Procedures can differ a great deal among the different states within the same country. Fortunately, there are references to assist involved parties to register vehicles in the proper manner. For instance, the Polk's *Motor Vehicle Registration Manuals,* Volumes I and II, provide descriptions of US procedures. Volume III incorporates procedures for Canada, a selection of European countries, Mexico, Costa Rica, Hong Kong, and Japan.

In Hong Kong, for example, the car registration occurs once in the car's lifetime. A registration mark is given, and the car is put in the appropriate class. Licensing a vehicle gives it the right to be operated on the road [10].

15.3 SECURITY DOCUMENTS

15.3.1 Principle

Vehicle registration documents belong to the group of security documents, because they need to be protected by security features against fraud. As mentioned earlier, the sale of a stolen car to an unsuspecting customer today requires supporting vehicle documents. If those documents were not (or badly) secured, it would be easier to reproduce or alter them to accomplish the transaction.

The technical examination of the document provides information about its genuineness, including the variable information on it. Several processes are involved in the life cycle of a security document: manufacture, personalization, issuance, and disposal. Fraud can occur at any time. Criminals tend to look for the weakest link in the security chain, and therefore countermeasures need to be taken at all levels.

The authorities controlling the registration of vehicles are responsible for developing and issuing secured registration documents, in cooperation with security printers. Based upon experience in use and abuse, depending on the availability of techniques and financial means, the level of security in a new generation of documents is usually higher. Figures 15-3 through 15-6 show two examples of vehicle registration documents secured against fraud. In addition to the traditional security features (among others, watermark, ultraviolet [UV] fluorescent printing, and fibers), the new Hungarian and the new Polish vehicle registration documents show some innovative features. The newly introduced Hungarian vehicle registration document is secured against counterfeiting with a hologram strip applied along the right border on the back side of the document (Figure 15-4). The new Polish vehicle registration certificate was introduced in late 2004 [11]. A 2D barcode including all personalization data is printed on the document (Figure 15-5). Also, a laser perforated serial number is found under title "D" in the front of the document and can be seen on the back of the document (Figures 15-5 and 15-6).

The security chain of a security document begins at the manufacturing level. Basic materials, such as substrates (e.g., paper), some kind of inks, and security features used for the manufacture of security documents, are not publicly available on the market. Only security printers have access to those.

The design of security documents is characterized by lines and structures generated by customized software packages to prevent reproduction with widely available software. Generally, the colors of the design are intentionally chosen to make difficult the imitation by conventional (four colors) offset or digital printing techniques. The security printer and the governmental authority responsible for issuing registration documents often work together for the choice of the security features, keeping an eye on the balance of them. Regarding registration documents, it is evident that the security features need to be clearly recognized by law enforcement agencies, which are trained to this task, as well as by the common citizen, who is rarely aware and experienced enough in checking the genuineness of security documents. To this extent security features are categorized in three levels:

First level: The genuineness of the security feature can be checked using the naked eye and without any tools. An average person should be able to check those features based upon public documentation distributed through flyers, broadcasted, or available on official websites.

Second level: The checker uses tools, such as a magnifier or a black light (UV lamp), for the inspection of the security features. UV features are very popular, and many shops and financial institutions are equipped to check those security features.

Third level: The inspection of those security features requires special know-how and/or equipment. Laboratories check those features.

15.3.2 Manufacture

Paper or polymer-based materials are substrates used for the manufacture of vehicle registration documents. Paper manufactured for security documents is optically dull (no fluo-

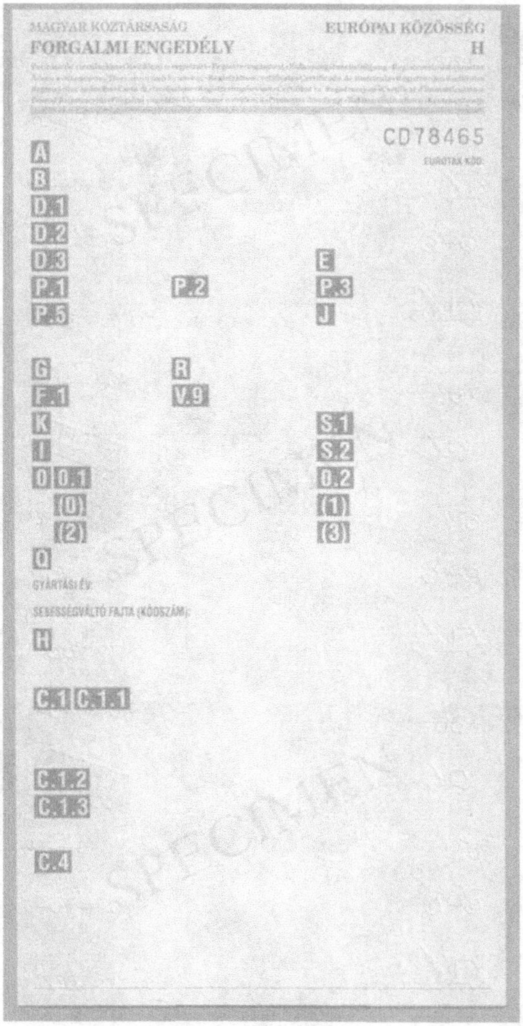

Figure 15-3

The vehicle registration document of Hungary (front). See Color Plate.

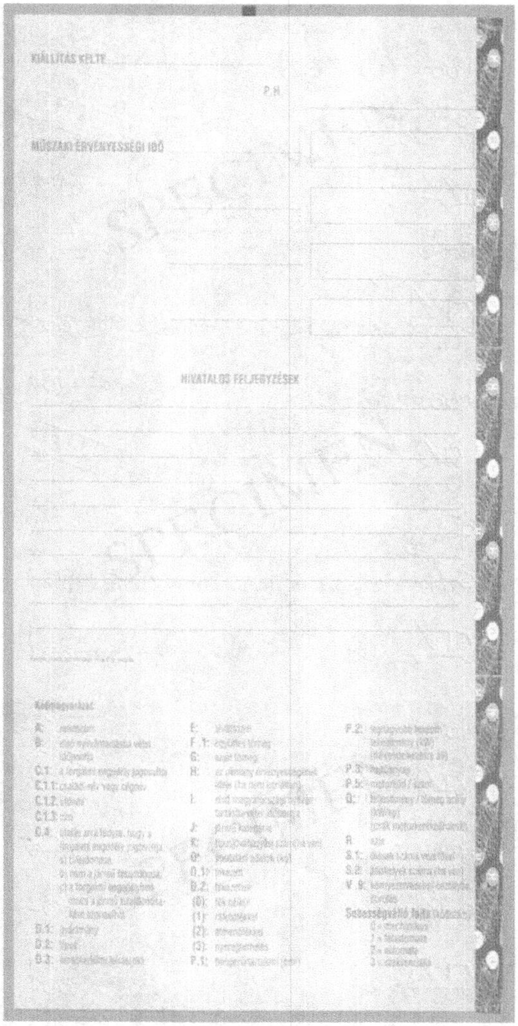

Figure 15-4

The vehicle registration document of Hungary (back). See Color Plate.

rescence under black light) and often presents a watermark, as shown in Figure 15-7. This security feature widely appears on bank notes and is well known by the public [12]. It can be easily checked by looking at the document with transmitted light. The paper could also include fluorescent items such as fibers or threads, as seen in Figure 15-8.

Four printing techniques are used and combined for the design of security documents: lithography, intaglio, silk-screen printing, and letterpress. The background printing composed of structures and lines is applied by lithography. This printing technique leaves a smooth amount of ink on the substrate where needed by the design. In some cases the inks

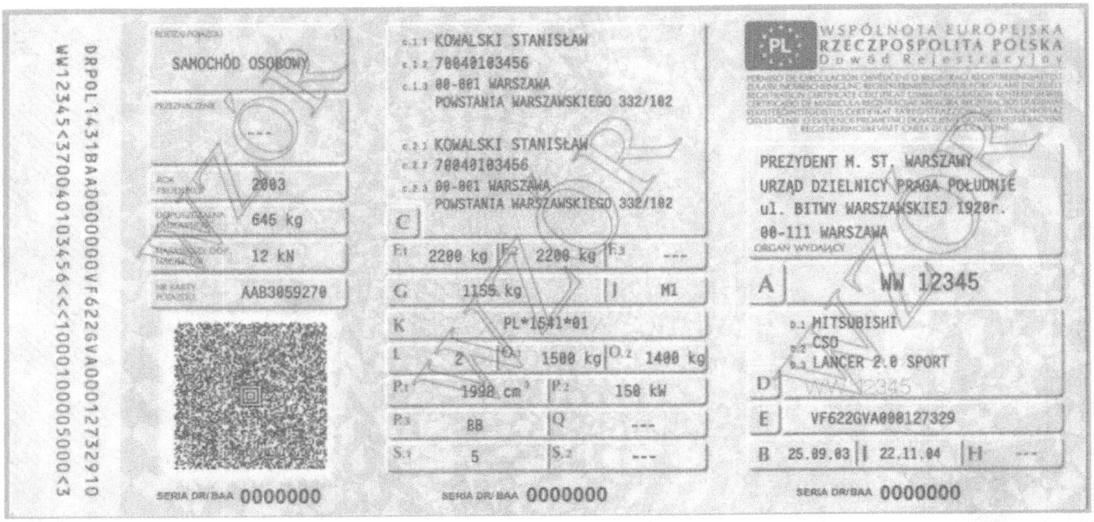

Figure 15-5

The vehicle registration document of Poland (front). See Color Plate.

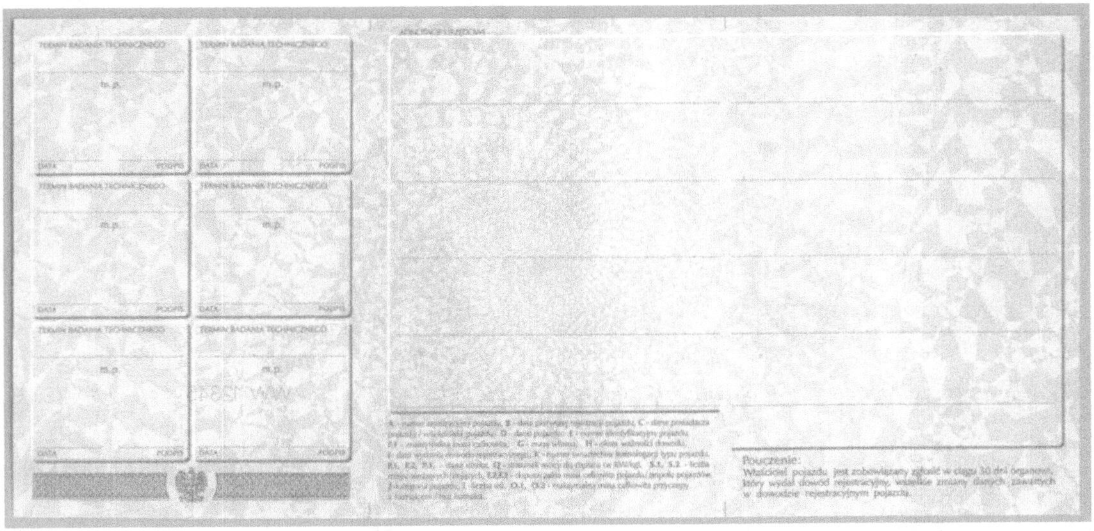

Figure 15-6

The vehicle registration document of Poland (back). See Color Plate.

used have fluorescent effects under UV light. Also, this technique allows the integration of additional security features such as see-through register, rainbow effect, and microtext [13]. An example of microtext from the Dutch vehicle registration document is shown in Figure 15-9.

Other design elements could be printed by other techniques such as intaglio [13]. This technique leaves a tangible amount of ink on the substrate, which can be seen as a relief when using an oblique light, as demonstrated in Figure 15-10. The design of the intaglio print could also include lines of microtext.

Figure 15-7

The repeated letters RDW and bars with four holes present in the background are from the watermark of the Dutch vehicle registration document at 0.7× magnification. (Courtesy of RDW, The Netherlands.) See Color Plate.

Figure 15-8

Blue, yellow, and pink fibers integrated in a paper sample at 2.5× magnification. See Color Plate.

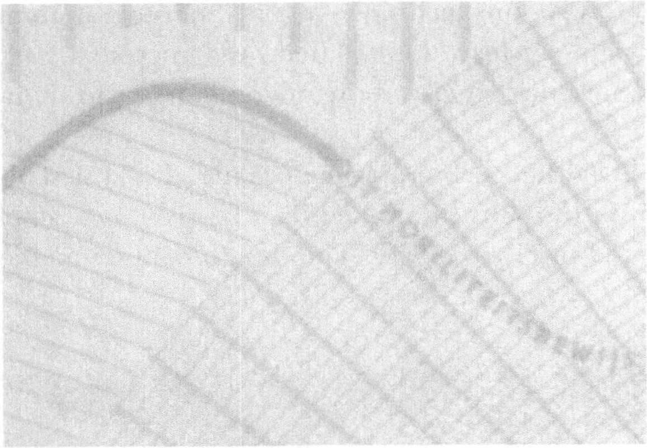

Figure 15-9

Example of microtext found on the Dutch vehicle registration document and seen at 7× magnification. The red line contains some microtext printed using the lithographic technique. (Courtesy of RDW, The Netherlands.) See Color Plate.

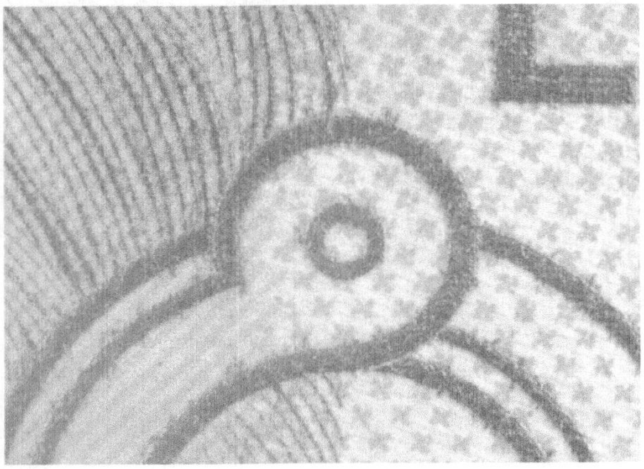

Figure 15-10

Example of a design element printed in intaglio on a Le Mont sample banknote at a 10× magnification. (Courtesy of KBA-GIORI, Switzerland.) See Color Plate.

Iridescent inks, such as optical variable ink (OVI), can be applied by silk-screen printing. This printing technique is recognizable when using a magnifier. The design element is composed of tiny ink spots, which fill the internal part of the element but are distinguishable at the borders by the characteristic ladder pattern, as shown in Figure 15-11. This pattern is due to the structure of the mesh, comparable with a fabric weft.

Prenumbering of documents is mostly performed by letterpress [13]. This printing technique can be recognized by the aspect of the borders of the characters (which can be either letters or numbers or a combination of both), as seen in Figure 15-12. The ink used for the numbering could also be fluorescent under a black light.

Other types of security features are the iridescent optically variable devices (OVDs) and, more specifically, the diffractive optically variable image devices (DOVIDs) [14]. Those are the generic names for design elements that can present different images and iridescent effects depending on the incident light beam (Figure 15-13). This group of security features

Figure 15-11

Green lines printed with a silk-screen printing technique at 2× magnification.

Figure 15-12

Numbering in letterpress of the Dutch vehicle registration document at 9× magnification. Note the border of the characters, showing slight inverted embossment, typical of letterpress technique. (Courtesy of RDW, The Netherlands.) See Color Plate.

contains not only holograms, but also devices under the patented names such as Pixelgram, Exelgram, and Kinegram. Recently, DOVIDs have been integrated into vehicle registration documents. The Dutch vehicle registration document shown in Figure 15-1 and the Hungarian vehicle registration document shown in Figure 15-4 are examples of documents containing such security features. However, OVDs are subject to imitation. The Kinegram has been reported to be the most secure DOVID [15].

Other features that involve an optical approach are printed structures hiding invisible information, which can be seen with special lenses or decoders (screen-decoded images). Last but not least, the security documents can be protected against photocopy with printed structures that become visible on the copy of the genuine document (scan and screen traps) [16].

This brief inventory of security features is certainly not exhaustive and should be considered as an indication of what an examiner could find in security documents.

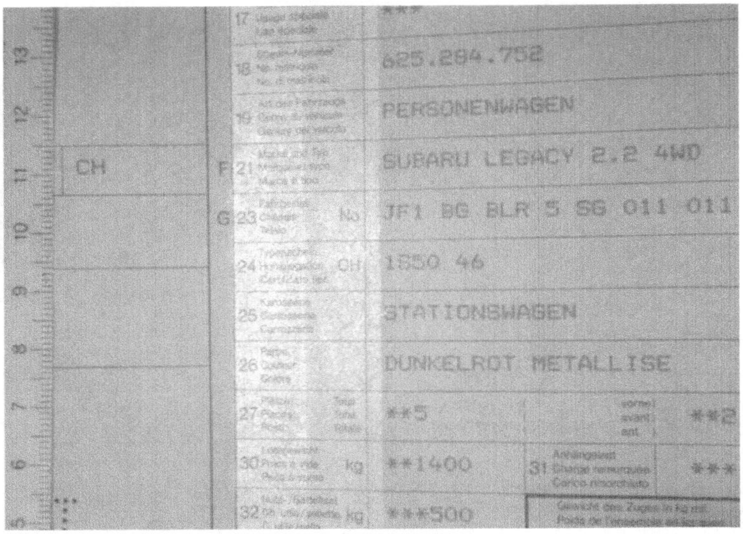

Figure 15-13

View of the iridescent effect present on the Swiss vehicle registration document. The two bands (green and purple) are only visible with the incident light and the observation at specific angles. See Color Plate.

As soon as the manufacturing process is completed, the finished blank security document is shipped for personalization to either a central place or satellite locations. Also, this logistic process needs to be secured, because a blank document could have an important commercial value in the criminal business. After the theft of blank documents, experienced forgers apply the variable information imitating the genuine personalization method. That is the reason why more and more secured personalization techniques are chosen.

15.3.3 Secured Personalization

Personalization is the technique used to fill in the variable information such as VIN or vehicle characteristics to the blank document. The choice of the best personalization technique is strictly related to the situation in a given environment. Aspects such as the issuing procedure (central/decentral), the availability of financial means, and the desired security level are considered in the decision-making process. Table 15-2 presents a summary of the characteristics of the most commonly used personalization techniques.

Alteration is known as a major fraud risk for the variable data added by the personalization. To limit the risk of the theft of blank documents, it is preferable to apply a personalization technique hardly available to the common citizen based upon one or a combination of different techniques to strengthen the security chain. Moreover, foils with

Table 15-2

Most commonly used personalization techniques and their characteristics.

Technique	Location	Cost[a]	Security level	Comments
Handwriting	Decentral	Very low	None	Authenticity is hard to verify by nonprofessionals
Typing	Decentral	Low	Very low	Can be easily imitated and altered
Ink jet	Decentral	Low	Low	Can be easily imitated
Dot matrix	Decentral	Medium	Medium to low	Higher ink penetration than ink jet
Laser printing	Decentral	Low	Very low	Can be easily imitated and altered
Dye sublimation	Decentral/central	Medium	Low	Difficult to recognize and it can be easily imitated
Thermotransfer	Decentral/central	Medium	Low	Difficult to recognize and it can be easily imitated
Laser engraving	Decentral/central	High	High	The information is far under the top, very characteristic aspect
Laser perforation	Central	High	High	Substrate is partially removed, very characteristic aspect

[a] Cost includes purchase and maintenance costs.

optical effects (e.g., retro-reflectance, UV fluorescence, etc.) are used to protect the data after the personalization process.

In the future, paper-based documents will be substituted with plastic cards bearing an electronic chip storing the information related to the vehicle. This will increase the security level of the documents and the variable data as well. The EU has already established rules for the issuance of such a document [8].

15.3.4 Secured Issuance and Disposal

Authorities responsible for the issuance of security documents, particularly identification and travel documents, are aware of the fraud risk related to their task. Fraud prevention means for them a secured environment (physical security), controlled procedures (personal and work security), and stock control (control on quantities). The fraud types related to issuance are described in Subsection 15.4.2.

In the final stage of their life cycle, documents are disposed. This relates not only to semi-manufactured products and finished blank products from the production, but also to personalized documents from the issuing stage. The producer disposes the first group of products according to strict reconciliation procedure, where quantities are counted and a report is archived after disposal. In the second group there are documents that need to be destroyed before issuance, for whatever reason (e.g., error or poor quality during personalization), or are returned after expiration. An invalidation and disposal procedure should be implemented at collection locations to minimize the risk of illegal "recycling" of genuine security documents by criminal organizations, that steal them from the collection location.

15.4 DOCUMENT FRAUD

15.4.1 Fraud Motives

There are multiple reasons and needs for trickery with a registration document. The main reason is the trade of stolen vehicles. The vehicle is stolen, the registration document is modified accordingly, if necessary, and the vehicle is put on the market for sale (mostly to a private person). The new owner or keeper is registered, and the thief disappears forever. In the event that the original registration document is available to the thief, the name of the keeper or owner appearing on the vehicle registration documents is the weakest link. The buyer should check the identity and the genuineness of the information about the vendor's identity on the registration document. More often, the original registration document is missing, whereas the thief needs to forge or produce a counterfeit document to sell the vehicle. In some cases, document fraud is committed for fiscal reasons. For example, in the Netherlands, cars aged 25 years and older are exempt from paying road taxes.

15.4.2 Types of Fraud

There are many different methods used to achieve the desired fraud. Table 15-3 presents a summary of these types of document fraud. The fraud can take place at two levels: the basic materials and/or the methods of personalization of the variable information. In general, authorities use preprinted forms for the application and issuance of registration documents. The personalization of the final vehicle registration document can take place at a local place (decentral way) or at a central place (central way). The examiner needs to check both the docu-

Table 15-3

Different methods used to create fraudulent documents and their characteristics.

Types of fraud	Genuine material	Correct personalization technique	Correct information
Counterfeit	No	No	No
Forgery/alteration	Partial	Partial	Partial
Fantasy document	No	No	No
Unauthorized issuance	Yes	Yes	No
Stolen blank	Yes	Usually not	No
Issuance based upon false information	Yes	Yes	No

ment and the personalization to establish the genuineness of the vehicle registration document. Also, the legitimacy of the variable information needs to be addressed by a cross-check with a database.

A/ Counterfeit

Counterfeiting means that the forger uses materials and techniques to reproduce a given document. Mostly, the genuine document is used as a model and the forger aims to simulate the aspect of the original document (Figures 15-14 and 15-15). In very exceptional circumstances, forgers have access to the genuine materials or are able to reproduce the original material. In case of doubts, the forensic examination of the document includes a detailed analysis of the physical and chemical composition of materials and inks used for the manufacture of the document. Instrumental analysis methods can also be applied.

B/ Forgery and Alteration

By altering an existing document, the forger changes the appearance of the document, typically modifying information appearing on it. The document will show security features related to genuineness of the basic materials. The examiner will need to observe very carefully the variable data to detect any manipulation.

C/ Fantasy Document

Sometimes the forger creates a document based upon his or her own fantasy. The materials used as well as the aspect of the document differ from the appearance of the genuine docu-

Figure 15-14

Detail of the genuine Dutch registration document (Part IA) at 1.5× magnification. See Color Plate.

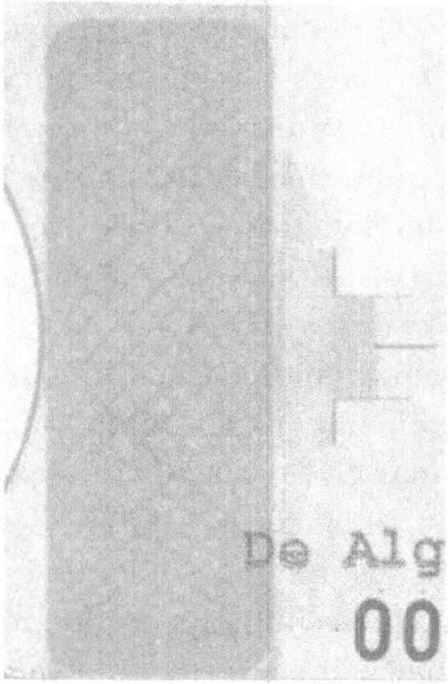

Figure 15-15

Detail of a counterfeit Dutch registration document (Part IA) at 1.5× magnification. See Color Plate.

ment. Because of the lack of standardization, this kind of fraud may be very attractive for criminals.

D/ Unauthorized Issuance

If a (corrupted) civil servant issued a registration document illegally (e.g., putting false information on it), the document would show the correct security features and personalization technique. Therefore, this type of fraud is difficult to detect. Based upon observation, the examiner will probably conclude that the document is genuine from the technical point of view. A detailed investigation about the circumstances of issuance and a possible cross-reference in the database, if the information in question has not been illegally updated, could supply evidence or point at this type of fraud.

E/ Stolen Blank

Documents that are stolen before they are personalized are called blank documents. The theft of blank documents happens more frequently in countries that personalize security documents in decentral way.

F/ Issuance Based upon False Information

Issuance based upon false information occurs when the applicant intentionally misleads the authority. The fraud could be based upon a false identification mark of the vehicle (and VIN) and false registration documents and/or license plates.

15.4.3 Fraud Trends

Digital printing techniques are widely used to imitate a security document in general and registration documents in particular. Even an occasional forger can obtain an amazing result with a high-resolution scanner and an ink-jet printer at home. Criminal organizations are using less often the traditional printing techniques such as offset and silk-screen printing for counterfeiting the entire document.

First-level security features are mostly used by forgers trying to imitate the primary effect of them. For instance, the fake watermark is printed on watermark-free paper. Second-level security features are often ignored by forgers, with the exception of UV features, which are widely known and checked in commercial circumstances or at the borders because of the large availability of UV lamps (black light) at these checkpoints. Therefore, this kind of feature may give a false sense of security if replicated effectively by the forgers. It is seldom necessary to examine the third-level security features. This would be done in a situation where the examination of levels 1 and 2 is inconclusive. Most forgers do not have the know-how and the technology needed to perfectly imitate or alter level 3 security features. More often, they will leave traces of the manipulation or will not use the correct materials and techniques for the creation of a document that will not be deceptive enough to mislead the examiner. Figures 15-16 to 15-22 show examples of counterfeit security features, along with the authentic version.

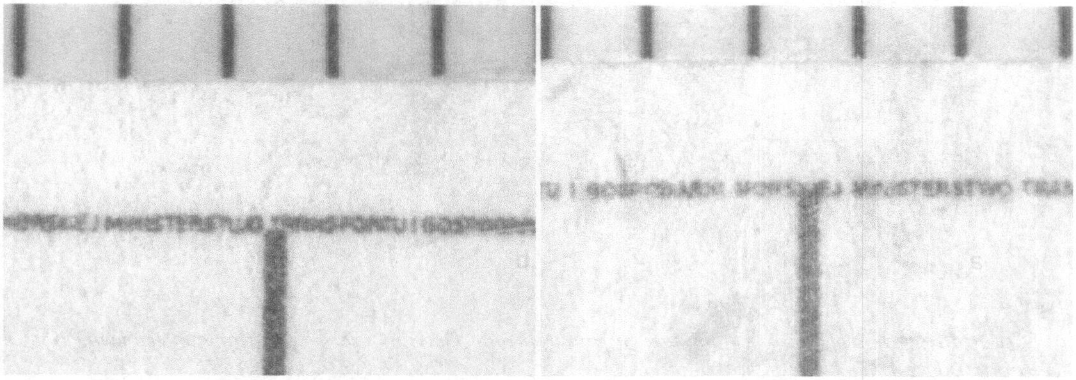

a b

Figure 15-16

(a) View of the microtext on an authentic Polish vehicle registration document. Note that the microimpression is not of a very good quality and could be misinterpreted as not being an original.
(b) View of the microtext on a counterfeit Polish vehicle registration document. Note the quality of the microimpression, which is not too poor compared with the authentic version (a). (Courtesy of The Swiss Border Guards [Corps Suisse des Gardes-Frontières].) See Color Plate.

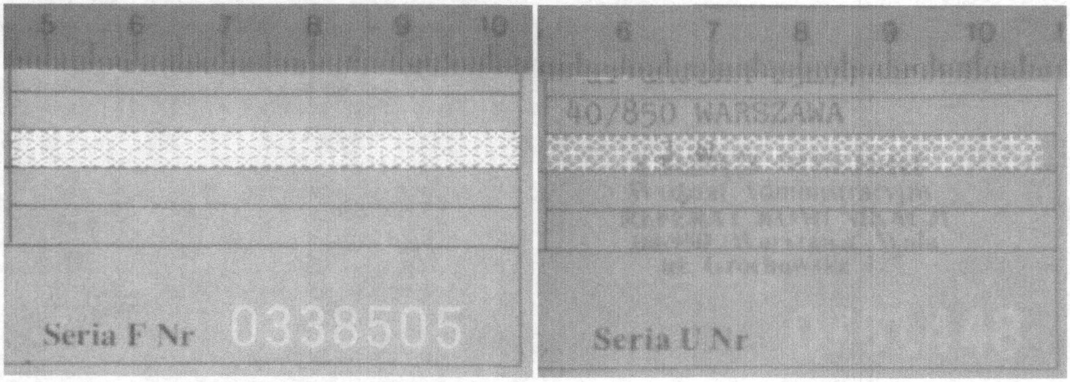

a b

Figure 15-17

(a) View of the UV fluorescent pattern and serial number on an authentic Polish vehicle registration document.
(b) View of the UV fluorescent pattern and serial number on a counterfeit Polish vehicle registration document. Note that the presence of a fluorescent pattern is not enough to guarantee the genuineness of the document; the pattern itself must also match. Also, note the difference in the color of the serial number and the fluorescence of the background of the document. (Courtesy of the Brigade de Police Technique et Scientifique, Geneva police, Switzerland.) See Color Plate.

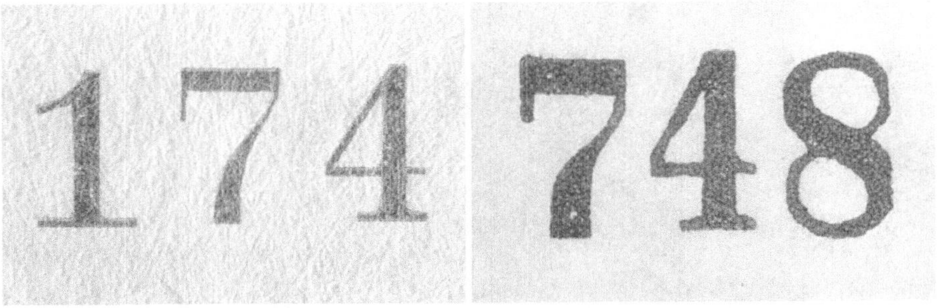

a b

Figure 15-18

(a) View of a portion of the serial number on an authentic German vehicle registration document. Note the 3D aspect of the ink.

(b) View of a portion of the serial number on a counterfeit German vehicle registration document. Note the difference in the border of the numbers, how the ink smears in the paper rather than presenting a relief. This is an example of a different printing technique being employed on a counterfeit document as compared to the genuine document. (Courtesy of The Swiss Border Guards [Corps Suisse des Gardes-Frontières].)

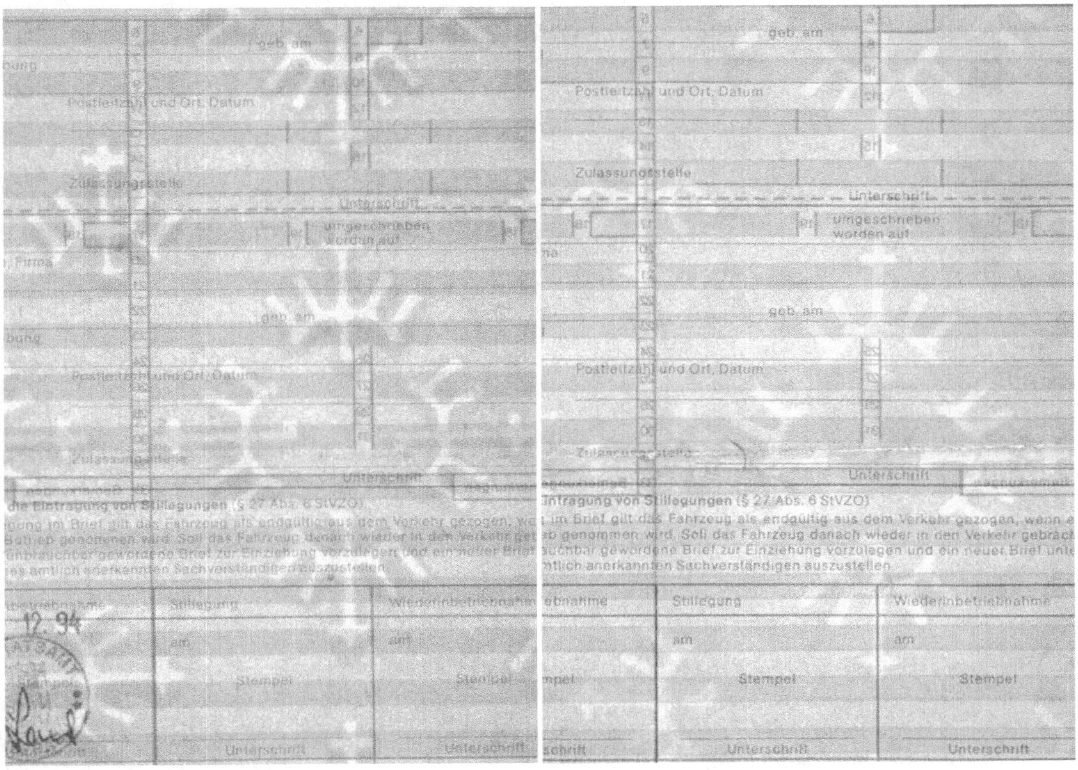

a b

Figure 15-19

(a) View of the watermark on an authentic German vehicle registration document.

(b) View of the watermark on a counterfeit German vehicle registration document. The mere presence of watermark is not enough to assess the genuineness of a document; the pattern itself must also match, which is not the case in this example. (Courtesy of The Swiss Border Guards [Corps Suisse des Gardes-Frontières].)

a b

Figure 15-20

(a) View of a fiber included in the paper on an authentic Lithuanian vehicle registration document.
(b) View of an imitation of a fiber on a counterfeit Lithuanian vehicle registration document. A line was actually drawn on the document to make it look like an included fiber. (Courtesy of The Swiss Border Guards [Corps Suisse des Gardes-Frontières].) See Color Plate.

a b

Figure 15-21

(a) View of an authentic Lithuanian vehicle registration document under UV light.
(b) View of a counterfeit Lithuanian vehicle registration document under UV light. Note the difference in the amount and characteristics of fibers, the serial numbers fluorescence, and the general fluorescence of the document. (Courtesy of The Swiss Border Guards [Corps Suisse des Gardes-Frontières].) See Color Plate.

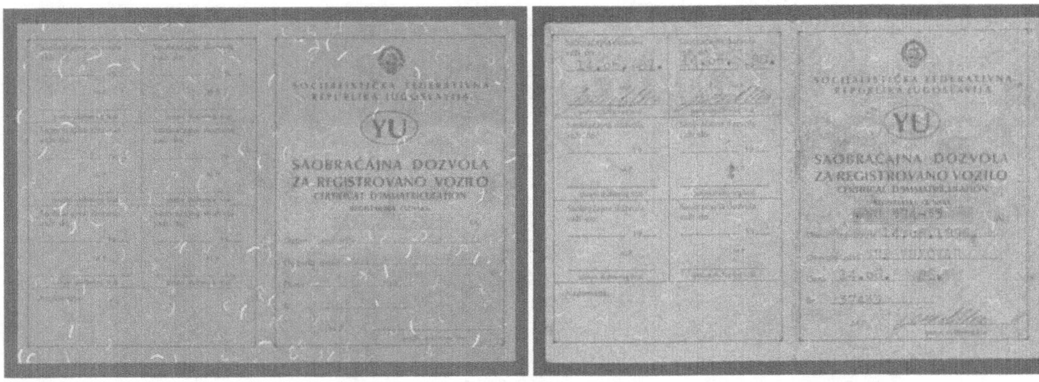

a b

Figure 15-22

(a) View of an authentic Yugoslavian vehicle registration document under UV light.
(b) View of a counterfeit Yugoslavian vehicle registration document under UV light. Note the difference in the amount and characteristics of fibers and the general fluorescence of the document. (Courtesy of The Swiss Border Guards [Corps Suisse des Gardes-Frontières].) See Color Plate.

15.5 FORENSIC EXAMINATION METHODOLOGY

15.5.1 Collection of Registration Documents

Vehicle-related documents could be found during the examination of stolen-recovered or crime-related vehicles. The collection of evidence occurs according to the standard forensic routine. The examiner wears gloves and safely stores the document in an appropriate container, such as a plastic bag or a paper envelope. If the document is wet, it should be air-dried before being stored in a plastic bag. In any case, a forensic document examiner should be consulted before manipulating any document, particularly if destructive examinations will be performed. Documents that are partially burned or carbonized require special care, as explained later in this section.

For documentation purposes, photographs of the documents should be taken and/or photocopies made. Gloves should be worn during this process. The reproductions are compulsory in the event of destructive tests, such as fingerprint examination or other manipulations, which will change the appearance of the evidence.

15.5.2 Equipment

The examination of registration documents requires several pieces of equipment, including different light sources, filters, and magnifications. Equipment manufacturers have integrated many of these examination tools in compact machines. Also, there is equipment that produces light at different wavelengths, such as Polilight and Crimescope for the detection of fingerprints.

15.5.3 Preliminary Visual Examination

Daylight is the preferred lighting for the preliminary examination of the document. The examiner (still wearing gloves) describes the document and measures its dimensions. The visual inspection includes the use of other light sources, sometimes in combination with color filters. The commonly called black light is a UV light source that is commercially available as a long-wave (366nm) and short-wave (254nm) version. Frequently, printing inks with a UV luminescent reaction can be observed with the naked eye under a long-wave UV light. The use of filters is needed to observe the effects obtained by infrared luminescence, where a high intensity blue-green light induces a luminescence effect visible on the infrared part of the spectrum [17].

15.5.4 Database Checks

The information shown on the registration document also needs to be checked with the issuing authority, especially if there is doubt of irregularities in the personalization or the data. As mentioned earlier, the SIS, the Interpol database, and the EUCARIS network can be accessed by authorized governmental users. Some countries provide public access to databases, such as Italy, which has a public Internet site to check numbers in the blank stolen documents database, as well as license plates, and VINs of stolen vehicles [18].

15.5.5 Comparison Process

All information regarding the substrate, the printing techniques, the colors, and the security features are noted. Based upon this information the examiner assigns the document to a given group (e.g., vehicle registration document from a given country) and looks for reference models of this country. A reference model is a document issued for information purposes by the competent governmental authority using the correct personalization technique and generally known as a specimen. Unfortunately, the variety of documents issued all over the world prevents law enforcement agencies to get complete collections of specimens. In some cases, it is possible to obtain the specimen from the issuing authority. However, this is time consuming. As a second choice, the examiner should look for a reliable detailed description of the specimen in the literature. Keesing Reference Systems has published descriptions of vehicle registration documents, driving licenses, and vehicle registration plates in the *The Interpol Guide to Vehicle Documents* [19]. Law enforcement agencies and registration authorities often manage collections of specimens and describe models from all over the world for information, fraud prevention, and training purposes.

Based upon the reference models/specimens or descriptions, the examiner compares the observed features and at the same time captures similarities and differences to establish the exact model of the document. It is important to stress that countries regularly change the appearance of (registration) documents. Establishing the genuineness of a questioned document depends on the availability of information about the different models, possibly in circulation at the same time. The known security features as well as the personalization features are compared in dimension, aspect, effect, and quality with the naked eye under different light sources, a magnifying glass, or a low-magnification microscope. Based upon this analysis, the examiner can determine the genuineness of the physical document. In addition to the examination of evidence, checking the variable information with the issuing authority is strongly recommended.

If no models are available for the comparison with the questioned document, an assessment of the security features can be done to approach tentatively the genuineness of the document. An experienced examiner inventories the security features and categorizes them according to their level. Other evaluation methods are described in the book *Optical Document Security* [16]. The results can be very diverse. In the most positive case, security features of all three levels are available and easily recognizable using routine inspection procedures. On the contrary, the examiner could find very poor security, which is often comparable with forgeries and counterfeit documents. Ideally, a frequency study of security features appearing in security documents could help to establish the value of a feature found in a questioned security document.

15.5.6 Examination Checklist

Table 15-4 presents a checklist of features that should be observed during the examination process of a document [20].

Table 15-4

Checklist for the examination of documents.

Type and title of the document	Country of issuance
Document number Numbering technique	State/province/territory of issuance
Format (height, width) in mm/inch	Validity
Substrate A: paper (optically dull?) Security features in substrate: watermark (description), planchettes, fibers, thread	Substrate B: polymer (optically dull?)
Background printing: lithography? Security features integrated in background printing	
Other printing techniques, e.g., intaglio, silk-screen printing	
UV fluorescence of the printing	
Optically variable devices, e.g., holograms, optical variable ink, etc.	
Other security features	
Personalization technique	

15.5.7 Electrostatic Detection Apparatus

The electrostatic detection apparatus (ESDA), manufactured by Foster & Freeman, is a device used in questioned documents examination to detect indented writing (imprint of writing on a material lying underneath the visible written substrate). The inspection by means of ESDA allows for the detection of invisible pressure lines of a handwritten text in the materials that has been used as a support (e.g., notebook). Because the examination procedure by ESDA requires the questioned document to be exposed to humidity, a negative interference with a subsequent examination of fingerprints is possible.

15.5.8 Burned or Carbonized Documents

The difference between burned and carbonized documents is given by their color: Burned documents are reddish-black, whereas carbonized are gray. Burned and carbonized documents should be transported from the crime scene to the laboratory in a box. The manipulation of burned documents can be facilitated by applying a thin layer of acrylic lacquer for watercolors [21]. First, a visual examination using different light sources is performed. Prior to the examination, the carbonized documents can be placed between two glass plates [22]. According to the literature, printed zones on carbonized documents can be revealed by observation with oblique light [23]. The fragility of the document could be minimized by exposing the pieces to water steam for up to six hours [23]. After the visual inspection, burned documents can be placed between two glass plates and "baked" in an oven to accomplish the carbonization. The substrate would lighten up, whereas dark text or writing would not change in color [24].

15.5.9 Instrumental Analysis of Paper and Inks

In some cases, by completion of the visual examination, the need for more detailed information about the materials used can rise. Material characteristics from both questioned and known sources are determined by physical and/or chemical analysis and compared [25]. Instrumental analysis requires expensive equipment and is carried out in specialized laboratories, experienced in the complex interpretation of the result of the analysis. Compared with the visual examination, which is considered nondestructive, most of the instrumental techniques require collecting a sample from the questioned document. By doing this, the document is permanently modified. With some techniques called semidestructive, the sample can be reanalyzed if necessary. Other techniques use the sample once such as the chromatographic techniques and are called destructive techniques.

The analysis of the paper can reveal characteristics such as the wire pattern or the composition of the fibers (e.g., percentage of cotton) [25, 26]. The examination of inks can include the determination of their color by spectrophotometry or their composition by thin-layer chromatography [25, 27, 28]. For toners, the analysis is carried out by Fourier transformed infrared spectroscopy, a semidestructive technique that allows for the discrimination between samples of toners from different producers [29].

15.5.10 Fingerprint Examination

The question about who was in contact with the questioned registration document could be asked during the investigation. To this purpose, the examiner submits the evidence to a fingerprint examiner. Visible fingerprints could have been already detected by the visual inspection of the document. A chemical processing is required to reveal possible latent fingerprints. The method for processing the document depends on the type of substrate. Porous substrates absorb fluids, and nonporous substrates do not absorb fluids [30].

Porous substrates are processed mostly by means of DFO (1,8-diazafluoren-9-one), ninhydrin, and, recently, 1,2-indanedione. All three reagents react with the amino acids present in the mark [31]. Afterward, the document could be processed in physical developer (PD), which reacts with water-insoluble components of the latent fingerprint deposit. Therefore, PD could be effective even if the surface has been wet [30].

Nonporous substrates, typically plastic cards or laminated papers, should be processed by the cyanoacrylate (superglue) fuming technique. If contrast is low, the revealed fingerprints can be enhanced by a colored or a luminescent stain [30].

ACKNOWLEDGMENTS

Many thanks to Johann Koning of the Rijksdienst voor Wegverkeer (RDW) for his precious support. Special thanks to Jacques van Zijp for his feedback.

BIBLIOGRAPHY

[1] European Commission (1996) Commission interpretative communication on procedures for the type-approval and registration of vehicles previously registered in another Member State, *Official Journal of the European Union,* **C 143,** I/3, pp 4–16.

[2] European Commission (1998) Opinion of the Economic and Social Committee on the "Proposal for a Council Directive on registration documents for motor vehicles and their trailers", *Official Journal of the European Union,* **C 19,** § 1.6, p 17.

[3] European Commission (1996) Commission interpretative communication on procedures for the type-approval and registration of vehicles previously registered in another Member State, *Official Journal of the European Union,* **C 143,** V/1, pp 4–16.

[4] United Nations Economic Commission for Europe (UNECE) (2001) *Annual bulletin of transport statistics for Europe and North America,* Volume LI, United Nations Publications, Geneva, Switzerland.

[5] European Commission (1998) Opinion of the Economic and Social Committee on the "Proposal for a Council Directive on registration documents for motor vehicles and their trailers", *Official Journal of the European Union,* **C 19,** p 17.

[6] United Nations Economic Commission for Europe (UNECE) (1968) Convention on Road Traffic, Vienna, Austria.

[7] European Commission (1999) Council Directive 1999/37/EC of 29 April 1999 on the registration documents for vehicles, *Official Journal of the European Union,* **L 138,** pp 57–65.

[8] European Commission (2004) Commission Directive 2003/127/EC of 23 December 2003 amending Council Directive 1999/37/EC on the registration documents for vehicles (Text with EEA relevance), *Official Journal of the European Union,* **L 10,** pp 29–53.

[9] European Commission (2004) Opinion of the European Economic and Social Committee on the "Proposal for a Regulation of the European Parliament and of the Council amending the Convention implementing the Schengen Agreement of 14 June 1985 on the gradual abolition of checks at common borders as regards access to the Schengen Information System by the services in the Member States responsible for issuing registration certificates for vehicles" (COM(2003) 510 final—2003/0198 (COD)), *Official Journal of the European Union,* **C 110,** pp 1–2.

[10] RL Polk & Company (2001) *Vehicle Registration Manual Volume III,* R.L. Polk & Company, Southfield, MI.

[11] Nowakowski D. (2005) New Polish vehicle registration certificates, *Keesing Journal of Documents & Identity,* Issue 11, Keesing Reference Systems, Amsterdam, The Netherlands.

[12] De Heij HAM. (2002) A method for measuring the public appreciation and knowledge of banknotes. *In: Proceedings of The International Society for Optical Engineering (SPIE) Conference—Optical Security and Counterfeit Deterrence Techniques IV,* **4677,** San Jose, CA.

[13] Haslop JM. (1993) Security printing techniques. *In: Optical document security,* ed Van Renesse RL, 1st edition, Artech House Publishers, London, United Kingdom.

[14] Van Renesse RL. (1993) Iridescent optically variable devices. *In: Optical document security,* ed Van Renesse RL, 1st edition, Artech House, London, United Kingdom.

[15] Andrade AA and Rebordão JM. (2000) Assessing DOVID security: A system approach. *In: Proceedings of The International Society for Optical Engineering (SPIE) Conference—Optical Security and Counterfeit Deterrence Techniques III,* **3973,** pp 248–256, San Jose, CA.

[16] Van Renesse RL. (2005) *Optical document security,* 3rd edition, Artech House Publishers, London, United Kingdom.

[17] Pfefferli PW. (1993) Forensic aspects of documents counterfeit. *In: Optical document security,* ed Van Renesse RJ, 1st edition, Artech House Publishers, London, United Kingdom.

[18] Ministero dell' Interno (2005) *Service III—system information interforce—information of public utility,* Office for Coordination and Planning of Police Forces, available at http://coordinamento.mininterno.it/servpub/ver2/principale_ing.htm, last access performed on May 28, 2005.

[19] Keesing Reference Systems B.V. (year unknown) *The Interpol guide to vehicle documents,* Amsterdam, The Netherlands.

[20] Naber M and Schuurmans J. (1999) *Handleiding documentenonderzoek,* 2nd edition, Elsevier, The Hague, The Netherlands.

[21] Meier J. (1964) Neues Hilfsmittel zur Ausservierung und Auswertung von versengtem und verkohltem Material, *Kriminalistik,* **8,** p 380.

[22] Garcia EC. (1964) Déchiffrage et conservation de documents brûlés, *Revue Internationale de Criminelle,* **176,** pp 70–81.

[23] Guenat M and Mathyer J. (1980) Lecture de documents carbonisés et brûlés. Un cas intéressant. *Revue Internationale de Criminologie et Police Technique,* **33,** pp 409–414.

[24] Santacroce G. (1999) The Forensic Examination of Fire and Water-Damaged Documents, *International Journal of Forensic Document Examiners,* **5,** pp 72–82.

[25] Brunelle RL. (1982) Questioned document examination. In: *Forensic Science Handbook,* ed Saferstein R, 1st edition, Prentice-Hall, Englewood Cliffs, NJ.

[26] Bijl LJ. (1990) *Zo goed als echt,* Koninklijke Vermande, Lelystad, The Netherlands.

[27] Kelly JD and Cantu AA. (1975) Proposed standard methods for the ink identification, *Journal of the Association of Official Analytical Chemists (AOAC),* **58**(1), pp 122–125.

[28] Olson LA. (1986), Color comparison in questioned document examination using microspectrophotometry, *Journal of Forensic Sciences,* **31**(4), pp 1330–1340.

[29] Roux C, Khanmy A, and Margot P (1989) Une nouvelle génération de contrefaçons : les photocopieurs couleurs: leur danger, leur identification, les moyens de lutte et de prévention à leur encontre, *Revue Internationale de Criminologie et de Police Technique,* **42**(3), pp 351–366.

[30] Champod C, Lennard C, Margot P, and Stoilovic M. (2004) *Fingerprints and other ridge skin impressions,* 1st edition, CRC Press, Boca Raton, FL.

[31] Wallace-Kunkel C, Roux C, Lennard C, and Stoilovic M. (2004) The detection and enhancement of latent finger marks on porous surfaces—a survey, *Journal of Forensic Identification,* **54**(6), pp 687–695.

CHAPTER 16

VEHICLE SEARCHING PROCEDURES

Stéphane Kummer

16.1 INTRODUCTION

Whether a vehicle is simply stolen, used to escape police, smuggle goods or drugs, or transport criminals, police investigators always have an interest in searching it. Criminals who just committed a crime often hide their take in the vehicle they are driving. They typically hide these objects in a manner that would prevent a simple search from discovering them. Indeed, narcotics, illegal weapons, counterfeit money, explosive devices, or even human beings can be hidden in a vehicle depending on the criminal activity. Modern vehicles present more possible dead volumes that could be used as caches. This is particularly true for vehicles equipped with many accessories.

There is a difference between the routine search of a vehicle, performed in the field during a common police traffic control procedure, and the methodical in-depth control that involves the disassembly of some parts of the vehicle, often requiring special tools. The basic principle of the search method is to use a systematic methodology by dividing the vehicle in different partitions. It is necessary to verify all possible dead volumes and other spaces that could contain particular objects.

The goal of this chapter is to present the complete and technical searching procedure of a vehicle to ensure that all caches have been found and controlled. This chapter mainly concentrates on passenger vehicles, including SUVs as they are searched in the same manner. Searching procedures in commercial tractor trailers, utility vehicles, and motorcycles are briefly discussed. The search procedures proposed in this chapter allow the reader to unveil most of the hidden places arranged in a vehicle. However, it is important to keep in mind that there will always be professional mechanics able to create particular locations, such as inside an engine, that may escape the trained eye of the forensic technician.

16.2 EQUIPMENT

In addition to the regular toolkit of a car mechanic, which typically contains wrenches and sockets of all sizes, screwdrivers, pliers, and files, it is necessary to get a complete set of tools and equipment adapted to the search of a vehicle. The equipment presented in Table 16-1 is recommended.

Table 16-1

Recommended equipment used to perform an in-depth search of a vehicle.

Equipment	Comments
Metal wire	The wire can be curved and cut at different lengths.
Flashlight	Different sizes of flashlight and a flexible-head flashlight are desired.
Mirrors	Different sizes are desired, including a flexible mirror that can be oriented to look through small holes, hidden locations, and hidden spaces.
Small video camera or optical device such as endoscope	This is not always necessary but very useful when looking inside hidden spaces that are not directly visible and hard to access.
Measuring equipment	A tape measure or a ruler is helpful, which is used to measure volumes and determine whether they have a hidden bottom.
Containers	They are used when a fluid circuit is purged to check its content.
Power tools	When different parts need to be taken apart, power tools can be very handy.

The list is not exhaustive because there are always cases requiring different equipment more appropriate to the particular circumstances. X-ray visualization systems adapted to the scanning of big objects, such as passenger cars and tractor trailers, can also be used. Chapter 18 presents some examples of such systems used to scan shipping containers at harbors. Smaller portable x-ray systems can also be used to examine smaller and more specific compartments inside a vehicle. These systems are developed to scan objects such as suspicious suitcases; however, they can be adapted and used to scan the inside of a vehicle.

Finally, it is important to always keep in mind that during the searching procedure, the forensic technician might encounter traces such as fingerprints, biological fluids, and toolmarks. This searching procedure is not to be confused with the forensic crime scene examination of the vehicle but still needs to take into account the potential presence of traces. Thus, the forensic technician needs to act accordingly.

16.3 SEARCHING METHODS

16.3.1 Definition

To control a vehicle, several methods can be used. These methods include visual observation, probing, and compression (touching). Other methods, such as x-ray radiography and other visualization systems, can be used on a small and large scale, such as border controls or in harbor or in particular cases. However, these are very expensive pieces of equipment that are not readily available and that require special training to be successfully used. They are not covered in this chapter.

A/ Visual Observation

Visual observation is the most used method to control the content of a certain volume. In instances where direct observation is not possible, it is necessary to use a mirror combined with properly oriented lighting to achieve this step. An endoscope is also an extremely useful visual observation tool that should be considered when spaces cannot be directly visualized.

B/ Probing

The control by probing is generally performed in cavities and spaces where full visual observation is not possible due to spatial configuration issues such as the presence of baffles or brackets. The probing technique typically consists of using a metal wire bent according to the space being searched. The goal is to direct the probe in all corners and directions inside the volume being checked. In this manner, any objects present in the volume can be detected. When circumstances permit it, it is also possible to use an endoscope instead of a metal wire probe, thus allowing for visual observation.

C/ Compression

The control by compression (touching) is performed on soft objects, such as upholstery, headrest, and trimming, or on hard surfaces covered with a softer material, such as textile, vinyl, leather, or soundproof foam. This control is performed with bare hands by touching and compressing the surfaces to ensure that no object is hidden.

16.3.2 Suspicious Elements

During the examination of a vehicle, the preliminary search sometimes reveals tools used to create caches inside the vehicle or to manipulate openings and compartments. Often, these are merely screwdrivers. It is therefore essential to pay attention to any tools or equipment that could have been used for such functions. Their discovery should bring suspicion and serious awareness.

Also, when looking through the vehicle, it is essential to pay extreme attention to screw or bolt heads. If these heads present scratches or unusual wear, if the paint around the head is chipped, or if different heads are present where only one type of head should be present, it might be evidence of recent removal. A recent removal of some elements could be due to the presence of a possible cache. Besides screw and bolt heads, the examiner should pay particular attention to any traces of dirt, shifts in paint color, differences of cleanliness, scratches, and stains. All these elements could be evidence of an item such as a plate or cover that has recently been changed, moved, or placed at this location, which might cover a hidden compartment. It is important to always be attentive during the entire searching procedure to elements that appear foreign to the particular location.

16.3.3 Partitions

A vehicle is composed of a complex shape with several compartments and hidden spaces. To ensure that a systematic approach is used and that the totality of the vehicle is searched,

Table 16-2

Different partitions used to perform a systematic search of a vehicle.

Partition number	Coverage
1	Engine compartment: from the front bumper to the dashboard support panel (firewall)
2	From the dashboard support panel (firewall) to the back of front seats (included)
3	From the back of the front seats to the back of the rear seats (included)
4	From the back of the rear seats to the rear bumper
5	Every space not in the engine compartment or the passenger compartment, including the undercarriage of the vehicle (chassis, wheels, wheel wells, tanks, etc.) and top part of vehicle (roof top box, etc.)

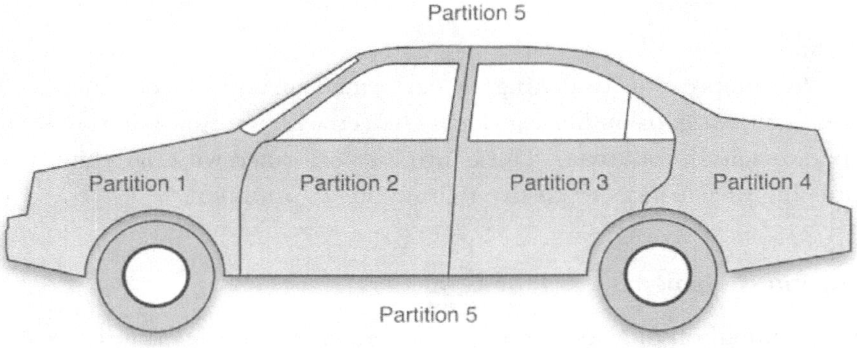

Figure 16-1

The five partitions into which a passenger vehicle is divided to ensure a systematic search. (Diagram courtesy of Eric Stauffer.)

it is necessary to divide it into different partitions. This allows for several people to work simultaneously on the vehicle without missing any compartments. A passenger vehicle should be divided into the five partitions, shown in Table 16-2. Figure 16-1 is a graphical representation of these five partitions in a passenger vehicle. For each of these partitions, there are particular searching procedures and controls that need to be performed.

A/ Partition 1

The following operations should be performed when searching the first partition.

- Look inside the volume of the bumper from its sides, top, and bottom. Use a mirror when necessary.
- Look in the space between the front grill and the radiator through the holes of the front grill and through the air intake(s), usually located aside or under the front grill.
- Look at the space between the reflector and the outside glass (or plastic) inside the headlights, park/turn lights, and all other lights.

- Look under the hood and more particularly control the soundproof material (typically foam) by probing, compression, and/or disassembly.
- Look inside the air-cleaner assembly after removing and controlling the air filter. Also, check by visualization or probing the snorkel tube or air intake.
- Control the battery by removing it and checking for a hidden compartment under it or inside it.
- Visually control the washer fluid reservoir. Probing might be necessary in some instances, particularly if there is a hidden bottom.
- Control the coolant expansion vase by purging the cooling system and removing the vase or by probing it.
- Visually control the electrical components found in the engine compartment such as horn, ventilation fan, and ignition coil.
- Visually control the compartment where the windshield wiper motor, gearing, and linkage are located.
- Look though the air intake of the heating/cooling ventilation system. Probing might be used for this operation.

B/ Partition 2

The following operations should be performed when searching the second partition.

- Look at the front doors and door panels. First, lower the window all the way down. The window should lower completely without any problems. Shed light in the opening between the window and the door panel to search for hidden objects. Looking from inside the door usually provides better results. The use of an endoscope might also be extremely pertinent to carry out this operation. Check the door panels, including the armrest and other compartments.
- Search the dashboard, glove box, speakers, and speaker recesses by disassembly, probing (with hand or wire), and visual observation. Figure 16-2 shows an object discovered hidden in the dashboard behind the glove box. Figure 16-3 shows some tools that were found hidden in a speaker recess. These tools were used in several burglaries.

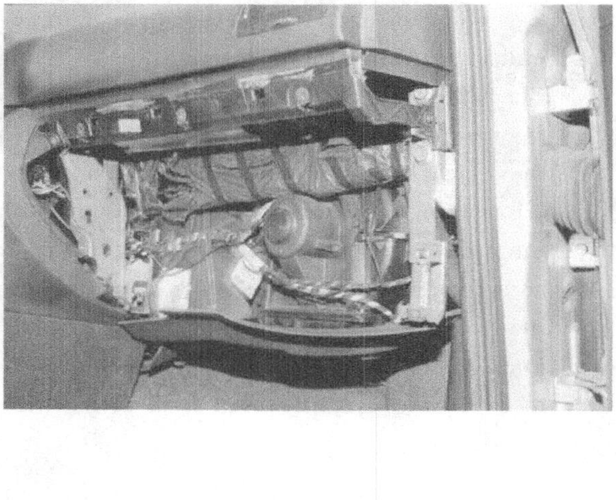

Figure 16-2

View of an object (wrapped in dark plastic and taped to the dashboard with black electrical tape) found hidden in the dashboard behind the glove box of a Skoda Octavia.

Figure 16-3

Tools used by a burglar that were found hidden in a speaker recess.

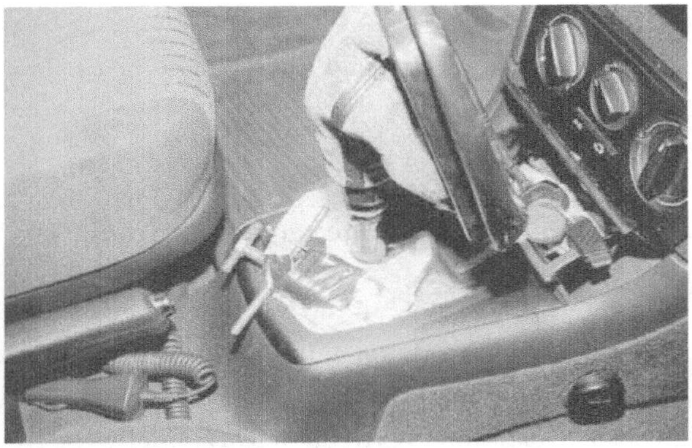

Figure 16-4

Tools used by a burglar that were found hidden under the gearshift lever cover in the center console of a Skoda Octavia.

- Control the ventilation ducts through the proper functioning of the fan and the probing of the ducts with the metal wire.
- Control the volume behind the A-pillar by compression. Disassembly might be required.
- Control visually the center console, hand brake lever, and gearshift lever. It might be necessary to remove the ashtray and other compartments to light the center console or at least the gearshift lever cover and hand brake lever cover. Figure 16-4 illustrates the discovery of burglar tools hidden under the gearshift lever cover, inside the center console.
- Check the ceiling by compression and touching with the palms of the hands. If the ceiling is made of a hard cover rather than a soft cover such as cloth or tapestries, it should be disassembled. When the vehicle has a sunroof or moonroof, it is checked by fully opening and closing it to make sure it is working properly and that no unusual resistance is present. In the most convenient position, the space between the sliding roof and the ceiling is checked by slightly prying open the ceiling and probing with the hands or a metal wire while visually observing the opening. Figure 16-5 shows the example of an object that was hidden between the ceiling cover and the moonroof.
- Look under the floor carpets, mats, and other coverings by removing the floor coverings, including the soundproof material, which is usually glued. Sometimes, it is possible to find holes

Figure 16-5

Cache in the ceiling. An object was visible in the opening between the ceiling cover and the moonroof. It necessitated prying open the ceiling cover once the moonroof was opened.

Figure 16-6

Object found under the carpet on the passenger's side front floorboard of an Opel Corsa.

in the bottom part of the car's body that are located in the area of the floorboards. In such instances, the area under the floor coverings should be probed with a metal wire. Figure 16-6 presents an example of an object that was hidden between the carpet and the floorboard. Figure 16-7 presents an ingenious cache professionally built into the floorboard of a Mercedes A170. Only by lifting the passenger's seat (Figure 16-7a) and removing the carpet is it possible to notice the metal plate attached to the floorboard by four screws (Figure 16-7b). Once the metal plate is removed, a cache of approximately $30 \times 30 \times 15\,\text{cm}$ is accessible (Figure 16-7c). In this particular instance, this cache was hiding a device used to neutralize department store anti-theft system.

- Through visual observation and probing, examine the space under and around the front seats. By compression, control the seats and any other upholstery around the seats.
- Check the rear storage pocket of the front seats.
- Control the horn on the steering wheel by removing the steering wheel cover/horn assembly. When in presence of vehicles equipped with air bags, this operation should only be performed by experienced personnel. A specific training course should be taken before disassembling any parts that are related to the air bag system. Air bags can be lethal when deploying. It is also

a

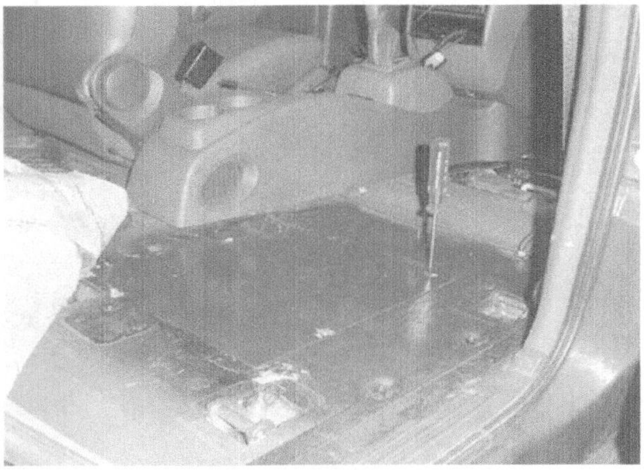

b

c

Figure 16-7

Example of a hidden compartment located under the front passenger seat in a Mercedes A170. (a) The seat is lifted to gain access to the carpet. (b) Once the carpet is removed, it is possible to distinguish the presence of a plate secured by four screws onto the floorboard. (c) The removal of the plate allows access to a cache of approximately 30 × 30 × 15 cm. (Photographs courtesy of Albane Bioley, Corps Suisse des Gardes-Frontière.)

possible to have a mechanic or the dealer to take apart this element under the surveillance of a forensic technician or investigator.

C/ Partition 3

The following operations should be performed when searching the third partition.

- Proceed identically to partition two with the rear doors and door panels. On some vehicles, rear windows might not lower all the way down as front windows would. In this particular instance, try to find any holes on the side of the door that are usually plugged with plastic cover and perform the observation through them.
- Proceed identically to partition two with the floor coverings and floorboards.
- Control the volume behind the B-pillar through the hole of the safety belt. Probing method or endoscope would work best in this instance. Disassembly might be required.
- Proceed identically to partition two with the ceiling. Nevertheless, if the car is a station wagon, it should be possible to pry open the ceiling cover from the ceiling through the opening of the trunk.
- Check the ashtray and other compartments of the center console. Disassembly might be necessary to visually observe the inside of the console.
- Check any extra ashtray or compartments located in partition 3.
- Remove, lift, or tilt the rear seats to check under them. Figure 16-8 shows money and a pistol found after removing the rear seats.
- If the gas tank is located under the seats, remove the firewall panel and control the tank through a gas gauge. An absolute control can only be performed by removing the tank and emptying it. An endoscope designed to be used in liquids such as gasoline can also be used for this control.
- Check the back of the rear seat and the central armrest.
- When dealing with three-door vehicles, check the interior side panels covering the rear fenders and accessible from the rear seats. Disassembly is required because the panels must be removed

Figure 16-8

Money and a pistol found after removal of the rear seats.

or opened to gain access to the space between them and the body. This space is of important volume and is often used.

- Remove the speakers and check them and their recesses. Check the rear dash under the window. Ensure that there is no hidden compartment under the deck. Some vehicles use that space to store first-aid kits or other items.

D/ Partition 4

The following operations should be performed when searching the fourth partition.

- Access the inside of the rear quarter panels through the trunk by removing or prying open the interior side panels. Also, it is sometimes possible to gain access through the side compartments (with or without doors) or by the ventilation ducts. Figures 16-9 and 16-10 show two examples of caches discovered in the volume delimited by the rear left quarter panel and the interior side panel.
- If the gas tank is located inside the trunk (or accessible through an opening in the trunk), remove the firewall panel and control the tank through a gas gauge. An absolute control can only be

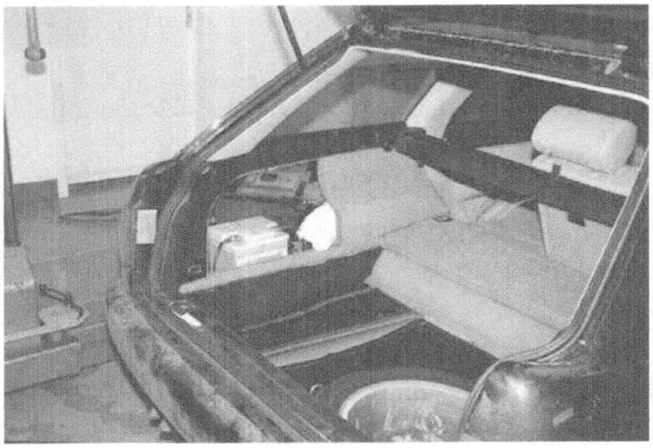

Figure 16-9

Cache discovered between the rear left quarter panel and the inside panel of an Audi A4.

Figure 16-10

View of the different spaces that can be used as caches and that located in the volume behind the rear inside panel of a Ford Fiesta.

performed by removing the tank and emptying it. An endoscope designed to be used in liquids such as gasoline can also be used for this control.

- The spare wheel should be removed and controlled by visual observation and by weight. If evidence of recent manipulation of the tire or the rim is present (such as remainder of fresh soap on the tire and wheel), it is necessary to proceed to the disassembly of the wheel to check its content. Figure 16-11 shows a spare wheel into which narcotics were found. The spare wheel well should also be controlled. Make sure there are no hidden bottom and compartment in the well.
- Control the rear taillights by removing their interior cover plate and the light bulbs from inside the lights. Check the volume inside the lights.
- If the vehicle is equipped with a liquefied petroleum gas (LPG) system, carefully check the tank. The welds should be checked on the tank and between the tank and the pressure regulator and valve.
- Carefully check the trunk's floor for the presence of hidden bottoms, recent welds, and fresh paint. If the trunk's floor presents any small openings, they can be checked by probing with a metal wire or by using an endoscope.
- Check the space between the trunk and the rear seats for the presence of hidden plates and dead volumes. Figure 16-12 shows the different steps that led to the discovery of a hidden compartment between the rear seats and the trunk of a Mercedes E220. At first (Figure 16-12a) there are no real indicators of such a compartment, except for a slightly reduced trunk space, which could easily escape the untrained eye. When the carpet is removed from the back of the rear seats, the indication of the presence of a door is obvious (Figure 16-12b). Behind the door, an important amount of drugs is found (Figures 16-12c, d, and f). The fake compartment can be removed from the vehicle to study the manner in which it was built (Figure 16-12e).
- Control the underside of the trunk by probing under the soundproof material or disassemble. Also, verify the content of any toolkit or roadside emergency kits located on the underside of the trunk lid or anywhere else in the trunk.

Figure 16-11

Upon control of this spare wheel from an Opel, which weight appeared unusually heavy, the tire was cut open and revealed the presence of drugs.

Figure 16-12

Example of a hidden compartment located between the back of the rear seats and the trunk in a Mercedes E220.
(a) When opening the trunk, no particular signs are observable, except for the slightly reduced trunk space.
(b) When the carpet is removed, indication of the presence of a door is directly noticed.
(c) Behind the door, a significant amount of drugs is discovered.
(d) The compartment extends from side to side.
(e) The disassembly of the cache allows for the investigator to understand better how such compartments are built.
(f) The seizure of illegal drugs found in the cache of the Mercedes. See Color Plate.

E/ Partition 5

The following operations should be performed when searching the fifth partition.

- Carefully observe the front fenders and wheel wells. It is necessary to disassemble or pry open the wheel well liner (usually made of plastic) inside the wheel wells. A very important volume is usually available behind the fenders. On some vehicles, there is an opening between the door and the body where the door hinges are located. In such instances, it is possible to control, with the proper lighting, the volume behind the fender through this opening.
- The rear quarter panels are also checked similarly to the front ones, which includes the removal of wheel well liner.
- The chassis or frame is carefully observed to notice any fresh welds and/or paint. The chassis or frame offers large dead volumes, particularly in its tubular structure and body rails. These volumes are very conveniently used by criminals to hide objects. Often, a cover plate from the chassis or frame is removed and objects are placed inside the dead volume. The plate is then welded back in place.
- Remove the wheel covers and hubcaps to check the volume inside.
- Remove the wheels and check them as for the spare wheel, by visual observation and by weight.
- Also, check inside the rims as well as the brakes and bearings for any signs of abnormal overheating.
- If the roof has a rack equipped with a cargo box, check its content. Also, if the roof rack is made of tubes with a certain dimension, make sure the tubes are empty. This can be done by either knocking them with a small hammer or probing them from one opening with a metal wire.
- Carefully control by visualization the body rails and door steps. When openings are available (usually covered by small plastic caps), probe the dead volumes using a metal wire. Figure 16-13 shows a cache inside the dead volume of rear left body rail toward the rear wheel well. This cache was discovered because of evidence of fresh welds and paint on the wheel well and body rail.

Figure 16-13

Cache in the body rail of a passenger vehicle that was located after the observation of fresh weld and paint inside the wheel well on a Renault.

16.4 OTHER VEHICLES

16.4.1 Tractor Trailers

Heavy commercial trucks and tractor trailers are very attractive to smugglers. They can transport important quantities of material and goods and offer a number of caches of important volumes. For example, imagine how much drugs can be placed in a hidden bottom covering the whole surface of a trailer's floor. These vehicles are searched using a methodology similar to that presented for passenger cars. However, there are some particulars specific to this kind of vehicle. They offer particular caches, both inside and outside the cargo area. These areas may include

- Inside compressed air tanks used for the braking system;
- In and around the coupling system of the trailer to the tractor;
- Inside fuel tanks, which are of a much higher capacity than the ones found in passenger vehicles;
- In accessories such as crane and other lifting tools, which comprise many dead volumes.

It is difficult to present a comprehensive list of all possible caches on heavy commercial trucks. Beside the engine compartment and the cabin, which are treated as partitions 1 and 2 of passenger cars, everything else has several different possible configurations. There are trucks used for many different applications such as carrying vehicles, pulling cargo trailers, transporting liquids (tank), or transporting dirt (dump truck). Thus, it is not possible to present a methodology applicable to all types of heavy commercial trucks. The important concept behind the search lies in the division of the vehicle in different partitions to ensure a good systematic approach to the searching procedure. Each partition is then searched with extreme rigor.

16.4.2 Utility Vehicles

These vehicles present very similar front ends to passenger vehicles. Thus, the same partitions 1 and 2 can usually be defined. However, behind the driver's seat the cargo area varies from the configuration of passenger cars. The cargo area should be inspected similar to the trunk of a passenger vehicle. In addition to checking the cargo area, the merchandise and goods present in the cargo space should be carefully inspected. This might involve the opening of boxes and crates.

16.4.3 Motorcycles

Dead volumes and compartments available on motorcycles are very small when compared to those found in a passenger vehicle. However, this does not mean they should be neglected when searching procedures are carried out. For example, the frame is typically made of metallic tubes, which are hollow in their center. These could carry certain items such as narcotics or money. The gas tank is also of a certain volume, and items could be hidden in it. Compartments containing toolkits and other small items could also be used to carry different items. The seat of the motorcycle is typically made of upholstery and could contain a hidden compartment. It is also important to check the space between the frame and the body panels. This operation might require disassembly. These spaces allow for hiding thin objects, but of consequent areas. Finally, if the motorcycle is equipped with carrying cases, much larger volumes are available and should be checked. False bottoms and hidden compartments could be present.

ACKNOWLEDGMENTS

The author would like to thank the Corps Suisse des Gardes-Frontière III (Swiss Border Control Authority, District III) for most of the photographs displayed in this chapter. The author would like to thank Eric Stauffer for translating this chapter into English.

CHAPTER 17

EXAMINATION OF VEHICLES INVOLVED IN TERRORISM

Greg Terp

17.1 INTRODUCTION

The purpose of this chapter is to understand the role auto theft investigations may play in fighting terrorism. Many criminal acts are solved based upon the direct investigation of the specific offense. A homicide investigator is able to process a crime scene, interview witnesses to a murder, and then identify and arrest the perpetrator. Information and leads from other law enforcement activities and investigations solve just as many cases. A uniform officer on patrol notices a suspicious suspect fleeing. After the suspect is caught, it is determined that he just committed arson. Another example is that of a burglary suspect who is arrested and quickly volunteers to provide information on a narcotics organization.

In the war on terrorism, auto theft investigation can be an important link. This chapter is designed to help investigators and forensic analysts understand how they might be able to contribute. The lead or key piece of a terrorist plot may come as a result of evidence recovered from a stolen vehicle or from the efforts in identifying a vehicle used in a bombing.

The most popular weapon of a terrorist continues to be an explosive device. Using a timing device allows the person setting the blast to be a safe distance away and even avoid detection. Since humans found they could make things explode and use this as a weapon, terrorists have favored improvised explosive devices (IEDs). These IEDs have been disguised in many ways, shapes, and forms to hide their deadly contents. The most deadly and destructive container is a vehicle. Vehicle bombs are the largest nonmilitary device and have shown they can bring down large buildings. One device could have the capability to kill hundreds of people.

Vehicles have become the favorite container of IEDs throughout the world, allowing terrorists to move large quantities of explosive and place them close to targets. The destructive force of a vehicle bomb has clearly been demonstrated time and again on international news reports. Hotels, federal buildings, embassies, and other large buildings have been heavily damaged or destroyed with a large loss of human lives.

Vehicles have other ties with terrorists. Some vehicles are stolen and sold for profit to help fund their organizations. Vehicles are used to carry operatives while planning and

Figure 17-1
Auto theft investigators are trained to identify vehicles and vehicle parts. Many vehicles are found in varying stages of dismantlement like this vehicle, which is missing its fenders, front end, hood, seats, and much more.

conducting surveillance of victims and targets. Vehicles are also used to transport weapons (including nuclear weapons) and smuggle narcotics.

Vehicles represent an excellent crime scene to gather vital information about terrorists and their organizations. A forensic examination may be called for to find clues, evidence, or just to identify a vehicle so it may be traced to a suspect. The vehicle may be totally intact, partially stripped as shown in Figure 17-1, or completely blown apart. Even a vehicle that is used as a vehicle bomb and blown into many pieces may be processed for valuable information and evidence. The van used in the bombing of the World Trade Center in New York City (USA) in 1993 was identified and led investigators to a suspect [1].

Stolen vehicles are becoming more involved in vehicle bombs, and forensic investigations of major bombings around the world have shown the ability of law enforcement to identify vehicles even after they have been blown into pieces. This leads to the identification and arrest of those involved. Rosie Cowan, an Ireland correspondent, wrote an article on January 28, 2002 regarding the Omagh (Northern Ireland) bombing that killed 29 people in 1998 [2]. Her headline, "Car thief 'clue' to Omagh bombing," shows the direct link in this terrorist incident in Northern Ireland. According to the story, an informant told the police that the day before the bombing he had been approached to provide a stolen vehicle for a Real Irish Republican Army (RIRA) terrorist group operation. He did not provide the vehicle but another car thief did, and 500 pounds of explosives were placed in the vehicle.

At any time it is crucial for investigators and forensic specialists to work closely and communicate as much as possible about a criminal act. A vehicle bombing requires this and even much more. The lead investigator of a vehicle bombing will most likely be a terrorism or major crimes detective. There will probably be local and national investigations. Specialists like bomb technicians and auto theft investigators will need to work closely together with crime scene technicians identifying all possible evidence.

17.2 USE OF STOLEN VEHICLES BY ORGANIZED CRIMINAL GROUPS AND TERRORIST ORGANIZATIONS

17.2.1 Organized Crime and Terrorist Organizations

According to the Emergency Response Research Institute (ERRI) Daily Intelligence report of October 3, 1997 [3], "Appearing before the House Committee on International Relations on Wednesday, FBI Director Louis Freeh warned that Russian organized crime poses a risk to U.S. national security and added that the risk of a nuclear attack by a terrorist/guerrilla group is now greater than at anytime during the Cold War by the old Soviet Union [. . .]. Russian organized crime groups hold the uniquely dangerous opportunity to procure and traffic in nuclear materials."

Organized crime by itself represents a serious threat to all developed countries. In addition, with the growing relationships between organized crime groups and terrorist organizations, the threat is a real and present danger to many communities of the world. On October 30, 2003, Grant D. Ashley, Assistant Director of the Federal Bureau of Investigation (FBI) Criminal Investigative Division, provided testimony before the Subcommittee on European Affairs of the US Senate's Committee on Foreign Relations on issues related to Eurasian organized crime [4]. Mr. Ashley stated the following: "Organized criminal enterprises are no longer bound by the constraints of borders. Such offenses as terrorism, nuclear smuggling, organized crime, computer crime, and drug trafficking have spilled from other countries into the United States. Regardless of origin, these and other overseas crimes directly impact U.S. national security and the interests of our citizens. Eurasian organized crime, because of its size, wealth and international reach, poses some of the greatest threats in this regard."

He further stated [4], "The FBI currently has 245 ongoing cases dealing with Eurasian organized crime. Fraud, transnational money laundering, extortion, drug trafficking and auto theft are the most frequent violations cited in FBI Eurasian organized crime cases". Figure 17-2 shows a vehicle used for such purposes. Mr. Ashley addressed connections to

Figure 17-2

This photo shows a Jeep Cherokee that had been stolen from Florida and located in Russia. Eurasian organized criminal groups are a major threat globally and auto theft is one of their funding sources.

organized crime groups in the Balkans [4]: "European police organizations now estimate that Balkan organized crime groups control upwards of 70% of the heroin market in some of the larger European nations, and are rapidly taking over human smuggling, prostitution and car theft rings across Europe."

A strong tie between significant organized criminal groups and stolen vehicles is clearly established. To link organized crime groups and terrorist organizations is a logical step to follow. On September 27, 2004, Cassandra M. Chandler, Assistant Director of Public Affairs at the FBI, made the following remarks in a speech at the National Air and Space Intelligence Center, Wright-Patterson Air Force Base, Ohio (USA) [5]: "What has changed over the years is that criminal and terrorist threats increasingly have an international dimension". While expanding her statement she was clear about the relationship: "The dark side of globalization is the dangerous convergence it has encouraged between terrorist, intelligence, and criminal groups, which all operate to some extent over the Internet and through interconnected, sophisticated networks. In this environment, the traditional distinctions between organized crime, cyber crime, espionage, and terrorism have broken down."

Strong evidence is found in specific cases where organized crime groups have dealt for stolen vehicles, surface-to-air missiles, and even bartered for nuclear weapons. In March 1995, the US Customs Miami Office of Investigations initiated an undercover operation involving the exportation of stolen vehicles. Using a Miami Beach police officer fluent in Russian, a deal was made to ship stolen high-end vehicles from South Florida to former Soviet Bloc countries. The vehicles would be placed in shipping containers, as shown in Figures 17-3 and 17-4, and smuggled past the borders. While negotiating with the two prime suspects, an inquiry was made whether the undercover officer would be interested in Soviet military weapons. The suspects thought the undercover agent was a member of a drug cartel from Colombia. The deal for 40 shoulder-fired missiles, capable of bringing down an aircraft, including helicopters, was made. The case then took an even more threatening twist when the suspects offered to sell a tactical nuclear weapon. The federal prosecutor brought the case to a close when the suspects were arrested before the missiles could be shipped [6].

Figure 17-3

Miami-Dade Task Force detectives opened a shipping container that appeared to be loaded with Xerox paper goods. Criminal organizations use different items to hide stolen vehicles.

Figure 17-4
After removing the paper goods seen in Figure 17-3, officers found two stolen vehicles in this shipping container at the Port of Miami in Florida.

More evidence of this link between organized crime and terrorist groups is found in the testimony by James Mack, Deputy Assistant for International Narcotics and Law Enforcement Affairs, US Department of State before the House Committee on International Relations Subcommittee on the Western Hemisphere on October 10, 2001 [7]. This testimony occurred just one month after the attacks on New York City and Washington, DC of September 11, 2001. In relation to organized crime and auto theft he stated the following [7]: "There often is a nexus between terrorism and organized crime." He added [7]: "Migrant smuggling, document fraud, arms trafficking, auto theft, contraband, and illegal financial transactions are tools for terrorists as well as narcotics traffickers". He mentions organized crime and auto theft specifically. In addition, auto theft suspects are often directly involved in narcotics trafficking, migrant smuggling, and financial crimes.

17.2.2 Use of Stolen Vehicles to Help Fund Terrorism

The ability for countries to fund terrorist groups has been dramatically impacted since September 11, 2001. Terrorist organizations have turned to economic crimes and narcotics trafficking to fund their operations. From credit card fraud to cigarette smuggling, funds are being diverted to further carry out the agendas of extremist groups worldwide.

At the 16th Annual Interpol Symposium on Terrorism, in Lyon, France, October 22–23, 2001, Willy Deridder, Executive Director of Interpol, addressed Interpol's position in contributing to the law enforcement efforts aimed at combating global terrorism [8]. In his remarks, he discussed their focus: "First, we need to prevent and dismantle the financing of global terrorism." He further stated [8]: "Just as a reminder of how complicated this can be, global terrorism is now funded differently as compared to its funding mechanisms 25 years ago. Terrorists now use complicated networks that are not specific to them". The pressure being placed on nations providing financial support to terrorist organizations is having an impact. To continue operations and maintain their growth, they must find alternative sources of income. Traditional organized criminal acts that have provided substantial profits to criminal organizations for many years are in a natural process of transition.

FBI Assistant Director Cassandra M. Chandler directly tied terrorist groups to organized crime and the funds derived from illegal activities [5]: "Credit card fraud is being perpetrated by the Russian mafia and by al Qaeda operatives. [. . .] Organized crime is laundering money for terrorists. Al Qaeda operatives are some of the world's largest heroin dealers. And earlier this year we busted yet another cigarette smuggling operation that was funneling money to terrorist organizations."

The Interpol website clearly supports this concept. Their section on terrorism and financing states the following [9]: "We must combat terrorism on all fronts. As with most crimes, the terrorist conceals or attempts to conceal his activities in preparing his crime to avoid detection; the terrorist's weakest link is the fact that he requires funding. To finance their activities, terrorist organizations engage in various criminal activities, not unlike other organized criminal groupings."

Interpol continues to stress the importance of addressing all revenue sources [9]: "The frequency and seriousness of international terrorist acts *are often proportionate* to the financing that terrorists might get. It is critical that law enforcement target the financial sponsors of terrorist activities, not just the actual perpetrators of the terrorist acts". Interpol estimates that the illegal trafficking of vehicles is a form of organized crime that produces USD 19 billion [10].

From Europe, to the United States, to Canada, and throughout the world, law enforcement agencies are finding ties to the financing of terrorist groups. As an example, in 1998 the Royal Canadian Mounted Police (RCMP) was involved in an auto theft investigation named "Project Mermaid" in the Ontario and Québec provinces. The vehicles were stolen in Montréal, Ottawa, Toronto, and Halifax by a Lebanese-based auto theft organization. A series of raids by Canadian law enforcement agencies (RCMP, Canada Customs, and Sûreté du Québec) working together led to the dismantling of this organized crime group. Police seized 55 luxury vehicles, such as Mercedes, Jaguars, and Jeep Cherokees, at a total value of USD 2.2 million. The cars were being smuggled in shipping containers to Russia, Western Europe, Kuwait, the United Arab Emirates, and Africa. The cars were sold for an average of USD 40,000, with 10% of the profits funneled to a terrorist organization [11].

17.2.3 Use of Stolen Vehicles for Explosives Transportation

Explosive devices have always been a favorite means for terrorists to carry out their agenda, putting fear into their targets. IEDs packed in varying containers give individuals the opportunity to blend the bomb in with the surrounding features and to reduce the likelihood of being discovered. Using a timing device gives the suspect the opportunity to escape and not be identified on the spot. Vehicles are large containers enabling the IEDs to be more powerful, causing more damage and fatalities [12]. Vehicles are able to park close to intended targets or on their routes through city or urban streets. It has only been in recent years that extensive security measures are being taken to limit vehicle access to vulnerable targets such as airports, government buildings, and military bases [13].

Vehicle bombs have been utilized in many areas of the world with high degrees of success. Europe, Asia, South America, and the Middle East have felt the effects of this dangerous tactic of terrorist groups since the early 1970s. Frankfurt, Germany was the target of a car bomb placed outside the US Army Headquarters on May 11, 1972, killing one person [14]. In July 1972, the Irish Republican Army (IRA) used 10 car bombs in a single day. Then on May 17, 1974, two car bombs exploded in Northern Ireland, one in Dublin and one in Monaghan, killing 33 people [14].

17.2.4 Use of Stolen Vehicles for Surveillance and Transportation

Vehicles are important in other ways to terrorist groups. As they plan and plot their activities, being identified is a major risk. When entering a country using commercial transportation modes, they risk that their identity is matched with a database indicating a heightened security alert. To lessen their chances of being discovered, terrorists can travel in private vehicles instead of using commercial airlines or trains, thus avoiding the security check lists as they move around a country. Even private vehicles pose a risk, so an altered/stolen vehicle or a rental vehicle stolen through a counterfeit credit card and a false driver's license reduces the ability of law enforcement to follow traditional tracking methods.

There is ample evidence to show that in planning major acts of violence, terrorist groups conduct extensive surveillance of possible targets. Each time, they run the risk of being identified by a curious neighbor, a suspicious police officer, or a security guard on duty. As terrorists are discovering in the aftermath of the attacks of September 11, 2001, in the United States law enforcement agencies were quickly able to identify their movements through even traffic stops by local and state law enforcement officers where citations were issued. As with organized criminal groups, it is a constant learning experience for all. Each mistake or detail learned from a previous operation will result in a modification to enhance future operations.

Vehicles, due to their nature, have many means by which they can be identified, thus enabling law enforcement to obtain evidence. In 1997, the North American Export Committee (NAEC) worked on the electronic reporting system (see Chapter 18) to help address the significant problem of stolen vehicles being exported. An FBI agent on the NAEC recognized the value of processing all vehicles being exported for more than just the fact that they were stolen vehicles. If the vehicles could be captured in a database for major investigations, it could possibly provide a lead to a homicide or terrorist case. The program was called VINNY (not an acronym) and was created to establish this connection at the FBI.

In 2004, the VINNY system was responsible for indicating three terrorist-related vehicles being exported from Miami and one vehicle linked to a homicide investigation. One of the terrorist alerts resulted in evidence being collected that showed movement of a suspect in other states.

If there is any possibility that a vehicle (see Figure 17-5), even if not reported as stolen, is linked to a terrorist suspect or action, it is important for law enforcement to notify the counter-terrorism investigators in their country. There is a possibility that evidence may be

Figure 17-5

When stolen vehicles are located as seen in this Miami chop shop, they should be treated as a crime scene. The complete scene should be processed for all possible forensic evidence.

in the vehicle. Even if vehicles have changed ownership, there could be some things that fell into hidden areas or cracks in seats. A complete examination should be conducted, including for possible forensic evidence related to DNA, explosive material, fingerprints, and/or trace evidence as presented in Chapter 4.

17.3 PROCESSING STOLEN-RECOVERED VEHICLES

17.3.1 Principle

All stolen vehicles should be processed for evidence and information that may identify the suspects involved. Because of the volume of stolen vehicles that are recovered, some police agencies do not process these vehicles unless there are specific reasons to do so. For some agencies, it must be one of their stolen vehicle cases; if the vehicle comes from another jurisdiction, unless specifically requested to do so, it will not be processed. The technical part of the forensic examination of a stolen-recovered vehicle was presented in detail in Chapter 4 and the values of the different possible evidence in Chapter 5. This section just presents some traces that can be recovered and how they apply in the collaboration between the different services involved in the investigation and the possible links to terrorism.

17.3.2 Forensic Evidence to Identify Suspects or to Cross-Link Vehicles

This is an area where command staff and investigators need to be made aware of the value in processing all auto theft. First, many auto theft suspects are involved in other criminal activities. It can vary from armed robberies to burglary and credit card frauds. The Los Angeles Police Department realized in 2003 that by making auto theft a priority, they could also have an impact on other crime statistics [15].

Second, many violent and/or major criminal suspects have a past for auto theft-related crimes. The earlier they are identified and put into law enforcement databases, the more likely they be caught through fingerprint or DNA analysis. For each stolen-recovered vehicle that is not processed, there is a risk that a current auto theft suspect will not be identified. This suspect will become wanted for a more serious crime in the future, but there will be no evidence available for comparison.

When a vehicle is identified as being involved in a terrorist investigation, the processing should be done by qualified crime scene investigators and/or forensic scientists. Many stolen-recovered vehicles are processed very effectively by uniformed officers or auto theft investigators; however, when the seriousness of the case rises to a level involving national security there should be no chances taken. It is critical that auto theft investigators work closely with forensic scientists to help determine what areas may have fingerprints and DNA, for example, and how to best recover this evidence. Nevertheless, the auto theft investigators' knowledge of chop shops and other related operations is crucial in showing obscure places a suspect may have touched. The power of teamwork is invaluable and should never be neglected.

17.3.3 Processing Suspicious Vehicles

Because vehicles may be used to transport suspects and explosives, it is paramount that stolen-recovered vehicles with terrorist ties are checked for indications of a weapon of mass destruction and explosives. This involves technology and expertise such as presented in Chapter 4 Part II beyond the scope of most auto theft investigators. The facts and circumstances of each case determine what type of examination is conducted. In late 1995, information was received in Miami, Florida that a Ryder truck rented in another state was involved in a suspicious incident. The vehicle was located at a service station where unknown suspects had dropped it off. Based upon facts known to the investigators involved in the case, the Bomb Disposal Unit was requested to examine the empty truck. An explosive detection canine team assisted the bomb technicians. The canine team conducted a cursory search of the truck. The dog indicated an interest in a rear portion of the vehicle where the bomb technicians noticed a distinct stain from an unknown substance. That portion of the truck was then cut out and sent to the lab for further examination and testing for possible explosive materials. The results of the forensic examination in the crime laboratory were then provided to the investigators. The case being still open, the results cannot be disclosed. Terror investigations can be frustrating to auto theft investigators and forensic crime scene personnel. In the interest of national security, the results of their evidence collection and the subsequent investigation may never be known to them. It is important to realize that the specific assistance that specialists like auto theft investigators provide in such a case can be a crucial link in a major case.

This case is an excellent example of a situation where a vehicle was recovered and processed for forensic evidence on a potential terrorist case. The investigators and technicians worked closely with each other, sharing specific information and using their training and knowledge to further the investigation from a possible lead. Although the vehicle had not been reported stolen, it was wanted for a suspicious incident occurring in another state.

The vehicle was rented to be returned to a different city and state than where it was located.

17.4 INVESTIGATION OF A CAR OR TRUCK BOMB CRIME SCENE

17.4.1 Principle

Auto theft investigators have a significant role to play in assisting in the aftermath of a car or truck bomb. Their knowledge of vehicles helps the primary terrorism investigators and the forensic scientists determine what pieces of debris may be part of a vehicle, which specific parts should have identifying numbers (Figures 17-6 and 17-7), and the different methods of tracking vehicles through the paper trails (Figure 17-8) of titling and registrations.

In addition to vehicle bombs, this expertise has been used to identify vehicles and victims in other acts of terrorism and disasters. When the two planes struck the World Trade Center in New York City on September 11, 2001 and the towers collapsed, hundreds if not thousands of vehicles were crushed below in the parking garages and adjacent streets. Each of these vehicles needed to be identified, and auto theft investigators were used in this process. On August 7, 1997, a cargo plane crashed into a commercial warehouse district just after takeoff from Miami International Airport, crushing and destroying vehicles in the parking lots [16]. Homicide investigators requested auto theft investigators to assist in identifying the vehicles involved so that a presumed identity of the victims could be obtained.

Figure 17-6

Auto theft investigators can play an important role in processing a vehicle bombing scene. Their knowledge of vehicle identification numbers is vital. This photo shows a vehicle identification number located in a confidential location.

Figure 17-7

Close-up view of the digits "3599" of Figure 17-6. Auto theft investigators also are trained to identify altered numbers. In this photo, what appeared to be two 9s are actually two 7s.

Figure 17-8

Identifying the vehicle is only part of the investigation; following the paper trail is equally important. During a search warrant an investigator found a box full of blank manufacturer's certificates of origin. They included some for General Motors, Toyota, and Lexus.

17.4.2 Understanding the Effects of Explosions on Vehicles

Many different explosive materials and possible combinations can be used to prepare a vehicle bomb. Traditionally, there are two categories of explosives, high and low [17]. High explosives are those that essentially detonate when initiated, meaning that they create an explosion whose shockwave travels at a very high speed. Low explosives are combustible materials and basically burn or deflagrate. Low explosives burn unless they are present in a confined space, in which case they explode. A common material used in pipe bombs is black powder. When black powder is sprayed on the ground and lit with a match, it burns very quickly but does not explode. If the same powder is placed inside a closed pipe and a spark is provided, an explosion ensues.

High explosives do not need to be in a contained environment to detonate. High explosives such as C-4 (a plasticized composition of cyclotrimethylenetrinitramine), trinitrotoluene (TNT), and pentaerythritoltetranitrate (PETN) are just some examples that have been used in terrorist activities. Many of these explosives are more difficult to obtain than the devices used in Bali and Oklahoma City that contained ammonium nitrate fuel oil (ANFO) [18, 19].

The effects on materials and human bodies of the different types of explosives can be used in the analysis of a bombing incident. A low explosive may sever a hand, whereas a high explosive may shred the same hand. An explosion has different effects with different properties. The sudden increase of pressure, due to the generation of gases and heat from the explosives, leads to a shockwave traveling outward in all directions from the seat of the explosion [20]. This shockwave is preceded by a massive positive-pressure front and followed by a negative-pressure wave [21]. Objects, bodies, or other containers close to the blast are forced violently away and fragmented. These fragmented pieces or fragments impact other objects in their paths. When a grenade explodes, the sides are specifically designed to break apart into tiny pieces or fragments that then fly at high speed, killing and maiming those in close proximity. When a vehicle is used as a container, its parts become fragments, as shown in Figure 17-9. Blasts can flip vehicles upside down and destroy some vehicle's parts to the point of complete disappearance, leaving a tangle of metal to

Figure 17-9

A vehicle bomb is a lethal and favorite weapon of terrorist groups worldwide. This photo clearly shows the effects of fragments being propelled at high speed. The passenger door is the largest visible piece, but all can be lethal. See Color Plate.

the investigators [22, 23]. Also, the negative-pressure wave creates a vacuum, which is filled by atmospheric air traveling back toward the seat of the explosion. This effect can move objects toward the seat of the explosion, in the opposite direction of the shockwave created by the explosion.

The pressure increase generated by the shockwave can even be more destructive than the flying of fragments. Consider that a protective bomb suit can keep fragments from killing a person, but the shockwave and the accompanying overpressure may still damage and destroy internal organs. It is the shockwave and its overpressure that brought down the federal building and damaged many others nearby in Oklahoma City on April 19, 1995. Figure 17-10 shows the effects of a charge of 2,268 kilograms prepared with calcium ammonium nitrate and confectioner's sugar and placed about 15 meters away from the vehicle [23]. The vehicle, a Toyota Celica, underwent heavy mechanical damage, which renders its identification much more difficult. Figure 17-11 shows the effects of a different charge placed 2.1 meters away from the vehicle [22]. In this instance, only few parts are left, and the identification of such a vehicle is extremely difficult. Many parts have been expelled at much further distances. These two photographs provide a good idea to the investigator of the condition in which a vehicle might be recovered post-blast.

17.4.3 Specialists Involved in Bombing Investigations

The comprehensive knowledge and understanding of these effects take specific training and years of experience. Bomb technicians or explosive ordnance disposal (EOD) personnel are essential to the processing of a bombing scene. In addition, the forensic analysis of

Figure 17-10

Effects of a charge of 2,268 kilograms of calcium ammonium nitrate mixed with sugar. The vehicle is a Toyota Celica and was located about 15 meters from the seat of the explosion. Identification of such a vehicle is rendered difficult, but the main structure of the vehicle is still present. (Copyright British Crown 1999/DERA. Reproduced with permission of the Controller of Her Britannic Majesty's Stationary Office.) See Color Plate.

Figure 17-11

Effects of another explosive charge located at 2.1 meters from the vehicle. In this case, most parts disappeared and the vehicle is barely recognizable. In such instance, the identification of the vehicle is a very difficult process, if not impossible. (Copyright British Crown 1999/DERA. Reproduced with permission of the Controller of Her Britannic Majesty's Stationary Office.) See Color Plate.

materials obtained from the site is critical to the investigation. Communication between crime scene examiners and those in the laboratory is critical not just when an incident occurs, but as an ongoing process. A team must be organized to use the talents of many specialists, including auto theft investigators.

EOD personnel are trained in the detection, identification, and characterization of explosives. They are taught the parts of an explosive train: the detonator, the booster, and the main charge. Each has parts and characteristics that may be found during the post-blast investigation. The bomb technician works to re-create the device used by finding the pieces that will help in the final analysis.

The crime scene of an explosion can be small, limited to a few feet, or as large as a hundred square miles. When an airliner exploded in the sky over Lockerbie, England on December 21, 1988, the evidence from the blast scattered over many square miles [24]. The plane was reconstructed inside a hangar, and the collection of evidence was extensive. From the investigation, they were able to determine the type of explosive, in what cargo container it was located, what suitcase was used, and what clothing was in that suitcase [25]. In addition, the timing device was identified. This all led to the identification and eventual prosecution of the terrorist subjects involved.

In June 1991, a small truck was driving in southwest Miami when a powerful explosion destroyed the vehicle and killed the driver. The vehicle rolled to a stop along the side of the roadway. The blast shattered the vehicle windows, and fire that followed burned the vehicle and the driver. The preliminary examination of the crime scene revealed debris scattered for hundreds of feet. The body was also heavily burned, to the point that the gender could not be immediately determined. The bomb squad responded and initiated the processing of the crime scene. One of the bomb squad technicians located a leg wire from a blasting cap on the roadway. Knowing that this was part of the explosive train and not just a piece of debris from the street, it was collected and submitted to the laboratory for forensic analysis. Although this case remains an open homicide investigation with no arrest to date, the then federal Bureau of Alcohol, Tobacco and Firearms used the technician's testimony in a New York bombing case: The forensic analysis of the leg wire determined that the same tool was used to cut the wire in both bombings. The victim of the Miami bombing was a member of Brigade 2506, a group of Cuban exiles involved in the Bay of Pigs, the failed US Central Intelligence Agency–sponsored invasion of Cuba in 1961 [26].

17.4.4 Identifying Explosives

Explosives identification is a major priority in the investigation of a vehicle bombing. In many instances, explosives can be traced to their manufacturer or country of origin [27]. Commercially produced explosives, for civilian or military purposes, have trace identifiers that can help determine when and where it was produced and in what shipments it was distributed. Even when explosives have been stolen, they can be traced back to the theft and leads are obtained. In the instances of explosives manufactured by suspects, as in ammonium nitrate devices, their mixture can be analyzed and thus the components used and their amounts can be identified. An identification of the explosives can also be used to link different incidents together. This will help identify groups or suspects involved in multiple bombings. The FBI Explosives Unit-Bomb Data Center compiles information in various databases concerning explosives and bombings to aid in the post-blast investigation [28]. Other law enforcement and military agencies throughout the world have compiled similar reference points. Law enforcement and forensic examiners worldwide use these databases when analyzing vehicle bomb incidents.

17.5 CASE STUDIES

17.5.1 World Trade Center, New York City, United States, 1993

On February 26, 1993, the United States experienced its first international terrorist incident involving a vehicle bomb [29]. The device exploded in an underground parking area just after noon. The explosives were placed in a rental van and parked in a location where the terrorist thought it would topple one of the World Trade Center towers into the other.

The planning of this event was extensive. The terrorists were able to obtain building plans that they studied to determine where to park the van. Fortunately, the device did not bring either tower down, but six people died from the explosion. The bomb crater was a hole over 30 meters in diameter, and over four floors of the parking garage were penetrated. It was a very powerful explosion that created an extensive crime scene to process.

The response was immediate with New York City Police Department officers and federal agents from several agencies gathering at the scene to participate in the post-blast investigation. The van used in the bombing was mixed in with the other vehicles and debris from the building. The coordination of the processing of the scene was a difficult and labor-intensive task. Bomb technicians, crime scene analysts, and auto theft investigators worked to identify the vehicle that contained the explosives. Bomb technicians relied on their knowledge and experience to determine the vehicle parts belonging to the suspect vehicle from other vehicles. Auto theft investigators helped identify vehicle parts from other debris. All the destroyed and damaged vehicles needed to be identified. Auto theft investigators were able to use vehicle identification and part numbers to characterize the different vehicles. Forensic specialists at the laboratory helped to identify the type of explosives and other evidence collected from the scene.

A number on a specific vehicle part enabled the identification of the rental van. This became an important aspect because it then allowed investigators to begin following the lead and to find where the van had been rented. As a result, an arrest was made in a very short time, and other suspects were identified. This allowed for more suspects to be arrested in New Jersey, just across the river from New York City. These suspects were actively preparing to detonate more vehicle bombs in tunnels and at a federal building. Many more lives were saved from the outstanding work of a team of investigators, technicians, and forensic specialists [30].

17.5.2 Federal Courthouse, Oklahoma City, United States, 1995

Domestic terrorism can be just as deadly and dangerous as its international counterpart. At 9:02 am on April 19, 1995, a rental vehicle that had parked in front of the Alfred P. Murrah Federal Building in Oklahoma City exploded. The devastating blast crumpled half of the building from top to bottom, as shown in Figure 17-12. The carnage was broadcast live on major television stations within minutes of the attack, showing the response by police and firefighters trying to save people still trapped inside. This terrorist act claimed the lives of 168 people and further shocked Americans who were not used to bombings of this magnitude.

As in the World Trade Center bombing, the crime scene was massive. The work of trained experts in their field would come together again and quickly identify the vehicle involved, another rental. A rear axle was located over a block from the explosion, and an FBI agent who had specific knowledge of auto theft investigation played a key role. He found a vehicle identification number derivative on the axle and had this number checked through the National Insurance Crime Bureau (NICB) to collect information about the vehicle. The NICB was quickly able to determine it was a Ryder rental vehicle based out of Miami, Florida, where the Ryder Corporate headquarters are located. Armed with a solid

Figure 17-12

View of the scene of the Alfred P. Murrah Federal Building in Oklahoma City after the explosion. Processing a vehicle bomb crime scene can be an enormous task and involves numerous agencies and personnel. Debris must be examined to locate vehicle parts that will enable investigators to determine many elements, including the identity of the vehicle and traces of the explosive used. (Source: Federal Emergency Management Administration.)

lead, investigators discovered who had rented the truck in time to learn that the suspect, Timothy McVeigh, had been arrested by a state police officer for an unrelated traffic offense within an hour of the incident while he was leaving the Oklahoma City area [31]. Forensic analysis of the evidence collected from the scene determined that the explosive used was ammonium nitrate and nitromethane, a very powerful explosive mixture that is readily available when its constituents are purchased individually [18].

17.5.3 Spain Assassination Case, Madrid, Spain, 2000

Some terrorist actions are specifically designed to impact the public, whereas others have specific targets such as military and government figures. Timothy McVeigh and his supporters had an anti-federal government agenda; in fact, the Oklahoma City bombing was on the anniversary date of the Waco incident.

In Spain, the Basque Separatists have a terrorist group that has fought a long war against the federal government and military. This group, the Euskadi Ta Askatasuna (translates from the Basque language as "Basque Country and Liberty"), known as ETA, uses car bombs as their favorite weapon. On January 21, 2000 two car bombs were detonated in Madrid, Spain. The first targeted an Army officer waiting to be picked up on the street. Within 30 minutes of the first device going off, a second car bomb exploded in the same neighborhood. This area was primarily housing military families.

Some of ETA's methods of operation include the use of stolen vehicles and remote control systems. The first car bomb's vehicle was stolen approximately three weeks prior to being detonated, in November 1999 [32]. Any time a car bomb is detonated in Spain, the first inclination is that ETA is responsible. It is important for investigators and forensic analysts to conduct a detailed examination of the bombing scene to determine whether signatures of specific groups are present. Stolen cars and remote devices coupled with a military/government target may indicate ETA's involvement.

Investigators must be careful and open minded so as not to jump to conclusions. A copycat may use other signatures to hide their criminal act. It may also take cases away from the actual suspects. The Madrid train bombings in 2004 are an excellent example of this. Initial reports speculated ETA responsibility, even though some attributes of the incident, such as heavy civilian casualties, did not point to ETA [33].

On October 30, 2000 another car bomb exploded in Madrid during early morning rush hour traffic. The target was a Supreme Court judge who handled military cases. The vehicle was parked on the side of the street and was detonated as the judge's car passed by, indicating a remote device more so than a timing device. A timing device would not be a good option because the judge's route could vary by a few minutes each day. An observer, making sure it detonated at the exact moment the target was passing by, could operate a remote device more effectively. A very similar scenario was used to assassinate the anti-Mafia Judge, Giovanni Falcone, outside of Palermo in Italy on May 23, 1992 [34].

In Madrid, although the target and three other people were killed and 30 wounded, the device was specific as a weapon to commit an assassination. Had the desire been to create fear in the general public, the car bomb could have been set to go off where more people would have been affected [35]. Again, early reports cited possible ETA involvement. The post-blast investigation later determined the vehicle was stolen several months earlier in Madrid, Spain. Just knowing that a particular group is involved is as important as making a criminal case against the actual suspect who committed the act. Counter-terrorism measures must not be limited to criminal judicial proceedings. As Interpol states in their website, law enforcement must also go after the terrorists' financial holdings and sources.

17.5.4 Sari Club and Paddy's Pub, Bali, Indonesia, 2002

Terrorism is clearly a global issue, and the use of vehicle bombs as a favorite weapon is equally worldwide. The previous case studies involved the United States and Europe. Taking an example from the other side of the globe is the October 12, 2002 incident on the island of Bali, in the town of Kuta. This terrorist act killed 202 people and injured over 200 more [19].

Indonesia suffered a major blow in their financially productive tourist center. Although Indonesia has suffered many violent acts in some areas of the island country, Bali had been excluded until this. Besides the geographical location, this was a true international incident with victims from over 22 countries counted among the casualties [19, 36].

As in the planning of the attacks of September 11, 2001 on the World Trade Center in New York, the plotters carried out extensive research and surveillance to conduct this operation. This included knowledge of the demographics of the people working or staying in the buildings targeted. In Bali, it was a nightclub neighborhood with the timing of one device programmed to have the greatest toll of human lives.

Two devices were used in a coordinated attack, aimed at killing more people. A general view of the scene, identifying the two explosion seats is presented in Figure 17-13. The first IED, a charge made of TNT, was carried inside a bar, the Paddy's Irish Pub, strapped onto the chest of a suicide bomber. Then, a van parked in the adjoining street was also an IED designed to explode as victims of the first explosion and spectators gathered outside. Some reports even indicated that the suspects actually blocked traffic on the street with additional vehicles as the first device was placed and the van parked. This attack was calculated and well planned to have as many fatalities as possible [37]. The scene investigation represented a tremendous amount of work, particularly because of its size. Identifying vehicles parked on the street was one of the tasks of the crime technicians and forensic scientists. This task was not easy because of the mechanical damage sustained by the vehicles, as shown in Figure 17-14.

What signatures are present in this brief summary of the Bali incident that might lead to a possible terrorist group? First, it was an act outside the normal acts of violence, similar to the Madrid train bombings in 2004, where ETA was at first suspected but later the focus

Figure 17-13

Aerial view showing damage sustained to the Sari Club (middle right area, marked with an S), Paddy's Pub (middle left area, marked with a P), and the surrounding buildings in JL Legian, Kuta, Bali during the Bali bombings of October 12, 2002. (Photograph courtesy of Australian Federal Police Forensic Services.)

Figure 17-14

View showing debris and damage sustained to vehicles and buildings near the Sari Club and Paddy's Pub, JL Legian, Kuta, Bali during the Bali bombings of October 12, 2002. (Photograph courtesy of Australian Federal Police Forensic Services.)

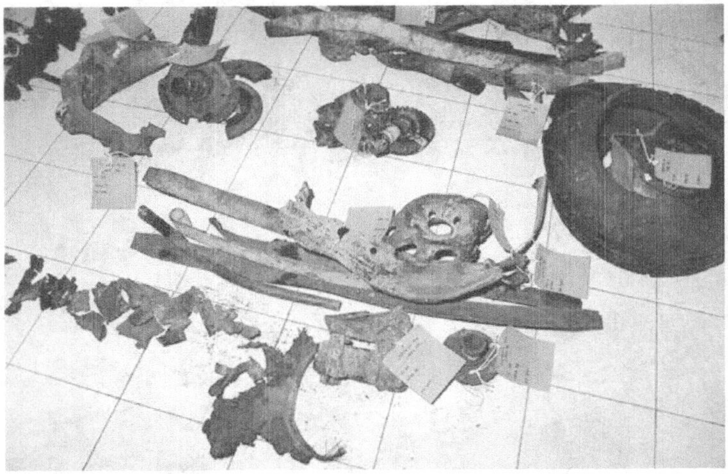

Figure 17-15

Components of the bomb vehicle used in the Bali bombings of October 12, 2002 that were recovered from the scene by the Indonesian National Police forensic members. (Photograph courtesy of Australian Federal Police Forensic Services.)

turned to al Qaeda: this bombing event was outside the norm. As stated earlier, it targeted an international site, similar to the World Trade Center. It is a financial center for the country, and it was done to kill many people, ensuring it received the international attention and public fear the perpetrators desired. The early focus of investigators had to be toward al Qaeda.

Authorities from Indonesia, Australia, and the United States were confident there was enough evidence linking a suspect with regional ties to al Qaeda. Four subjects were charged with the incident and convicted. One of these subjects filed the serial numbers on the vehicle to hide any possible links to the terrorists involved [19, 37].

Forensic analysis was helpful in finding out two key pieces of evidence in this case. The explosive used was first thought to be C-4, a powerful military explosive material. Later, it was determined that a chlorate-based mixture was used, similar to the Oklahoma City bombing and much easier to acquire. Second, the Indonesian National Police forensic members extensively searched the debris to recover a large number of the bomb vehicle components, as shown in Figure 17-15. The recovered components, including the chassis rails of the vehicle, were reassembled by the Indonesian National Police forensic members. The deliberately obliterated vehicle identification number was restored, which ultimately led to the identification of the vehicle and the current owner at the time of the bombings.

BIBLIOGRAPHY

[1] Federal Bureau of Investigation (1994) *1993 Bomb Summary,* US Department of Justice, Washington, DC.

[2] Cowan R. (2002) *Car thief 'clue' to Omagh bombing,* Guardian Unlimited, available at http://www.guardian.co.uk, last access performed on June 4, 2005.

[3] Macko S. (1997) FBI Director warns Russian organized crime threatens U.S. national security, *ERRI Daily Intelligence Report—ERRI Risk Assessment Services,* **3**(276), available at http://www.emergency.com/rusn-mob.htm, last access performed on June 4, 2005.

[4] Federal Bureau of Investigation (2003) Testimony of Grant D. Ashley, Assistant Director, Criminal Investigative Division, FBI before Subcommittee on European Affairs, Committee on Foreign Relations, United States Senate, October 30, 2003, "Eurasian, Italian, and Balkan Organized Crime", *Congressional Testimony,* available at http://www.fbi.gov/congress/congress03/ashley103003.htm, last access performed on June 4, 2005.

[5] Federal Bureau of Investigation (2004) Remarks prepared for delivery by Cassandra M. Chandler, Assistant Director, Public Affairs, Federal Bureau of Investigation, National Air and Space Intelligence Center, Wright-Patterson Air Force Base, Ohio, September 27, 2004, *Press Room Major Speeches,* available at http://www.fbi.gov/pressrel/speeches/chandler092704.htm, last access performed on June 4, 2005.

[6] Terp G. (2004) *Personal Communication with former Assistant Special Agent in Charge Keith Prager, US Customs Office of Investigations,* April 2004, Miami, FL.

[7] Mack J. (2001) *Providing support to counternarcotics and other anti-crime efforts,* Testimony before the House Committee on International Relations Subcommittee on the Western Hemisphere, Washington, DC, October 10, 2001, US Department of State, Bureau for International Narcotics and Law Enforcement Affairs, available at http://www.state.gov/g/inl/rls/rm/2001/sep_oct/6215.htm, last access performed on June 4, 2005.

[8] Derrider W. (2001) *Opening remarks,* 16th Annual Interpol Symposium on Terrorism, October 22–23, 2001, Interpol, Lyon, France, available at http://www.interpol.int/Public/Terrorism/financing.asp, last access performed on June 5, 2005.

[9] Interpol (2002) *The financing of terrorism,* Interpol, Lyon, France, available at http://www.interpol.int/Public/Terrorism/financing.asp, last access performed on June 5, 2005.

[10] Interpol (2003) *Vehicle crime,* Interpol, Lyon, France, available at http://www.interpol.int/public/vehicle/default.asp, last access performed on June 4, 2005.

[11] Criminal Analysis Branch (1998) *Organized crime and automobile theft,* Intelligence report of the Criminal Analysis Branch, Criminal Intelligence Directorate, Royal Canadian Mounted Police, available at http://www.rcmp.ca/crimint/sparkplug_e.htm, last access performed on June 4, 2005.

[12] Bureau of Alcohol, Tobacco, Firearms and Explosives (2004) *Vehicle bomb explosion hazard and evacuation distance tables,* ATF I 5400.1, US Department of Justice, Washington, DC.

[13] Department of Homeland Security (2003) *DHS advisory to security personnel, no change in threat level,* Press Release of September 4, 2003, available at http://www.dhs.gov, last access performed on June 10, 2005.

[14] Wikipedia (2005) *Car Bomb,* http://en.wikipedia.org/wiki/car_bomb, last access performed on June 4, 2005.

[15] Los Angeles Police Department (2003) *Zero tolerance crime task force focuses on San Fernando Valley,* News release of August 11, 2003, available at http://www.lapdonline.org, last access performed on June 10, 2005, LAPD Media Relations Section, Los Angeles, CA.

[16] Candiotti S. (1997) NTSB: *Doomed Miami cargo jet not overweight—first victim positively identified,* CNN, available at http://www.cnn.com, last access performed on June 10, 2005.

[17] Tenney LD. (1943) *The chemistry of powder & explosives,* GSG Associates Publishers, San Pedro, CA.

[18] Wikipedia (2005) *Oklahoma City bombing,* available at http://en.wikipedia.org/wiki/oklahoma_city_bombing, last access performed on June 4, 2005.

[19] Wikipedia (2004) *2002 Bali bombing,* available at http://en.wikipedia.org/wiki/2002_Bali_Terrorist_Bombing, last access performed on June 4, 2005.

[20] Akhavan J. (1998) *The chemistry of explosives,* The Royal Society of Chemistry, Herts, United Kingdon.

[21] DeHaan JD. (2002) *Kirk's fire investigation,* 5th edition, Prentice Hall, Upple Saddle River, NJ.

[22] Phillips SA, Lowe A, Marshall M, Hubbard P, Burmeister SG, and Williams DR. (2000) Physical and chemical evidence remaining after the explosion of large improvised bombs. Part 1: firings of ammonium nitrate/sugar and urea nitrate, *Journal of Forensic Sciences,* **45**(2), pp 324–332.

[23] Cullum H, Lowe A, Marshall M, and Hubbard P. (2000) Physical and chemical evidence remaining after the explosion of large improvised bombs. Part 2: firings of calcium ammonium nitrate/sugar mixtures, *Journal of Forensic Sciences,* **45**(2), pp 333–348.

[24] Wikipedia (2005) *Pan Am Flight 103,* available at http://en.wikipedia.org/wiki/Pan_Am_Flight_103, last access performed on June 4, 2005.

[25] Oliver M. (2001) *Lockerbie: the crucial evidence,* Guardian Unlimited, available at http://www.guardian.co.uk, last access performed on June 10, 2005.

[26] Swarns RL. (1991) Truck blast kills driver, *Miami Herald of June 8, 1991,* Miami, FL.

[27] Bureau of Alcohol, Tobacco and Firearms (1993), *Firearms and Explosives Tracing Guidebook, ATF P 7520.1 (9/03),* Department of Treasury, Washington, DC.

[28] Federal Bureau of Investigation (2000) *FBI Laboratory—Explosives Unit,* available at http://www.fbi.gov/hq/lab/org/eu.htm, last access performed on June 4, 2005.

[29] Williams D. (1998) The bombing of the World Trade Center in New York City, *International Criminal Police Review,* pp 469–471.

[30] Martella J. (1993) *Management of the World Trade Center Bombing Investigation,* FBI Bomb Commanders Conference, October 4–8, 1993, Quantico, VA.

[31] United States District Court (1997) *United States of America vs. Terry Lynn Nichols,* Criminal Action No. 96-CR-68, Reporter's Transcript of November 4, 1997, Volume 61, available at http://www.cnn.com, last access performed on June 10, 2005.

[32] Hellenic Resources Network (2000) *Voice of America, 00-01-21,* available at http://www.hri.org, last access performed on June 4, 2005.

[33] Larson J and Bar-on S. (2005) *Lessons from Madrid bombing—What can U.S. learn about terrorist profiles and planning?,* Dateline NBC, available at http://www.mmsnbc.msn.com, last access performed on June 10, 2005.

[34] Romolo FS. (1999) *Falcone bombing: a forensic overview,* Conference, Institut de police scientifique et de criminologie, University of Lausanne, Lausanne, Switzerland.

[35] Staff and agencies (2000) *ETA blamed as Madrid car bomb kills three,* Guardian Unlimited, available at http://www.guardian.co.uk, last access performed on June 4, 2005.

[36] Lennard C. (2003) *World view: The Bali bombing,* Third Annual TWGFEX Symposium for Fire and Explosion Debris Analysis and Scene Investigation, Orlando, FL.

[37] Moore J. (2002) Car bomb security: Bali was a wake up call, *National Law enforcement Trainers Association Online Magazine,* available at http://www.nleta.com, last access performed on November 26, 2004.

CHAPTER 18

INVESTIGATION FROM THE PUBLIC SIDE

Greg Terp

18.1 INTRODUCTION

"It's just auto theft, we only investigate for the insurance companies anyway!" This quote was from a lieutenant with a major metropolitan police organization discussing the communications bureau's plan to handle all auto theft calls over the phone. This was in direct conflict with the auto theft section within the same department that was not only trying to make sure all reports were taken by an officer in person, but also that they require a stolen vehicle affidavit. There is an enormous misunderstanding on what an "auto theft" means, not only among the general public but also with law enforcement agencies. This lieutenant was saying what many others believe: Why should public agencies allocate manpower and resources to investigate something that private industry is interested in?

Auto theft is not just a "property crime." It has a tremendous impact on each citizen (see Chapter 1). Primary victims feel an economic loss, even with insurance because a deductible usually has to be paid. All citizens who pay car insurance feel the cost of auto theft through higher rates. Auto theft also affects the sense of security for members of a community. Finding one's vehicle stolen can be a very personal experience.

On April 1, 1976 a joint investigation by the Florida Highway Patrol (FHP) and the former Dade County Public Safety Department, presently the Miami-Dade Police Department (MDPD) Auto Theft Section, focused on a corrupt clerk in a Florida Department of Highway Safety and Motor Vehicles (DHSMV) Driver's License Bureau operating in Miami Beach, Florida [1]. As investigators were making illegal purchases at the DHSMV office, several blocks away two detectives noticed a suspicious car in a motel parking lot that later was found to be stolen. The detectives, two highly experienced auto theft investigators, approached the motel clerk to determine in which room the person driving the car was. As they approached the suspect's room, they were ambushed by the suspect with a shotgun and mortally wounded. The other investigators at the DHSMV office heard the shots and responded. A third auto theft detective was shot and killed by the suspect. To this day, this incident remains the most tragic loss for the MDPD in its history.

Over the years many law enforcement personnel and civilians have been killed over stolen vehicles. On March 3, 2005, four uniform officers of the Royal Canadian Mounted Police were killed while investigating stolen vehicle parts and marijuana growing in Alberta, Canada [2]. One of the most dangerous weapons of terrorist groups involves vehicles, some of which are stolen. Countless other criminal acts are directly and indirectly related to auto theft crimes, including a substantial involvement of organized criminal groups worldwide [3]. There is a strong need for public law enforcement agencies to investigate auto theft. There is an even stronger need to investigate these crimes effectively and to share information between investigators, crime scene technicians, and other criminal investigators.

18.2 INITIAL REPORTING AND INVESTIGATION OF A STOLEN VEHICLE

As with any crime, the investigation should start at the scene of the crime. Unfortunately, in most initial calls reporting a stolen vehicle, the crime scene is the vehicle and its current location is unknown. Even with the vehicle taken, some evidence may still be found and processed from the location of theft (see Chapter 4). Broken glass, tire tracks from a second vehicle, shoeprints, and blood are just some of the possible elements to be considered by the initial responding officer. Although auto theft may be the primary criminal action, some vehicles are used in other criminal acts at a later time; thus, evidence not collected at the theft scene can be lost, which can lead to a very frustrating situation if the vehicle becomes the focus of a more significant crime such as murder, sexual battery, or armed robbery. But this theft scene is usually located in a public area, and its evidence might have been destroyed or contaminated beyond usefulness. Although this makes it very difficult to process "the crime scene," it should not stop the initial investigative process. There is a key witness who needs to be interviewed: the person reporting the vehicle stolen (see Chapter 2).

There are estimates that 20–30% of vehicles reported stolen in the United States are insurance fraud cases or owner give-ups [4]. In some locations and/or during difficult economic periods these percentages may even be higher. These people arrange for their vehicles to be taken and dumped in canals, rivers, or lakes; set on fire; or exported out of their country. They then report the vehicles stolen and collect the insurance claim.

It is essential for the initial officer taking an auto theft report to document what the person is saying about how the crime occurred. The date, times, place, and how the theft occurred are the normal primary interests for law enforcement. Observations of the person making the report could be key as well. Did they appear nervous? Were their statements sincere? Did they contradict themselves at any point? In addition, do they still have their original keys to the vehicle? This is a vital question that can help determine fraudulent intent. A person reporting a theft from a shopping mall parking lot should still have their keys in their possession. If they were at a movie, what movie did they see and when did it start? Good follow-up questions by the officer taking the report, based upon the statements of the person, the environment/location of the theft, and the officer's own observations could only be beneficial to the investigation.

As police agencies try to more effectively serve the community, a measure sometimes used is calls-for-service. Agencies are rated on how quickly they can respond to citizen calls reporting suspicious activities, criminal activities, or general complaints. Hence, many departments have instituted reports written by officers based upon information received via phone calls. Specific criminal actions that do not require an officer to respond to the complainant's location are handled with information provided by the person without ever having someone seeing the scene or the victim. Some jurisdictions have decided to include auto theft cases in their phone reporting system. This not only sacrifices the opportunity for an officer to directly and properly evaluate the person, but it also favors the perpetration of fraudulent reporting. Agencies that are serious about reducing auto theft and effectively investigating criminal acts will require an officer to respond at the scene. In some special cases, people can make reports in person at police stations.

Another effective tool in the initial reporting process is the vehicle theft affidavit. This has been an instrumental document in many investigations. First, it forces the reporting person to make a formal declaration as to what they allege to have occurred. The affidavit is a legal document that has the effect of a sworn statement. Should contradictions be later identified, the person may be prosecuted for perjury or for filing a false police report.

Second, affidavits are immensely helpful in the prosecution phase of the investigation. Some jurisdictions require the victim to make a formal statement to the prosecuting attorney's office before a case is filed. This means that the victim or reporter must make a visit to the prosecutor's office, taking time out from work or school. People, however well intended they might be, sometimes do not show up. This requirement leads to a difficult situation if the victim is not local and was only visiting or if the car is recovered in another jurisdiction far away from their home. The affidavit allows the prosecutor to have a formal statement so they can proceed to prosecute or "file" the case. Although it does not eliminate all court-related activities, it can reduce at least one visit, and this is the most important visit. Most auto theft cases do not go to trial but are pleaded out in negotiations between defense attorneys and prosecutors. The affidavit gives the prosecutor evidence of a prosecutable case and helps to win a better plea. Affidavits are an invaluable tool that should be considered by all agencies.

18.3 RECOVERING STOLEN VEHICLES

18.3.1 Principle

The importance of recovery cannot be underestimated, yet one third of stolen vehicles are never recovered [5]. In some cases, only parts of the vehicle are found, such as in chop shops,[1] as shown in Figures 18-1 and 18-2.

---

[1] Chop shops are locations where stolen vehicles are dismantled for their parts. Chop shops can be in a commercial building, a residential garage, or even a yard area. In general, finding two or more stolen vehicles being dismantled and/or major component parts from two or more stolen vehicles constitutes a chop shop by statute.

Figure 18-1

This pickup truck was found in a Miami, Florida chop shop and had been stripped of most major component parts. Investigators worked to identify the vehicle, which was a difficult task due to its condition.

Figure 18-2

View from another angle of the same truck in Figure 18-1, which shows the extent of the work done by the auto thieves. Many times they have specific "part orders" to fill for a local body shop that has a similar wrecked vehicle. Note the cup of soda located under the steering column. The cup belongs to an "employee" of the chop shop and might bear fingerprints and DNA traces that could lead to his or her identification. Such evidence should not be overlooked while processing a chop shop.

Finding and recovering stolen vehicles is a challenging and interesting task. It can be as simple as a police officer checking the license plate of an abandoned vehicle against a database of reported stolen vehicles or as difficult as the forensic examination of a vehicle where an engine number that has been ground off must be restored.

18.3.2 Clues to Finding Stolen Vehicles

The officer's power of observation is critical to identify a stolen vehicle from the hundreds of thousands of vehicles encountered everyday, whether as a patrol officer or as an investigator. Knowing where to look for stolen vehicles is the first step. Then, knowing what to look for is the next logical step in finding stolen vehicles.

A/ Crime Trends

Knowledge of crime trends helps officers locate stolen vehicles. In earlier days of law enforcement, pin maps were used to show patrol officers what types of criminal activities were occurring in their patrol zones. This information was gathered from incident reports and follow-up investigations. The police officer could visually see from where stolen cars were being taken and where they were recovered. Today, computers aid in the analysis by providing more detailed and timelier maps of that information (see Chapter 21).

B/ Cool-Off Zones

Using their own experience, information from other officers, and common sense, officers can put themselves in position to find suspicious vehicles. "Cool-off" zones are to an experienced officer what fishing holes are to an experienced fisherman. Because of the rise of tracking technology (see Chapter 20), thieves have started leaving "hot" (recently stolen) cars in locations that allow them to "cool off". Basically, cars are parked in public areas such as apartment buildings, hospitals, and commuter lots, where they will not attract attention and will not provide a direct link to the thief. The vehicle is allowed to stay there for a certain period of time to see whether it is equipped with a tracking device. Sometimes the vehicles stay only for a few days, whereas other times they may stay for months. The thieves even move the vehicle from spot to spot within the same parking lot to reduce suspicion. If the vehicle is equipped with a tracking device, the police will eventually find it and recover it. After a certain period of time, the thief comes back to the cool-off zone. If the car is missing, it means that it was likely equipped with a tracking device and recovered by the police. If the car is still there, it can be safely moved again because no tracking devices are present. When such a location is found, it can lead to the discovery of many more vehicles if the pond is properly "fished" by officers. Just recovering all the stolen vehicles present eventually leads to a dry "fishing hole" as the thieves find a new, less risky location.

C/ Dumping Sites

Dumping sites are locations where vehicles are disposed after they are no longer wanted or needed. Sometimes just the shell of a vehicle is dumped after its parts have been stripped. These locations become important in discovering more stolen vehicles. By documenting times, dates, and locations, patterns can show when surveillance teams would be most effective.

D/ Indicators on Vehicles

For officers on patrol, there are many clues to indicate whether a vehicle being operated on the roadways or parked may in fact be a stolen vehicle. With additional training specific to auto theft investigation, patrol officers can find more sophisticated stolen vehicles and organized theft groups. Investigators have unique training and experience to identify altered-stolen vehicles through numerous means. Some of the common indicators that a vehicle may be stolen are as follows:

- Rear side window missing (broken to gain entry while minimizing damage);
- License plate (tag) missing;
- Dirty plate on clean vehicle or clean plate on dirty vehicle;
- Accumulated debris such as leaves and dirt on the vehicle;
- Parked away from other vehicles or in a guest spot;
- Damaged or missing door or trunk lock;
- Driver appearing nervous;
- Parts stripped off vehicle.

18.3.3 Vehicle Identification Officers

Vehicle identification officers (VIO) are patrol officers who have been given specific advanced training on auto theft. This program enables regular patrol officers to identify more sophisticated auto theft-related activity. Training consists of basic vehicle identification, title fraud, secondary identifiers, stolen-altered indicators, and other specific information.

The Miami-Dade County Auto Theft Task Force showed the importance of this training during the execution of a search warrant in a warehouse area within the City of Miami. The Task Force officers were waiting in the area of the target warehouse they had identified as a chop shop operation. While waiting for the signed warrant, they observed two City of Miami uniformed police officers driving through the area repeatedly. The warrant arrived and the Task Force members requested uniformed officers from the City of Miami Police Department to assist. The two officers who had been driving through the area responded. The search warrant was executed and several stolen vehicles (Figure 18-3) in the process of being dismantled were found.

The two uniformed police officers then related that they knew a chop shop was in the area because they found stripped frames dumped nearby. They had used the knowledge to

Figure 18-3

By providing patrol officers with more knowledge of auto theft, more stolen vehicles can be recovered. In this case, specific knowledge allowed investigators to identify a stolen vehicle operation that did not catch the attention of uniformed officers who had visited the shop earlier.

define the location to be searched but did not have the next level of knowledge of what to look for. In fact, they had entered the target location and looked at several stolen vehicles without knowing it. With the VIO advanced training, these officers would have found the chop shop before the detectives.

18.3.4 Technology

Technology continues to evolve and improve on a daily basis. The first and most important factor to remember about it is that criminals also have the ability to use technology. Also, typically they do not have the same budget restrictions or bureaucracy limitations than public agencies in obtaining the latest and most advanced equipment.

A/ Computer Systems

The computer is a technology utilized extensively by investigators, patrol officers, supervisors, managers, and the criminal element. Today, computer systems make it easier not only to replicate documents, identification stickers, and vehicle titles, but also to enable subjects to share software and templates. During a search warrant in Miami relating to a subject counterfeiting vehicle titles, it was found that he was also manufacturing counterfeit concert tickets, social security cards, and much more. The subject did not speak English and arrived only recently from Cuba as an immigrant. Yet with a little instruction, a top of the line computer system, and shared software, he was able to start a successful counterfeiting operation (until his arrest).

Investigators need to understand how the forensic examination of a computer is conducted. Many law enforcement agencies are establishing a forensic information technology section specifically to handle this growing field. There are many legal issues associated with the collection of digital evidence. In addition, computer technicians can best advise on how to prevent the loss of evidence when seizing computers. Computer systems are frequently involved in many cases today; thus, investigators have a duty to either know how to handle this important technology or where to find the forensic support needed.

B/ Databases

Understanding computer systems and databases is important in the process of finding stolen vehicles. In the United States, investigators can compare databases from vehicle registration departments with those from the National Insurance Crime Bureau (NICB), Insurance Bureau Canada (IBC), or Oficina Coordinadora de Riesgos Asegurados (OCRA).[2] These companies work as a liaison between the insurance industry and law enforcement in the United States, Canada, and Mexico, respectively. The NICB database can provide

[2] The Oficina Coordinadora de Riesgos Asegurados, created in 1994, is the Mexican equivalent to the US National Insurance Crime Bureau and has for its goal to coordinate operations of localization, identification, and recovery of stolen vehicles.

information on vehicles previously exported, salvage claims, nonconforming numbers, and reported as stolen. Learning how to "mine" these databases will enable investigators to follow the electronic or real paper trail.

C/ Tracking Devices

The evolution of tracking devices continues to enhance law enforcement ability to locate stolen vehicles. Tracking systems now include wireless transmissions and global positioning systems working with cell phone capabilities. These tracking devices (see Chapter 20) provide many opportunities not only to recover stolen vehicles, but also to get intelligence information and aid in surveillance operations. Investigators need to be familiar with tracking devices and stay abreast of advances in this evolving technology. The success of tracking devices has not gone unnoticed by criminals. Their first action to counteract tracking devices was the development and use of cool-off zones, as explained previously.

Another tactic criminals use is to learn where tracking devices are installed and then look in these locations when they steal a vehicle. This is why tracking companies try to keep their installation process secretive and why law enforcement should be careful in discussing tracking devices around civilians and subjects. The use of tracking devices should not be divulged more than is necessary for the courts, including in written reports. It may make a great press release to talk about the success of one case involving a tracking device, but it may harm future efforts.

Criminals have also found technology that will alert them to a tracking device on a vehicle. If the investigator suspects this, then the tracking device can be set up to only transmit at specific times or upon activation.

D/ Bar Code Readers

Today, all products have an attached bar code, from cereal boxes to batteries, including vehicles and vehicle parts and titles. Bar code readers are important and effective tools for the field investigator to help identify true vehicle identification numbers (VINs) [6]. Some even have an integrated wireless capability so the number can be downloaded and checked for theft messages and other vehicle information. Officers can quickly scan and inventory or check a large quantity of vehicles in a relatively short time period. Although many criminals have been able to counterfeit VINs, many times they do not get the bar code correct. This helps to identify stolen-altered vehicles.

Many titles and even some other documents use bar coding to provide a security feature. There are different styles of bar codes, such as two- and three-dimensional. Investigators need to be aware of what type they will encounter during their examination.

E/ License Plate Readers

License plate reader (LPR) systems have been developed for many purposes, and their operation can be a tremendous asset to law enforcement agencies. There are fixed systems used at land border crossings and entrances to specific sites such as harbors or airports [7].

Figure 18-4

LPRs enable law enforcement officers to identify stolen vehicles through a check of the license plate. This system can be mounted on marked or unmarked vehicles as shown in this photo. The cameras mounted on the roof capture tag images that are checked with an on-board database. Systems continue to improve and become more and more covert in appearance.

Mobile systems enable officers to check locations where stolen vehicles are being stored, cooling off, or parked. Some of these systems can process up to 1,000 plates an hour being driven at speeds ranging from 8 to 40 kilometers per hour (5 to 25 miles per hours).

There are companies throughout the world developing LPRs, and each has advantages and drawbacks [8]. Some companies are looking for police agencies to purchase the system, whereas others are working with private industry to fund the systems for law enforcement. The systems may be used by marked patrol vehicles or covertly used on unmarked vehicles, such as the vehicle shown in Figure 18-4. The operators of the system may not be investigators or even sworn officers; in Toronto, Canada, they use police aides to operate their systems. It is important that whoever is operating the LPRs, as seen in Figure 18-5, understands the importance of not just recovering stolen vehicles, but whenever possible to first notify investigators. There are several factors to consider whether to recover or notify. They depend on the availability of a surveillance team, a tracking device, and whether or not the vehicle looks abandoned.

18.3.5 Processing Recovered Vehicles

A stolen vehicle that is located, no matter what the circumstances or situation are, should be considered as a crime scene. This is an important point for all investigators and patrol officers to consider, because their actions from this point forward not only dramatically

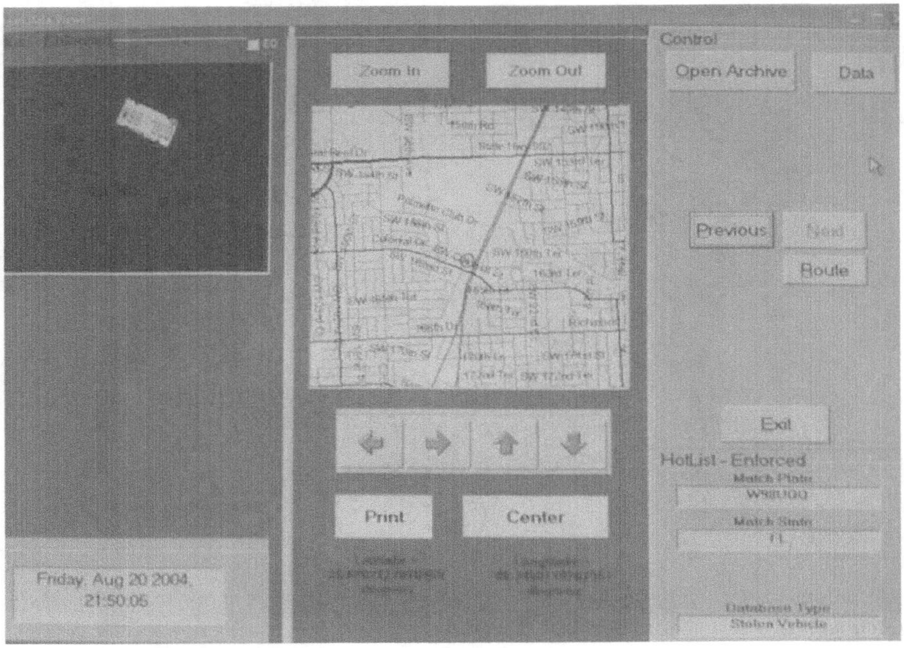

Figure 18-5

The LPR system shown in Figure 18-4 works with a computer located in the vehicle. This system captures a picture of the license plate (upper left corner), marks the position of the vehicle on a map (upper center) through global positioning systems technology, and indicates the results of the query (right). This all transpires in seconds and allows the operator to immediately take a proactive measure if desired.

affect the outcome of the primary investigation of a stolen vehicle, but also could have much more serious consequences.

Stolen vehicles are many times used in other criminal activities, including narcotics trafficking, burglaries, robberies, homicides, and even terrorism-related activities. All rules of preservation, search, and collection of evidence from crime scenes should apply. Law enforcement agencies should have a policy in place for the recovery and processing of stolen vehicles.

Some agencies have officers or investigators processing vehicles by themselves. In these cases, the officers involved should have received specific training in the recognition and collection of evidence. This training needs to be an ongoing process so that these people are aware of new techniques and procedures. Communication between the officers processing vehicles and the laboratory scientists evaluating any submitted evidence is essential. Laboratory scientists can provide the feedback necessary to improve the collection process. Also, knowing what evidence is important to the forensic scientists allows more effective time management for all involved personnel.

If crime scene technicians process the vehicle, officers must ensure they effectively communicate information specific to the investigation and any additional aspects. An example is the recovery of a stolen vehicle that fits the method of operation of a serial rapist. This knowledge would enable the technician to collect more samples of possible trace evidence (such as fibers, glass, paint, and hair) and body fluids (for DNA) from specific areas within the vehicle (see Chapter 4 Part I).

With the rise in attention to terrorist activities, trace evidence may not constitute the only items collected from a stolen vehicle. A stolen vehicle may actually be an improvised explosive device (see Chapter 4 Part II). Any suspicion that explosives are within a stolen vehicle will modify all protocols, with explosive ordinance personnel (bomb disposal) called on to first make sure the vehicle is safe before it can be processed. Safety comes first and evidence comes second.

18.4 INVESTIGATING AUTO THEFTS

18.4.1 Principle

Auto theft investigations are handled in many different ways by the responsible agencies depending on several factors, including the extent of the problem, the community response, and an understanding of the role auto theft plays in other criminal activities. Some major metropolitan police agencies have actually disbanded their auto theft investigation units, treating auto theft as either another property crime and/or as organized crime. Auto theft is more complex than most police administrators want to understand; therefore, it is important for investigators to educate their command staff as to the types of cases they are working.

Vehicles are stolen for a variety of reasons. The easiest manner of dividing auto thefts for investigative purpose is in the following two categories: those stolen for profit and those stolen not for profit. Professional thieves or groups are interested in making a profit. It takes highly trained and experienced investigators to identify and recover stolen-altered vehicles. The primary role of auto theft investigators is the for-profit side, which is discussed in detail in the next subsection.

Conversely, it is much easier to identify stolen vehicles that have not been altered. By providing crime analysis data to patrol officers and other investigators, cases where vehicles were stolen for transportation or to be used in other criminal activities can be solved. For example, if vehicles are being used in armed robberies, crime analysts can provide the robbery detectives with the locations from where vehicles are stolen, names of known suspects, and locations where vehicles are being abandoned. Thus, they can focus their investigation to these locations and individuals.

18.4.2 Types of Thefts

A/ Export

The NICB highlights the effects of stolen vehicles being exported in a 1995 study [9]. The study clearly shows a correlation between communities in the United States with high vehicle theft rates and port/border communities. With the growing export problem and high national auto theft rates, US Customs and the NICB created the North American Export Committee (NAEC) to find ways to ebb the flow of stolen vehicles overseas [10]. The NAEC was established in 1995 with key stakeholders from the United States and Canada, representing federal, state, and local law enforcement agencies as well as private industries. Later, the NAEC expanded to include Mexico in an official capacity and other international partners from Europe in an unofficial capacity. The NAEC's goal is to find ways and means to address the problem of exportation of stolen vehicles for all countries, and more specifically North America.

With the growing world economy, stolen vehicles have become an important commodity. With international laws and treaties weak at repatriating stolen vehicles back to the victim's country, it is a high-profit and low-risk enterprise. Even when vehicles are located in a foreign country, many governments are reluctant to take a vehicle from their citizens. In fact, stolen vehicles from a particular country are sometimes found being used by government officials in another country.

The best course of action is to stop the exportation of stolen vehicles before they cross international borders. This is a difficult and complex effort. Customs officials are the primary law enforcement agency responsible for stemming the flow of stolen property from one country to another. In most countries, customs officers are more involved with in-bound processing than out-bound processing. With the rise in terrorism worldwide, hampering the export of stolen vehicles is even less a priority for customs officials.

Many countries also lack effective laws to help reduce the exportation of stolen vehicles. In the United States, all vehicles being exported must be presented to US Customs 72 hours before leaving. This provides law enforcement with time to check the status of a vehicle. Various databases can be used to identify stolen and stolen-altered vehicles being exported.

In January 1997, the NAEC, the US Customs, the NICB, and the MDPD Auto Theft Task Force initiated an electronic reporting system at the Port of Miami in Florida [11]. This system took all the vehicles being presented for export and sent a batch list to NICB's database to check for potential stolen vehicles. The vehicles were checked against the National Criminal Information Center theft files and NICB's previous salvage, previous export, and nonconforming files. A list of suspect vehicles was then sent back to Miami the next day, all of this occurring within the 72-hour period. On a daily basis, approximately 200 to 300 vehicles were checked electronically, with about 15 to 25 suspect vehicles identified for further inspection. This system has now expanded to the top 70 ports in the United States.

As the development of the electronic reporting system was progressing, another critical area was the identification of the smuggling of stolen vehicles hidden in containers. There will always be regular smuggling methods where vehicles are taken across a land border or placed on a cargo ship without being properly manifested, but the bulk of smuggling is done through shipping containers. In 1997, the NAEC looked at various means of identifying vehicles inside these containers. X-ray and gamma ray technology was used to scan shipping containers (see Figures 18-6 and 18-7) [12]. After several demonstrations and

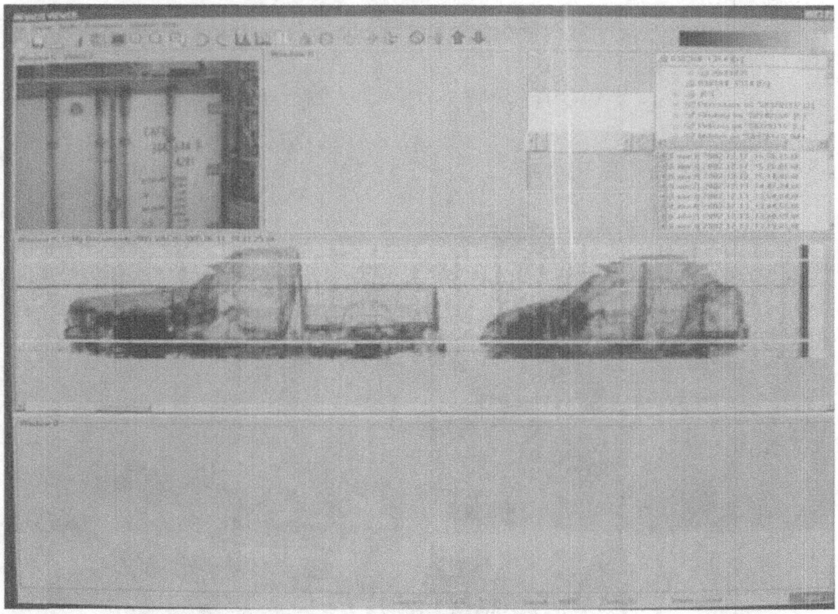

Figure 18-6

Figure 17-3 showed a shipping container with paper goods hiding two stolen vehicles. This figure is a gamma ray photograph of the same container. Container imaging systems such as the Stolen Auto Recovery working at the Port of Miami in Florida clearly shows law enforcement personnel that two vehicles are inside the container.

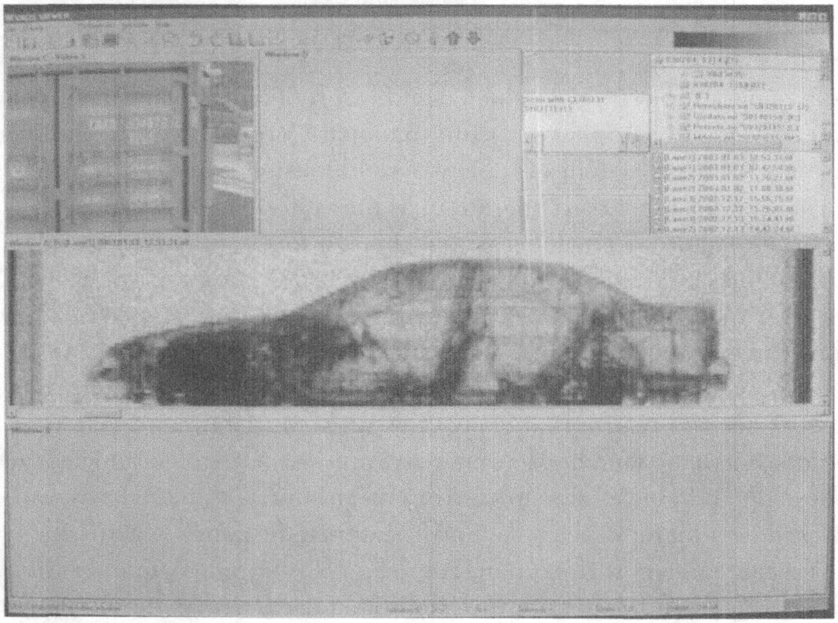

Figure 18-7

The Stolen Auto Recovery image is similar to an x-ray image. In the upper left corner is a picture of the container number for the purpose of tracking. The upper right corner indicates the lane the container was scanned in and the date and time it entered the Port of Miami. The bottom image allows the investigator to quickly identify the presence of a vehicle.

Figure 18-8

With the Stolen Auto Recovery system, Miami-Dade Task Force members no longer have to guess when opening a container—they know in advance that a vehicle is present inside. Some of the operators are able to determine the year, make, and models from the image obtained on the screen as shown in Figures 18-6 and 18-7.

presentations, a pilot project was conducted at the Port of Miami where a gamma ray system manufactured by Science Applications International Corporation was modified to create the Stolen Auto Recovery system [13]. This system located six stolen vehicles in shipping containers the first week it was in operation, as illustrated in Figure 18-8.

The important factor to remember is that this is a for-profit enterprise, albeit an illegal enterprise. As efforts are initiated to catch stolen vehicles, the criminal element will adjust and change their modus operandi. As these systems became more effective, criminals tried alternative methods to export their stolen vehicles, such as altering paperwork (titles, etc.), altering VINs, or shipping out of ports that had little or no outbound enforcement efforts.

Investigators need to realize that export investigations begin and end away from ports or land borders. Ports and land borders are like funnels where stolen vehicles can be located and leads developed. Vehicles are stolen from communities to be delivered to other communities. It is important for officers and investigators to be aware of activities in their areas that might indicate an export operation. For example, shipping containers in a residential or business area that is normally not involved in export activities should be considered as suspicious and should deserve further investigation. A report of vehicles being placed in shipping containers, especially with flatbed wreckers, is a key factor. When discovering a chop shop, the presence of stolen parts being packaged or wrapped in plastic may indicate that the shop is related to export activities.

B/ Domestic Investigations

Stolen vehicles are sold to local buyers (within the same community or country) with their value determined by the ability to obtain a clear title or ownership. Vehicles sold as is with no altering of the VIN or counterfeit documents are worth a small percentage of their true commercial value. A vehicle with altered numbers and a clear title can be worth the full value and sold to unwary customers.

Understanding the nature of business helps guide investigations. As with any business, auto theft is driven by supply and demand. High demand for specific types of vehicles or parts results in similar vehicles being stolen. In addition, like any commercial operation, there are suppliers, brokers, and buyers. Targeting only one area will result in mediocre results; all areas need to be addressed.

The suppliers or thieves and suspects altering identification numbers and documents can be identified through various means. When a subject is apprehended for stealing a vehicle, the interview by an auto theft investigator is vital to learn more about the process. Knowledge of crime trends and known vehicle theft suspects will help determine the value of information from a cooperating subject.

C/ Vehicle Parts Market

The cost of automotive replacement parts drives another illegal facet of auto theft: the parts market. Vehicle repair and body shops can maximize their profits and compete unfairly with other businesses when using stolen parts. It is important for communities to have specific laws or ordinances that regulate this industry. In addition, law enforcement officers should be granted the ability to inspect records, parts, and vehicles to ensure no illegal activities are taking place. Regular inspection of repair shops should be a part of any auto theft investigation unit.

Major component parts in the United States require identification numbers to be affixed by anti-theft labels or stamps into the part (see Chapter 6). The absence of such labels or the grinding off of numbers is a criminal offense. Thus, when law enforcement officers inspect a repair shop, body shop, or salvage yard, they have the legal tools to take action, even without clear indication of theft.

Many manufacturers have argued in favor of reduced parts marking as they have introduced anti-theft initiatives such as immobilizer ignition systems. These systems may initially deter thefts, but criminals will still find a way to steal vehicles by using original or duplicate keys. Stolen parts will still be obtained but less will be identifiable, making the investigation and prosecution more difficult.

D/ Title Fraud

Fraudulent documents are key elements to high profits for stolen vehicles (see Chapter 15). In the United States alone there are over 50 different titles issued, whereas in Canada there is only one registration form issued. An important number of different valid documents lead to a situation where organized theft rings can steal vehicles in one jurisdiction and transfer documents from another to show a clear title. Title fraud is used in both domestic and export cases. It is difficult for personnel at vehicle titling departments to have knowledge of all the different forms the various countries use to prove ownership of a vehicle.

For example, an Aston Martin Vanquish leased in Switzerland (valued at CHF 390,000, equivalent to approximately USD 300,000) and that was not allowed to leave Europe was exported to the United States. The Swiss registration was provided as a proof of ownership to the Florida Department of Motor Vehicles. An examiner used a reference manual to verify that the Swiss registration was valid. The translation of French words was not available for the indication that the vehicle was leased and a Florida title was issued. The vehicle was then traded in for a Porsche Cayenne (about USD 85,000) and USD 65,000 in cash. As a result of cooperation between Swiss and Floridian investigators, the subject was arrested and the vehicle was recovered [14].

An investigator from the MDPD Auto Theft Task Force initiated Operation Cyber Search, where all State of Florida Department of Motor Vehicle records for active registrations were matched against NICB's databases. Over 95,000 suspicious vehicles were identified statewide [15]. Vehicles were divided into four categories: straight stolen, prior export, salvage, and nonconforming. Investigators on Operation Cyber Search responded to the suspicious vehicles and were able to determine whether they were altered or not. The altered vehicles were seized, and their "owners" were questioned. The vehicles were then processed to identify their original VINs and to obtain information regarding the theft. Suspects were identified through interviews with the "owners" and through the documents submitted to obtain a clear title. Among the four lists, the prior export was the most productive for finding stolen-altered vehicles. Suspects used the VINs of exported vehicles on similar vehicles that were stolen to reduce the ability of law enforcement to identify them. The nonconforming list concerns the VINs that do not meet the algorithm created for US vehicles (see Chapter 6). Most vehicles on this list were actually entered improperly due to administrative errors. Any true nonconforming vehicles were generally stolen vehicles that were subsequently altered. The salvage vehicle list was the second most productive group of vehicles. Detectives found both stolen vehicles and vehicles with titles being washed to hide accidents and flood damage.

Title fraud investigations in and of themselves are important and require specific knowledge and training (see Chapter 15). Partnerships with licensing agencies increase the level of knowledge needed to identify forged documents.

The United States has developed a program, the National Motor Vehicle Title Information System, to enable each of the 50 states to verify title transactions to reduce title fraud [16]. Other systems in Europe, Mexico, and Canada are being developed to provide the same service and eventually links with each other. This will greatly assist investigators in identifying fraudulent titles.

E/ Insurance Fraud

The first factor to consider when addressing insurance fraud is the manner in which the agency takes vehicle theft reports. In the United States, because vehicle thefts are property crimes and agencies are trying to better serve their citizens, many have started allowing phone reports. This enhances the ability to commit insurance fraud. Without a face-to-face meeting, a police officer is not able to judge the validity of the reporter's statements of fact and to ask the proper questions. The vehicle theft affidavits mentioned earlier are an important part of addressing insurance fraud. Having a sworn statement locks the reporter of the theft into a story that can later be questioned and disputed. Affidavits should be a part of all vehicle theft reports.

Law enforcement investigators need to develop relationships with insurance company investigators and special investigation units personnel. By working together, they can develop leads and verify data that indicate a potential fraud has occurred. It is important to let the reporters of thefts make voluntary statements to the insurance industry. Many times these statements provide the evidence needed to make an arrest.

Sometimes, vehicles that have not been reported stolen are discovered under suspicious circumstances, such as in a shipping container or in a cool-off zone. In such situations, investigators need to proceed cautiously, not immediately notifying the owner that they have the vehicle until enough time has elapsed for the owner to file a complaint. When a vehicle is found on Friday at the port and the owner states it was stolen from their driveway on the following Monday, the investigation has strong elements to conclude insurance fraud.

F/ Identity Thefts

As with other economic crimes, vehicles are stolen through the theft of identity and personal information. With more and more anti-theft equipment being installed on vehicles, criminal suspects may now take vehicles costing thousands and thousands of dollars by just using a pen to sign a lease or sales agreement under someone else's identity. Credit cards are used to steal vehicles from rental companies. Using stolen credit card information and false identification, vehicles are easily obtained.

Auto theft investigators need to work closely with economic crime units. Many times, criminal groups are involved in multiple economic criminal activities with some suspects stealing credit card information, others running credit card laboratories, and others making purchases, including renting vehicles. The sharing of information is vital to linking auto thefts and other financial crimes.

G/ Theft of Specialty Equipment

Investigative units should either have experts in specialty equipment or know where to find information relevant to it. Heavy equipment (as shown in Figures 18-9 and 18-10) is a prime example that knowledge of what a product identification number is and where it can be located on the equipment is important. The National Equipment Registry is a private industry company that provides law enforcement with additional information when dealing with suspicious heavy equipment [17]. Other special equipment includes motorcycles, boats, all-terrain vehicles, personal watercraft, rental vehicles, dump trucks (Figure 18-11), semi-tractors, and trailers.

Figure 18-9

The image obtained from the Stolen Auto Recovery system clearly shows the presence of heavy equipment in a container being exported from Miami. Such equipment is valued very high on the black market list for shipping to other countries.

Figure 18-10

This backhoe found in a shipping container has specific identifying information available through the manufacturer, such as the product identification number. The shipping container makes it more difficult to find the numbers.

18.4.3 Long-Term Versus Short-Term Cases

Administrators of law enforcement agencies need to understand the benefits of long-term cases versus short-term cases: Using both will have the greatest impact, whereas ignoring one for the other will produce marginal results. Long-term investigations are time consuming, expensive, and do not quickly produce a number of arrests. They have a greater impact on theft rates over a longer period of time and produce longer sentencing of suspects. These cases involve identifying organized criminal rings and the people involved. Cases may take months or even years to investigate and prosecute.

Figure 18-11
Even large vehicles such as dump trucks, shown here being recovered from a shipping container at the Port of Miami, are stolen for both the domestic and foreign markets.

Short-term cases can have an immediate impact and are the result of directed activity. Prosecution rates are less effective and rarely end up in lengthy sentencing. These investigations are focused on the street-level thefts by identifying the individuals involved. Suspects arrested on short-term cases may sometimes be used for long-term cases. Theft rates can be affected temporarily but eventually start to rise again without a long-term approach being integrated.

18.4.4 Utilization of Informants

Information is the key to any investigation, and informants play an important role in developing leads and identifying major players in organized crime organizations [18]. It is important to remember that information always flows in two directions; informants may provide intelligence, but they also absorb intelligence. Informants are always looking for counter-intelligence, so that they and their friends can avoid future law enforcement efforts. Investigators must always be careful when using an informant. Most agencies have specific protocols for using informants to make buys or deliver stolen property, vehicles, or fraudulent documents. Informants can provide inside information that enables investigators to collect the most evidence. Informants can be classified as formal and informal.

A/ Formal Informants

Formal informants are properly documented by a law enforcement agency. This documentation ensures that informants are checked for background, pending judicial cases, motivation, and personal history. Documenting an informant allows agencies to pay for some information and allows investigators to use that information in formal court proceedings.

B/ Informal Informants

Informal informants are not documented for various reasons and many times are called "sources." These sources provide valuable information but generally are not used in court proceedings and are not compensated.

18.4.5 Undercover Operations

No matter what crime is being investigated, undercover operations are dangerous. Auto theft investigators should never take for granted that they are just dealing with a routine auto theft case.

Undercover operations are a necessity, specifically when doing long-term investigations. To learn as much as possible about a group of criminals and to make the best prosecutable case, undercover operations are exercised. Informants and/or officers portray other criminals involved in the illegal business. Officers should be well versed in the type of case they are working.

The decision to conduct an undercover deal should weigh the benefit to the case. Safety considerations should be paramount with the planning of alternative courses of action and an emergency plan. Everyone involved in the deal should be briefed and known to each other. As much background as possible should be learned about the subjects involved, including past actions and a criminal history check. If informants are used to set up the meeting, their motivation and reliability should be determined.

18.4.6 Search Warrants

One of the best tools for gaining information and evidence of criminal activity directly from known suspects is a search warrant. This judicial order allows investigators to enter property, vehicles, and computers to identify and seize records and instruments of criminal activity. In the United States, investigators must establish probable cause that a crime has been committed and that evidence is currently located at the location to be searched [19]. The information provided in the search warrant must be timely and relevant to the investigation. Trash or garbage pulls remain an excellent means for finding fresh evidence of certain activities at a given location. A review of the search warrant request should be conducted by supervisory personnel, management, and a prosecutor before being presented to a judge for approval.

The execution of a search warrant should be considered in the same light as an undercover operation; they all have the potential to be dangerous. A checklist should be established by every agency to ensure that all necessary steps are taken to identify the right location and that it is executed in the safest manner. A tactical team should be used to execute the search warrant any time there is the potential that armed or dangerous subjects are involved. The moment it is served should be decided as the safest time that will enable investigators to secure the most evidence. Safety should remain the primary factor.

The search team should be decided in advance with a recorder, photographer, and search officers clearly established. If a computer is suspected of being involved, it should be articulated in the warrant. Some jurisdictions require an additional warrant to examine a computer. After completion of the search warrant, an after-action report should be completed indicating the outcome of the serving of the warrant.

18.4.7 Bait Car Systems

Technology has provided law enforcement with the ability to place bait car systems in areas of high theft that will capture suspects and record them on video. These systems use several integrated technologies to alert law enforcement when a vehicle is entered and stolen. The vehicle location is provided as it moves through traffic. When law enforcement approaches the vehicle, the latter can be disabled and the doors locked, sealing the suspects inside. The suspects are recorded with audio and video during the entire incident.

These real-life videos have a dramatic effect when played on television, making other auto theft suspects leery of getting caught as well. Also, they have enormous interest from the media and provide a visual outcome of a criminal being apprehended, which is usually appreciated by the public.

Bait car systems have limitations for investigative purposes. Many suspects arrested are driven by a crime of opportunity motive rather than being professional auto thieves. The use of these systems necessitates the knowledge of the most popular vehicles being stolen, the method of theft, and the best locations to put them out. Even then, the potential for the bait car to be stolen is not necessarily high. It takes time and research to be most effective. Bait cars can be part of an effective campaign to reduce high auto theft rates in a targeted area. They have a short-term impact and the video evidence will enhance the prosecution aspect.

18.5 PROSECUTION OF AUTO THEFT CASES

The success of an investigation is not determined by the arrest as much as by the prosecution. It is important for all investigators to understand that the communication between investigators and key stakeholders involved in their cases is a crucial factor in the outcome of the case. This includes the prosecutorial aspect of the case.

As mentioned earlier, auto theft cases can be categorized as long-term or short-term. Street-level cases are very basic and do not require an extensive knowledge of the specifics for a prosecutor to handle the case. The key factors in a simple case where a suspect is found driving a stolen vehicle do not require a deeply involved investigation and rarely extensive processing of the scene by a forensic technician. However, the crime scene processing should not be completely left out. The collection of physical evidence, even in case of an auto thief caught red-handed, is always an extra layer of evidence that should never be neglected. Furthermore, it may lead to the discovery of trace evidence from an accomplice, not present in the car at the time of the arrest but who participated in the theft, because some auto thieves do not act alone. Indeed, other trace evidence that might link the criminal to other crimes might be present in the vehicle. Finally, it is always a good experience for the investigator and the forensic technician to examine a vehicle where the modus operandi and the thief are known.

Many auto theft cases require much more involvement, from an understanding of the intricate paperwork flow to how VINs that have been ground off can be restored and used to identify a stolen-altered vehicle. For these cases, prosecutors need to have a greater understanding of the auto theft field. In jurisdictions with high auto theft rates, it is most productive to assign a specific prosecutor to handle all major cases. These prosecutors should also attend auto theft investigation training courses, seminars, and intelligence meetings.

Long-term cases, specifically dealing with major targets and organized groups, need vertical prosecution.[3] As soon as a major target is identified or a complex case is recognized, a prosecutor should be brought into the investigative team. This allows the investigators, crime scene technicians, and managers to make clear decisions on the direction of the case. Case management is critical to a successful prosecution and the advice of the prosecutor is invaluable.

Another consideration for the prosecution is the question of which venue should the case be brought under. The option to take a case under a federal or local court system depends on where the case will have the biggest impact. Knowledge of what is expected at each level will help the decision-making process.

18.6 TYPES OF AUTO THEFT INVESTIGATIVE UNITS

Auto theft units are not the top priority of any law enforcement agency, federal or local, unless the theft rates have become so high that the politicians have made it the crime du jour. Violent crimes, narcotics, and terrorism have, as they should, higher priority.

Agencies establish auto theft investigative units based upon their respective concerns and knowledge of how to best address this specialized criminal enterprise. Some agencies understood the nexus between auto theft and other crimes. One of the most interesting analyses was given by the Los Angeles Police Department where it was decided to process every single stolen vehicle. Explanation was that by reducing auto theft, it helps solve other criminal activities and that overall crime rate falls.

Reactive auto theft units respond to the criminal acts, assigning cases to investigators based upon potential leads. In high theft areas, the number of cases can be overwhelming, allowing little free time for investigators to be proactive. Most cases are short-term in nature.

---

[3] Vertical prosecution is a term used to characterize a situation where the prosecutor is involved since the very early stages of the investigation.

Investigators need to have specific knowledge and training of auto theft crimes or their impact will be limited.

In some jurisdictions, general property crimes detectives handle auto theft investigations. The scope of the vehicle theft problem dictates the resources allocated. When reactive investigators develop leads into organized groups or identify major targets, if there are no proactive investigative avenues available, opportunities to have significant long-term effects are lost.

Proactive investigative units do not respond to specific criminal acts. They address a significant issue such as vehicle theft by initiating cases and operations designed to identify major targets and organized groups. In addition, proactive measures are used to force compliance as when inspections are conducted of salvage yards, shipping lines, body and repair shops, and dealers.

The most effective auto theft units combine proactive and reactive investigations. The flow of information between reactive and proactive cases is imperative to being the most successful.

The multiagency approach provides the best forum for a criminal investigation where suspects routinely cross jurisdictional boundaries. Having representatives from federal and local law enforcement, customs officials, department of motor vehicle personnel, and private industry investigators combines various resources and experiences in one location. As auto theft continues to be more of a global crime, multiagency task forces will become the most effective means to address this costly illegal business.

BIBLIOGRAPHY

[1] Buchanan E and Montgomery L. (1976) 3 Metro policemen killed, *Miami Herald of April 2, 1976,* Miami, FL.

[2] Oakes WA. (2005) *UPDATE—Shooting leaves four RCMP officers dead,* Royal Canadian Mounted Police Newsroom, available at http://www.rcmp-grc.gc.ca/news/n_0508_e.htm, last access performed on June 12, 2005.

[3] Macko S. (1997) FBI Director warns Russian organized crime threatens U.S. national security, *ERRI Daily Intelligence Report—ERRI Risk Assessment Services,* **3**(276), available at http://www.emergency.com/rusn-mob.htm, last access performed on June 12, 2005.

[4] Turnberger GA. (1994) Recovered stolen-recovered burned—Investigation of suspicious motor vehicle fires, *The APB,* **2**(4), pp 21–22.

[5] Scafidi F. (2005) *Hot wheels: now you see them, now you don't!,* National Insurance Crime Bureau press release of February 28, 2005, available at http://www/nicb.org/public/newsroom/hotwheels/pressrelease.cfm, last access performed on June 12, 2005.

[6] Bender Enterprises (2004) *VINtrack,* available at http://www.vintrack.com, last access performed on June 12, 2005.

[7] Tremblay L. (2002) *National Insurance Crime Bureau update by COO Jim Spiller,* Notes from the North American Export Committee Meeting of May 13–15, 2002, Halifax, Canada.

[8] Insurance Bureau of Canada (2003) AutoVu program—seeing success on Toronto streets, *Challenges and Champions,* **5**(3), p 6.

[9] Perin M. (1997) Auto theft: Costs are high for both insurers, victims, *Houston Business Journal of February 17, 1997,* available at http://www.bizjournals.com/houston, last access performed on June 15, 2005.

[10] Conn L. (2000) Car thefts top cop, *Canadian Underwriter,* March issue, available at http://www.canadianunderwriter.ca, last access performed on June 16, 2005.

[11] Staff writers (1997) Exports program prevents stolen vehicles from cruising, *Spotlight,* National Insurance Crime Bureau, Issue II, pp 4–5.

[12] Crepeau PJ. (2004) Waging war against stolen vehicle exports & fraud, *Insurance Advocate,* **115**(3), pp 26–27.

[13] Zollars R and Adams J. (2000) *SAIC's stolen automobile recovery system (STARS) to play major role in recovery of stolen automobiles at major seaport locations,* Science Applications International Corporation news release of October 2, 2000, available at http://www.saic.com/news/oct00/news10-02-00.html, last access performed on June 12, 2005.

[14] Horgan K. (2004) *Case number 494537-C,* Miami-Dade Police Department Report, Miami, Florida, pp 1–2.

[15] Florida Motor Vehicle Theft Prevention Authority (1999) *Grant Awards,* 1999 Annual Report, p 12.

[16] American Association of Motor Vehicle Administrators (2004) *Vehicle Registration & Title Committee summary of activities,* AAMVA, Arlington, Virginia, available at http://www.aamva.org/committees/comVRTmasterplan.asp, last access performed on June 13, 2005.

[17] National Equipment Register (2005) *Combating heavy equipment theft,* available at http://www.nerusa.com, last access performed on April 10, 2005.

[18] Miami-Dade Police Department (1997) *Confidential informants,* Departmental operations manual, standard operating procedures 4–44, section 6 undercover operations, Part II, pp 277–285.

[19] Miami-Dade Police Department (1997) *Searches,* Departmental operations manual, standard operating procedures 4-44, section 7 undercover operations, Part I, pp 298–315.

CHAPTER 19

INVESTIGATION FROM THE PRIVATE SIDE

Part I: The European Perspective

Marc Stauffer

19.1 INTRODUCTION

Switzerland, unfortunately, appears to have one of the leading statistics in the number of stolen vehicles per capita. According to the 1999 Interpol statistics, cited by Junghans, there were 1,130 stolen vehicles per 100,000 inhabitants in Switzerland [1]. This figure is 33% greater than that of the United Kingdom and more than 50% greater than those of other European countries [1]. This figure is even more impressive when expressed per number of registered vehicles rather than per number of inhabitants: It jumps to 1,700 stolen vehicles per 100,000 registered vehicles, which would mean that nearly one of every 59 vehicles is stolen every year in Switzerland (see Subsection 1.3.3 for a brief discussion of the reservation around these figures). In any case, these critical numbers seriously influence the insurance industry, as Switzerland is among the world's most insured countries.

According to the European Report, there are about 1.2 million cars stolen each year in the European Union, which represents one car stolen every 26 seconds, for a total cost of about EUR 15 billion (about USD 18 billion) [2]. Among these thefts, an important part (about 20%) is not actual theft but rather insurance fraud [3]. Therefore, insurance companies have an important role to play in the investigation of auto theft claims.

This chapter explains the process followed by the insurance company to handle an auto theft claim. The information presented in Part I is a direct reflection of the Swiss experience, which is designed to respond to the Swiss insurance laws and regulations. However, this approach can be adapted to foreign insurance companies since many insurance companies are present on an international market and because the basis of all auto theft insurance claims is similar around the world.

19.2 INFORMATION IN POSSESSION OF THE INSURANCE COMPANY

During the establishment of the policy, the insured has to provide the insurance company with the information presented in Table 19-1. As shown, this information is very limited and concerns both the insured and the vehicle.

The insurance company provides the insured with a temporary proof of coverage. This document is necessary to get the vehicle registered through the department of motor vehi-

Table 19-1

Information required by the insurance company to establish a vehicle insurance policy in Switzerland.

Information regarding the insured		
Last name	First name	Date of birth
Citizenship/Origin	Date of issuance of driver's license	Photocopy of identity card
Information regarding the vehicle		
Make and model	Type	Vehicle identification number
Date of first time on public road	Cylinder displacement	Price as new
Price of accessories	Number of seats	Number of doors

cles (DMV). Once the administrative part of the registration process has been performed, the insurance company receives further information directly from the DMV. This information consists of the license plate number and the color of the vehicle.

19.3 COLLECTION OF BACKGROUND INFORMATION IN CASE OF THEFT

19.3.1 Insured's Notification of Theft

As soon as the insured is aware of the theft of his or her vehicle, he or she is required to immediately report it to the police and to the insurance company. "Immediately" is a vague term and does not necessarily mean within a few seconds. It is commonly required that the insurance company must be notified within the 24 hours following the theft or knowledge of the theft. This preliminary filing can be done by telephone. It starts the first necessary measures taken by the insurance company.

19.3.2 First Measures Taken by the Insurance Company at Time of Theft Notification

A/ Notification to the DMV

The insurance company is required to notify the DMV, because the report made to the police is not forwarded there. According to Swiss road traffic laws, the company has to maintain the coverage on the vehicle (only for damages created to third parties) for 60 days after the notification of the theft [4]. Since the driving record of the thief or thieves is not known to the company, it is important for the company to be alleviated as soon as possible from the duty of paying possible damages created by the vehicle to third parties. These damages could reach considerable amounts depending on the types of activities or accidents engendered by the vehicle.

B/ Notification to the Manufacturer

Depending on the vehicle's make, it is also very important to notify the vehicle's manufacturer. In Switzerland, neither the police nor the DMV informs the manufacturer of the theft. More vehicles, particularly luxury ones, are equipped with global positioning systems (GPS) and other tracking systems (see Chapter 20) and can be tracked by the manufacturer or other private companies. In some instances, it is even possible, as with some Mercedes-Benz models, to have the vehicle blocked no matter where it is located around the world [3]. With new electronic systems used to open doors and to start vehicles, some of these systems can be operated remotely and, more interestingly, can be used to remotely deactivate essential components of the vehicle, preventing it from being driven any further.

19.3.3 Questionnaire

A questionnaire, particularly designed around the report of a stolen vehicle, is sent to the insured by the insurance company. The questionnaire typically includes, but is not limited to, the following questions:

- Who parked the vehicle immediately preceding the theft (hereafter referred to as "the driver" as it might be a different person from the insured)?
- At what date and time was it parked?
- List the names of people that were accompanying the driver at that date and time. *This information is used to obtain further information regarding these people, who will be interviewed at a later date to corroborate or contradict the statement from the insured.*
- Who discovered that the vehicle was stolen (hereafter referred to as "the discoverer")?
- At what date and time was the theft discovered?
- Where did the theft happen (description of the location)?
- What were the reasons for being at this location? Where was the driver coming from and going to? Was the driver going to meet someone?
- Who was going to drive the vehicle after it was parked?
- At what date and time was it going to be driven again?
- Describe, in chronological order, what the discoverer did immediately after discovering the theft of the vehicle.
- To which police station did the discoverer go to serve notice of the theft?
- In which language did the complaint filing occur? *It is important to know if the conversation with the police occurred in the insured's native language to prevent the situation where he or she could later state that the police did not clearly understand the facts around the complaint if no translator was available.*
- Was the vehicle properly locked (doors, windows, moonroof, etc.)?
- Was the steering column lock engaged?
- Was the vehicle equipped with an anti-theft system?

- If so, was the system original equipment manufacturer (OEM) or was it installed by a third party after purchase of the vehicle?
- If this system was not OEM, does it correspond to the norms of the manufacturer of the vehicle?
- Was the vehicle equipped with an alarm system?
- If so, was the alarm system engaged?
- Where were the keys of the vehicle?
- How many keys did the owner receive when the vehicle was purchased?
- Have any copies of the keys been made?
- If so, how many? Where and when did the owner (or anyone else) make these copies? *It is possible that the insured made copies of the keys, which are then provided to a third person who steals the car. Also, the owner might give the original keys to that person, missing an original key or a set of original keys after the facts. These are very common fraud scenarios. See Chapter 10 in this regard.*
- Was the vehicle's registration card in the vehicle?
- Did anyone have any personal items in the vehicle? If so, please establish a complete list of these items and provide the receipts. *It is very important to control the receipts that are provided to the insurance company. The insured might try to take the opportunity of this theft to get reimbursement for items that he or she never possessed. Also, some insureds do not have the item's receipts and might just manufacture fake receipts. See Subsection 19.4.4 in this regard. The presence of many objects in a vehicle that are of very high values, such as several leather jackets or many cameras, should be considered with suspicion and must be controlled with extreme details.*
- Who is the primary driver of the vehicle (name and address)?
- Has this person been the primary driver since this vehicle was purchased?
- How many kilometers or miles was the odometer of the vehicle reading at the time of the theft?
- Does the insured have a homeowner's or renter's insurance policy?
- If so, provide the insurance company's name and policy number.
- During the last 10 years, did the driver, discoverer, or insured have any owned or driven vehicles that were stolen? If so, where, when, and under which circumstances did these occur (including exact description of the vehicle with license plate numbers). *It is possible to obtain more information from law enforcement agencies if the person is known as an offender or as a victim in other crimes.*
- During the last 10 years, did the driver, discoverer, or insured obtain any compensation for stolen vehicles or for items that were present in a vehicle that was then stolen? *In case of negative response, it is important to still request information from other insurance companies whether compensations were paid or not.*
- If so, which insurance company(ies) handled the cases (address, policy and claim numbers)?
- How did the driver get back to his or her residence after the vehicle was stolen? What type of transportation was used? Please attach any receipts (train, bus, or plane tickets, rental company, hotel night receipts, etc.).

- Did the driver, primary driver, or insured have any other vehicle insurance policies? If so, with what insurance company(ies) (include the policy number(s))? *It has happened that the insured had the same vehicle insured through several insurance companies to obtain compensation more than once.*
- Did the driver, primary driver, or insured already have any claims in liability with the vehicle? If so, with what insurance company and when?
- Did the driver, primary driver, or insured have any claims in partial or full coverage for any vehicles? If so, with what insurance company and when?

It is clear that some questions will change depending on the circumstances. For example, the first question would not be applicable if the vehicle was carjacked. The questions presented here provide an idea to the reader of the types of question asked to the insured.

The questionnaire comprises several pages. The questions are not asked in order, which is done so that the interviewee may not be able to provide consistent answers if the story was fabricated. Also, the questions might be repeated or different and additional questions might be asked depending on the circumstances. Also, different people might be interviewed. Thus, it is important to have the page number and the total number of pages appearing on each page of the transcript. At the bottom of each page, date, location, and insured or interviewee's signature must be present.

In addition, if the theft occurred in a foreign country, the following comment must appear at the end of the questionnaire: "If the theft occurred in a foreign country, you are requested to notify the police of your location of residence. This is an important step, which ensures that the stolen vehicle will be searched in our country. If this step was already performed, it is important to know which police department was notified. If not, the police department must be notified immediately."

Until all the necessary documents are provided by the insured to the insurance company, the time period allowed by law for the latter to pay the insured does not commence. Therefore, it is in the insured's best interest to cooperate and to properly answer the questionnaire sent by the insurance company as quickly as possible.

19.3.4 Procuration

Aside from the questionnaire, the insurance company requires the insured to provide a procuration, giving consent to the company to gather information from whoever (people or entities) it would deem necessary to process the claim. Figure 19-1 shows an English translation of such a procuration.

This document allows the insurance company to obtain information from several different companies such as banks, phone companies, and law enforcement agencies. Obtaining phone records can provide very useful information, such as the commonly dialed phone numbers, along with dates and times of phone calls, and also phone calls performed just before and after the theft. Also, this document allows the insurance company to request

Policy No	0000-XX
Claim No	0000-XX
Date of Loss	December 31, 2004
Insured	Last name, first name, date of birth, address
Vehicle	Make and model, license plate number, VIN

Procuration

The aforementioned person authorizes *XXX insurance company* to request all information deemed necessary by *XXX insurance company* to any other insurance companies, department of motor vehicles (including driver's license records), all administrative entities, telecommunication companies, vehicle manufacturers, and other private institutions. This document revokes any physical person's, corporation's, administration's, or institution's obligation to protect personal data. In addition, this document authorizes *XXX insurance company* to forward any information in its possession to third parties.

Date and Location **Signature**

Figure 19-1

Example of a procuration that the insured is required to sign for the insurance company. This document allows the insurance company to request and collect information from phone companies, banks, and insurance companies. It is a release of information from the personal data protection regulations.

information from other insurance companies to obtain claim history for the insured without any risks of being in contempt of personal data protection regulations. It is clear that if the insured refuses to sign this procuration, the insurance will claim that it is not in possession of all necessary elements regarding the handling of this claim. In such an instance, the company will typically deny the claim.

19.4 INVESTIGATION

19.4.1 *Verification of the Questionnaire*

It is very important for the insurance investigator to have a clear understanding of the circumstances surrounding the disappearance of the vehicle. This is the reason why it is crucial to obtain as many details as possible regarding the whereabouts of the insured or the person that was in control of the vehicle at the time of its disappearance. After a careful control of all the statements and information collected, the investigator should visit the site of theft and make all necessary verifications regarding the statements of the people involved, directly and/or indirectly, in the theft. The keys of the vehicle should be provided to the insurance company and their authenticity should be verified as well as whether or not copies have been made; this is addressed through proper forensic examination.

19.4.2 *Investigation Methodology*

It is very important to conduct the investigation in a methodological manner and to act from general to particular.

A/ Contact with Law Enforcement

Contact must be established first with the police from the location where the theft occurred. The investigator should meet with the police officers handling the theft while he or she is visiting the site of theft to verify the information provided by the insured in the questionnaire. If the theft occurred in a foreign country, it is important to determine how the system works in that country and how auto theft crime is handled. There is usually a very dense network among insurance companies, which facilitates exchange of information. If the insurance company is also present in that foreign country, obtaining information is even easier. Also, another very fruitful alternative in obtaining information is to go through the membership of international associations. The International Association of Auto Theft Investigators (IAATI) and the International Association of Special Investigation Units (IASIU) are the most useful in obtaining information in foreign countries regarding auto theft cases [5–7]. The author successfully answered many questions regarding claims in foreign countries by contacting fellow members of the IAATI.

Also, it might be pertinent to contact the police department from the location of residence of the insured. This may reveal other interesting information, such as prior criminal records or suspicious activities in the neighborhood in which the insured might be involved.

B/ Contact with Other Insurance Companies

It is important to contact other insurance companies to verify that the insured did not already file a claim for the same theft. This type of fraud is sometimes practiced by insureds who, if successful, obtain compensation from both insurance companies. Insurance companies in Switzerland have developed among them, and in accordance with the Swiss Insurance Association (SIA), a network of people authorized to provide information [8]. This network, called the Zentral Information Stelle (ZIS), is operated by the SIA. Because of the Swiss law on personal data protection, the communication of this information is done in a confidential manner; otherwise, legal consequences would arise. If this information is to be used against the insured, it must be communicated in writing. However, if the investigator leading the interview does an excellent job, this information does not have to be requested in writing, because the interviewer would be able to lead the insured to admit this information by him or herself. Once this information provided by the insured is in the hands of the insurance company, it can be used without any problem.

C/ Use of a Private Detective

If it becomes necessary, a private detective is hired to follow the insured for a certain period of time. This allows the company to determine with which bank(s), repair shop(s) and car dealer(s), and, perhaps, other insurance company(ies) the insured is involved. In some instances, the repair shops or car dealers are known from the police or from the insurance industry as being part of some sort of fraud scheme. When suspicion arises regarding the genuineness of the claim, it is good to have an idea of the insured's whereabouts.

D/ Forensic Examination of Keys

In the scope of the investigation, the whole set of keys collected by the insurance company should be forwarded to a forensic examiner and/or to the manufacturer of the vehicle (particularly when dealing with keys including transponders) as soon as possible. The forensic examination of keys (as explained in Chapter 10) may reveal whether copies of the keys were made. In such instances, it might contradict the statements of the insured, who was requested to provide all keys (and copies) of keys in his or her possession and who signed a document attesting that no (other) copies were made.

Only the manufacturer of the vehicle can verify that the keys belong to the vehicle that was stolen. Also, some keys include the mileage of the vehicle at last start recorded on their electronic components [3]. This element can be extremely pertinent, particularly with leased vehicles that are stolen right before the end of the lease. In some instances, these vehicles present a mileage much higher than allowed by the leasing contract, which would engender an important amount to be paid by the vehicle's insured at the time of the restitution of the vehicle, thus creating a motive. For example, an insured leases a vehicle for CHF 1,800 (about USD 1,400) per month. The vehicle's value is above CHF 100,000 (about USD 77,000) and had a lease term allowing a maximum of 10,200 km per year for four years. Thus, a maximum of 40,800 km is allowed at time of the restitution, with every extra kilometer charged at CHF 0.89 (about USD 0.70). If the insured drove only an extra 10,000 km (about 7 km more per day), he or she would have to pay an extra CHF 8,900 (about USD 7,000) at the time of the restitution. Thus, in this instance, it would seem much easier to "sell" the vehicle to the insurance company by having it "stolen".

E/ Control of Phone Listings

It is very important to scrutinize the different calls received and dialed by the insured prior to, during, and after the theft. It is possible to request a list of the phone calls received and dialed from a given phone number from the phone operating companies. The phone companies must, by law, comply with this request when the proper procuration is provided to them [9]. However, it is not presently possible to know the names and addresses corresponding to the phone numbers, unless a judge requests the discovery of this information. Indeed, only a judge can request the information regarding the exact location of a cell phone during the phone calls, or at any time for that matter. However, the list of phone numbers can provide a very good indication of the insured's communications in many cases, as shown with the example in Subsection 19.4.4.

19.4.3 Checklist

To ensure that no element is left behind by the insurance claim specialist or special investigator, a check list was developed as shown in Figure 19-2.

Suspicions are not always obvious. However, some events or particularities of the case may allow the investigator or claim handler to believe that the case might be suspicious in nature.

Stolen Vehicle Checklist

Claim No 0000-XX **Name of Insured**

Location of Claim Switzerland | Other — Dates of notification/creation

Document Requested — Received: yes / no

- ⇨ Procuration
- ⇨ Questionnaire
- ⇨ Vehicle registration
- ⇨ Vehicle keys — number
- ⇨ Service/maintenance booklet
- ⇨ Emission control booklet
- ⇨ Title
- ⇨ Bill of sale
- ⇨ Leasing agreement
- ⇨ Vehicle accessories proof documents
- ⇨ Stolen items proof documents
- ⇨ Travel costs proof documents
- ⇨ Police report of theft location
- ⇨ Police report of residence location

Follow-up — Completed: yes / no

- ⇨ Police report from another country (translation, if necessary)
- ⇨ Police report of Swiss police
- ⇨ DMV notification
- ⇨ Credit verifications
- ⇨ Control of keys' authenticity and copies
- ⇨ Cancellation of vehicle search, if recovered
- ⇨ Other requests

Controls — Completed: yes / no

- ⇨ Policy (accessories, contract, ..)
- ⇨ Other insurances
- ⇨ Premiums payment history
- ⇨ Prior claims
- ⇨ Insurance companies information center (ZIS)
- ⇨ Have all the documents requested to the insured been received?

Transmission to Head of Service

- ⇨ In any instances, if irregularities are noticed
  In case of suspicion of fraud

- ⇨ **In any instances, before claim settlement**

Head of service approval — Decision: **yes** / **no**

Figure 19-2

Example of a checklist used to ensure that the handling of the auto theft claim is complete.

In general, it is a good practice to be wary of an insured who had a claim right after the establishment of his or her policy. Also, if the frequency of claims is significant (for example, three stolen vehicles in two years), there is ground to undertake serious verifications of the authenticity of the thefts. If the theft occurred before the insurance premium is paid, right before the expiration of the policy, or just after the payment deadline expired, it would require further investigation. By law, Swiss insurance companies must cover the policy up to 15 days after the collection notice. For example, if the collection letter for the premium was sent on July 15, the premium was paid on August 1, and the theft occurred on July 25, a serious investigation should be carried out. Section 19.7 of this chapter presents some fraud indicators, which are used to determine whether the claim is suspicious or not. There are many fraud indicators, and they vary among countries due to differences in culture, social behavior, laws, and regulations.

19.4.4 Case Examples

Figure 19-3 shows an example of a fake receipt that was provided to the insurance company as a proof of items present in a Volkswagen Polo that was stolen in Geneva, Switzerland. The car was stolen with allegedly two coats, which value totaled CHF 5,500 (about USD 4,200). When the insured was asked to provide a receipt, the document in Figure 19-3 was received.

This document appears to be a receipt for two shawls rather than coats from a store in Srinagar, India. Interestingly, the rate column shows the letters "C.H.F."(international abbreviation for Swiss francs). After interview with people familiar with the region of Srinagar, India, it was determined that it would be impossible to pay with Swiss francs there, that it is not reasonable to believe that the store's owner would know the Swiss francs abbreviation to be CHF, and that the price of two shawls in Srinagar could never reach such a ridiculous amount. The investigation from the insurance company revealed that the cost of two shawls in Srinagar would more reasonably correspond to INR 5,500 (Indian rupees) (about USD 130) and that the letters CHF were likely added at a later date on the receipt. The claim was denied. If further forensic examination was to be conducted, this document would be submitted to a questioned document examiner to determine whether the ink used to write the letters CHF was the same as the ink used to write the remaining text.

In another example, a client notified the theft of his Porsche 911 Turbo around 22:00 in Geneva, Switzerland. While looking through the list of phone calls made from his cell phone provided by the phone company, it is possible to find the call made to the police that evening at 22:00:34. In order to justify his presence about 60 km (40 miles) away from his residence that evening, he provided the insurance company with a restaurant receipt, where he had dinner with his wife. The restaurant is located downtown, about two minutes on foot from where the vehicle was allegedly parked. The time printed on the restaurant receipt is 21:56, thus four minutes before the call was made to the police. In such conditions, it was necessary to go to the site and retrace the activities of the insured before the discovery and notification of the theft. It was important to determine whether the receipt was brought to his table right after being printed, if he paid for it right away or not, if he left the restaurant normally or in a hurry, and, of course, if the restaurant's time on the cashier machine matches the time

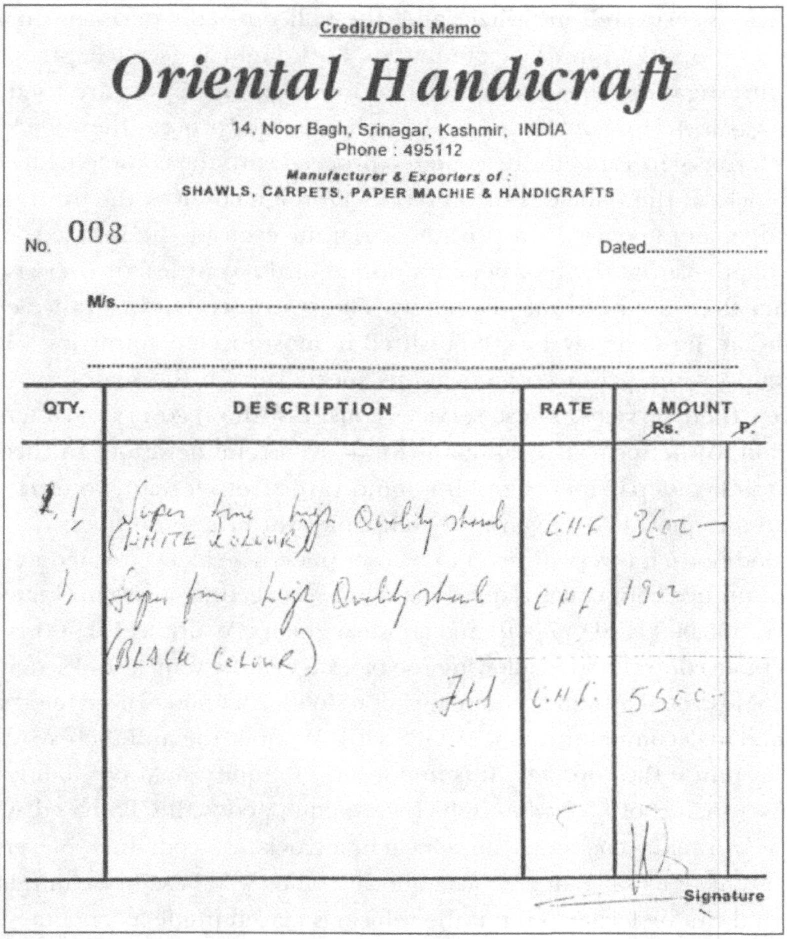

Credit/Debit Memo

Oriental Handicraft

14, Noor Bagh, Srinagar, Kashmir, INDIA
Phone : 495112

Manufacturer & Exporters of :
SHAWLS, CARPETS, PAPER MACHIE & HANDICRAFTS

No. 008 Dated.......................

M/s...

QTY.	DESCRIPTION	RATE	AMOUNT Rs.	P.
1	Super fine high Quality shawl (WHITE COLOUR)	C.H.F	3600	—
1	Super fine high Quality shawl (BLACK COLOUR)	C.H.F	1900	
	Total	C.H.F.	5500	—

Signature

Figure 19-3

Receipt received by an insured to justify the reimbursement of two expensive coats allegedly present in a vehicle that was stolen. This receipt was counterfeited. The letters "C.H.F." were very likely added to the receipt to increase the value of the two shawls (not coats) from 5,500 Indian rupees (USD 130) to 5,500 Swiss francs (USD 4,200).

from the phone company records. After this investigation, it was determined that this timeline was extremely suspicious. In addition, the keys to the vehicle were sent to Porsche for verification. Porsche responded to the insurance company, indicating that the keys provided could not, in any case, start or even open the vehicle allegedly stolen. The keys did not correspond to the vehicle identification number (VIN) of the vehicle. The claim was denied.

19.4.5 Recovery of the Vehicle

It is primordial to know under which circumstances the vehicle was retrieved. In Switzerland, a stolen vehicle can be taken very quickly across the border and disappear. This strongly contributes to the low proportion of vehicles recovered in Switzerland.

If the vehicle is recovered in Switzerland, the police usually perform the investigation, including a very thorough forensic examination of the vehicle (see Chapters 3 and 4). The crime scene investigation report and criminal investigation report are used by the insurance company as is. If, for some reason, the police do not process the vehicle, it is advised that a private forensic investigator be called to proceed with the complete forensic examination of the vehicle. If the vehicle is recovered in a foreign country, the insurance is notified by a law enforcement agency, by a private citizen, or even by the insured. As a matter of fact, it often happens that the first person notified of the vehicle's recovery is the owner or insured, rather than the insurance company. The reason is that insurance companies are not considered at the same level as the insured in most foreign countries.

More recently, some private organizations specialized in the search of stolen vehicles started to offer their services. These services consist of the recovery of stolen vehicles and their restitution to the insurance company for a given remuneration. In this fashion, they regularly visit police departments and impound yards from foreign countries, check VIN, and then send a request to all insurance companies with these numbers.

In these conditions, it is very difficult to recover physical evidence related to the theft itself. More often, insurance companies require for vehicles exceeding a certain value (for example vehicles of CHF 100,000 [USD 77,000] and above at Phenix Assurance) that they are equipped with both a remote control system allowing the blockage of the vehicle and systems giving their locations by GSM (Global System [or Standard] for Mobile), a standardized international system for digital mobile telecommunication, or GPS [10]. Without the anti-theft systems, insurance companies may refuse the coverage. It is important to require dual possibilities of localizing the vehicle. As a matter of fact, when vehicles are equipped with GPS-based systems, thieves counteract this by rapidly enclosing the vehicle in a truck or a container, preventing the positioning of the vehicle by satellite (see Chapter 20). The GSM-based system does not present this problem and can be located even if the vehicle is placed inside a container or truck [10].

19.5 CLAIM SETTLEMENT

19.5.1 Payment of Claim

According to the general conditions of Phenix Assurances, if the vehicle is not recovered within a month of its disappearance or theft and if the insurance company is in possession of all necessary information and documents, the insurance is required to pay the claim to the insured. The amount of the claim is estimated by the insurance company, either via an internal insurance adjuster or via a public adjuster. When referring to all necessary information and documents, these include all the documents requested to the insured and other companies as described earlier in this chapter as well as the results of all investigations and technical examinations performed on the vehicle. Hence, it is extremely rare that these documents and information are collected within a month, thus preventing the settlement of the claim. A time span of three to four months is not unusual before it is possible to settle the claim that did not present any suspicion. The main lines of the process described in this chapter do not vary greatly between the different insurance companies in Switzerland and in Europe in general.

19.5.2 Discovery of the Fraud

If the theft is discovered to be a fraud at a later date, the insurance company can file a legal, both criminal and civil, complaint through the authorities. This allows for recuperating the amount of the claim payment and the administrative proceedings of claim handling and court proceedings.

In most instances, if the fraud was performed by the insured because of a serious need for cash flow, the amount paid through the insurance settlement has already been spent and is not available anymore. If, in addition, the insured is sentenced to jail time, it is impossible for him or her to earn enough money to reimburse the insurance company. The insurance company has very little chance of recouping its losses. For this reason, before settling any claims, it is important to make sure that all dispositions have been taken and that the investigation was performed in a very thorough and comprehensive manner. Discovering a fraud after the claim has been settled is a situation in which an insurance company does not want to see itself.

ACKNOWLEDGMENTS

The author would like to thank Eric Stauffer for translating this chapter into English.

BIBLIOGRAPHY

[1] Junghans P. (2004) *Halte au trafic: La mafia des voitures,* JC Lattès, Paris, France.

[2] Author unknown (2004) Justice and home affairs: Car stolen every 26 seconds in the EU, *European Report* of February 7, 2004, available at http://www.eb-iaati.org, last access performed on July 10, 2005.

[3] Wolff H. (2004) *Diebstahlschutz für Pkw: Vom Lenkradschloss zur elektronischen Wegfahrsperre,* Allianz Suisse SIUs semiannual meeting, Bern, Switzerland.

[4] Recueil Systématique (1958) Attestation d'assurance, suspension et cessation de l'assurance, *741.01 Loi fédérale sur la circulation routière (LCR),* Article 68, Alinéa 2, p 31.

[5] International Association of Auto Theft Investigators (2005) *What is IAATI?,* available at http://www.iaati.org, last access performed on July 10, 2005.

[6] International Association of Auto Theft Investigators European Branch (2005) *Who we are,* available at http://www.eb-iaati.org, last access performed on July 10, 2005.

[7] International Association of Special Investigation Units (2004) *IASIU mission,* available at http://www.iasiu.org, last accessed performed on July 10, 2005.

[8] Swiss Insurance Association (2003) *About us,* available at http://www.svv.ch, last access performed on July 17, 2005.

[9] Recueil Systématique (1992) Droit d'accès, *235.1 Loi fédérale sur la protection des données (LPD),* Article 8, pp 3–4.

[10] Reco Infomobility (2004) *Redco S3 satellite location and protection terminal product data sheet,* available at http://www.redcoinfo.it, last access performed on July 17, 2005.

CHAPTER 19

Part II: The US Perspective

Glenn Wheeler

19.6 INTRODUCTION

Most insurance companies today have special investigation units (SIUs). Historically, SIUs have been in place since the early 1970s and were predominantly on the property side of the insurance business. That was primarily driven by the fact that every state in the United States had arson against property statutes and each state carried a requirement that suspected arson to property is to be reported to the appropriate division of law enforcement. All those statutes have some form of immunity for the party reporting the alleged arson. Because of these reporting requirements and immunities, there was assurance that such information would be reported and properly investigated by the authorities and, where appropriate, prosecuted. Before SIU development in the insurance industry, suspected arson investigations were conducted by independent companies hired by the policy-issuing company. Gradually, the insurance companies began to identify some of their employees as specialists in arson investigations. These early SIUs were poorly organized due in large part to apparent lack of knowledge and support from the management of the insurance companies.

In the late 1970s and early 1980s, SIUs began taking on a whole new complexion. Auto theft was on the rise, and insurance premiums were following that same pattern. The insurance-buying public demanded that the insurance companies do something about the price of insurance. States began to pass statutes requiring companies writing insurance within their state to staff an SIU to address the auto theft and fraud problems. Those demands and requirements created an explosion in the number of SIUs across the country. They are now in place to investigate claims in all lines of insurance, including property, casualty, and workers compensation.

Auto theft, arson, and insurance fraud are no longer the sole responsibility of law enforcement. SIUs have assumed the primary role of investigation, and this resulted in advanced training for the insurance claim representative and law enforcement, which in turn has improved communication between the private and public sectors. In many cases, efforts to combat fraud have resulted in assigning prosecutors dedicated to the prosecution of these types of crime.

Generally, fraudulent claims are made because of financial stress. For example, in most cases it is far easier and more profitable to settle a theft claim with an insurance company than it is to sell a used vehicle. Investigations of possible frauds explore the financial background of those who are involved in suspicious claims. Once an event occurs (a loss is reported), the wheels are set in motion to evaluate the claim for damages. The claim representative receives the notice of claim from different sources depending on the configuration of the insurance company and its reporting procedures. The information gathered from the report determines the kind of loss that occurred, the kind of coverage involved, and the applicable limits.

This chapter presents the systematic and thorough investigation performed by the insurance company in the United States, which leads to the most appropriate decision on a claim. That decision may be payment or denial. Either way, the correct decision is the best result for all policyholders.

19.7 INDICATORS OF FRAUD

Indicators of fraud are sometimes referred to as "red flags" or "suspicious loss indicators." Over the years, the National Insurance Crime Bureau (NICB) has developed a very complete list of suspicious loss indicators [1]. Copies of those indicators can be obtained by simply contacting the NICB [2]. Every claim representative and special investigator should be trained in the recognition of these indicators, because they eventually determine the need for a thorough investigation. Without recognition of these indicators, possible fraudulent claims will go undetected.

A single indicator would not necessarily mean a fraud. Part of the insured behavior and some circumstances that are described as indicators can actually be perfectly justified by the insured outside the scope of a fraud. However, clusters or multiple indicators should trigger a thorough and systematic investigation by an SIU or other trained claim representative.

In the automated world, it is now possible to electronically identify suspicious losses through the use of rule-based systems. ISO Properties, Inc. has developed ClaimDirector, a predictive modeling tool that identifies potentially fraudulent claims when certain events occur in the presentation and investigation of the loss [3]. SAS has developed Enterprise Miner, a software package with models that allow a company to implement their own suspicious loss indicators [4]. These types of products have been in use by credit card companies and banks for some time and are growing rapidly in the property and casualty arena. Where these products are currently in use, they prove themselves to be efficient and accurate. Regardless of the method used, it is the indicators that trigger an in-depth investigation, not the ethnicity, geographic location, or occupation of the insured party. It is important to always maintain objectivity and focus during an investigation.

The NICB auto theft indicators include not only some general indicators, but also some concerning the insured, related to the vehicle, coverage, and reporting [1]. The following is a very brief list of some of these indicators:

- Vehicle has a good anti-theft device.
- Vehicle is a leased vehicle with very high mileage.
- Vehicle has a (recent) history of mechanical problems.

- Vehicle was recently for sale.
- Coverage was obtained via walk-in business to agent.
- Insured claims expensive contents in vehicle at time of theft.
- Insured recently called to confirm and/or increase coverage.
- Insured is behind in loan payment on vehicle and/or other financial obligations.
- Recovery is made with no ignition or steering lock damage.
- Vehicle is recovered with seized engine or blown transmission.
- Vehicle is recovered close to scene of theft, by the insured or a friend of the insured.

It is recommended that any insurance claim specialist obtain a complete listing of these indicators from the NICB.

19.8 INVESTIGATION

19.8.1 Principle

First, one must ensure that a systematic procedure is established to handle the investigation. A thorough review of the reported loss must be carried out. The report must contain all the required information. If anything is missing from the report, the reporting agent should first be contacted to see whether someone has the information and simply failed to report it or if this information must be sought through the insured or other third parties. If there is a police report, it should be included in the documents. To be in compliance with the insurance contract, the insured must report the theft to law enforcement as soon as possible. The essential point is to make sure that all questions have been addressed and that nothing was left to conjecture. It is vital to be thorough in the investigation. Reluctance to asking questions could lead to a lack of pertinent information.

19.8.2 Examinations

A/ Scene Investigations

By the time the investigator is notified of a vehicle theft loss, it is likely that several scenes will require examination. The loss site must be visited. The vehicle could have been reported stolen from a residence, a mall parking lot, an employee's parking lot, or any number of locations. An examination of that site will determine whether there is any evidence to aid the investigation. For example, if a vehicle was reported stolen from a residence, the owner should be asked to identify where the vehicle was last seen. By reviewing that location, a determination may be made that there was forced entry to the vehicle such as broken glass found on the ground, or forced entry in the garage, or no forced entry anywhere at all. While at the loss site, a canvass of the surrounding neighborhood should be conducted to identify potential witnesses to the loss (see Chapter 2).

The recovery site of the vehicle can also provide evidence important to the investigation. The manner in which the vehicle got to the site might be identified. If the vehicle was burned, there will be evidence of where it sat during the time it was burning. There may be evidence of what was used to burn the vehicle and how it was burned (see Chapter 12).

B/ Vehicle Examination

The recovered vehicle must be inspected. First, one must ensure that it is the vehicle insured under the policy. Then, an estimate of the loss must be performed to determine whether the vehicle is a total or a partial loss. In this last case, the owner's permission to move the vehicle to a repair facility must be obtained. While inspecting the vehicle it should be possible to quickly establish how the vehicle was stolen (see Chapters 8 and 9). Photos of the exterior and interior of the vehicle should be obtained to clearly identify the condition of the vehicle and if there is any preexisting damage. It always important to keep in mind that a full forensic examination of the vehicle can be conducted and can bring very pertinent information (see Chapters 3 through 9). If this examination is to be performed, the vehicle should not be touched until the forensic expert is on site and has control of the vehicle. If anyone would touch the vehicle prior to the arrival of the forensic examiner, crucial pieces of physical evidence might be contaminated or lost. After this examination, the insurance adjuster or claim representative may proceed to further examination. If, for any reason, no full forensic examination is performed, the insurance personnel can conduct a simple test of the vehicle, by trying to turn the steering wheel and move the shift lever (with automatic vehicles). This kind of examination reveals if either of those items is locked. It is also possible to use the services of a forensic expert to determine whether the ignition was defeated or a key was used to steal the vehicle (see Chapters 8 and 9).

C/ Collection of Evidence

During the scene investigation, any evidence discovered should be gathered and properly identified. If not preserved, the evidence may change or be removed. A chain of custody must accompany every item of evidence. It cannot be stressed enough that collection of evidence might lead to a claim of spoliation of evidence from an opposed party, particularly if the collection was not properly executed. It is always best to have recourse to the services of a forensic scientist when dealing with examination, recognition, and collection of evidence. However, this is not always possible. The steps below should be carefully followed to maintain the integrity of the evidence:

I Photograph the evidence in place before moving.

II Identify the evidence with an identity tag to include

- i Claim number;
- ii Date of loss;
- iii Name of insured party;
- iv Date of discovery;
- v Location of discovery (use photos and diagrams).

III Select a secure location to store the evidence.

19.8.3 Use of Expert Services

The recovery of a reported stolen vehicle often requires the services of forensic or other experts. If it is suspected that a stolen-recovered vehicle has suffered mechanical failure before or after the reported theft, a mechanical expert should be contracted to analyze the mechanical parts in question. If an engine failure is suspected, there are some preliminary tests (vehicle fluid evaluation) that the SIU investigator can conduct to help determine whether a mechanical expert should be used (see Chapter 11). Ignition lock experts with knowledge in manual locking mechanisms as well as the more sophisticated computerized ignition systems should be considered whenever there is a suspicion that the vehicle in question may not have been stolen but was an arranged disappearance (see Chapters 8 and 9).

Before hiring an expert, it is important to make sure that the experts have proper credentials and, where required, proper certification and/or licensure. Certifications and licensures require experts to periodically attend training courses and to keep up with the development in their field of expertise. At least once a year, it is good practice to verify the expert's attendance to training sessions and their current licensing requirements. Nevertheless, many disciplines do not necessarily offer a certification program or a respectable certification program. A noncertified expert does not necessarily mean that he or she does not have competency in their fields. It is imperative not to be blinded by the "certification" label and look with more detail at the resume of a potential expert.

19.8.4 Documentation and Authorizations

A/ Activity Log

An activity log must be kept, briefly describing what is being done in the investigation. The log should provide the date and time of the activity and should be initialed by the investigator. There must not be any gaps in the entries. For future reference, the log will provide a clear outline of what was done during the investigation and what, if any, information was gathered.

B/ Authorizations

Automobile insurance policies contain language giving insurance company representatives the right to inspect the vehicle to verify that it is the vehicle insured and to verify damages. But just how far does that right go? Logic would dictate that anything in clear view is not private and the owner has no right to assume privacy. However, it might be a privacy issue to open a briefcase found in a vehicle and to inspect the documents in that briefcase. It is imperative to know what the case law decisions about privacy are and if a search of documents is a violation of privacy issues.

To avoid conflicts with privacy issues, it is good practice to obtain permission from the property owner to inspect the property in question and to remove any evidence found in or on the vehicle. Permission can be gathered verbally; however, a recorded or written permission statement is preferred. A defense attorney familiar with insurance contracts and privacy case law issues is the most qualified to help determine whether an authorization is needed.

19.8.5 Sources of Information

A/ Insurance Records

Underwriting records of the current insurance company may contain information critical to the investigation. A copy of the original application should indicate previous insurers, licensed drivers in the household, loss histories, and additional information. Misrepresentation in the application could be an intentional act to disguise a fraud or simply a mistake on the part of the applicant. If the insured had a prior insurance, the request of information from that insurance company can provide information on his or her loss history, which can show a pattern of possible fraudulent losses. A good example is repetitive vehicle theft losses.

If it is believed that there is misrepresentation in the application, a determination must be made if the misrepresentation is material to or has a direct effect on the claimed loss. If it is determined that the misrepresentation is material to the claimed loss, two options other than payment are available. An insurance policy can be rescinded (void ab initio), which simply implies that the insurance contract is canceled retroactively, as if it was never initiated. This is done because the applicant made misrepresentations in the application that, had the truth been known, would have caused a rejection of the application. A good example is an applicant indicating they had never suffered any losses with a previous insurance company, but a review of the underwriting records of the previous company shows three vehicle theft losses. This is something the applicant knew or should have known but did not reveal. Because of current underwriting rules, the new insurance company would not have accepted the new applicant if the previous loss history had been known. Voiding coverage under a policy is triggered when an insured makes a misrepresentation (fraud) in the presentation of the claim that is discovered during the investigation of a claim. A good example is when an insured indicates that their vehicle was stolen and a subsequent investigation determines that no theft occurred but the owner participated in the vehicle's disappearance. For a misrepresentation to trigger a rescission or a voidance, the misrepresentation must be material or must have a direct bearing on the issue. A decision to rescind or void a policy should be made with the advice of counsel.

B/ Agents or Brokers

Agents or brokers who have produced the business should be questioned about their relationship with the parties to the claim. They might be personally familiar with the insured or not at all. Sometimes it is pertinent to know how they were introduced. For example, if it is through a referral, it would be interesting to know who made the referral and what is the claim history and relationship of the person making the referral.

C/ DMV Records

Motor vehicle records (MVRs) can be obtained through a DMV or Secretary of State offices. These records contain information on vehicle accidents and thefts. It is not uncommon to find that an insured suffered a vehicle accident on the same date as a reported theft. MVRs may contain information missed in the underwriting process and can contain information leading to a rescission.

D/ Title

Title history should be a routine part of every auto theft investigation. First, it must be ensured that the vehicle insured is owned by the party(ies) making the claim. There might be a lien holder or other owners involved. The vehicle might have been branded as salvage, is a rebuilt vehicle, or the real mileage does not match the mileage claimed by the owner (given the time of ownership and probable miles driven). Also, the identification of the vehicle might be questioned after reviewing title information (see Chapter 6).

E/ Public Records

Public records are available to everyone over the Internet. A search of public records should include verification of addresses of the parties to the loss. Telephone numbers should match the address given for the phone number. A phone number search may help identify other parties listed to that phone who have experienced similar losses as those claimed by the parties in the investigation. There are literally dozens of companies providing the service of public records searches. Equifax and TransUnion are two of the largest companies offering automated searches of public record and can be a very efficient way of conducting searches [5, 6].

19.8.6 Interviews

A/ Principle

Interviews and statements can be recorded or handwritten. Recorded statements are more efficient and are generally more thorough, capturing every word that was spoken. Regardless of the method used, opportunity to obtain information through interviews and statements should not be neglected. In some instances, an interviewee's attitude changes over time, as well as the recall of his or her memories. A reconstruction of the loss can be made by asking what the facts of the loss are, when and where the loss occurred, why was the vehicle there, who was at the scene of the theft, where the parties to the loss were going, how they got there, and by which route.

Anyone identified as an insured under the policy should be interviewed, and when appropriate, a statement should be obtained. Any parties with a monetary interest in the claim should be contacted to provide a statement or, at the very least, an interview. For example, a lien holder on a stolen vehicle should be able and willing to provide a record of monthly payments and whether the payments are current. This can reveal a default or a pending repossession.

B/ Statements Under Oath

SUOs are provisions provided for in all property and casualty insurance policies and can generally be found under the duties of an insured. The language is styled as follows: "an insured will answer questions under oath when asked by anyone we name, as often as we reasonably ask and sign copies of the answers" [7]. In most instances, an attorney selected by the investigator should conduct the SUO. SUOs and depositions are similar in nature because they both have a court reporter present and recording the interview. However, unlike a deposition, an attorney representing the party providing the statement cannot examine the witness or instruct the witness on a response to the question. Because SUOs are a contractual obligation, if the insured party refuses to provide a statement, they have created a breach in the contract and the contract can be voided. It is essential to be aware of the position of the courts on voiding a contract.

C/ Other Considerations

Joint owners to the property will help to determine who has an interest in the loss and, if so, what that interest is. Payments or denials must be conveyed to all parties with an owning interest in the property. Laws differ by jurisdiction. The party reporting the loss, if other than an insured, should be contacted and interviewed. Quite often, observations believed as unimportant could help with the investigation.

Neighbors are a good source of information. They hear "rumors" and see evidence of problems, and every neighborhood seems to have at least one party who knows just about everything that is going on. Verification of all information must be made whether factual or rumored.

Effort should be made to become acquainted with key people in law enforcement. A good connection to law enforcement may be through the NICB. An opportunity may exist to exchange information about parallel investigations. Also, firefighters are a possible source of information. They could be the first parties on the scene of a stolen burned vehicle, and their observations could be helpful to the investigation.

Eyewitnesses are obvious contacts. Those contacts should occur as soon as possible because time can fade memory. When conducting witness interviews, it is important not to overlook the possibility of children and their observations. Generally, the statements from someone under the age of seven years is considered incompetent; however, their statements may reveal additional information to verify.

19.8.7 Timelines

All statements and interviews should be included with any notes in order to develop a timeline of the loss. This should determine whether all parties interviewed agreed on the time the event occurred until the loss was reported and whether all previous and subsequent intervening events occurred as reported. It is essential to identify any gaps that would have provided someone with an opportunity to participate in the loss or if the reported loss was a fabrication with substantial discrepancies.

The following example is a simple timeline that clearly indicates a discrepancy that needs further clarification. This provides a good opportunity to the insurance company to ask who, what, where, when, why, and how.

April 7, 23:00: Vehicle discovered burned by police.

April 8, 00:30: Time at which the owner parked the vehicle in front of his home after dropping his best friend at a friend's home, according to his statement.

April 8, 10:00: Police notifies the owner of the location of vehicle.

April 8, 11:30: Owner's best friend takes owner to see vehicle.

19.8.8 Financial Background

The investigation should include the gathering of information related to the income of the party(ies) making the claim. When the theft of a vehicle has occurred, the owner should state when the vehicle was purchased and from who, as well as the purchase price and whether it was financed or paid for in cash. Regardless of how the vehicle was purchased, a verification of where the money came from should be made.

What is the income and liabilities of the party making the claim? If cash was paid, does the income seem to support such a large cash outlay? If there was a lease or lien, were payments current or was there an eminent repossession? What other liabilities are there?

Part of the financial worth of an individual is the amount of money he or she can or should anticipate in the payment of a claim, so damages to the vehicle must be evaluated as soon as possible. Even if there is a strong suspicion of fraud, established claim procedures must be followed to prevent accusations that there was a predetermination to deny the claim.

19.9 CLAIM EVALUATION

19.9.1 Reservation of Rights or Nonwaiver

A/ Reservation of Rights

A reservation of rights is a formal letter addressed to the parties of the loss explaining that an investigation of discovered discrepancies will not prevent the insured(s) or the insurance company from exercising their rights under the contract and continued investigation will not imply that there is coverage. The discrepancies remaining unresolved should be spelled out in the letter and tailored to fit the circumstances.

The letter is mailed, certified with a return receipt request, to the last known address of the party(ies) to the claim. The letter should be created and sent as soon as discrepancies have been found and a resolution cannot be made of the discrepancies.

Timing of a reservation of rights letter can be critical to the claim. Some argue that this item should be issued as soon as there is suspicion that the claim may be fraudulent. Others support the philosophy that a letter should not be issued until every possibility for making payment has been explored. Issuing the letter too quickly may imply that there was a predetermination that the claim was a fraud and a complete investigation would have clarified any discrepancies.

B/ Nonwaiver

A nonwaiver, although similar to a reservation of rights, is generally in a standard printed form with blank spaces to fill in with the discrepancies and is obtained in person. This document must be signed and dated.

C/ General Considerations

Both methods require separate notice to each party involved in the loss. A reservation of rights or nonwaiver can also be recorded, requiring a full explanation and having the insured party attest to the statement. Regardless of which method is used, as soon as there is solid information that there is a question regarding the coverage, rights must be reserved.

19.9.2 Fraud Determination

Now that all the apparent evidence has been gathered, it must be evaluated to determine whether there was a motive, means, and opportunity to commit a fraud. To this point, there has been a team effort. Now, the team has to evaluate all the pros and cons. The SIU investigator is the primary person gathering and sorting evidence, and the investigator's supervisor should be providing input. Forensic experts have been providing their input on the evidence they analyzed, which will play an important role in the decision making process. Any attorneys used during the investigation provide their opinion of the loss. Multiple disciplines help the investigator arrive at the best possible resolution.

Does the evidence support an accidental loss to the insured party or a loss that was intentional? If the evidence supports a fraud, a denial of payment will be made as soon as possible. The denial should be conveyed in a letter by registered mail with a return receipt request. The denial letter will spell out why the claim is being denied and will provide the denied party a last opportunity to explain the discrepancies that led to the denial.

19.9.3 Disposition of the Claim

If the final evaluation determines that a fraud has not been committed, payment of the claim is made as soon as possible based upon the earlier evaluation of the damages. After payment is made, one should never overlook subrogation possibilities. If the investigation demonstrates who the responsible party was, the damages may be collectible.

If the decision is to pay the claim, it should be done promptly based upon the amount of damage, which was established early in the investigation. If the decision is to deny the claim, that decision should be conveyed as quickly as possible with an explanation of why the denial was made.

Compromise is not uncommon in the insurance industry; however, when first-party property claims are made and fraud is discovered, there should be no room for compromise. The insurance industry requires a total absence of fraud to allow for full payment of damages.

BIBLIOGRAPHY

[1] National Insurance Crime Bureau (2001) *NICB interactive indicators,* available at http://www.sedgwicksiu.com, last access performed on July 11, 2005.

[2] National Insurance Crime Bureau (2005) *About NICB,* available at http://www.nicb.org, last access performed on July 11, 2005.

[3] ISO (2005) *ClaimDirector,* available at http://www.iso.com, last access performed on July 21, 2005.

[4] SAS International (2005) *SAS Enterprise Miner 5.1 fact sheet,* available at http://www.sas.com, last access performed on July 11, 2005.

[5] Equifax (2005) *Equifax products,* available at http://www.equifax.com, last access performed on July 11, 2005.

[6] TransUnion (2005) *Business solutions,* available at http://www.transunion.com, last access performed on July 11, 2005.

[7] State Farm Mutual Automobile Insurance Company (year unknown) *Car policy form 9822.7,* Bloomington, IL.

CHAPTER 20

VEHICLE TRACKING

Greg Terp

20.1 INTRODUCTION

The advancement in technology has provided law enforcement with more means to recover stolen vehicles, track suspects, and conduct surveillance. This technology has also aided law enforcement managers to reduce personnel costs for overtime while increasing effectiveness. It is important for investigators and management to understand what technology is available, its applications and limitations, and how to fund it.

The most important source of evidence to vehicle theft is the vehicle itself. The sooner a stolen vehicle is recovered, the greater likelihood that evidence will be available. Tracking devices now make it possible to locate stolen vehicles immediately or within hours after a theft. They also increase the possibility of capturing suspects while still in the stolen vehicle.

Traditional investigations of stolen vehicles are time consuming and labor intensive. One of the biggest costs in time and money is conducting a surveillance of an unoccupied stolen vehicle, waiting for somebody to arrest. Tracking devices now enable surveillance teams to respond after a vehicle is started, reducing the time spent waiting. The vehicle may then be followed until it is decided to make the arrest.

Tracking devices are also used to monitor informants and suspects actions. These uses generally require a court order but provide additional intelligence previously only available from a surveillance team. Investigators can also maintain a greater degree of separation from counter-surveillance maneuvers by suspects.

Tracking devices should be used by law enforcement to gain knowledge and information to identify suspects and organizations. If the technology is only utilized with the intent to recover stolen vehicles, it may actually drive auto theft up in a community. If the recovered stolen vehicles are being taken for profit, then another vehicle will be stolen to fill the order. An emphasis must be placed on furthering investigations when possible.

Investigators and crime scene processing personnel should become knowledgeable of the technology of tracking devices and equipment being used in their country, state, and/or region. Because of a concern for suspects learning what the technology looks like and where it might be installed, this information is and should be very guarded. Most companies have a law enforcement liaison that can be consulted when needed. Bomb disposal units have been called to check "suspicious" devices attached to vehicles, only to learn they were law enforcement tracking devices.

20.2 TYPES OF TRACKING DEVICES

20.2.1 Lojack

One of the most popular commercial tracking devices is the Lojack. In 1989, Lojack was authorized by the US Federal Communications Commission to operate their system on a regular basis. By October 25, 2004, Lojack was deployed in 22 states and the District of Colombia and was licensed in 25 additional countries. Lojack estimates that over 100,000 vehicles worth an estimated USD one billion have been recovered using their technology [1]. Systems in other countries include Boomerang in Canada and TRACKER in South Africa [2].

This technology is based upon a wireless system installed in a vehicle. The system is inactive until a police report is made and the stolen vehicle identification number (VIN) is entered into the law enforcement computer system. This sends a signal to the vehicle and activates a transmitter. A signal with a unique identifier is emitted from the stolen vehicle or property.

Lojack has provided law enforcement agencies with receivers that pick up the signals from stolen vehicles. An example of such a receiver is shown in Figure 20-1. These receivers have a direction finding capability that allows officers trained by Lojack in recovery operations to find the vehicle identified by the specific signal [3].

These receivers are placed in vehicles and aircrafts. Because they operate on radio wavelengths and are in line of sight (the radio signals go directly from the source to the receiving antenna), aircraft have less interference from buildings and trees affecting their reception, thus giving them a greater range of signal reception. When an officer first observes a signal, other officers are then notified of the signal. Police communications operators are able to provide information from the signal identifier as to the vehicle's make and style. Officers coordinate their movements until the vehicle is visually observed. If occupied, the officers coordinate their approach and make the arrest. If unoccupied, officers make a determination as to whether to call in a surveillance team or immediately recover the vehicle.

Figure 20-1

Lojack provides equipment to law enforcement such as this receiver being used in an undercover vehicle. Officers are trained by Lojack how to use the equipment to find stolen vehicles.

Lojack signals are effective even when the vehicle is inside a building, warehouse, parking garage, or shipping container. In some cases, when officers locate the stolen vehicle being tracked by the Lojack signal, they find other stolen vehicles.

Lojack is cautious on divulging how and where their equipment is installed in a vehicle or on other property. Heavy equipment, generators, and other property have found it beneficial to have Lojack installed.

With success has come awareness by criminals. Vehicle theft suspects now park stolen vehicles for a period of time in a location such as a parking garage, apartment building parking lot, or on the street away from their operations. They wait to see if the police respond, thus letting a hot car cool off (see Subsection 18.3.2B). If the police do not respond within a certain time period, they then move the vehicle to where they will work on it.

Lojack does not provide street-by-street locations. It is a radio direction finder system designed to recover or locate vehicles. From a consumer perspective, it has been very successful in recovering stolen vehicles and reducing the cost of auto theft to its customers. Many insurance companies provide discounts in premiums with recovery systems like Lojack on an insured vehicle [4]. From a law enforcement perspective, it provides another excellent tool to locate and recover stolen vehicles and property.

One benefit from the introduction and popularity of Lojack, along with other tracking systems, has been the reaction of auto thieves by using cool-off zones for recently stolen vehicles. This has provided law enforcement with knowledge that can be used to further investigations into major targets and organized groups.

20.2.2 Global Positioning Systems

The global positioning satellite (GPS) system was developed for military objectives [5]. US President Reagan directed GPS to be made available for civilian purposes after the Russians mistakenly shot down a Korean Airliner in 1983 [6]. This system can provide a user with a location within 50 feet or less anywhere in the world. It also has the capability to log positions at various time increments determined by the user. A hiker in the woods no longer needs bread crumbs to follow the path home but can see foot by foot where the path is. A favorite fishing spot in a lake, river, or ocean can be exactly marked, including the path through treacherous shoals or reefs, so that a person can return there safely time and again.

GPS technology has changed navigation forever and has revolutionized services in vehicles. No longer do drivers have to be lost without a map: A GPS device, either portable or fixed in a vehicle, can show on a map where the vehicle is presently located and even provide detailed instructions on how to arrive at the desired location. GPS technology is integrated with cell phones to provide the information to the customer or service provider.

GPS tracking devices also help recover stolen vehicles. As the technology became available for commercial development, vehicles were quickly identified as a potential revenue source. These systems can be part of factory-installed systems, like General Motors' OnStar, or can be aftermarket installed.

When a vehicle equipped with a GPS tracking system is stolen, law enforcement may obtain the tracking history and the vehicle's present location from the service provider, if authorized by the customer. Officers can be directed to the exact location the vehicle is at or the last location where the vehicle was driven to. In some cases, subpoenas or even court orders are necessary.

Some limitations occur with buildings, parking garages, and shipping containers, because the satellite signal is blocked. When the vehicles emerge from cover, the signal can be detected again.

When a vehicle is moving, GPS systems can provide the exact street, the direction of travel, and the speed, as shown in Figure 20-2. Most systems are monitored by a service provider who must communicate to law enforcement agencies, which in turn relays the information to officers out on the street. This delay can be crucial during surveillance operations. As technology improves, law enforcement will have the capability to monitor the activity directly from a computer in their vehicle.

The consumer version of a GPS system is monitored at all times. In fact, if a GPS-equipped vehicle is involved in an accident with air bag deployment, the service provider

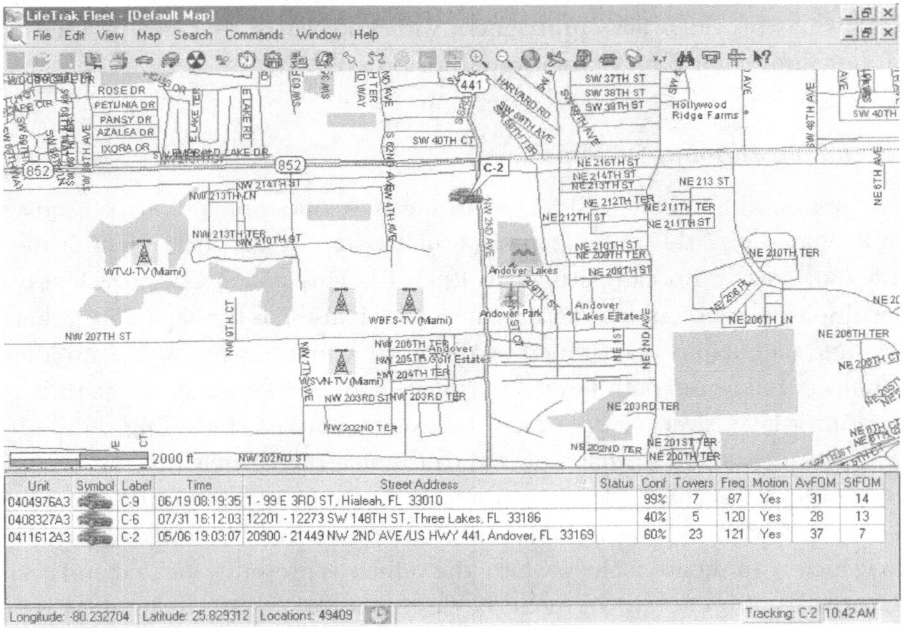

Figure 20-2

Mapping software as seen in this screen shot enables investigators to track stolen vehicles and suspects. Both GPS-based and systems such as Ituran (as shown here) provide the exact location, time, and even speed of a vehicle being tracked. The bottom portion of the screen indicates three active tracks being monitored at one time, C-2, C-6, and C-9. The vehicle C-2 is clearly shown in the center of the map. See Color Plate.

is automatically notified and emergency first responders are dispatched to the right location. Unlike Lojack, where the police must be officially notified before the system is activated, a consumer may notify the service provider immediately if the vehicle is stolen and law enforcement authorities are quickly involved.

Auto theft suspects are very familiar with GPS systems and how they operate. Professional thieves immediately deactivate GPS systems through various means. Law enforcement GPS systems are used for covert installations where suspects are not aware of a tracking system. Stolen vehicles found cooling off, vehicles part of undercover deals, and suspect or informant vehicles are tracked to obtain intelligence. Many rental vehicles now have GPS tracking capabilities that have been used to recover stolen vehicles [7].

The GPS system capabilities allow law enforcement to check a vehicle history, seeing exactly where a vehicle has traveled during a specific time frame. Investigators may then respond to areas indicated to conduct further surveillance or intelligence gathering. Real-time tracking enables investigators to actively follow vehicles and suspects. Surveillance teams can be directed to positions that will not risk compromising the investigation. Counter-surveillance maneuvers are less threatening and easily compensated for.

As tracking devices become more commonplace on vehicles, counter-measures by professional thieves will expand as well. One method of countering tracking devices is an appropriate apparatus that signals if a transmitter is sending out signals. GPS systems can be programmed to provide updates at specific times or only when a vehicle is in motion. These updates can include the vehicle history and the current location. This reduces the ability of criminals to know if a tracking device is operating on a vehicle.

20.2.3 Ituran System

Ituran is currently operating in Argentina, Brazil, Israel, and the United States. Future plans are in development to offer this service in more countries in Latin America and Eastern Europe. Similar to GPS systems in their ability to map locations, Ituran are radio tower-based tracking systems [8]. This system also may triangulate a radio signal transmitting from the vehicle. It is not affected as much as the GPS system by parking garages and shipping containers. The coverage area of the towers limits the service area of the system, while the GPS system operates anywhere in the world where it has a clear signal to a satellite and a cell phone capability.

This radio signal-based system also has the ability to program the number of locales, from 20-second intervals to longer time periods. The 20-second intervals are most pertinent to follow moving targets or vehicles; a vehicle can travel several city blocks in a minute, and this could mean the difference in keeping or losing a vehicle in a dense urban area.

20.2.4 Cell Phones

Cell phones have been tracked using their electronic signals with various technologies. These electronic devices use a triangulation from cell towers to help locate the phone. This provides investigators with a general area from where the cell phone is transmitting. Then, a more sophisticated device could be used to find the specific location of the cell phone. Most of this technology is very confidential.

New cell phones have GPS capability that can be used to locate a suspect and/or vehicle in addition to the above method. Electronic devices can pinpoint specific cell phone transmissions, enabling investigators to positively identify suspects. Cell phones have become popular tools for criminal suspects for doing business and illegal transactions. Some jurisdictions require court orders to track through this means [9].

20.3 INSTALLING TRACKING DEVICES

20.3.1 Vehicles Equipped With Tracking Devices

A/ Vehicles Factory Equipped

As more tracking companies are formed, it is difficult to stay abreast of the various systems. Some vehicle manufacturers even use different GPS companies on different models. Investigators should learn to know the companies most common in their area and obtain contact information for them. Conducting an Internet search provides a relatively complete and updated listing.

B/ Vehicles Post Factory Equipped

Some tracking systems are designed for installation after being sold to the consumer, either for commercial purpose or personal use. Many companies have systems for fleet management such as delivery trucks or cars, taxis, emergency vehicles, and rental vehicles. Heavy equipment and even generators have had tracking devices installed because of high theft rates [10]. When a police report is taken, officers should ask whether or not it was equipped with a tracking device.

20.3.2 Installation in the Scope of an Investigation

The installation of tracking devices requires a technician who understands the system being installed and the vehicle wiring. In major agencies, technical support units perform this function. In a smaller unit or agency without a specialized technical support unit, the investigators need to have the training and knowledge to do it by themselves.

Auto theft investigators should work closely with their technical support personnel. Vehicles similar to those targeted for covert installations should be provided so that training and practice can be done in preparation for in-the-field operations.

Technicians need feedback from the investigators as to how effective each installation is during the course of the tracking operation. This enables them to fine tune their installation procedures. Some technicians even keep a logbook of each installation to be more effective in future installations.

20.3.3 Covert Installations

A/ Principle

The covert installation in the field is risky and dangerous, but when successful it is very enlightening. When the decision is made to conduct a covert installation in the field, it takes a lot of teamwork to achieve the desired goal: tracking a vehicle without its driver's knowledge.

The first part is finding a stolen vehicle cooling off or a suspect's vehicle that has been targeted. A surveillance team must be deployed around the vehicle until the technician has approved the final installation. The safety and security of those involved in the operation depends on the team's level of alertness. While the information is relayed to the technicians as to the type of vehicle and the location, other personnel can be working on the court order if required.

When the technicians arrive in the area of the vehicle, they must stay in constant communication with the surveillance team. As the installation is in progress, another member of the investigative team needs to monitor the system to make sure a good and clear signal is being received. After the installation is complete and the technicians have left, the surveillance team should remain in place for a short time period to see if anyone shows interest in the vehicle.

An investigative team member should visually check the vehicle on a frequent basis, at least several times per week. Some vehicles have been towed, and the tracking system was not engaged because the vehicle was never started. Other systems can be programmed so that any movement of the vehicle will turn the system on.

B/ Legal Guidelines

Investigators should become familiar with federal and local laws and court decisions that are applicable to tracking vehicles [11]. Blank court orders should be obtained and readily available to personnel within the investigative unit. A specific prosecutor should be briefed on vehicle tracking installations and should be on-call for covert installations. This helps to expedite the process, reducing the time spent on surveillance.

All legal guidelines should be discussed and approved by agencies legal staff prior to the implementation of a tracking program [12]. Proper documentation needs to be maintained on every installation. All notes and reports should be confidential in nature while the case is ongoing.

C/ Court Testimony

As with any information about a case, officers should be reserved in their testimony, keeping the use of tracking devices quiet unless necessary. An officer may clearly state that a vehicle was under surveillance without indicating that a tracking device aided it. Each court session provides the suspects with an opportunity to gain insight into investigative methods of operation and officers should avoid compromising future cases if possible.

20.4 THE USE OF TRACKING DEVICES FOR INVESTIGATIONS AND RECOVERY OF VEHICLES

20.4.1 Principle

As technology becomes more advanced, investigators will continue to find ways in which to use tracking systems to further their investigations and recover stolen vehicles and property. The criminal elements will also learn from their mistakes and use technology to counter law enforcement efforts. It is very important for investigators to limit the publicity and discussion with suspects and informants about tracking devices.

Most auto theft cases in the United States rarely go to a full trial; they are often plea bargained. Unless lawful discovery legally requires disclosure, information in reports should be limited. In some other countries, the use of such investigation's techniques must be disclosed, even if the case does not go to full trial.

20.4.2 Verifying Sources of Information

When using informants and sources of information, investigators need to realize that information flows in two directions. Many informants are double agents playing both sides. Tracking devices can be utilized to verify information received from an informant and to know when a source is not being truthful.

Placing a tracking device in a vehicle being used by an informant, a source, or even a suspect can provide an immense amount of intelligence. Frequented locations and times can be recorded for surveillance assignments. Meetings with other sources or suspects can be discovered.

There are different means to placing a tracking device on a suspect's vehicle. Once the court order has been approved, a decision on how to do a covert installation needs to be made. The environment where the suspect lives, works, or spends time dictates if it can be done without contact. If not, there needs to be a planned event that will provide an opportunity without suspicions being raised.

For informants and sources, it is possible to tell them directly and get their approval. This depends on the relationship with the investigator and the situation. Although the reason might be for an undercover operation, the tracking information can be obtained 24 hours a day. This method has risks, especially because it is compromised immediately by the source knowing about it.

A vehicle with a tracking device installed can be provided to the source. This method needs careful analysis by agencies legal affairs personnel as there can be other liabilities incurred.

A source can also be treated as a suspect with a covert installation. It is easier to set up an event to provide an opportunity with an informant or source if the environment is not conducive for an installation at home or work.

20.4.3 Aiding the Surveillance of Suspects

Whenever it is feasibly possible, tracking devices should be placed on suspects' vehicles. Major targets and organized criminals are very aware of law enforcement surveillance and use many counter-surveillance measures. A tracking device enables investigators to maintain a loose tail when a suspect moves from location to location. The history of movement recorded when a suspect is not under active surveillance helps identify locations the suspect frequents. This gives the investigators the ability to estimate trends and obtain knowledge of where a suspect may be heading so that surveillance teams can actually setup ahead of the suspect's arrival.

When using the tracking device for active surveillance, at least one investigator or support person needs to be in direct communication with the entire surveillance team and actively monitoring the computer showing the track. Any delay going through a service provider diminishes the effectiveness and possibly requires a closer surveillance. This may compromise the operation and should be avoided at all costs.

The surveillance team should be aware when possible interference of tall buildings or parking garages may affect their ability to monitor the tracking device. The knowledge of the type of system being utilized and its capabilities is thus primordial. Some systems can be hard wired into the vehicle, for long-term use with a permanent power source. Other battery-operated systems limit the time span of their effectiveness.

20.4.4 Undercover Operations

Tracking devices add an additional layer of security when conducting an undercover operation. Officer safety should always be the primary factor considered. Any vehicle being used in an undercover operation, especially if an officer is inside, should have a tracking device in case the unexpected occurs.

Tracking devices should be used in all vehicles being sold in undercover operations. This allows the investigators to follow the vehicle through the network, whether the vehicle is destined for export, the domestic market, or the chop shop for parts. In addition, when officers are acting as buyers, tracking devices can be installed when the vehicle is taken for a test ride or borrowed to later be returned. This allows the investigators to reject the vehicle for a bogus reason yet gain the intelligence that would be much more difficult and costly to obtain from an active surveillance.

Tracking devices are not just for vehicles. Heavy equipment, personal watercrafts, cargo shipments, and other expensive property such as generators can also have tracking devices installed.

20.4.5 Installation on Vehicles Cooling Off

As discussed earlier, a considerable benefit of tracking devices has been the response by auto theft suspects in cooling off vehicles. This provides investigators an excellent opportunity to gain intelligence.

Investigators need to communicate with officers on the street as to where vehicles may be found cooling off and to not approach them when found. Officers in plain clothes and unmarked vehicles are best suited for checking cool-off sites, plate numbers, and VINs without drawing attention to them. If a stolen vehicle is located, based upon the observations, a surveillance team should be called in until a tracking device can be covertly installed.

Even though there should not be any expectation of privacy for a stolen vehicle, many jurisdictions still require a court order. This process can be expedited with documents, prosecutors, and a procedure to obtain a judge's approval in place.

Once the installation is completed, the surveillance teams can then be withdrawn until the vehicle moves. Vehicles should be checked visually on a routine basis, because the suspects sometimes move the vehicles from parking space to space to avoid drawing attention. When the vehicle finally starts moving, a surveillance team needs to respond. The team documents and records all activity.

During surveillance for export cases, vehicles are followed to locations where they are loaded into shipping containers. This will lead to other stolen vehicles being placed in the same and additional containers.

Vehicles being altered are taken to a location where they are processed. If investigators believe the VIN and other vehicle numbers have been changed, at some point the new numbers must be recorded. There is a possibility that the tracking device will be found during the process. Other vehicles will be followed to chop shops.

There is a risk that a stolen vehicle that is tracked could be lost and never recovered. This risk must be weighed with the fact that for each vehicle tracked, the probability of finding additional stolen vehicles and identifying more suspects is enhanced.

20.4.6 Bait Car Operations

Bait car systems are designed to capture suspects stealing vehicles using tracking systems. These systems record the suspects on video as they steal the vehicle, their actions while driving, and their reactions when police move in to arrest. These operations have immense interest by media outlets and news agencies. They make potential auto theft suspects leery about stealing a vehicle [13].

Bait cars are selected based upon what type of vehicle is most likely to be stolen in a specific area, as discussed in Chapter 21. Authorization to work with private industry vehicles can be obtained from insurance companies. These vehicles are then outfitted with the proper system.

The system allows a suspect to enter the vehicle and drive it away. A tracking device indicates to the law enforcement agency that the vehicle is moving and police officers are directed to its location. When officers advise that they are in position for a take down, the system enables, for example, the engine to be turned off and the doors locked. This helps capture the suspects with little or no chance of a pursuit [14].

20.5 CASE STUDIES

The following cases are examples of various tracking cases conducted by the Miami-Dade Auto Theft Task Force in Miami, Florida, United States.

20.5.1 Operation Company Car

In 1996, investigators located a stolen Toyota Land Cruiser in the northern part of Miami-Dade County, parked in an apartment complex parking lot. The vehicle appeared to be cooling off and prime for a tracking device. The investigator, James Woodruff, radioed the office with the vehicle information, and a supervisor faxed a copy of the tracking order to Assistant State Attorney William McGee. Additional investigators responded to the apartment complex to assist with the surveillance.

While McGee was getting a judge to sign the order, a call was made to the Technical Support Unit. They were advised of the type of car, the location, and that a court order was being processed. Technicians then scheduled to do a covert field installation and headed to the site.

With people walking around and cars moving in and out of the parking lot, technicians parked next to the stolen vehicle and began working on the installation. Within a short period of time they had opened the vehicle and installed the tracking system from Teltrac (during this investigation, the tracking system was owned and operated by Teltrac, which was then sold to Ituran, who continues to operate the system in south Florida). The signal was strong, indicating a successful operation. The surveillance team cleared later, and investigator Woodruff waited for the vehicle to move.

During the next two weeks, two more stolen vehicles, a pickup truck and another Land Cruiser, were located in other parts of Miami-Dade County and tracking devices were successfully installed on them. Other investigators had their vehicles to monitor, as did investigator Woodruff.

Each time a vehicle moved, investigators responded to photograph and record all activities and suspects observed. A gas station by Miami International Airport was the first clue that all three vehicles might be involved with the same auto theft suspects. When discussing the cases, it was noted that all three vehicles had been tracked to this gas station.

Later during the investigation, one of the vehicles was taken to a residential neighborhood and parked in the rear of a house. The house belonged to a known suspect who alters VINs and produces counterfeit labels. After several days the vehicle was driven to another house approximately two miles away. An investigator was able to sneak up to the vehicle and get the altered number now placed on the vehicle [15].

The stolen pickup then started moving from a home it was parked at to the house of the known suspect. It too went into the rear yard where it could be altered. Meanwhile, investigators working title fraud cases intercepted the false title application for the first stolen Land Cruiser that had been altered. The vehicle was to be sold once a clean title was obtained. The altered number was from a vehicle that had already been exported. The number was reborn on a stolen vehicle.

When the third vehicle was driven to another house and into a garage in Hialeah, a suburb of Miami, the lead investigator prepared a search warrant for the home. Upon executing the search warrant, a vehicle-altering laboratory was discovered being operated from within the house. Two more stolen vehicles were found in the yard. Many evidentiary items were found that clearly linked the suspects involved to a large organized group. The intelligence gained from using tracking devices clearly helped the identification and prosecution of a prominent organized auto theft group.

20.5.2 Lojack Recovery on the Miami River

When a stolen vehicle report was entered into the police computer system, it initiated the signal in the vehicle. An officer working near the Miami River immediately received this signal. As he and other officers equipped with Lojack responded to the area, the signal became stronger and indicated it was possibly on one side of the Miami River and was no longer moving.

An officer entered one of the many shipping yards located on the approximately seven miles of waterway called the Miami River. The stolen vehicle was found parked in a building on the shipping yard waiting to be placed on a coastal freighter (Figure 20-3) headed to the Caribbean. Officers had US Customs inspectors respond, and a search of the ship docked at the facility found nine more stolen vehicles.

Figure 20-3

A Lojack signal from one stolen vehicle led police to a shipping yard on the Miami River where other stolen vehicles were being loaded onto a coastal freighter.

Figure 20-4

After officers found a stolen vehicle being loaded (see Figure 20-3) onto the ship, nine more stolen vehicles were found already loaded under mattresses. In this case, one tracked stolen vehicle led to the recovery of nine more.

The vehicles had been placed on the ship with mattresses spread over them to conceal them, as shown in Figure 20-4. Without the Lojack signal coming from the tenth car, all of the cars would have sailed and not been recovered.

20.5.3 Tracking a Toyota Land Cruiser to Chicago

Most operations do not go according to plan, and one must always be ready to take a loss. Even when things go badly, intelligence can be learned. With the success of Company Car and other investigations, the Task Force members continued to find stolen vehicles in cool-off locations and kept the technical support officers busy. A stolen Toyota Land Cruiser was found in an apartment complex just west of Miami International Airport. A tracking device was installed successfully, and the monitoring of the system was initiated.

At the time, the tracking program did not have the capability to alert when a vehicle was moving. The vehicle moved several times over the next week, including a stay for several days at a residence in the northeast section of Miami-Dade County.

On a Friday night, one of the Task Force members noticed the vehicle was northbound on the Florida Turnpike in St. Lucie County, near Ft. Pierce, Florida, approximately 110 miles north of Miami. The vehicle was traveling at 90 miles per hour. It left the mapping coverage area for Teltrac and only the longitude and latitude were displayed.

A call was made to the Aviation Section, and a pilot was able to provide the location from the coordinates being broadcast on the monitor. The vehicle was heading toward Orlando. The Florida Highway Patrol was called and provided with the description of the vehicle and its approximate location. They were unable to find the vehicle, and it disappeared off of the computer screen. Through further investigation, it was determined that the Land Cruiser had the VIN altered at one of the homes in northeast Miami-Dade. It was then used to possibly transport narcotics to an unknown location north of Florida.

Approximately two weeks later, the tracking device showed up in Chicago, Illinois. At the time there were only three areas in the eastern United States with Teltrac coverage. The vehicle was observed to move from a particular location to another. The Chicago Police sent a team of investigators to locate the vehicle. Before it could be located, the signal ended and was not seen again.

Within a few days of the Chicago incident, a Toyota 4Runner was observed parked in the same parking lot in which the Land Cruiser had first been observed. It was a stolen vehicle from the same neighborhood in Chicago where the Land Cruiser signal had last been seen in. Even though the tracked vehicle was never recovered, it ended up providing intelligence information and suspects involved in other criminal activity.

20.5.4 Out to Sea

A stolen vehicle was found in a parking lot in Miami Lakes. A tracking device was placed on the vehicle. After several days, the vehicle was finally observed moving south into Miami. A surveillance team was dispatched to find the vehicle. Before the team could catch up with the fast-moving vehicle, it entered the Port of Miami.

There was one big problem: The vehicle was not driving on its own but was in a shipping container. The container was placed in the shipping yard with thousands of other containers. This occurred late on a Friday afternoon. The surveillance team arrived at the port and began checking the shipping documents of the containers that had arrived in the same time span. Over 20 suspect containers were opened without finding the vehicle.

The signal remained strong, but it could not pinpoint the vehicle's exact location. For two days, different methods of searching the container with the vehicle were tried with negative results. On Monday morning at 1:30 am, the signal showed it was moving, only it was no longer at the Port of Miami; it was in the shipping channel heading out to the ocean. The signal went due east to the edge of the Gulf Stream and turned southeast in the shipping lanes headed for the Caribbean (Figure 20-5). It obviously had been loaded during the night aboard a ship and was headed to another port of call.

Immediately, the Task Force was able to determine the ship that left at the time the signal left Miami. A review of the manifest was completed, and based upon specific information, relative to previous shipments of vehicles, including weights, type of product, and destination, 10 possible containers were identified. These containers were ordered to be returned to Miami. Upon their return each was searched, but the stolen vehicle was not located.

This case showed both law enforcement and the private industry the need for scanning technology to see inside shipping containers, as presented in Chapter 18. The cost of this

Figure 20-5

Some stolen vehicles get away on ships like this headed to the Caribbean, South America, Central America, Europe, or Asia.

particular investigation to the shipping company, the freight forwarders, their customers, and law enforcement was tremendous. This led to excellent future cooperation by all stakeholders when gamma ray technology was later tested at the Port of Miami, knowing each would benefit from this effort.

BIBLIOGRAPHY

[1] Goldberg H, Godles JA, and Stefani L. (2004) *Petition For Rule Making before the Federal Communications Commission,* Lojack Corporation Legal Document of October 25, 2004, Lojack Corporation, Westwood, MA.

[2] Crocker G. (2005) Lady Stripper Marks 5000 Arrests for Tracker, *The APB,* **11**(1), p 53.

[3] Cole J. (2005) Lojack Colorado Year-End Report 2004, *The APB,* **11**(1), p 51.

[4] Lojack (2005) *About Us,* available at http://www.lojack.com, last access performed on April 10, 2005.

[5] GPS World (2005) *GPS History,* available at http://www.gpsworld.com, last access performed on April 10, 2005.

[6] Rand Corporation (2004) *GPS History,* available at http://www.rand.org/publications/mr/mr614/mr614.appb.pdf, last access performed on April 10, 2005.

[7] CBS News (2004) *GPS Keeping Tabs on Car Rentals,* available at http://cbsnews.com/stories/2004/03/06/eveningnews/main604461.shtml, last access performed on April 10, 2005.

[8] Ituran (2005) *About Us,* available at http://www.ituranusa.com, last access performed on April 10, 2005.

[9] Koerner B. (2003) *Your Cell Phone Is a Homing Device,* available at http://www.legalaffairs.org/issues/july-august-2003/feature_koerner_julaug03.html, last access performed on April 15, 2005.

[10] Lojack (2003) *2003 Construction equipment theft study,* Lojack Corporation, Westwood, MA.

[11] Supreme Court of the State of Washington (2003) *State v. William Bradley Jackson,* Docket number 72799-6, available at http://www.courts.wa.gov/opinions, last access performed on June 3, 2005.

[12] Orion (2004) Hugh's hints—legal aspects of GPS, *Orion's Belt,* **5**(2).

[13] Brockman D. (1997) Bait vehicles, *The APB,* **3**(1), p 35.

[14] Phillips RA. (1999) Cops go fishing for car recoveries, *The APB,* **5**(2), p 51.

[15] International Association of Auto Theft Investigators (1997) Dade County Multi-Agency Auto Theft Task Force wins 3M/IAATI award: Anti-theft VIN labels instrumental in identifying "re-numbered" stolen vehicles, *The APB,* **3**(3), pp 45–47.

CHAPTER 21

VEHICLE CRIME MAPPING

Jerry Ratcliffe

21.1 INTRODUCTION

Although much of this book focuses on the investigation of individual vehicle thefts, this chapter seeks to place each crime within a broader context. This is done so that lessons learned from the investigation of multiple crimes can help the investigator build up a picture of the crime pattern in the local area. Armed with an understanding of the larger crime picture, crime prevention and detection resources can be more effectively applied to stop further car thefts, either by targeting organized groups of offenders or by tackling the problem in a vulnerable crime hotspot.

The chapter starts by explaining the basics of routine activity theory, which suggests that for a crime to occur there is a requirement for the meeting in time and space of a motivated offender and a suitable target, in the absence of a capable guardian. For the investigator, understanding the interplay of these three elements (motivated offender, suitable target, absence of capable guardian) can be a starting point from which to consider ways to prevent future thefts. All three elements of routine activity theory must come together in time and space, and one of the best ways to understand the spatial and temporal patterns is with a geographic information system (GIS). Car crime is an offense of opportunity, and GIS allows investigators to visualize the patterns of opportunity.

Next, the basics of how a GIS works are explained. This outline includes the generation of map coordinates through geocoding, and how the layers approach within a GIS allows for different pieces of information to be combined within a single analysis. Crime mapping is not just about looking at vehicle theft, but about understanding the context so that crime can be prevented and offenders can be arrested.

The chapter then goes on to examine vehicle crime in particular, with a look at the ways in which a GIS can spatially explore the characteristics of locations such as where a car is stolen from, where it is stripped or damaged, and where it is abandoned or arsoned. This includes understanding the social demographics of these areas with census data, mapping likely routes the vehicle was driven, and calculating distances driven between these crime sites. The chapter also explains a number of ways that modus operandi, vehicle model and characteristics, and offender crime scene behavior can be mapped to determine whether significant spatial patterns exist.

The chapter goes on by returning to the interplay of crime theory and GIS application by suggesting ways in which an understanding of the spatial dynamics of the journey from theft to abandonment can suggest ways to prevent future offenses. Two particular new technologies that benefit from careful positioning with GIS are explored: license plate reading systems and bait cars. The chapter concludes with a primer on mapping services, so that the investigator has an overview of what resources are available, and knows the data requirements for a GIS to function as an effective investigative tool.

21.2 VEHICLE CRIME: A CASE OF INFORMATION OVERLOAD?

Although the first chapter of this book covered a number of statistics, it is worth taking a moment to reiterate the sheer volume of car crime because it relates to a problem of information overload suffered by many who work in this field. Taking the United States as an example, every year over one million vehicles are stolen. As Figure 21-1 shows, this translates to a rate of over 400 vehicles per 100,000 people every year [1]. In metropolitan areas this rate can be even higher. In the Washington, DC metropolitan area the rate for 2002 was 679 cars per 100,000 people; in Miami, Florida the rate was a stunning 899 cars per 100,000 people; and in Phoenix, Arizona the rate of 1,320 (per 100,000 people) was over three times the national average. Although the rates in Figure 21-1 have, in recent years, not reached the levels of the early 1990s, this still represents a considerable drain on the resources of law enforcement and insurance companies.

Things are even worse in some other countries. According to the International Crime Victim Survey, American cars are fairly safe from theft compared with a number of other

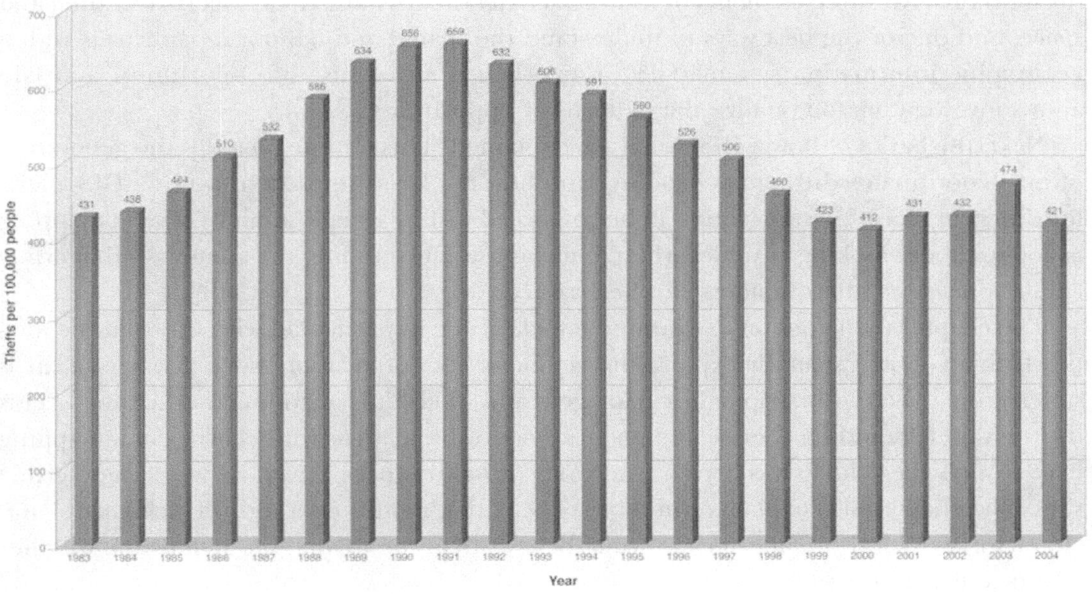

Figure 21-1

Vehicle thefts in the United States, per 100,000 people, 1983–2004 [1].

nations. In 2000, Americans reported that of every 100 vehicle owners, 0.6 experienced a vehicle theft. By comparison, the Scottish experienced car theft at twice that rate, in Australia cars were stolen at more than three times the US level, and in England and Wales the car theft rate of 3.0 per 100 owners was five times the US car crime rate [2]. Even though other countries experience higher rates of theft and the United States has apparently recovered from the high rates of theft experienced in the early 1990s, vehicle crime is an enormous problem.

When crime reaches this sort of magnitude, there is a tendency for investigators and other law enforcement officials to respond to vehicle crime with a sense of inevitability. The information overload caused by the magnitude of car theft numbers can make it impossible to discern any patterns without the aid of information technology. When research was undertaken in Nottingham (UK) to discover why police officers were not generally aware of the vehicle crime hotspots (areas of highest offense density) in their patrol areas, one officer remarked that although both burglary and vehicle crime locations from the previous 24 hours were written on a board [3], "You can come on in the morning and find 20 auto crimes, and only two burglaries which are much easier to remember". With 20 new locations every day, it was understandable that officers were left with the impression that vehicle crime was a contagion that had spread across the whole police district.

To get a better understanding of high volume crime, law enforcement has turned to information technology. If crime data are entered onto a computer system, then it is often possible to use the power of the microprocessor to detect patterns that would not otherwise be perceivable to humans. Indeed, it is well known that spatial patterns are perceived differently from person to person [4], and in many cases people have a tendency to see patterns where none exists, believing they see a pattern in data that is in reality random. Given that vehicle crime has the highest reporting rate to the police—95% of all stolen vehicles in England and Wales are reported [5]—then it would seem essential that information technology is used to its utmost to extract the best information and intelligence possible in the fight against crime.

One recent addition to the information technology revolution is crime mapping. Mapping technologies are relatively new to the world of law enforcement but yet are being adopted at a fast rate [6–8]. Mapping is an ideal medium for an investigator to appreciate the larger picture of car crime. Although there is a tendency for investigators to view the investigation of auto theft on a case-by-base basis, mapping can allow the investigator to place an individual offense into a broader perspective and possibly see a larger context or pattern of offending. Before examining how mapping systems work, we should ask a crime-related question: Why bother to map crime? If, as many police officers believe, crime is essentially random, why go to the effort of mapping vehicle crime? To answer this question, we need a short primer on crime theory.

21.3 WHY MAP CRIME?

21.3.1 Routine Activity Theory

For police operational commanders trying to prevent the apparent deluge of theft, the sheer number of car crimes can be daunting. However, for an investigator examining a handful of high-value car thefts, the reverse may be true. There might not appear to be any sort of spatial or temporal pattern that can aid the investigation. In low volume investigations, it can help to understand the behavioral patterns of offenders. Although the world of criminology is full of academic theories, many of them are quite useless to the law enforcement practitioner. There are, however, a couple of ideas that can help an investigator understand better the mechanisms of crime and provide possible remedies. The following theories can be considered all a part of the general field of environmental criminology, and foremost of these ideas is that of routine activity theory.

In its most basic form, routine activity theory states that for a crime to occur, it is necessary for the meeting in time and space of a motivated offender and a suitable target, in the absence of a capable guardian [9, 10]. When these three things occur, we have all of the chemistry for a crime. In other words, the opportunity to commit a crime exists. To prevent a crime from happening, it is therefore necessary to remove one (or more) of the three components. From an investigator's perspective, it is probably most rewarding to remove the motivated offender. However, much of the crime prevention that takes place today is more focused on the other elements. The growth of closed circuit television (CCTV) systems and the rise in the number of private security guards are indicative of crime prevention strategies that seek to provide the capable guardianship that makes crime more difficult. Other ways to try to increase the guardianship of personal property include the use of neighborhood watch, increased street lighting, and doormen at nightclubs. All these features try to replace the capable guardian, the absence of which is a necessary component of a crime event.

The increase in technologies used to make vehicles theft-proof is evidence of policies that try to make a target less "suitable." Mechanisms such as electronic keys, remote entry gadgets, and engine immobilizers are all ways that manufacturers use to make their brand of car less attractive as a target (see Chapter 8). The aim is less to prevent all vehicle thefts, but simply to make their cars look less attractive to a thief in the hope that the offender will target another brand of cars. Individuals can do this for themselves by using physical mechanisms such as metal bars that lock to the steering wheel. These crime control strategies help to reduce the attractiveness of the target, because they increase the effort required to successfully steal the car.

Crime prevention mechanisms to increase the guardianship of property or to make the target less attractive have their origins in the "rational choice perspective" of crime control [11]. The rational choice perspective recognizes that most offenders make some sort of a decision to commit a crime. They weigh up some of the pros and cons and balance the potential rewards against the chance of being caught. In essence, the offender makes a rational choice to steal a particular car. If that is the case, then this provides a potential mechanism to prevent some car crime. If offenders recognize an opportunity, then it may be possible to influence their perception of that opportunity. "Situational crime prevention" has evolved from both routine activity and rational choice thinking and concerns itself with preventing the opportunities for crime. Situational crime prevention consists of a number of tactics to reduce opportunity, tactics that are crime specific and involve manipulating the physical environment in a systematic and permanent way. These ideas are designed to change the perception of offenders such that it appears the effort and/or risk has increased or the rewards have reduced [12].

There are a number of car crime-related examples. As said, engine immobilizers and electronic keys increase the perceived effort to steal a car, whereas car alarms and vehicle tracking systems that are connected to police cars dramatically increase the risk of being caught (see Chapter 20). For offenders wishing to steal from vehicles, the removal of valuable items from visible areas within the car reduces the perceived rewards of crime, as does the encoding of car entertainment systems so that they have to have a secret number keyed into them if they are disconnected from the vehicle's battery. All these are effective crime control strategies if they change the offender's perception of the opportunity.

21.3.2 Seeking Crime Opportunities

Given that the availability of an opportunity is central to offending, where are the opportunities? It would be tempting to believe that vehicle crime opportunities are everywhere, but if that were the case then there would not be car crime hotspots: Crime would be everywhere. As most investigators know, this is not the case. Car crime hotspots are a result of the search behavior of offenders.

When offenders are seeking out a car to steal, they use the same search strategies that the rest of us use for noncrime events. If we wish to purchase milk, we do not travel 10 miles to do so. It would seem pointless to travel such a long distance when the article in question is available from a shop much closer to home. Traveling over long distances increases the effort required, as well as the time taken to get the milk. From a criminal perspective, increased distance means more effort required, more time required, and a greater risk of capture. It also increases the chance of wandering into an unfamiliar area, where the escape routes and behavior patterns of local people and police are not known. Unknown areas increase the risk for offenders [13]. Like us, therefore, offenders seek out opportunities close to home.

Crime pattern theory [14] suggests that offenders not only commit crime close to home, but also in the vicinity of other "anchors" [15] such as work and school or the homes of friends and along the pathways between these places [16]. By mapping crime we can get an insight into the routine activities of offenders' lives, and by linking offenses to a particular offender or a group of offenders, we are able to use this knowledge to predict where the offender might live, a process sometimes referred to as "geographic profiling" [17]. The application of knowledge in this fashion requires mapping the spatial pattern of offenses, which demands for a crime mapping system.

21.4 HOW DOES CRIME MAPPING WORK?

21.4.1 Geographic Information Systems

Although it is possible to tack a map to a wall and to stick pins into it, the volume of car crime makes this an unrealistic approach to understanding crime patterns. Instead, crime mapping systems use the digital technology GIS.

A GIS is "a system for c apturing, storing, checking, integrating, manipulating, analyzing and displaying data which are spatially referenced to the Earth. This is normally considered to involve a spatially referenced computer database and appropriate applications software" [18]. GIS has its origins in the early attempts by geographers to automate the process of cartography but soon grew to include analytical functionality. Law enforcement was slow in getting aboard the GIS wagon, but that changed with the digital revolution of the 1980s that saw a dramatic reduction in the costs of hardware and desktop computing became mainstream.

Now many police departments have the ability to map crime and criminal activity and to compare the crime event locations with other features of the police district. The new computerized approach means that investigators can not only map vehicle crime, but can also generate specialized queries, such as only mapping the locations of stolen Honda Accords or all vehicles stolen on Saturday nights. These more refined queries can tackle questions of offender specialization or can help an investigator to better understand temporal patterns of auto theft. If the data are collected, it also provides the potential for investigators to analyze modus operandi patterns or to examine spatial patterns of thefts based upon the characteristics of a vehicle. For example, if a police or insurance company database records the value of a stolen car, then a spatial pattern of high value thefts may be apparent because a ring of thieves are targeting particular car parks or shopping malls. Alternatively, a cluster of thefts of a particular model may be present. Mapping systems can easily filter low value vehicles or certain models from the map, leaving only the vehicles of choice on the display. There are examples of this later in the chapter.

There are some basic processes that take place within crime mapping that are useful to understand. The following section explains geocoding, how layers function in a GIS, and the different types of queries that can be asked of a GIS.

21.4.2 Geocoding

GIS are particularly suited to vehicle crime data because of the inherent geography that exists in crime incidents—a crime usually occurs at a location. In the case of vehicle crime, it usually occurs on a street. Geocoding is the term given to the process of getting the address or location of a crime event on to a map. In the United States many years ago, the government constructed a digital file of every street in the country, and each record details the location of the street (mapped to coordinates), the street name and type, the zip code, and the house numbers that run along each side of the street. Many other countries quickly followed this approach because it provides a great deal of spatial knowledge about where we all live. Geocoding uses special software built into a GIS to convert a text string of an address into a point on the map. A few countries (such as the United Kingdom) have moved to the stage where individual street addresses and intersections are maintained in a massive database and can be used as lookup tables. In other words, when a police officer enters the street location of a car theft, the computer automatically enters the map coordinates of the offense location. Geocoding is not just a computer-based system, but a process that starts with the initial crime report, as can be seen in Figure 21-2.

The process of geocoding is not infallible. When reporting crime, there are many ways that the address of the offense can be entered incorrectly onto a records management system. Sources of error include the following:

- The victim does not know the exact location of the offense.
- The street name is spelled incorrectly.
- The wrong street type is recorded (e.g., "street" instead of "avenue").
- The address entered onto the system does not exist (such as an error in house number).

There are more sources of error (for a full list, as well as some potential solutions, see [4, 19]); however, the point is that it is rare for a crime mapper to successfully geocode all crime. At this early stage in an analysis, it is important that all information pertaining to the vehicle crime is recorded as accurately as possible and carefully entered into a database. An error at this stage remains in the system and reduces the efficiency of the analysis.

Figure 21-2

Geocoding from original crime report, through geocoding within a GIS, to final map production. (Source: http://www.jratcliffe.net. Reprinted with with permission of Dr. Jerry Ratcliffe.)

Address data relating to where the theft or crime occurred are particularly important and must be collected accurately and carefully. On the assumption that in many cases investigators do not map crime themselves but use the skills of a crime analyst, a useful question for the investigator is to ask the crime mapper what is the geocoding hit rate. The geocoding hit rate is the percentage of crime reports that are successfully geocoded and mapped, and it is simply (in a vehicle crime example) the number of geocoded and mapped crimes divided by the total number of recorded vehicle crimes in the police department database, multiplied by 100 to make a percentage:

$$Geocoding\ hit\ rate = \frac{Mapped\ crime}{All\ recorded\ crime} \cdot 1$$

This is a useful question to ask, because it allows the investigator to know how much crime is not being shown on any map and so allows the investigator an appreciation for the completeness of the picture. In other words, if the geocoding hit rate is 95%, then this shows the investigator that the crime mapping system was able to successfully place a dot on the map for all but 5% of the vehicle crime stored in the record management system (the standard crime recording database). Although this is not an infallible indicator (e.g., it does not indicate a problem with incorrect addresses that can be mapped), it does give the investigator some idea of how much crime is not shown on the map. Generally, geocoding hit rates in excess of 90% are considered acceptable, though one study has found that for high volume crime 85% is an acceptable limit [19]. Anything less than 85% should be viewed with suspicion and caution, as a significant proportion of the recorded crime is not being shown on the map.

21.4.3 Layers

Crime mapping is not simply about placing a dot on a map to show where a crime has occurred, though there is undoubtedly value in this. A GIS can also provide the capacity to view crime data alongside other useful information. GIS have the ability to layer different pieces of information so that the geographical relationship between disparate urban features can be visualized (Figure 21-3). For example, it may be that a group of offenders is suspected of stealing cars when owners are distracted while getting cash at an automated teller machine (ATM). If the locations of the ATM devices are known, then these can be added to the map as a second layer. This has been done in Figure 21-4.

Crime mappers can generate thousands of different layers of information, information that can be combined and analyzed together or independently. Example layers that police can overlay with crime data include

- Police district and beat boundaries;
- Local authority areas;
- Locations of known chop shops;
- Gas stations;

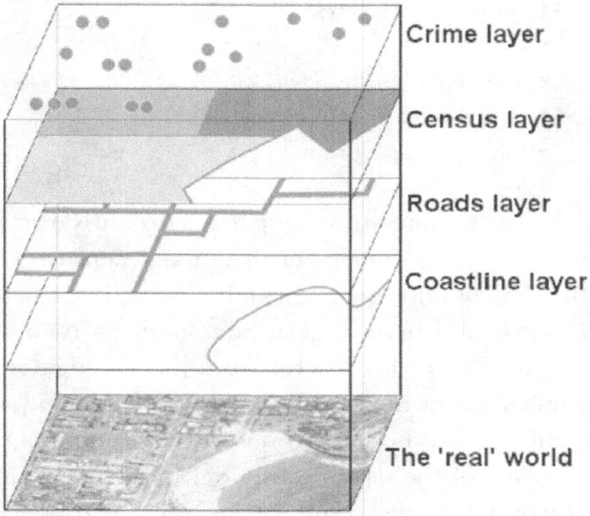

Figure 21-3

The layer principle, showing different layers of information related together by geography. In a GIS, layers can often be thought of as a series of spatially referenced transparencies [20]. (Source: Crime pattern analysis, available at http://www.jratcliffe.net. Reprinted with permission of Dr. Jerry Ratcliffe.)

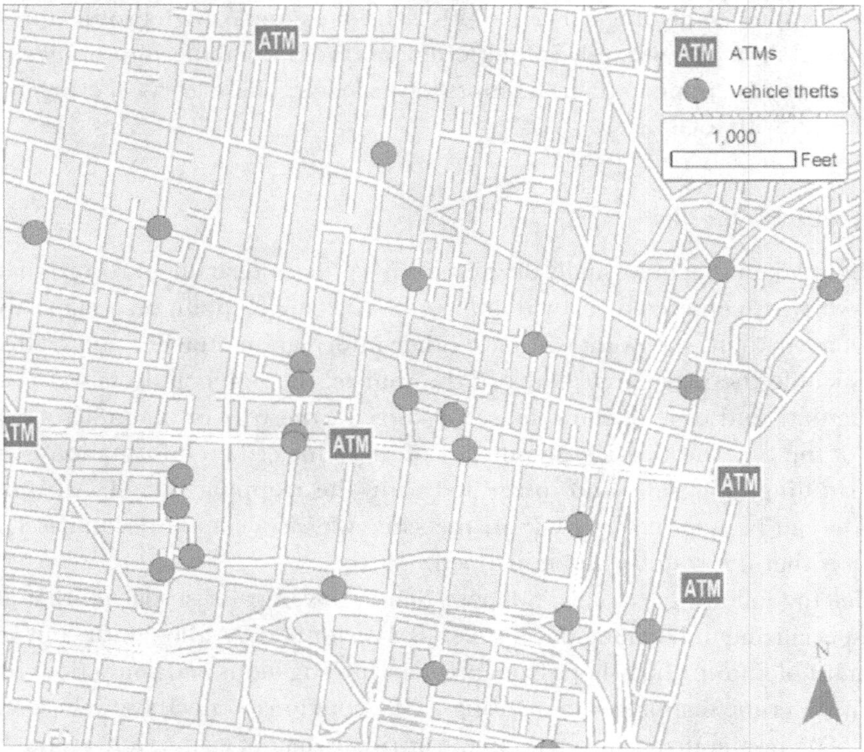

Figure 21-4

Car theft incidents and ATM locations in north Philadelphia.

- Roads and freeways;
- Banks;
- Late-night fast food outlets;
- Neighborhood watch areas;
- Public transport routes.

From a more strategic perspective, it is also possible to use a GIS to overlay socioeconomic information from a national census. This gives a different flavor to the crime problem by allowing a researcher to explore whether high vehicle crime rates are occurring in areas of different racial mix, lower economic status, or high unemployment. Although this may seem less useful from an investigator's perspective, it may also be possible to combat car crime by improving neighborhood cohesion or identifying areas for publicity campaigns and crime prevention advice in a particular language.

Some of the most useful layers that can be added to vehicle crime maps are the crime hotspots for other crime types, such as robbery and burglary. If a police department is planning a crackdown to combat a vehicle crime problem, it can be useful to select a target area that has other crime problems as well. The purpose of this is to try to achieve a "diffusion of benefits" so that the crime control benefits of, for example, a robbery strategy, could also spread to combating car crime [21–23]. For example, high profile uniform police presence in an area can not only inhibit offenders from breaking into homes or robbing citizens, but can also impede their attempts to steal cars. These areas of crime overlap can be easily found with a GIS.

21.4.4 Attribute Queries

GIS use digital data that are usually derived from a police department record management system or which are entered directly into the GIS. Although a point on a map can show the location of a car theft, the point can have other information attached to it. This attribute data can include the make and model of the vehicle, the home address of the registered keeper, the date and time of the offense (if known), the year of manufacture of the car, the color of the car, and the license number of the vehicle. The information is stored in a database running alongside, and integrated with, the mapping system. All this attribute information can be queried by a GIS in the same way as a standard database. The value with a GIS is that the results can be mapped.

Consider the following example. Suppose that an investigator suspected that a city chop shop was specializing in breaking down stolen BMWs for parts. Although the investigator has obtained a list of known chop shops from previous investigations and convictions, they are all over the city. A crime mapping system, using a GIS, could map only those vehicle crimes that involved BMWs by automatically querying the attribute data of each crime event. This would immediately reduce a huge series of dots all across the city to a smaller subset of points. We know from environmental criminology theory (introduced in an earlier section of this chapter) that offenders do not tend to travel further than is necessary to commit a crime. The investiga-

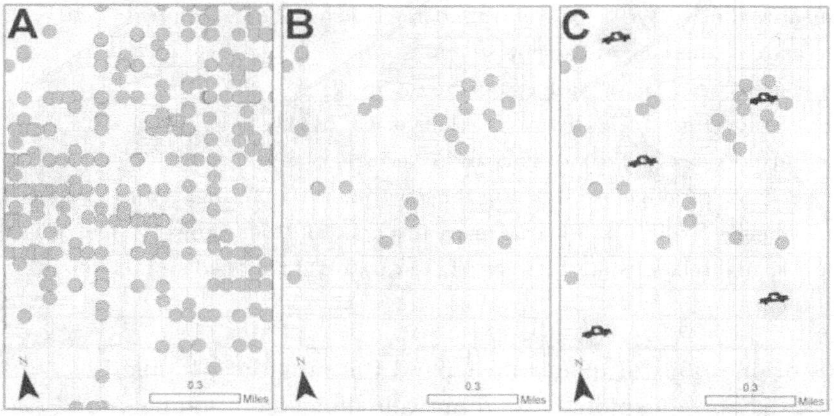

Figure 21-5

Three maps that demonstrate a querying and layer process in a GIS. Map A shows all vehicle theft locations in an area for a six-month period. Map B uses a query to only select and map those crimes that involved BMW vehicles. This query allows the investigator to focus attention on a certain type of vehicle; however, a similar query might also just map cars that had been the victim of arson or which had been stolen from locked garages. Map C adds a layer to map B, a layer that shows geocoded locations of known chop shops. This additional information may lead the investigator to suspect that one chop shop in particular, in the northeast of the study area, may be specializing in BMW parts.

tor could therefore surmise that if there are any areas of clustering of stolen BMWs, then it is possible that the chop shop may be near that location. At the very least, it helps to narrow the focus of inquiry to the more likely locations. With minimal extra effort, the addresses of the chop shops could be geocoded and then mapped alongside the theft locations for the BMWs and shown as different symbols. This has been done in Figure 21-5, where map A shows the locations of vehicle thefts for a six-month period. A query has been run to filter out all stolen vehicle that are not BMWs, so that map B now only shows BMW thefts. In the final image (map C) the locations of known chop shops have been added (as vehicle symbols in a gray circle). The result is a map of stolen BMWs and all known chop shops: a map that would narrow the spatial focus and advance the criminal inquiry and a map that, in the case of Figure 21-5C, suggests one location in particular as worthy of closer attention and possibly surveillance.

21.4.5 Spatial Queries

The previous example explored a spatial relationship between an urban feature (chop shops) and car crime but left it to the map reader (the investigator) to interpret the result. More complex queries can tap into the analytical power of GIS to determine the answers to many questions of crime and space. These are questions that are beyond a standard database's functionality. For example, it is possible to map not only the location of a car theft, but also the location where the stolen vehicle was recovered. From this it is possible to ask a barrage of potentially useful questions:

- How far does the average car travel from being stolen to being dumped?
- Are there areas where many cars are stolen?
- Are there areas where many stolen cars are dumped?
- Are the hotspots for vehicle theft near ATMs or banks?
- Are the hotspots near schools or bars and taverns?
- Are hotspots for cars that are stolen in the day different than night-time hotspots?
- What percentage of all the stolen cars in an area is stolen from within 500 feet of a subway station?
- Are there spatial relationships between where cars are abandoned and the modus operandi used to steal them?

These types of questions dramatically expand the range of possible lines of inquiry and allow the investigator to explore beyond individual incidents and begin to look at patterns. The essence of problem-oriented policing is that understanding patterns of criminal behavior may be the start of a more productive solution to a crime problem, a solution that solves the overall problem rather than a solitary offense [24–26].

21.5 RECOGNIZING PATTERNS AND TRENDS

21.5.1 Maximizing Investigator Effort

Where is the value in these types of questions for the average investigator? The answer lies in an examination of the role of the investigator in modern law enforcement. The influential Rand Corporation study examined the role of investigators within the criminal justice system in the United States [27]. The report calculated the use of investigator's time from observation of detectives in Kansas City, Missouri (USA). Forty-five percent of their time was spent on non case-specific general duties, such as administration, traveling, reading general reports, or conducting general surveillance; 22% was expended on cases that were never solved; 26% on paperwork and administration for already cleared cases; and only 7% was spent investigating unsolved cases that would eventually be cleared. Therefore, they spent 93% of their time engaged in activities that supported investigations rather than actively working on cases that had the potential to be cleared successfully. Given that investigators only have limited time available in their busy lives, it would appear prudent to focus their energies on the most promising leads and activities. GIS can help to provide a spatial focus to active investigations, especially those investigations that had potentially valuable leads, such as modus operandi patterns or suspect descriptions.

When Eck [28] and Brandl and Frank [29] studied detectives at work, they recognized that the most effective use of detectives' skills lay in the investigation of crimes that had moderate suspect information. Other offenses, such as when a suspect was known, would be solved irrespective of the effort put into the case, whereas offenses where there was weak suspect information would most likely never be solved, however hard the investigator worked.

In 2001, Singh concluded that the most valuable refocusing of detectives and investigators was toward a more problem-oriented focus for investigative policing [30]. By moving the attention of investigators away from individual incidents and toward the recognition of local crime problems, the investigative time of the detective could be more fully used to achieve greater crime reduction. In other words, by working on problems and crime patterns instead of individual cases, the detective can become more effective and efficient. GIS can play a crucial role in this, and the next examples show two ways in which GIS can aid the problem-oriented investigator.

21.5.2 Hotspot Mapping

One way that crime maps can effectively move the attention away from an individual case focus is to map problem areas geographically. Figure 21-6 shows two maps of vehicle theft in Philadelphia, Pennsylvania (USA): the first is a simple point map, whereas the second has a hotspot surface that uses the individual points to estimate a crime intensity surface in the same visual manner as a weather temperature map.

The advantage of the hotspot map is that it focuses attention on the geographical areas of concern rather than on individual points. These geographical areas—crime hotspots—are often a more useful problem to investigate. By solving the causes of crime in an area, law enforcement can prevent and reduce more crime.

The value of hotspot maps can be further enhanced, as shown in Figure 21-7. This image shows a hotspot map of car crime in the eastern suburbs of Sydney (Australia). Attached to five of the main hotspots is a chart showing the temporal pattern of car crime in each

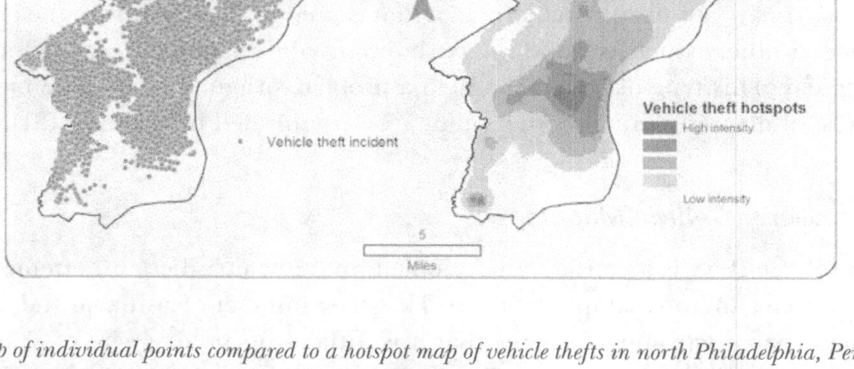

Figure 21-6

A simple map of individual points compared to a hotspot map of vehicle thefts in north Philadelphia, Pennsylvania. The sheer volume of offenses in map A shows the value in using hotspot techniques to show practitioners where the main problem areas are (map B). See Color Plate.

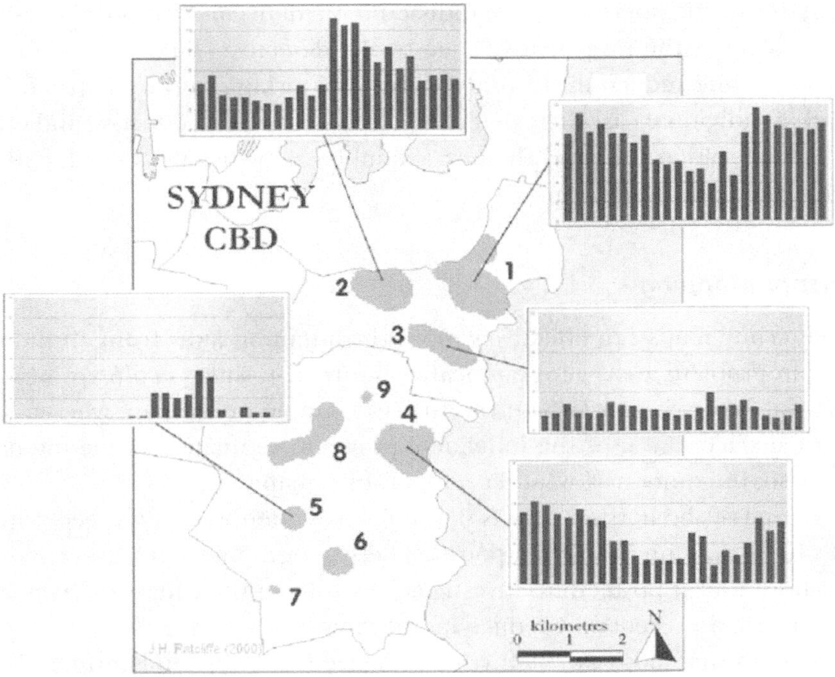

Figure 21-7

Vehicle crime hotspots for the eastern suburbs of Sydney, Australia [31]. Each vertical bar represents the volume of crime estimated in each hotspot for each hour of the day. Each graph starts with the hour from midnight to 1:00 am and continues through to the last hour of the day. The higher the bar, the greater the amount of crime. For example, hotspot 2 appears to have a sudden surge in crime during the middle of the day, just around 2:00 pm, whereas hotspot 5 is only victimized in the afternoon and evening. (Reproduced with kind permission of Springer Science and Business Media.)

hotspot area. Each bar in the graphs shows the proportion of car crime for each hour of the day, running from midnight to midnight. As can be seen, some crime hotspots (numbered 1 and 4) have a vehicle crime problem that is predominantly during the evening and night, whereas others (such as 2 and 5) are shopping centers and have car crime problems during the day. This type of image can help a problem-oriented focus. The methodology for this type of mapping can be found in an article published by Ratcliffe [31].

21.5.3 Recovered Vehicle Mapping

Although vehicle theft is a considerable problem in many jurisdictions, attempts to catch car thieves often run into a simple problem. The sheer number of available and unattended cars in the urban environment means that it is difficult to predict where a car thief will strike next. Although the application of environmental criminology theory can help in this task, there are other spatial properties of car crime that a GIS can help with.

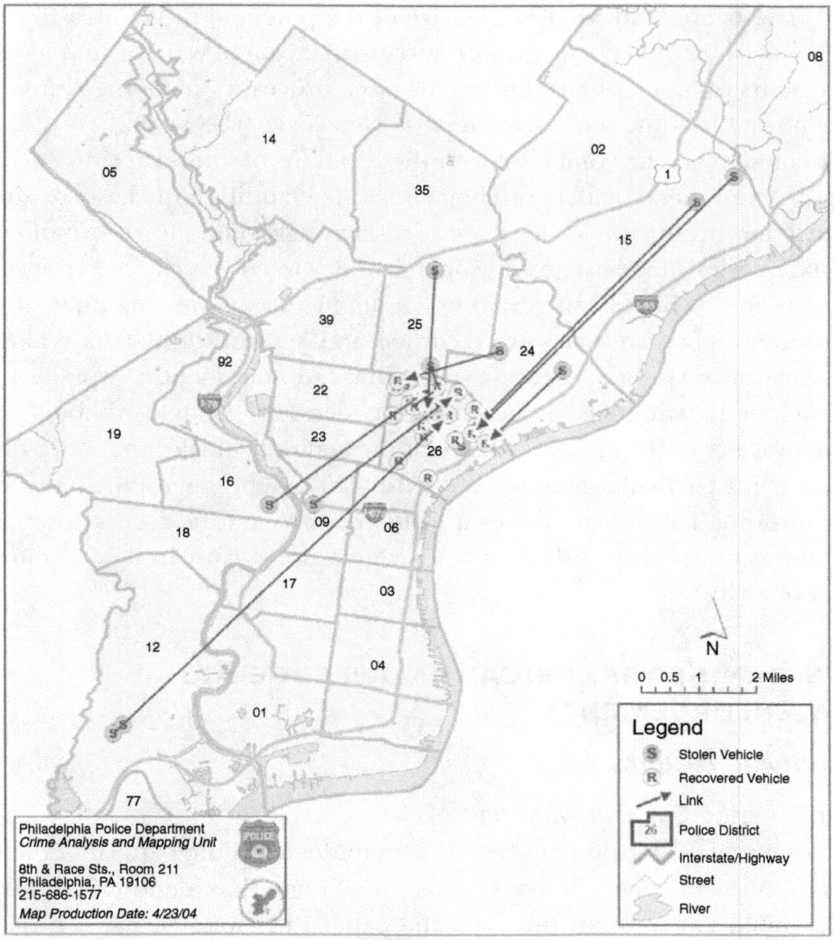

Figure 21-8

The point of theft and the point of recovery for stolen vehicles that were discovered in a police district in Philadelphia, Pennsylvania. Although most of the cars were stolen and recovered within the same police district, it is noticeable that some stolen cars that were recovered in close proximity were stolen from some distance away. In particular, two cars were stolen from many miles away in the southwest part of the city, whereas two other cars were stolen from the northeast. These vehicles may have been stolen to order or were used as transport to get to a particular place by the offender. See Color Plate.

In Figure 21-8, the focus of the map is less on where the cars have been stolen from and more on where the abandoned vehicles have been located. The value of this type of map was demonstrated by a police officer at an East London (UK) police station who went to considerable effort to determine that one offender was stealing cars from all over London, cars which would turn up on a Friday and Saturday night within less than a quarter mile of each other. He eventually discovered that a known offender was visiting nightclubs all over London and then stealing a car to get home. While an attempt to catch him stealing a car would have been fruitless given the size of the potential target area (a city of nearly 10 million people), the officers were more successful laying in wait for him to return home in the early hours of Saturday morning. This same process of mapping recovery locations has also been used to some success in Newark, New Jersey (USA) [32].

This type of map can be combined with the locations of known recidivist car thieves or with a map of scrap metal dealers or chop shops. The addition of these, or similar, layers can often help an investigator make sense of a seemingly random pattern of events.

Additional queries can assist an investigator when stolen vehicles are recovered. Examination of recovered vehicles can reveal much about the modus operandi of a car thief. If modus operandi characteristics are recorded and kept in the database along with the spatial information relating to location, then this can help identify a pattern of offenses that may have been committed by the same offender. For example, although cars may be stolen from a variety of locations, a cluster of burned-out vehicles that have been deliberately set on fire in a particular manner may exist. This might suggest to an investigator that attempting to prevent the thefts across a wide area may not be a good use of resources, whereas attempts to catch the offender destroying a stolen car can be concentrated into a smaller part of a city.

21.6 HOW CAN GEOGRAPHICAL KNOWLEDGE AID LAW ENFORCEMENT?

21.6.1 Identifying Patterns

A/ Problem-Oriented Crime Prevention

The key to comprehend a group of offenses is to understand the pattern of the crimes: the reason that links the incidents. It may be that the offender has identified a weakness in the security of a public car park. In this case, the pattern becomes clear when the events are mapped and the points cluster at the car park. In another pattern, a group of offenders may have discovered that a particular make of car is vulnerable to a certain new type of attack. In this case, the spatial pattern may be broader and less clustered; however, it still exhibits some clustering in the general area of residence of the offenders. In both cases, the offenders have identified systemic weaknesses and are exploiting opportunities to commit crime. The first is a weakness in building security, and in the second example the weakness is in the manufacture of a vehicle. In both cases, the investigation can be more effective if the investigator can identify and resolve the weakness, or in other words, prevent the opportunity. Most opportunities are inherently spatial, and this is where GIS use is fundamental in the working of problem-oriented policing.

B/ Hotspots

GIS can not only identify the most crime-ridden places, but also the most crime-ridden times. Understanding of these spatiotemporal patterns is essential if patrolling officers or crime prevention schemes are to be effective in both a spatial and a temporal sense. Indeed, one police director describes a vehicle theft CompStat meeting where "during the discussions, it became apparent that the time of day was a critical factor" [32]. Hotspot geography is quite flexible. It can therefore be used at a scale large enough for a police chief to determine broad patrol problems and to allocate resources across police districts, and it can also be used at the microlevel such that individual blocks and streets can be mapped.

C/ Potential Routes

With the capacity to map patterns of auto theft and vehicle recovery, it is possible to predict likely routes of an offender or group of offenders. Armed with this knowledge, police can direct their attention to the most likely times and places where offenders may operate. Indeed, with the added value of attribute data matched to crime events, it may also be possible to predict an offender's favorite vehicle type. Resources can be sited at pinch points along routes (such as freeway ramps or bridges) or in the vicinity of the vehicle dump locations.

21.6.2 Targeting Crime Detection Technologies

Once hotspots or potential routes for stolen vehicles have been identified, there are a number of different theft prevention technologies that can help reduce the incidence of car theft. This section focuses on two crime fighting resources that are most effective when deployed with a knowledge of spatial patterns: license plate reading technology and bait cars.

A/ License Plate Reading Systems

The latest technology that has become available is automated number plate recognition, also known as license plate reader (LPR) (see Subsection 18.3.4). These systems are composed of digital cameras that feed images to optical character recognition software. This software converts the digital image of a license plate into a digital string of letters and numbers. The final aspect of the system is a link to a (user-provided) database of stolen or suspicious vehicles. The system can also be used for access control to secure locations, as long as a database of approved vehicles exists, but the value to an investigator is in the detection of stolen vehicles. For example, a police department in Texas has managed to tap into the LPR system used by US Customs to scan the tags of cars crossing the border with Mexico. The police department runs the license number through a database of stolen and flagged license numbers, and cars that come up with a hit are stopped at the checkpoint [33].

Mobile systems are now available, allowing police to patrol an area and scan parked and mobile cars from the comfort of a police cruiser. This type of vehicle check will be most effective when used in hotspots where offenders are known to steal or abandon stolen vehicles.

B/ Bait Cars

Bait cars are becoming more popular across the United States and Canada and have been used successfully in the United Kingdom for some time (see Subsection 18.4.7). These law enforcement vehicles range in sophistication, but the basic concept is to tempt an offender to steal a vehicle, which is in fact disabled in such a way that the offender can be caught. There are two main ways to use a bait car. First, it can be used to catch an offender as soon as they enter the car or, second, to follow the vehicle once stolen to locate potential chop shops where the vehicle may be broken down for stolen parts. The particular technology within bait cars varies. Some simply alert officers nearby when the car has been broken into. This requires police officers to be nearby on surveillance. More advanced models lock the offender in the vehicle until police arrives. Other forms of technology allow officers to track the location of the car, through the use of radio transponders fitted into the vehicle. When the vehicle is reported stolen, the tracking unit is activated and the police alerted. The unit broadcasts a signal that is detected by tracking computers, which are fitted in police cars or at fixed land sites [34]. Finally, if a vehicle is stolen and is moving, remotely operated ignition immobilizers and fuel cut-off switches can help police stop the vehicle safely.

The Taskforce for Regional Autotheft Prevention (TRAP) in Los Angeles, California (USA) uses surveillance officers to maintain surveillance of the vehicle. This has been referred to as a traditional watch-and-wait method [35]. TRAP installs videotapes in its vehicles to improve the conviction rate in court. The six TRAP teams make between 5 and 15 arrests, on average, in a four-hour period [35]. While the TRAP team conducts active surveillance, the use of global positioning system (GPS) technology allows cars to be parked unattended, letting police officers perform other duties because they are no longer needed for surveillance. As a result, there is less demand on police and law enforcement officers are only required once a crime has been committed. A number of forces have utilized this labor-saving approach, including the Ohio State Highway Patrol, who used 10 GPS-enabled bait cars in Cleveland, Columbus, Cincinnati, Elyria, and Toledo [36].

In all cases, the key to successful use of bait cars lies in the accurate targeting of the right make and model of vehicle and in the optimal sighting of the bait car location. As Sallybanks notes [34]: "Local level analysis should be carried out to establish the precise nature of the vehicle crime problem in the local area. Knowledge of the types, makes and models of vehicles targeted by thieves and the specific hot spot locations is crucial to the successful deployment of such an operation."

21.7 A PRIMER ON MAPPING SERVICES

If an investigator wants to map vehicle crime, the first place to look into is often the local police department. Unfortunately, it is often the case that many detectives are unaware of the resources available locally, and many police departments have developed local mapping systems, which already map vehicle crime. If no crime mapping systems are available within the police department and the investigator wants to be instrumental in starting a system, there are a number of possible routes. The first is to tap into existing systems outside of the police department. Many local municipalities run GIS services to support electric, gas, and communications facility management, and these systems can often be converted to mapping crime data.

Although local companies are often enthusiastic to offer mapping to police (at a cost, of course), it is advised to stick with a system from one of the two large companies: ESRI (http://www.esri.com) or MapInfo (http://www.mapinfo.com). They already have years of experience developing GIS, they are unlikely to disappear overnight, and they have a number of third-party software solutions that integrate with their systems.

A number of sources of help are available. Locally, most universities have a GIS capability in their geography department. Some also maintain a GIS capability in criminal justice departments. These university resources are often willing to help police in return for access to crime data, which they can use for teaching and research. A local collaboration can therefore be fruitful for both the university and the police. Furthermore, in the United States there are federal resources available. Advice can be sought from the Mapping and Analysis for Public Safety (MAPS) program of the National Institute of Justice (http://www.ojp.usdoj.gov/nij/maps); their website contains a primer on crime mapping and also a list of software resources, grant opportunities to pay for a GIS, and other related material. This site is also of value to readers from overseas. Their conferences are particularly valuable for people new to crime mapping.

State and regional law enforcement training programs now offer GIS and crime mapping training at reasonable cost. Training in crime analysis, which includes a discussion of crime mapping, is also available through the International Association of Chiefs of Police (IACP, http://www.theiacp.org). The Crime Mapping and Analysis Program of the National Law Enforcement and Corrections Technology Center (NLECTC) has been offering crime mapping training to US law enforcement for a number of years (http://www.nlectc.org/cmap). Further information on other training opportunities is available from the MAPS program, which also maintains a list server. The list server regularly carries training opportunities in different parts of the United States and, less regularly, overseas.

Finally, there are a number of texts available. The broader field of GIS is burgeoning with introductory books, but investigators may find that a search for crime analysis books may be more useful. Two recent additions to the field are *Introduction to crime analysis: Basic resources for criminal justice practice*, by Deborah Osborne and Susan Wernicke [37], and *GIS and crime mapping*, by Spencer Chainey and Jerry Ratcliffe [20].

21.8 DATA CHECKLIST

Investigators can help their crime mapper by accurately gathering the following information:

- The theft or crime location: This is often best recorded as the exact address outside which the car was parked at the time of the crime. Geocoding engines can usually map addresses with high levels of precision.

- The vehicle recovery location: Again, a street address is often the best information to gather.
- The details of the car: Make, model, year, color, and type of car will help a crime mapper search for patterns in crime.
- The time of the offense: If the exact time is not known, there are techniques (such as aoristic analysis [31]) that can estimate the offense time for crimes that are recorded with start and end times. These are the times when the vehicle was last seen unmolested (start time) and when it was first discovered as being the target of a crime (end time).
- Details of the theft location: For example, was the vehicle parked in a garage, or in a public car park.
- General modus operandi details: For example, was the vehicle locked at the time of the offense, how was the vehicle broken into, how was it defeated, if it was arsoned, or stripped of parts, or abandoned in a remote area. These characteristics can be queried in a database if recorded.

BIBLIOGRAPHY

[1] Federal Bureau of Investigation (2005) *Uniform Crime Reports: Crime in the United States,* http://www.fbi.gov, last access performed on May 27, 2005.

[2] Van Kesteren J, Mayhew P, and Neuwbeerta P. (2001) *Criminal victimisation in seventeen industrialised countries,* Wetenschappelijk Onderzoek-en Documentatiecentrum, The Hague, The Netherlands.

[3] Ratcliffe JH, and McCullagh MJ. (2001) Chasing ghosts? Police perception of high crime areas, *British Journal of Criminology,* **41**(2), pp 330–341.

[4] Harries K. (1999) *Mapping crime: Principles and practice,* US Department of Justice, Washington, DC.

[5] Dodd T, Nicholas S, Povey D, and Walker A. (2004) Crime in England and Wales 2003/2004, *Home Office Statistical Bulletin,* Home Office, London.

[6] LeBeau J. (2000) *Demonstrating the analytical utility of GIS for police operations: A final report,* National Institute of Justice, Report NCJ187104, Carbondale, IL.

[7] Mamalian CA, and LaVigne NG. (1999) The use of computerized crime mapping by law enforcement: Survey results, *National Institute of Justice Research Preview,* Washington, DC.

[8] Weisburd D. (2001) Translating research into practice: Reflections on the diffusion of innovation in crime mapping. In: *Proceedings of the Fifth Annual International Crime Mapping Research Conference,* National Institute of Justice, Dallas, TX.

[9] Cohen LE, and Felson M. (1979) Social change and crime rate trends: A routine activity approach, *American Sociological Review,* **44**, pp 588–608.

[10] Felson M. (1998) *Crime and everyday life: Impact and implications for society,* Pine Forge Press, Thousand Oaks, CA.

[11] Clarke RV, and Cornish DB. (1985) Modeling offenders' decisions: A framework for research and policy. In: *Crime and justice: An annual review of research,* ed Tonry M and Morris N, University of Chicago Press, Chicago, IL.

[12] Clarke R. (1992) *Situational crime prevention: Successful case studies,* Harrow and Heston, Albany, NY.

[13] Brantingham PJ, and Brantingham PL. (1984) *Patterns in crime,* Macmillan, New York, NY.

[14] Brantingham PL, and Brantingham PJ. (1993) Environment, routine, and situation: Toward a pattern theory of crime. In: *Routine activity and rational choice,* ed Clarke RV and Felson M, Transaction publishers, New Brunswick, NJ.

[15] Rengert GF, and Wasilchick J. (1985) *Suburban burglary: A time and place for everything,* Charles C. Thomas Publishing, Springfield, IL.

[16] Brantingham PL, and Brantingham PJ. (1993) Nodes, paths and edges: Considerations on the complexity of crime and the physical environment, *Environmental Psychology,* **13**, pp 3–28.

[17] Rossmo DK. (2000) *Geographic Profiling,* CRC Press, Boca Raton, FL.

[18] Department of the Environment (1987) *Handling geographical information: The report of the committee of enquiry,* Her Majesty's Stationary Office, London, England.

[19] Ratcliffe JH. (2004) Geocoding crime and a first estimate of an acceptable minimum hit rate, *International Journal of Geographical Information Science,* **18**(1), pp 61–73.

[20] Chainey S, and Ratcliffe JH. (2005) *GIS and crime mapping,* John Wiley and Sons, London, England.

[21] Clarke RV, and Weisburd D. (1994) Diffusion of crime control benefits. In: *Crime prevention studies,* ed Clarke RV, Criminal Justice Press, Monsey, NY.

[22] Green L. (1995) Cleaning up drug hot spots in Oakland, California: The displacement and diffusion effects, *Justice Quarterly,* **12**(4), pp 737–754.

[23] Painter K, and Farrington DP. (1999) Improved street lighting: Crime reducing effects and cost-benefit analyses, *Security Journal,* **12**(4), pp 17–32.

[24] Eck JE, and Spelman W. (1987) *Problem solving: Problem-oriented policing in Newport news,* Police Executive Research Forum, Washington, DC.

[25] Goldstein H. (1990) *Problem-oriented policing,* McGraw-Hill, New York, NY.

[26] Scott MS. (2000) *Problem-oriented policing: Reflections on the first 20 years,* Office of Community-Oriented Policing Services, US Department of Justice, Washington, DC.

[27] Chaiken JM, Greenwood PW, and Petersilia J. (1977) Criminal Investigation Process—A Summary Report, *Policy Analysis,* **3**(2), pp 187–217.

[28] Eck J. (1983) *Solving crimes—the investigation of burglary and robbery,* Police Executive Research Forum, Washington, DC.

[29] Brandl SG, and Frank J. (1994) Relationship between evidence, detective effort, and the disposition of burglary and robbery investigations, *American Journal of Police,* **13**(3), pp 149–168.

[30] Singh D. (2001) Community-oriented investigation at the North Miami Beach police department, *BJA Practitioner Perspectives,* US Department of Justice, Report NCJ 185367, Washington, DC.

[31] Ratcliffe JH. (2002) Aoristic signatures and the temporal analysis of high volume crime patterns, *Journal of Quantitative Criminology,* **18**(1), pp 23–43.

[32] Santiago JJ. (1998) The problem of auto theft in Newark. In: *Crime mapping case studies: Successes in the field,* ed LaVigne N and Wartell J, Police Executive Research Forum, Washington, DC.

[33] DeFranco LM. (2000) Share, and share alike, *Law Enforcement Technology,* **27**(3), pp 68–72.

[34] Sallybanks J. (2001) Assessing the police use of decoy vehicles, *Police Research Series Paper 137,* Home Office, London, England.

[35] Mertens J. (2003) Thieves tempted by bait, *Law Enforcement Technology,* **30**(4), pp 36–43.

[36] Morckel KL. (2002) GPS vehicle tracking improves auto theft enforcement, *Police Chief,* **69**(12), pp 39–42.

[37] Osborne D, and Wernicke S. (2003) *Introduction to crime analysis: Basic resources for criminal justice practice,* Haworth Press, Binghamton, NY.

CHAPTER 22

INTERNATIONAL COLLABORATION THROUGH INTERPOL

Alain G. Barbier

22.1 INTRODUCTION

Mr. YS, salesman for an important German car manufacturer in Lyon (France), had been in his office for about 20 minutes. The morning was going slowly with very few customers. Suddenly, Mr. YS heard a noise coming from the showroom. He ran out of his office to investigate and was confronted by a masked man brandishing a gun. In the showroom, Mr. YS could see a number of other armed men who were shouting at the staff, telling them to be quiet.

The gang asked for the keys of the five most expensive cars in the showroom, which the manager handed over. The men shouted at the staff to open the garage door. At this moment, Mr. YS tackled the gunman nearest to him and tried to overpower him, but a second member of the gang came over and fired his weapon into the ceiling. Mr. YS quickly understood that an insured car was not worth his life. In just 12 minutes, the six men had stolen five cars worth EUR 400,000 (about USD 500,000).

In such cases, what does Interpol, as a unique and global international police organization, put in place to prevent and fight such incidents, and auto theft in general, around the world? Interpol supports member countries to combat the proliferation of organized vehicle crime in their country, and training is also given to help develop an integrated and multidisciplinary approach. This chapter presents Interpol, its role in the worldwide battle against auto theft, and the different programs assisting law enforcement agencies in their operation.

22.2 HISTORY

The notion of tackling car crime is not a phenomenon of the early 1990s on the part of those seeking to adapt international police cooperation policy to major changes in the socioeconomic environment of various countries. As early as the 1970s, the idea of taking a more systematic and practical approach to such crime had already begun to emerge in international circles and especially at the International Criminal Police Organization (ICPO), or Interpol. Over the years, the project gradually took shape, and the stolen vehicles unit was established.

The first task of the Interpol's stolen vehicle unit was highly administrative; it consisted of simply keying information on stolen vehicles into Interpol's criminal database, along with details of criminals actively involved in vehicle trafficking. In the late 1980s and early 1990s, the unit gradually became more specialized, building up knowledge of the criminal environment and context, which highlighted the major problem facing police cooperation: the exchange of information on stolen vehicles or lack thereof.

In 1996, with the help of a brilliant engineer, Interpol developed the first global database for stolen vehicles, known as the Automatic Search Facility-Stolen Motor Vehicles (ASF-SMV). The rollout started for Interpol member countries on April 1, 1996 [1]. Alongside this important change, the first moves were made to establish contact with external partners, both public and private. From 1995 to 1999, these early initiatives, and the resulting developments, led to increasingly widespread use of Interpol's stolen vehicle database.

Interpol's program to combat illegal traffic of vehicles set up a three-pronged action plan consisting of the exchange of reliable qualitative information, training initiatives, and the transnational coordination of operations to combat organized vehicle crime. Theft detection methods are in constant evolution to adjust to the criminal activity. In the early 1980s, when vehicle theft began to increase, national authorities initially responded by creating a national database of vehicles reported as stolen. This tool was designed to facilitate the search for and identification of stolen vehicles by law enforcement agencies. Over time, the volume of information on stolen cars rose in inverse proportion to the number of police officers able to carry out these checks. The database, as it was, became obsolete because criminals managed to stay one step ahead. They transferred the vehicle identification number (VIN) of already registered vehicles to stolen cars, thus hiding their illegal origin.

In response to these two problems—limited police capacity and the registration of stolen cars with VINs of existing registered vehicles—countries developed a link between the database of stolen vehicles and that of registered vehicles. This was designed to prevent the re-registration of stolen vehicles listed in the police database within a same country.

Starting in the mid 1990s, police authorities also responded by setting up closer cross-country links, or even creating supranational databases, at regional level, such as the Schengen Information System (SIS) for the member countries of the European Union (see Chapter 15). On a transnational level, only Interpol had a truly international tool with the stolen motor vehicles database ASF-SMV. Taking a lead from the police, national vehicle registration services in Europe developed an international network for the exchange of information on registered vehicles, called EUCARIS (the EUropean CAR and driving license Information System), which is a network that allows participating countries to exchange data related to motor vehicles and driver's licenses via their vehicle registration authorities. It represents an on-line method of preventing the criminal laundering of stolen vehicles.

Is this the definitive solution? Of course not, although it does mark a considerable step forward in the international effort to combat illegal trafficking of vehicles. The basic prerequisite remains the participation of all countries, and this has not yet been achieved. There is evidence to show that criminals modify their behavior as a result of enhanced police cooperation and information exchange.

Many international organizations have in place a number of resolutions and agreements in relation to vehicle crime. These organizations include the United Nations, the World Customs Organization, the Council of Europe, Interpol, and the European Council of Ministers of Transport [2–5]. They have all recognized the urgent need for their member countries to engage in various activities such as the rapid exchange of information, the use of Interpol's ASF-SMV database, the cooperation with other parties concerned (vehicle manufacturers, registration authorities, insurance companies, etc.), and the systematic preregistration checks. In total, more than 75 recommendations or resolutions have been adopted by international organizations in relation to trafficking of stolen vehicles. However, following the events of September 11, 2001 in the United States, there was a dramatic shift in priority levels for different types of crime. Unfortunately, criminals can not only target countries where technological developments in crime fighting are not particularly advanced and where, if available, the Interpol's database is not properly used, but also countries where that crime priority has completely disappeared.

22.3 EXTENT OF THE PHENOMENON

22.3.1 Main Picture

There are more than 9,000 vehicles stolen every day! During the time it takes to read this chapter, between 70 and 140 vehicles will have been stolen worldwide. Around the world, a vehicle is stolen every nine seconds [6].

Vehicle theft and trafficking is a significant problem throughout the world in terms of economic costs and distress caused to victims. According to one study of nearly 60 countries in Europe, North America, Africa, and Asia, every year at least 3.4 million vehicles are stolen and approximately 1.6 million are recovered [7]. The same study estimates the current profits from such crime to be higher than the combined annual profits of two of German's leading car manufacturers, Mercedes and BMW. The study arrives at a final profit of EUR 21 billion (about USD 25 billion). These impressive figures provide some insight of the financial gains open to criminal organizations. Of course, it is necessary to bear in mind that this "dirty money" will be used by criminal organizations to finance other illegal activities such as terrorism, drug trafficking, bribery of officials, arms dealing, and trafficking of human beings or will be invested in the global economy.

22.3.2 Measuring Vehicle Crime

There is no standard measurement of the impact of vehicle crime. One of the most solid statistics is the cost of insurance claims, which is used by most countries in their threat assessment. Unfortunately, it is very difficult to calculate the full impact of vehicle crime. Various measurements exist, each of which taking different factors into account as presented hereafter.

A/ Direct Financial Cost of Vehicle Crime

The direct cost of vehicle crime is mostly borne by insurers. This cost does not include the direct cost for the justice, uninsured losses, and the indirect costs.

B/ Cost in Anticipation of Crime

Potential victims of crime take defensive and precautionary measures to reduce the risk of becoming a victim of car theft. These include increased vehicle security such as alarms, engine immobilizers, steering locks, and other anti-theft systems. Other precautionary measures also include parking vehicles in well-lit low-risk areas. In addition to this, individuals may take out insurance policies to offset many of the financial losses associated with a vehicle theft.

C/ Cost as a Consequence of Crime

When a vehicle is not recovered, the victim bears the cost of replacing it. In many cases, the victim is insured and is limited to the payment of the deductible, whereas the insurance company pays the remainder. However, if the victim is not insured or does not file a claim, then the entire cost falls on the victim.

D/ Cost in Response to Crime

Vehicle crime has a financial impact on the justice system and law enforcement agencies, including costs for police, prosecution, and any other attached legal services and aid. Of course, the costs incurred by the courts, jurors, and witnesses as well as the administration of the prisons should also be taken into account. In addition, in response to crime, insurance companies, having borne the bulk of the direct costs of vehicle theft and thus vehicle crime, will in turn increase premiums for both the victims and nonvictims of crime to compensate for the loss.

E/ Emotional and Physical Impact

Many opinion polls relating to crime and security show that the four major factors most often cited by citizens, in an order that differs from country to country, are terrorism, drugs, theft (in the home or of vehicles), and street violence. These factors are those which contribute most to the feeling of insecurity. Until several years ago, these four types of crime were dealt with in a different manner. Terrorism and drugs were, by definition, placed in the "organized crime" category, whereas the others tended to be classified as urban crime, with no international significance and no links at all with organized crime. However, we have to recognize that crime linked to vehicle trafficking requires structures, which are in some way organized, with theft being merely one stage in the criminal process. Hence, illegal trafficking of vehicles is perceived by the authorities as both general and organized crime. Notwithstanding this, crimes such as these, no matter at what level they occur, are upsetting and heighten our sense of insecurity. This feeling of insecurity reflects people's perception of crimes taking place on their doorstep or widely reported in the press.

Vehicle trafficking contributes to the sense of insecurity, because of the sheer number of vehicles stolen daily (one every nine seconds worldwide) and because each of us runs the risk of buying a second-hand car, which has been stolen, only to have it seized during a police check. Finally, the fact that criminals are now resorting to increasingly violent means to steal vehicles is another source of insecurity.

22.3.3 Where Do All These Stolen Cars Go?

The only finding to emerge from data on vehicles stolen and subsequently recovered is that North America and Europe are the biggest suppliers of such vehicles [8]. The demand for smuggled stolen vehicles remains high for a variety of reasons, including relative costs, availability, desirability, and as a method of payment for other types of criminality (such as exchange of drugs), thus avoiding the necessity for money to change hands. Intelligence received and developed by various law enforcement agencies demonstrates the existence of an important international market for stolen vehicles, notably in Eastern Europe, The Balkans, Africa, and the Middle East.

To establish with any degree of accuracy what routes are used by the criminals, a certain amount of basic information is required. The data we have at the present time relates to the places where vehicles are found (not conclusive, given the relatively small number of vehicles recovered as far as theft for financial gain is concerned), the places where vehicles are intercepted at some stage in the process (too rare), and the qualitative information resulting from operational analyses (too infrequent).

Given the established fact that a stolen vehicle is swiftly disguised to make it "resellable," it is then considered as a second-hand car and becomes part of a lawful process. The path it takes is the same as the path followed by any legitimate second-hand car, which is exported and imported.

The routes of these vehicles may be inferred from the economic logic of supply and demand. Large American cars with a fuel consumption rate too high for the European market are rarely found on the narrow roads of western Europe, where fuel-saving engines are a commercial asset for car builders. Similarly, off-road vehicles will find a buyer more easily in countries where road conditions necessitate their use. The distribution of stolen vehicles thus reflects consumer demand.

Vehicles are produced by a small number of manufacturers based in a handful of countries. New cars are thus transported to consumers over the world by road, rail, air, and sea. Use of vehicles is regulated in each country by a registration system. Once registered and on the road, they may be used locally, nationally, or internationally. During its life, a car can have several owners and users, thus giving rise to a global market for second-hand vehicles. "New" vehicles are offered for sale by dealer's networks, car markets, or directly by the owner. Many new or second-hand cars are exported by boat.

This vast business is vitally important to a country, both socially and economically. However, it offers tremendous opportunities for criminals seeking large profits. Hiding behind yet actively involved in the legal market, they take rational and strategic decisions. Regardless of the particular conditions or systems in place, the fact that the vehicle constitutes valuable "raw material" means that criminals will adapt to any circumstances and exploit the weaknesses of the system. One of the current flaws in international cooperation is no doubt the fact that customs units working in seaports and registration services, which carry out physical or document checks on a vehicle before its registration, are some of the weakest links in the criminal intelligence chain.

Dr. Cyriel Fijnaut, professor of criminology and penal law at the Katholieke Universiteit Leuven in Belgium and at the Katholieke Universiteit Brabant in The Netherlands, explained on a strategic level the extent of and the main routes taken by stolen vehicles [9]: "The phenomenon (of illegal traffic in vehicles) is an implicit illustration of global imbalances. In less developed countries, it is a question of survival. Transport is also a necessity here. Vehicle theft and trafficking are among the illegal activities which improve the distribution of wealth and well-being."

22.4 INTERPOL APPROACH

22.4.1 The Illegal Process

Working on the basis of the notion that a vehicle is any motorized device used to transport people or goods, Interpol considers that the criminal activity of vehicle trafficking encompasses various offenses such as theft, fraud, embezzlement, misuse of entrusted property, fraudulent alteration of documents or VINs, use of forged documents or VINs, possession of stolen property, misrepresentation, corruption, tax fraud, offenses committed with accomplices, and money laundering. Those offenses could be found in the penal code of Interpol's member countries, but the offense called illegal trafficking of vehicles does not exist per se. In the light of such offenses, vehicle crime has been defined as any activities that further the marketing of a stolen vehicle for illicit gain.

To tackle the phenomenon internationally, Interpol studied the "lawful" process of a vehicle from its manufacture to its sale. This allowed the international police organization to determine which stages are most vulnerable and need international attention to offer tailored support to the national police services of its member countries. Figure 22-1 shows the different stages undertaken by a vehicle from its birth to its final use. The stages for which Interpol provides support have international involvement.

Once the vehicle has reached the "use" phase of the lawful process, it may continue along one of several paths. The main possibility is of course, that of theft. There are different opportunities for criminals to launder a stolen vehicle before it is put back on the market by using the identification of a vehicle that has been involved in an accident, exported, sold, or abandoned. After the theft, offenders carry out various operations, including making the stolen vehicle unidentifiable to enable the future buyer to register the vehicle once it has been sold. Such activities may be broadly divided into three main stages: theft, alteration, and sale.

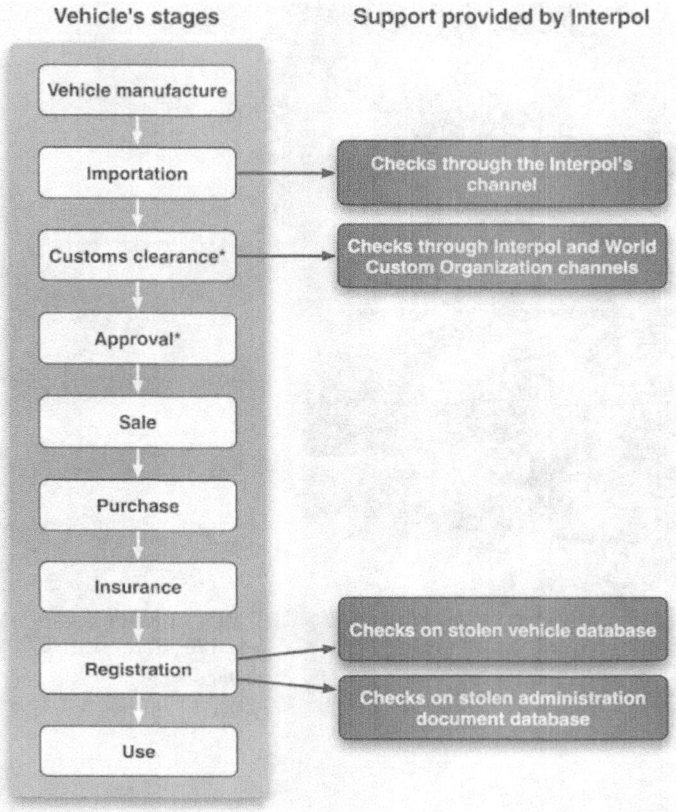

Figure 22-1

*The different stages a vehicle undergoes from its manufacturing to its use on public roads with the different supports provided by Interpol to prevent and fight the market of stolen vehicles. *Customs clearance and approval procedures depend on the regulations of each country, community, or union (e.g., in the European Union, approval is more flexible and systematic customs controls have been abolished).*

Figures 22-2 and 22-3 show a Mercedes crushed by a train, which was then registered at a later date. Was the car completely rebuilt or is it the result of the laundering of a stolen car? Indeed, it was a stolen car where the identification plate and chassis number had been changed based upon the ones from the damaged Mercedes.

22.4.2 Methodological Considerations

A/ Hypothesis

To date, no scientific studies, either national or international, have been conducted in an attempt to identify or define specific categories of theft or to provide a breakdown of crime figures in absolute and relative terms. There is, however, a general hypothesis according to which the relative number of so-called professional thefts is rising every year and criminal

Figure 22-2

Mercedes crushed by a train; its identification plate and VIN were used at a later date to launder another car of the same make and model (see Figure 22-3).

Figure 22-3

Stolen Mercedes re-registered using the identification plate and chassis number of the crushed car illustrated in Figure 22-2.

organizations or organized criminal groups are making increased net profits from their vehicle trafficking activities [10]. A number of key indicators support this hypothesis as described below.

B/ Trends in Vehicle Thefts and Carjacking

The days of stealing the neighbor's car just to go for a spin are now over. Criminals prefer to steal vehicles without damaging them and to obtain the original documents, which makes subsequent disposal quicker and more profitable.

Although it is not a new phenomenon, recent years have witnessed exponential growth in carjacking statistics in all countries around the globe. Carjacking is defined by Interpol as "the theft of a motor vehicle with violence or threatening behavior to the driver or one of the passengers" [11]. The key point in this definition is that it highlights the pattern that criminals do not hesitate to make contact with and use violence on their victim to obtain the desired goods. In the past, theft was largely committed by criminals who avoided all contact with their victims. Acting in this manner gave the thief something of a head start on the police; he or she can be far away before an investigation even begins. Although this type of theft is much more visible and affects the population to a much greater extent than traditional theft, in most countries the number of carjacking cases represents a tiny proportion of all thefts committed.

The reader should be made aware, however, of current developments in carjacking. This type of theft usually takes place on public highways, and the offenders are thus more likely to be intercepted by the police. This is why some of them prefer "homejacking," which is defined by Interpol as "the theft of a motor vehicle with violence or threatening behavior to the occupiers of a property" [11]. The thief obtains the vehicle keys and documents before stealing the car. If the offender has no contact with the victim in thefts of this type, Interpol then speaks of the "garage procedure" [11]. A further development is that of stealing directly from new car showrooms as presented in the introduction of this chapter. Small criminal groups arrive in a showroom with enough accomplices to drive the stolen cars from the showrooms to a secret place before export. Carjacking and homejacking are most common in the Sub-Saharan Africa, America, and Europe. Nevertheless, this type of crime is now beginning to emerge in Asia and Australia.

One of the likely causes of this type of theft is the more widespread sale of modern vehicles with effective built-in systems of protection (see Chapter 8), and the need for criminal organizations to meet the endless demand from potential buyers. Another cause is the stricter controls on registration from governments. Another reason may be inferred from the many advantages that such thefts offer; keys and documents can be obtained without damage to the vehicle, thus ensuring a high resale value and avoiding the need to replace keys and lock cylinders and/or produce false papers.

The theft of plant equipment conerns any equipment used in construction, demolition, and quarrying as well as agricultural machinery and trailers. Although it has a financial cost, it is generally speaking seen as a low police priority. This combined with a relative ease of stealing plant equipment, the remote and unsecured areas where it is kept, and the high profit margins available make it a very attractive target for organized crime. This is borne out by the recovery rate for stolen plant equipment, which is broadly estimated to be lower than 5% worldwide [12].

C/ Enhanced Vehicle Protection

Developments in car manufacturing are dictated mainly by customer demand, and car builders have responded by setting up engineering and design departments to devise efficient protection systems for new vehicles (see Chapter 8). The advent of individual and sectoral[1] motor insurance has prompted those manufacturers with the largest market shares and the highest number of vehicles stolen to build in effective systems of protection such as immobilizers or tamper-proof locking systems. In this manner, car manufacturers avoid being indirectly penalized by the actuarial measures taken by insurance companies.

D/ Recovery Rate

Analysis of vehicle theft and recovery statistics reveals that the number of thefts in absolute terms has fallen since the mid-1990s, whereas the proportion of unrecovered vehicles has slightly risen (from 45% to 60% in most industrialized countries) [12]. Nevertheless, this latter category includes an increasing number of vehicles less than five years old. These vehicles have the highest monetary value and are of most interest to criminal organizations.

E/ Involvement of Criminal Organizations

Over the last 10 years, large-scale investigations into criminal organizations or organized criminal groups have shown that auto theft is often used to help finance other activities (see Chapter 17). These activities can be criminal (terrorism, prostitution, drugs, weapons trafficking, burglary, armed raids, etc.) or noncriminal such as property investment.

F/ National Controls

In the aftermath of the events of September 11, 2001 in the United States, technological and political developments in national and international institutions have led to improved controls in the most concerned countries. These tighter controls, particularly on the importation and registration of vehicles, have prompted criminals to adapt their activities, with growing specialization and a more professional approach.

G/ Conviction and Sentencing

As each member country will confirm, the standard sentences for the various offenses associated with illegal traffic in vehicles are negligible, few criminals serve a full sentence,

---

[1] Sectoral means depending on the profile of the client (sex, age, professional activity), the type of car, and the region where the client lives.

and such penalties rarely represent the maximum possible. In some instances, exceptions to light sentences occur, such as the following example where a car trafficker was sentenced to 10 years of prison in Belgium.

Interpol received information from France that several cars stolen from tourists had been discovered for sale in a garage. Upon receipt of the message, the Interpol National Central Bureau in Brussels sent the information to a regional investigation unit in Liège, Belgium for further investigation. The first steps of the investigation indicated that seven different Belgian garages had sold the vehicles concerned in France, Luxemburg, Germany, The Netherlands, and Italy. A large-scale investigation was launched, and several months later an organized group led by three individuals was identified as being responsible for the thefts of more than 300 vehicles. The investigation revealed that a number of criminals were paid to forge the vehicles or to provide forged documents for the stolen vehicles. Others were responsible for the international distribution, and drug addicts were paid to sign commercial bills identifying them as garage owners. Interpol gave its support through its network, allowing the investigators from the countries involved to exchange information and to further the investigations in Belgium, France, Luxembourg, Italy, Germany, and The Netherlands. After two years of scrupulous searches, home searches, and accountability verifications, Mr. FH was arrested in Spain. However, the other two ringleaders were still free. He decided to protect his accomplices, took all the judicial charges, and was sentenced to 10 years by the Belgian Court.

22.4.3 Victims and Actors

In a study carried out for The Netherlands, Professor Cyriel Fijnaut wrote the following [13]: "One wonders how many people, both within and outside Western Europe, are involved. And what role do garages or scrap yards—legal and illegal—play? To what extent is the shipping industry involved in this business? How many papers must be altered or forged to ensure 'hassle-free' trade and transportation? How widespread is corruption among police and customs officers on vehicle trafficking routes?"

With regard to legal business, Dr. Fijnaut had the following to say [13]: "What does this crime mean for car manufacturers and insurance companies? We should first of all recognize the fact that most stolen cars will be replaced by new ones. The individual citizen then finds himself paying higher prices to reflect all the technical features required to counter theft—and there is a corresponding rise in insurance premiums."

From a police point of view, police officers could also be considered as victims due to the fact that vehicle crime is not a political crime priority even if the solving rate for vehicle theft is very low. Therefore, most international specialists in this field call for an organized approach to fighting such crime, approaching by proactive investigation mala fide dealers and setting up task forces or international operations in which the motor divisions of insurance companies, police forces, ministerial departments, customs authorities, and financial inspectors should actively participate.

22.4.4 International Operations

A/ Introduction

In the mid-1980s, police forces turned their attention to the transportation of stolen vehicles by boat. Following the initial success resulting from tighter controls, there were some remarkable changes in criminal behavior. Previously, stolen vehicles had been transported as they were; however, the tighter controls led to criminals making alterations to the car before its transportation. License plates from other countries were affixed to the cars or the identifying number on the chassis was changed, which of course made it much more difficult for police officers carrying out checks in ports to identify the vehicle. Later still, changes were observed not only in the chassis numbers of stolen vehicles, but also in engine numbers. Identification then became almost impossible.

The theft and systematic exportation of vehicles has evolved rapidly since the mid-1990s. Such changes occurred in response to the use of improved search techniques: Criminals have adapted their behavior to minimize the chances of being intercepted, in particular by:

- Transporting stolen vehicles in shipping containers filled with other materials (toilet paper, plastic buckets, etc.);
- Importing and transporting vehicles on boats along inland waterways or on trains, which enables criminals to escape police checks;
- Falsifying documents (of vehicle and/or container) or altering the vehicle itself (car ringing).

For their part, police authorities must respond to such trends by refining existing search techniques to increase the chances of intercepting offenders and stolen vehicles. It is indeed true that on the final leg of the journey out of its country or region of origin, a stolen vehicle often passes through a port. Supranational operations are therefore being mounted in certain parts of the world according to the need for improving the regional and international cooperation in the seaports.

B/ European Operation: An Example

International developments led to more focused activities in dealing with organized crime. Following the implementation of the SIS, politicians requested that a pilot project for all member states committed to the Schengen Agreement (most of the European Union countries at the time of those operations) be set up. The principal objective of the project, as formulated by the political body, was to get a good insight into criminal activities and activities of organized groups carried out in European Union territory or in transit through that area. The operation period was to last four months divided in three subperiods of three to five days each. During the operations, police officers and experts were exchanged.

Based upon the political principal objective, concretely, the activities consisted of joint teams from European countries to undertake physical control at the borders to detect stolen vehicles and arrest persons known as criminal in the Schengen database. Therefore, those teams were deployed overall on the European countries' borders and took part in several bilateral or multilateral operations focused on the north-south and east-west corridors and harbors. The operations were coordinated by national project leaders. The project leaders had prepared in secrecy the different time laps for the operation during preparatory meetings in 1996. The operations were carried out between April 1, 1997 and June 30, 1997. To reach a successful international operation, the project leaders had to prepare 24-hour service within their countries, contact with the car manufacturers, and use only the English language to exchange information between countries and operational teams. The result of the operation was as follows:

- 180,000 vehicles controlled;
- 200,000 persons checked;
- 15,000 police officers involved;
- 348 stolen vehicles discovered.

C/ African Operation: Two Examples

Interpol regularly helps African countries to establish joint operations in targeted regions. Mainly organized through roadblock operations, vehicles are controlled by mixed operation units from the involved countries. This allows for a faster determination of the real identity of the driver and of the validity of the vehicle's registration documents, which are checked via the Interpol database. This will reveal if the car in question is known as stolen in the international collection. Those operations led to some arrests but mainly to the discovery of drugs, weapons, and, of course, stolen vehicles. The Operation Makhulu Makhulu, initiated by the Southern African Regional Police Chiefs Cooperation Organisation (SARPCCO), aimed to tackle cross-border crime in some prioritized areas, which were of concern to member States under which theft and hijacking of motor vehicles was the first priority. The objectives of the operation were to

- Identify stolen vehicles from member countries;
- Collate and disseminate intelligence on cross-border crime syndicates and individuals;
- Gather evidence and arrest all criminals for eventual prosecution irrespective of country;
- Recover and repatriate stolen and fraudulently acquired cars to their country of origin;
- Gather intelligence to assist future joint operations.

The countries involved were Angola, Botswana, Malawi, Namibia, South Africa, and Zimbabwe. These countries put in place roadblocks and used the Interpol database, mobile radio, and operational data mobile to conduct efficient searches. The strategy of that specific operation was driven by the production of intelligence using each country's intelligence capabilities. But in that special operation, it was very difficult to work with "old" intelligence (even if it only dated weeks back) because of the speed of developments in this type of crime and the fact that criminals use the vast extent of the territories to regularly change their modi operandi. The final result of that operation was the recovery of 253 stolen vehicles discovered as follows:

- Botswana (84);
- Namibia (33);
- South Africa (74);
- Zimbabwe (62).

It was determined that these vehicles originated from:

- Botswana (3);
- Namibia (6);
- South Africa (101);
- Zimbabwe (8);
- Mozambique (1);
- Unknown countries (134).

In total, approximately 2,500 criminals were arrested for various offenses including vehicle theft, firearms, drugs, and illegal immigration.

Another operation in the southern region of Africa was launched because vehicle crime has continued to rise in that region despite efforts to thwart criminal activities. Elaborate syndicates are responsible for vehicle theft in southern Africa. The syndicates include businessmen running legally registered holding companies and auctions enabling stolen cars to be issued with the substitute documents and easy processing of change of ownership. These criminals operated from Tanzania, Mozambique, Botswana, South Africa, Zambia, and Zimbabwe. Stolen cars are driven to neighboring countries by locals and are sold to unsuspecting clients with documents showing them as second-hand vehicles or re-exported to countries in Central Africa, which include Burundi, Rwanda, and Democratic Republic of the Congo. Therefore, the aim of the operation was to provide a database for known cross-border criminals in that specific region. The operation led to the identification of 422 criminals known to police as a thief, a seller of stolen cars, in possession of stolen cars, or member of a vehicle crime syndicate. Those criminals were from Botswana (64), Lesotho (55), Namibia (25), South Africa (11), Swaziland (25), Tanzania (165), Zambia (23), and Zimbabwe (54).

D/ International Operations: The Example of Operational Focus on a Strategic Seaport in Antwerp, Belgium—Project "GIRAFE"

In 1998, as Head of the National Program, the author managed a project cofinanced by the European Commission to bolster European cooperation in tackling vehicle crime in the European region with support and cooperation from the US Federal Bureau of Investigation and the Royal Canadian Mounted Police. The project was focused not only on investigators but also on middle-management personnel. Over a five-day period, investigators and middle-management decision makers met and were trained to better understand the Belgian, European, and worldwide situation in relation to this specific phenomenon.

The investigators then went to the seaport of Antwerp to check a number of containers that had been highlighted by intelligence as particularly suspicious. The managers developed a sound methodology to gather and analyze data and produce intelligence. The project was notably based upon the work achieved by the Belgian Federal Police. The intelligence activities enabled the Belgian Federal Police to reduce the number of containers to be checked from 100,000 to just 25. The result was that from the 25 selected containers, 10 cars were recovered and their combined value was 10 times higher than the investment made by Belgium and the European Commission. In total, 100 police officers took part in this operational training, 10 stolen cars from 7 different countries were recovered, two arrests were made, and a network was established between the investigators and the decision makers from 17 different countries. This has led to other seaport operations organized by Great Britain, The Netherlands, Italy, Germany, the United States, and Canada.

In addition to the support given by Interpol during the joint operation (database, 24-hour support, specialized officers on the spot), the experts at the Interpol General Secretariat provided training to allow every country in every region to understand the international character of that crime and the structure of organized groups. Indeed, given the complexity of the illegal process, it is clear that such criminal activity demands a high degree of organization. Each stage calls for specific expertise, as outlined below.

22.4.5 Specific Criminal Expertise

A/ Thief

The thief often specializes in one make or type of vehicle as far as straightforward theft is concerned. Organized gangs have developed an in-depth knowledge of electronic vehicle protection systems and can neutralize them in a matter of minutes. Violent car thieves, however, have no technical experience of specific types of cars (in fact, the only qualification the thief needs in this case is a violent nature).

B/ Forger

Compared with the number of thieves or couriers, there are relatively few forgers. Their task is to develop and maintain a network of contacts enabling them to buy and sell documents (stolen or blank) and to use their expertise to forge and alter these papers. For this reason, Interpol offers to its member countries a database of stolen administrative blank documents (see Chapter 15). The database of stolen documents enables member countries to check a car and its related documents through Interpol.

C/ Ringer

A ringer is a criminal who performs the ringing or re-VINing of stolen vehicles. Ringers are often vehicle technicians or mechanics. Over the years, they have developed their technical knowledge and became more adept at foiling the efforts of the police and vehicle control services. They regularly use up-to-date techniques to give new identities to stolen vehicles.

D/ Courier

The courier may be anyone in need of money and has no real knowledge of the organization behind the vehicle(s) he or she is asked to transport.

E/ Seller

The seller is hired either for his or her salesmanship or because he or she has the requisite facilities for marketing the vehicles (e.g., garage owner or employee).

F/ Coordinator

The coordinator is hired to recruit or make contact with the thieves or forgers/ringers. He or she often acts as a buffer between the head of the organization and those involved in the criminal process. He or she may also carry out criminal activities (as thief, forger, ringer, or seller).

G/ Organizer

The organizer supplies the initial capital and takes a commission. He or she is often involved in other types of crime and uses vehicle trafficking to help provide liquid assets for these activities.

In conclusion, most of those involved are "freelancers." Analyses have shown the process to be based upon a highly coordinated and well-established structure [14]. However, each specialist or specialty is more or less independent of the others. This means that an organized group can simply hire and pay specialists according to its needs.

22.4.6 An Integrated Multidisciplinary Approach

Setting out the UK government's five-year strategy, Jack Straw, then Home Secretary, stated that car crime accounted for one quarter of all recorded crimes [15]. It costs more than GBP 3 billion (approximately USD 5.5 billion) per year in Great Britain and causes great distress and inconvenience to the victims. He said that was why the Prime Minister had made the issue a national priority, with the number one aim of reducing car crime by 30% in five years. The strategy focused on 14 areas, some of which are outlined below:

- Greater security in car parkings;
- Cooperation with vehicle registration services;
- Enhanced vehicle protection;
- New market regulations for scrap cars;
- More comprehensive information on vehicles;
- Checks on vehicles before registration;
- Controls on exports.

Analysis of the legal and illegal processes highlights the complexity of both the phenomenon itself and the measures required to effectively combat this type of organized crime.

Organized criminal activity linked to vehicle trafficking covers all regions of the world and affects everyone. No public body or private organization can fight this alone. Organized crime transcends national and international borders. Therefore, the approach to combating this phenomenon must be integrated, embracing all those involved, from crime prevention to law enforcement and victim support services. The second element of the strategy stems from the vital need for close cooperation between those concerned, hence the multidisciplinary aspect. There are two separate strands to implementing this strategy: first, following through the reactive aspects of police work, and second, finalizing the proactive approach. It is clear that the ultimate aim of these two elements is to arrest offenders and implement ad hoc measures designed to reduce this type of crime.

A/ The Reactive Approach

The reactive aspect is unavoidable for Interpol, yet it must be approached with a focus on the following three pillars: the fast and reliable exchange of high quality information (relating not only to criminals but also to the vehicle itself), training initiatives, and measures to target offenders. The type of information exchanged includes not only judicial information (notification of vehicle theft), but also administrative information from car manufacturers (vehicles produced), customs authorities (vehicles imported and exported), registration services (vehicles registered), and insurance companies (vehicles "written off"[2]). Hence, the first stage of the partnership between the various bodies involved is to bring together the different types of information available. Of course, enabling access to all this information does increase the need for training. Such training covers other areas such as investigative strategy and procedure, technical identification of a vehicle, large-scale operations, case law, international cooperation, and the sharing of information with the judiciary system for more effective prosecution of offenders, part of wider measures targeting the perpetrators of car crime, the third pillar in the reactive approach.

B/ The Proactive Approach

Finally, once the various basic elements of the integrated multidisciplinary approach are in place, the proactive aspect of the strategy can be developed. Linking up the databases and allowing key partners to access this information will make the registration of stolen vehicles difficult in any country connected to that system or platform. Police investigations can then move into a more proactive mode. This is part of the process of seeking, pooling, and storing basic information on criminal activities to identify the organized groups involved in vehicle trafficking. Processing this information, using strategic and tactical analysis techniques in particular, should enable operations to lead to the arrest of key criminals in such organizations.

---

[2] A vehicle is written off when the vehicle expert working for insurance companies considers the vehicle as a wreck that is not repairable.

CI VIRA 17

Interpol was supporting that approach when it developed the Project VIRA 17. The VIRA 17 Project started in January 2001 with a feasibility study and a promise from European insurance companies to cofund the project [16]. The project aimed to set up an information platform to enable member countries to consult vehicle- and crime-related information "supermarket style"[3] in a highly effective manner, making use of the latest techniques. At the same time, Interpol National Central Bureaus (NCBs) were responsible for working with their national partners to ensure effective support for operational investigators.

This tool, based upon a "one-stop shopping" strategy (checking several databases at the same time), was aimed at freeing up investigative resources to deal with criminal intelligence activities and proactive inquiries. In fact, 75% of investigation capacity is currently used to search for information already held in different unconnected databases. In addition, the tool would have offered national police services, a policy instrument at managerial level for identifying both national and international trends and developments in this field. Finally, the platform should have helped to inspire confidence and reinforce the sense of controlling car crime throughout the world, with particular benefits for member countries. Because of changes in priorities, the representatives of insurance companies could not pursue any further the cooperation and the project was stopped in 2002.

22.5 TECHNOLOGY TO HELP POLICE WORK

22.5.1 Principles

A theft response and vehicle recovery system is a system combining various items of equipment and human elements that communicate and interact through various interfaces in line with standard procedures to facilitate the recovery of a vehicle reported stolen by its owner. Technology is assuming an increasingly important role in our society. New technological inventions enable to protect an object, for example, such as if the object is used against the wishes of its owner (as in theft), a signal is transmitted immediately to a private alarm center. These new technologies are obviously of particular interest in motor vehicle applications.

Such developments have direct consequences for police services previously used to responding to emergency calls from alarms linked to fixed points (private residences, offices, etc.) and following a particular procedure. In contrast, an emergency call in response to an alarm coming from a moving object, in this case a vehicle fitted with a tracking system, requires a new and well-thought out approach to police intervention and deployment.

[3] Supermarket style: where any police officer could get any information related to a vehicle through Interpol I-24/7 network like shopping in a supermarket.

22.5.2 *Tracking Systems*

The tracking system consists of an electronic device, which enables the location, speed, and direction of a vehicle to be continuously monitored via a communications center (see Chapter 20). The system is triggered automatically by a signal from a theft detection device. It is rather like a silent alert system insofar that when an equipped vehicle is stolen, the police are kept informed of its location and movement via a private alarm center. Then, the most appropriate moment to intercept the vehicle and arrest the offenders is chosen. In this way, patrol takes no risks and the chances of locating the stolen car are high. At present, only few vehicles are fitted with such systems, but they will doubtless become more widespread in the future.

According to Working Group 14 of the European Union, Interpol, as an international organization, also has a role to play in the use of these "theft response" systems, because it is responsible for the exchange of criminal information and for the support of countries in international operations to combat organized crime [17]. Interpol is also guided by a progressive vision in its belief that a growing number of vehicles will be fitted with such systems, which will offer a great range of possible applications, such as with vehicle hire firms or debt recovery companies, although this would no longer concern an international organization such as Interpol.

22.6 THE INTERPOL TOOLS

Vehicle crime is not only tracing stolen cars but also arresting criminals. Interpol's ASF provides police-to-police interaction on all types of criminal information ranging from simple criminal history checks to more complex matters covering all areas of the criminal activity. The number of entries present in the database as of June 2005 are as follows:

- Criminal names (170,000 names including 8,300 terrorists);
- Fingerprints and DNA profiles;
- Stolen travel documents (7.1 million);
- Stolen administrative documents (1 million);
- Stolen motor vehicles (4.6 million);
- Credit cards.

The database related to stolen motor vehicles (SMV) came into operation in 1996. It now contains 3.4 million entries on vehicles reported as stolen across 88 countries. In 1998, the ASF-SMV was connected to the database of stolen vehicles, license plates, and boats operated by the US Federal Bureau of Investigation. This link enables over 32 authorized countries to check whether vehicles have been stolen in the United States or Canada, adding then 1.2 million stolen vehicles from the United States and Canada.

One hundred eighty-two countries can access the ASF-SMV, and 80 of them use it regularly for systematic checks. This database was among the first important successes of Interpol in terms of systematic police data exchange. This useful tool is consulted more than 1.2 million times per year with an average of 16,000 vehicles discovered on an international level in addition to the national and regional databases such as the SIS.

The relevancy of the support given by Interpol can be underlined through the notable example of stolen vehicles from South Africa: In the year 2004 (from January to August), the ASF system enabled police to trace 1,292 vehicles stolen in South Africa, located in 32 different countries. Taking into account the fact that those cars are right-handed drive cars, the investigators appreciated the support given by Interpol, which drew their attention on stolen vehicles being intercepted in countries where only left-handed drive vehicles are used (Figure 22-4).

Figure 22-5 presents the number of records, the number of queries, and the number of positive answers recorded during the last four years through the ASF-SMV. The latter means

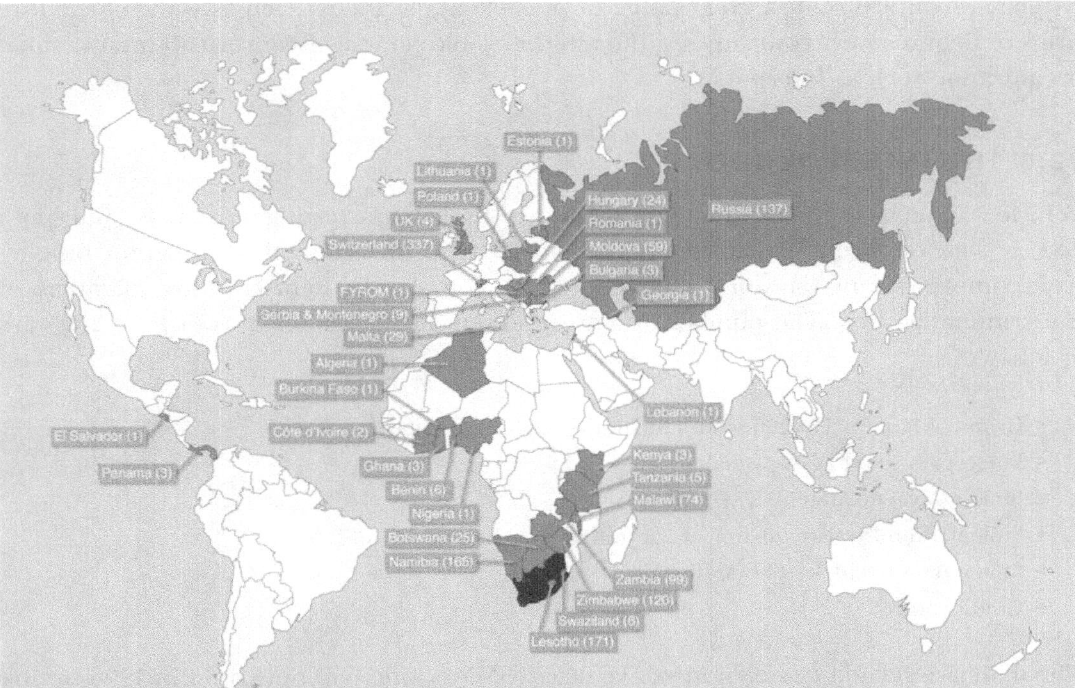

Figure 22-4

Graphical representation of some results obtained by using the Interpol ASF-SMV database. In this example, the ASF-SMV database enabled investigators to trace 1,292 cars stolen in South Africa in 32 different countries from January to August 2004. South Africa has 886,026 vehicles recorded in the database. (Diagram created by Eric Stauffer. FYROM stands for Former Yugoslav Republic of Macedonia.)

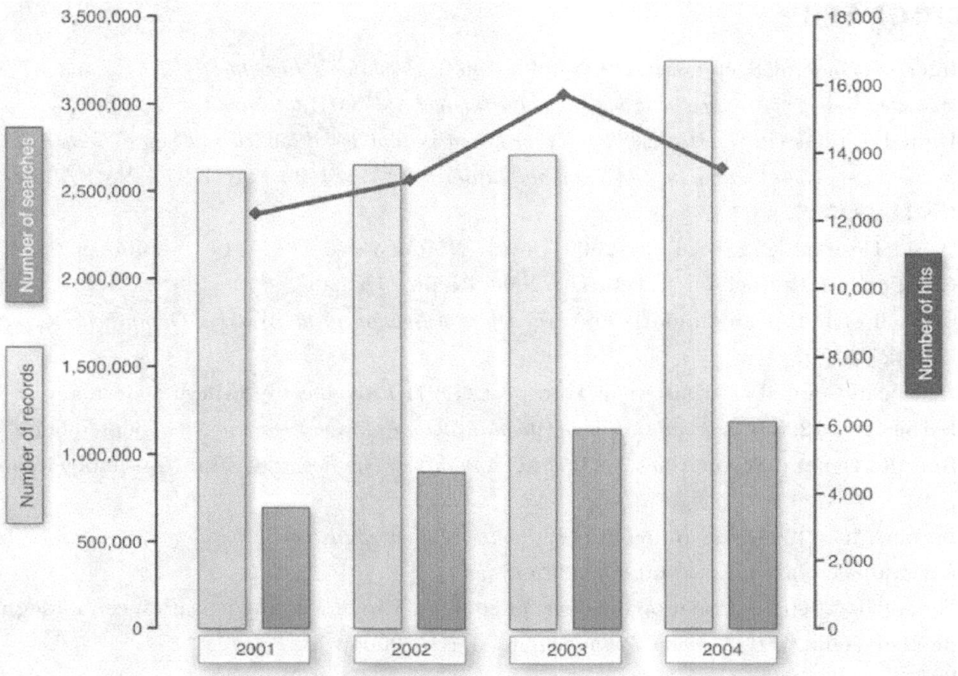

Figure 22-5

Graphical representation of the number of records, number of queries (searches), and number of positive answers (hits) obtained by using the Interpol ASF-SMV database from 2001 to 2004. (Diagram created by Eric Stauffer.)

that a search by a country in the Interpol database led to the identification of a stolen vehicle introduced by another Interpol member country.

22.7 CONCLUSION

Car crime is not restricted to the theft itself. It is but one part of a highly profitable area of criminal activity, where conviction is rare. This form of crime, organized on a worldwide scale, impacts other types of crime such as terrorism (financing), drug trafficking, arms dealing, and trafficking in human beings. Because this crime is also influenced by economic changes, special attention should be given to the development of emerging countries where an increase in vehicles has been identified or in regions where war or civil war offer an open market for such crimes.

Of course, Interpol does not claim to have the solution. However, it does recommend that countries improve their approach to the phenomenon by cooperating more closely and in a more organized fashion with other private and public sector bodies. Each of these partners is a key element in the strategy and can thus help not only to reduce the number of thefts but also to enable police search and investigation services to arrest the organizers or "ringleaders" of this crime, which is so profitable yet so rarely punished.

BIBLIOGRAPHY

[1] Interpol General Secretariat (1995–1996–1998) *Resolutions from the 64th, 65th, and 67th General Assembly Sessions of Interpol with regards to the Interpol ASF-SMV database,* Lyon, France.

[2] United Nations (1997) *UN resolution by the Economic and Social council made on 6th session of the Commission on Crime Prevention and Criminal Justice* (28th April–9th May 1997), No E/ 1997/30 E/ CN.15/1997/21, p 69.

[3] World Customs Organisation (2000) *Report N°EC0060E0, TE6-4,* 19th Session of the Customs Co-operation Council on 4th February 2000, Berlin, Germany.

[4] Council of European Union (1999) *The Dutch delegation in the Working Group Police Co-operation,* Doc 12715/99.

[5] European Council of Ministers of Transport (1997) *Resolution* **97/2**, Berlin, Germany, p 1.

[6] Barbier AG. (2000) *Personal Communication during a Press Conference in Sofia,* September 27, 2000.

[7] Interpol General Secretariat (2003) *2002 Annual Activity Report,* General Assembly, Report N°1, p 4.

[8] Barbier AG. (1999) Les routes. In: *Positionnement en matière de trafic illégal de véhicules,* Interpol General Secretariat, Lyon, unique edition, p 6.

[9] Fijnaut C. (2000) Transnationale Misdaad en de rol van de Verenigde Naties, een uitdaging voor de 21ste eeuw, *APB (Algemeen Politieblad),* **6**, March 2000, p 16.

[10] Barbier AG. (1995) *Internal Strategic Analysis of Vehicle Crime,* Internal document of the Belgian Gendarmerie developed for the National Program Against Vehicle Crime.

[11] Interpol General Secretariat (1999) *Interpol Guidelines on Vehicle Crime,* Lyon, France.

[12] Barbier AG. (1999) Description du phénomène. In: *Positionnement en matière de trafic illégal de véhicules,* Interpol General Secretariat, Lyon, France, pp 6–9.

[13] Fijnaut C. (1987) The internationalization of criminal investigation in Western Europe. In: *Police Cooperation in Europe,* Van den Brink, Lochem, The Netherlands, pp 32–56.

[14] Barbier AG. (1999) Analyse des auteurs suivant le processus criminel. In: *Positionnement en matière de trafic illégal de véhicules,* Interpol General Secretariat, Lyon, France, pp 11–12.

[15] Straw J. (1999) Foreword. In: *Tackling Vehicle Crime, Five Years Strategy,* The Home Office Communication Directorate, p 2.

[16] Interpol General Secretariat (2002) *Project VIRA 17,* 71st General Assembly, Report N°23, Lyon, France.

[17] Working Group 14 (2001) *Report from the Technical Committee on "After-Theft System for Vehicle Recovery"* European Committee for Standardization.

ABBREVIATIONS

2D	Two-dimensional
3D	Three-dimensional

A

AAMA	American Automobile Manufacturers Association (USA)
AAMVA	American Association of Motor Vehicle Administrators (USA)
ABS	Acrylonitrile-butadiene-styrene
AC	Alternating current
AFIS	Automated fingerprint identification system
AFP	Australian Federal Police
AIAM	Association of International Automobile Manufacturers
ALPCA	Automobile License Plate Collectors Association
ANFO	Ammonium nitrate fuel oil
ASA	American Standards Association
ASF-SMV	Automated Search Facility-Stolen Motor Vehicle (Interpol)
ASPT	Assembled from parts
ATF	Automatic transmission fluid
ATM	Automatic Teller Machine

B

BCM	Body control module
BKA	Bundeskriminalamt
BMW	Bayerische Motoren Werke (Germany)

C

CCD	Charge-coupled device
CCTV	Closed circuit television
CD	Chemiluminescence detector
CD	Compact disc
CFR	Code of Federal Regulations
cm	Centimeter
CODIS	Combined DNA Index System
CSS	Component set strategy
cSt	Centistoke
CVMA	Canadian Vehicle Manufacturers' Association

D

DAB	Diaminobenzidine
DART	Diagnostics and Reprogramming Tool
DC	Direct current
DIV	Direction des Immatriculations des Véhicules (Belgium)
DFO	1,8-diazafluoren-9-one
DfT	Department for Transport (Great Britain)
DGPS	Differential global positioning system
DGT	Dirección General de Tráfico (Spain)
DHSMV	Department of Highway Safety and Motor Vehicles (Florida, USA)
DMV	Department of motor vehicles
DNA	Deoxyribonucleic acid
DNT	Dinitrotoluene
DOVID	Diffractive optically variable image device
DPA	Diphenylamine
DST+	Digital Signature Transponder Plus
DVD	Digital videodisc or digital versatile disc
DVLA	Driver and Vehicle Licensing Agency (Great Britain)

E

EC	Ethylcentralite
ECD	Electron capture detector
ECM	Electronic control module
ECU	Engine control unit
EDX	Energy dispersive x-ray analyzer
EGDN	Ethyleneglycoldinitrate
EOD	Explosive ordnance disposal
EPG	Enhanced protective glass
ERRI	Emergency Response Research Institute
ESDA	Electrostatic detection apparatus
ETA	Euskadi Ta Askatasuna (Spain)
ETD	Electronic Transponder Duplicator
EU	European Union
EUCARIS	EUropean CAR and driving license Information System
EuVID	European Vehicle Identification Database
EVD	Explosive vapor detector

F

FBI	Federal Bureau of Investigation (USA)
FHP	Florida Highway Patrol (USA)
FYROM	Former Yugoslav Republic of Macedonia

G

GC	Gas chromatography
GIS	Geographic information system
GM	General Motors Corporation
GMC	GMC Trucks (a division of GM)
GPS	Global positioning system
GRC	General rifling characteristics (FBI database)
GSM	Global system (or standard) for mobile
GSR	Gunshot residue
GVWR	Gross vehicle weight rating

H

HEAT	Help Eliminate Auto Theft (Virginia State Police, USA)
HPLC	High-performance liquid chromatography
HSG	High security glass

I

IAATI	International Association of Auto Theft Investigators
IACP	International Association of Chiefs of Police
IASIU	International Association of Special Investigation Units
IBC	Insurance Bureau of Canada
ICPO	International Criminal Police Organization (Interpol)
IED	Improvised explosive device
IMS	Ion mobility spectrometry
IPC	Instrument panel cluster
IRA	Irish Republic Army
ISN	Immobilizer serial number
ISO	International Organization for Standardization

K

KBA	Kraftfahrtbundesamt (Germany)
kHz	KiloHertz
KLPD	Korps Landelijke Politiediensten (Netherlands)
km	Kilometer

L

LCV	Leucocrystal violet
LIV	Landelijk Informatiecentrum Voertuigcriminaliteit (Netherlands)
LPG	Liquefied petroleum gas
LPR	License plate reader

M

m	Meter
MAPS	Mapping and Analysis for Public Safety (USA)
MATS	Mechanical anti-theft system
MCO	Manufacturer's certificate of origin
MCOT	Multi-car opening tool
MDMA	Methylenedioxymethamphetamine
MDPD	Miami-Dade Police Department (Florida, USA)
MGT	Modified Griess test
mm	millimeter
MSO	Manufacturer's statement of origin
mtDNA	Mitochondrial DNA
MVR	Motor vehicle record

N

NAEC	North American Export Committee (USA, Canada, and Mexico)
NC	Nitrocellulose
NCB	National Central Bureaus (Interpol)
NCIC	National Crime Information Center (FBI, USA)
NG	Nitroglycerine
NGS	New Generation Star
NHTSA	National Highway Traffic Safety Administration (USA)
NICB	National Insurance Crime Bureau (USA)
NIJ	National Institute of Justice (USA)
NLECTC	National Law Enforcement and Corrections Technology Center (USA)
nm	Nanometer
NMVTRC	National Motor Vehicle Theft Reduction Council (Australia)

O

OCRA	Oficina Coordinadora de Riesgos Asegurados (Mexico)
OEM	Original equipment manufacturer
OVD	Optical variable device
OVI	Optical variable ink

P

PCM	Power control module
PD	Physical developer
PDQ	International Forensic Automotive Paint Data Query database (RCMP, Canada)
PETN	Pentaerythritoltetranitrate
PPE	Personal protective equipment

ppm	Parts per million
PVB	Polyvinylbutyral

R

RCMP	Royal Canadian Mounted Police
RDW	Rijksdienst voor Wegverkeer (Netherlands)
RDX	Cyclotrimethylenetrinitramine
RFID	Radio frequency identification
RIRA	Real Irish Republican Army
ROV	Remotely operated vehicle

S

SAE	Society of Automotive Engineers (USA)
SAN	Styrene-acrylonitrile
SARPCCO	Southern African Regional Police Chiefs Cooperation Organisation
SDD	Silca Diagnostic Device
SEM	Scanning electron microscope
SIA	Swiss Insurance Association
SIS	Schengen Information System
SIU	Special investigation unit
SPR	Small particle reagent
SUO	Statement under oath
SUV	Sport utility vehicle

T

TATP	Triacetone triperoxide
TNT	Trinitrotoluene
TRAP	Taskforce for Regional Autotheft Prevention (Los Angeles, USA)
TÜV	Technischer Überwachungsverein (Germany)

U

UCR	Uniform Crime Report
UK	United Kingdom
UNECE	United Nations Economic Commission for Europe
US	United States
USA	United States of America
USFA	United States Fire Administration
UV	Ultraviolet

V

VATS	Vehicle anti-theft system
VDS	Vehicle descriptor section

VIN	Vehicle identification number
VIO	Vehicle identification officer
VIS	Vehicle identifier section or vehicle indicator section

W

WMI	World manufacturer identifier
WPMI	World parts manufacturer identifier

Y

YAG	Yttrium aluminum garnet

Z

ZIS	Zentral Information Stelle (Switzerland)

INDEX

Page numbers noted with "f" or "t" indicate related figures and tables.

Figure 3-7

Scene of a hit and run motor vehicle accident where numeric and alpha markers were used to identify the location of physical evidence such as scuff marks, tire tracks, and clothing items. (Photograph courtesy of AFP Forensic Services.)

Figure 3-8

Aerial view of the scene depicted in Figure 3-7. Yellow markers and white outline indicate scuff marks caused by the movement of the victim on the ground. Black and white markers indicate tire tracks created by the backward and then forward movements of the vehicle. Markers 1, 14, and 16 identify dental stone of tire tracks. (Photograph courtesy of AFP Forensic Services.)

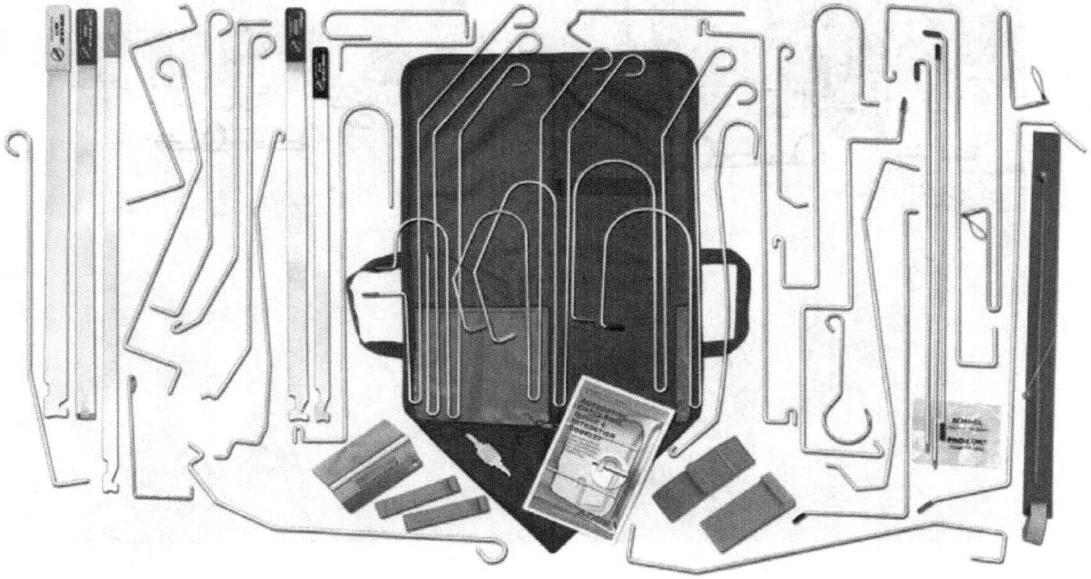

Figure 4-13

Example of a lockout toolkit (an auto opening kit Pro-lok Complete kit 2000 containing 47 pieces) used by professional locksmiths to penetrate vehicles. Auto thieves also use this kind of kit to illegally penetrate vehicles. (Source: Complete Kit 2000 (AKCOM00), http://www.pro-lok.com. Reprinted with permission of Pro-lok, Orange, California.)

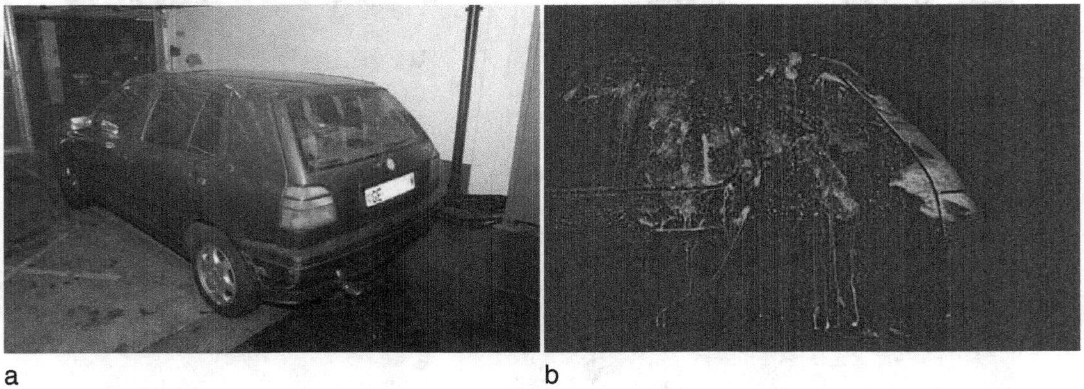

a b

Figure 4-20

(a) This VW Golf was used in a homicide, and it was suspected that traces of blood were present on the body around the trunk. Unfortunately, due to the color of the vehicle, the bloodstains are not visible to the naked eye.
(b) When luminol is applied to the vehicle, the bloodstains become readily visible. Then, it is possible to sample the different areas using the technique presented in Figure 4-19.

Figure 4-20 Continued.

(c) View of the same vehicle as in Figure 4-20a before application of luminol.

(d) View after application of luminol.

(e) View of a vehicle's rear seat whose top has been cut off for laboratory examination. No apparent stains are present.

(f) View of the same seat as in Figure 4-20e after application of luminol. Blood is still present in the seat cushion.

(g) A particular photographic technique consisting of exposing the film to a short flash right after its exposition to the luminescence produced by the luminol provides a better visualization of the blood stains. Compare this image with Figure 4-20f.

(h) By inverting the color of the image using computer software, it is possible to make the white-blue luminescence appear dark red, closer to the natural color of blood. This technique is particularly appreciated when presenting evidence to a jury.

(Photographs (e) through (h) courtesy of Sgt. Stewart Mosher, Broward Sheriff's Office.)

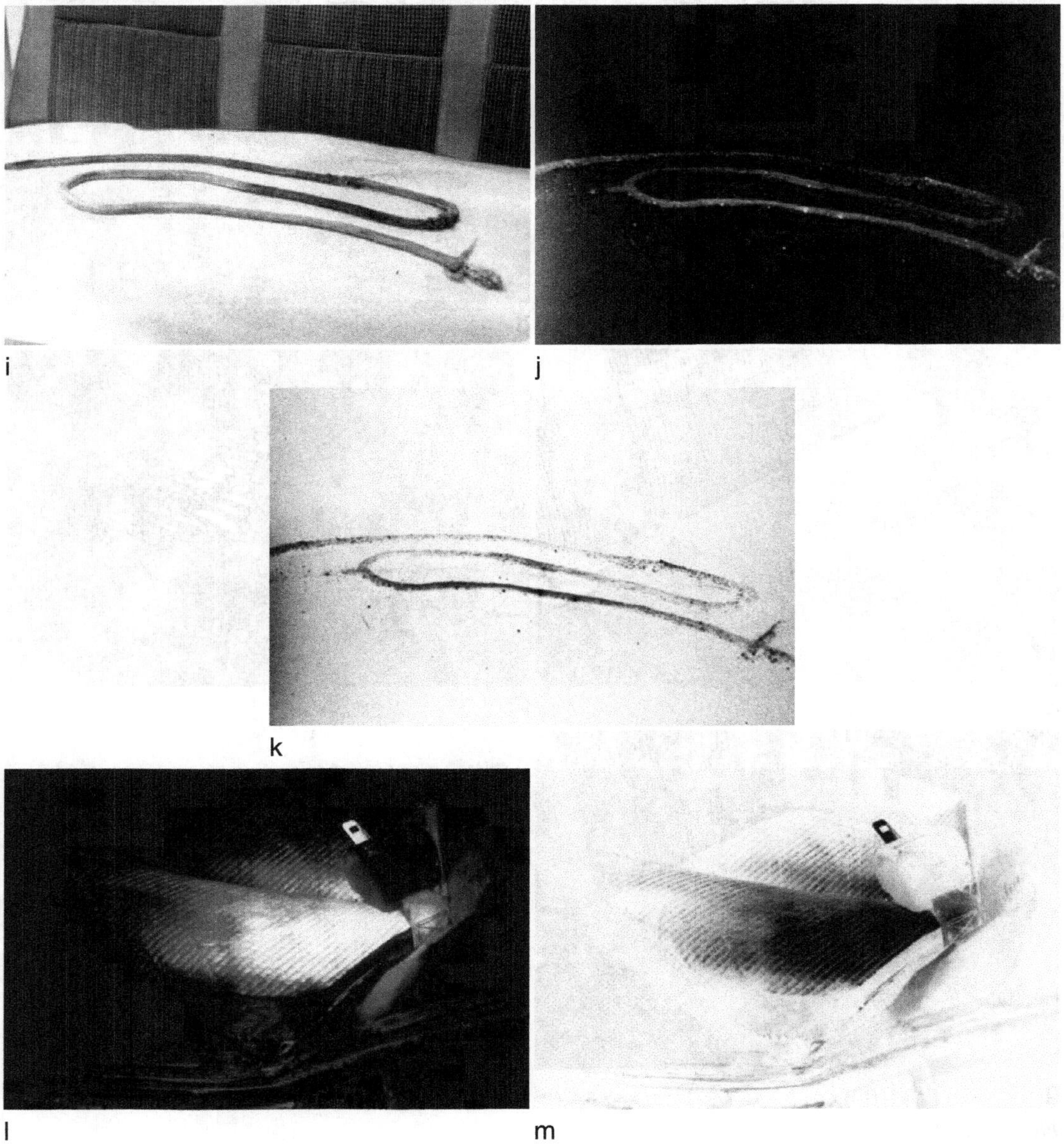

Figure 4-20 Continued.

(i) This rope was found tied on the leg of a car seat and was allegedly used to tie a person who was bleeding. A naked eye examination did not reveal any blood traces.

(j) Same view after application of luminol.

(k) Same view after inverting the colors in computer software.

(l) View of a car seat after application of luminol, revealing important possible stains of blood.

(m) Same view inverted with computer software. The color rendering provides a clearer picture of the bloodstain.

(Photographs (i) through (m) courtesy of Sgt. Stewart Mosher, Broward Sheriff's Office.)

Figure 6-36

Retroreflection security feature present on 3M anti-theft label. (Source: 3M Security Systems Division (1996) Automotive Security Labeling System, p. 1. Reprinted with permission of 3M Security Systems Division.)

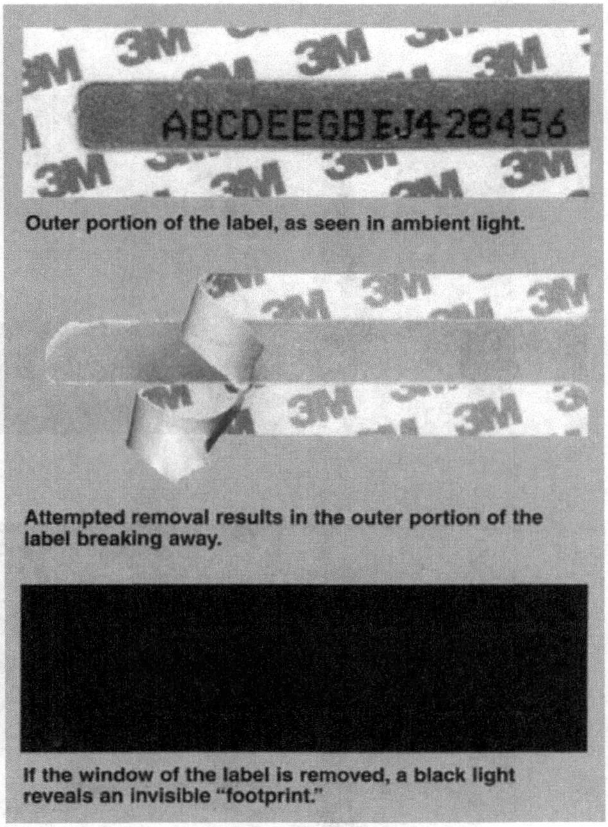

Figure 6-37

Illustration of the breaking away of a 3M anti-theft label when removed from vehicle and illustration of the footprint left after removal of the label and seen under ultraviolet light. (Source: 3M Security Systems Division (1996) Automotive Security Labeling System, p. 2. Reprinted with permission of 3M Security Systems Division.)

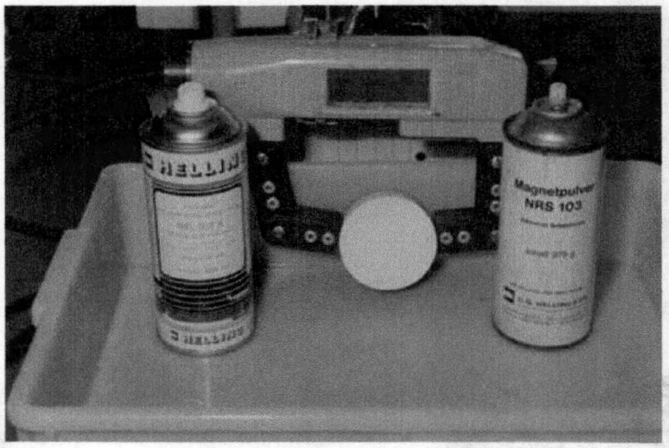

Figure 7-20

Laboratory device using magnetic particle method.

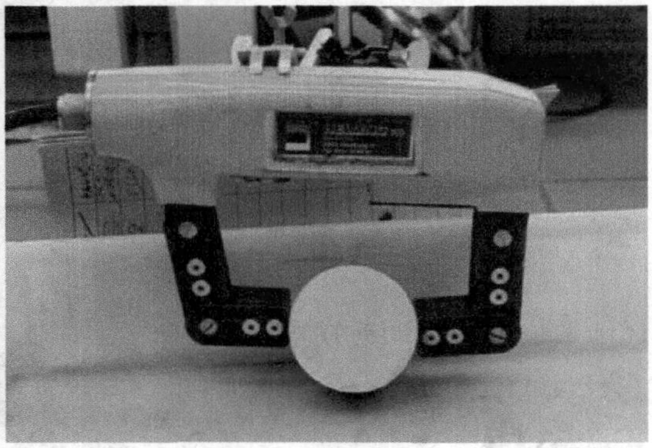

Figure 7-21

Magnetic metal (ferritic steel). Erased and restored characters after magnetization and spraying with fine magnetic particles. For contrast enhancement, it may be useful to paint the surface with white colors first.

Figure 8-16

The foil is securely wrapped around the key head to block transmissions and inserted into the lock for testing. Note the theft light in dash that is illuminated as shown with the arrow.

Figure 9-28

Driver's side floor debris area of a 2002 Suzuki Aerio, which was a stolen vehicle recovered burned. Note the steering column remains (T) and the debris area directly below.

Figure 9-29

The clump of metal being recovered is the melted ignition lock—column lock—starter switch assembly housing. The 2002 Suzuki Aerio is equipped with a standard single-component assembly.

Figure 9-31

The ignition lock on the left displays severe thermal damage to the housing from direct exposure to the fire, which indicates that the lock was secured to the column at the time of the fire. Note the undamaged lock wafers embedded in the melted housing. The lock on the right was placed on the floor below the steering column of a vehicle before a test burn. After the vehicle burned to completion, the lock cylinder was recovered. The lock housing had some discoloration, however there was no thermal damage or distortion. The evidence clearly demonstrates that the lock cylinder had been removed from the column and placed on the driver floor before the fire.

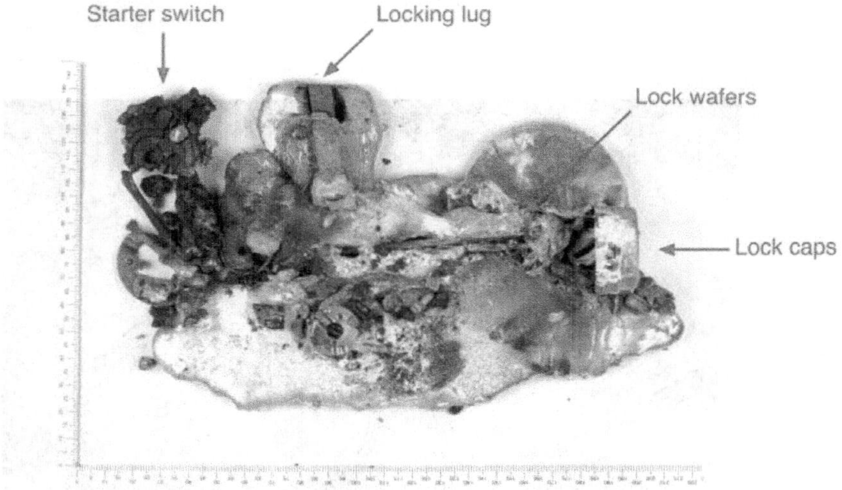

Figure 9-32

View before x-ray of the melted remains of a single component ignition lock—column lock—starter switch assembly as it would be recovered from a burned vehicle driver's floor area. It is possible to distinguish some parts such as the starter switch, the locking lug, and the lock caps. However, it is not possible to distinguish the lock wafers. (Photograph courtesy of Eric Stauffer.)

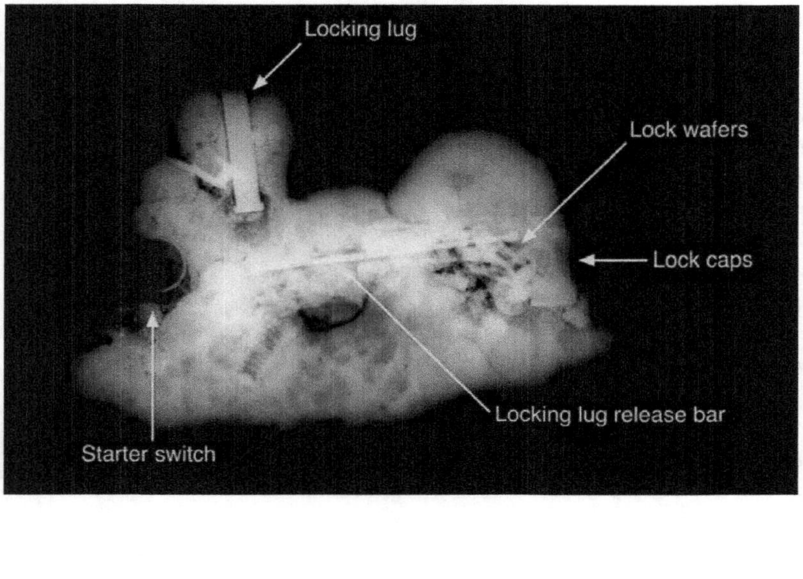

Figure 9-33

Same assembly viewed under x-rays. The x-ray clearly reveals the ignition lock components and wafers, the locking lug release bar, and the column locking lug embedded in the melted housing. The starter (ignition) switch is more difficult to see due to its lower density. (Photograph courtesy of Eric Stauffer.)

a b c d

Figure 12-2

(a) Photograph of a vehicle that accidentally caught fire on the highway. The fire started in the rear part of the passenger compartment and had been burning for a couple of minutes before this view. T = 0:00 [min:sec].
(b) T = 0:54.
(c) T = 1:56.
(d) T = 6:23.

Figure 12-6

Example of a deliberately set fire of a 1995 Lexus SC300. A plastic jug filled with gasoline was found on the front seat, partially burned. The vehicle did not undergo flashover, as the perpetrator closed the doors and windows. The fire died by starvation of oxygen. (Photograph courtesy of Mark D. Culver, Key Fire Investigations, Inc.)

Figure 12-7

View of the interior of the 1995 Lexus SC300 showing fire damage on the upper part of the passenger's compartment. Note the heavy soot deposit throughout the vehicle, due to an incomplete combustion reaction. (Photograph courtesy of Mark D. Culver, Key Fire Investigations, Inc.)

Figure 12-8

View of the trunk of the 1995 Lexus SC300, which was splashed with gasoline but not ignited. Notice the unburned match lying on the carpet of the trunk. (Photograph courtesy of Mark D. Culver, Key Fire Investigations, Inc.)

a

Figure 12-19

(a) View of the front left of a Chrysler 300, clearly showing a point of origin from outside the vehicle.

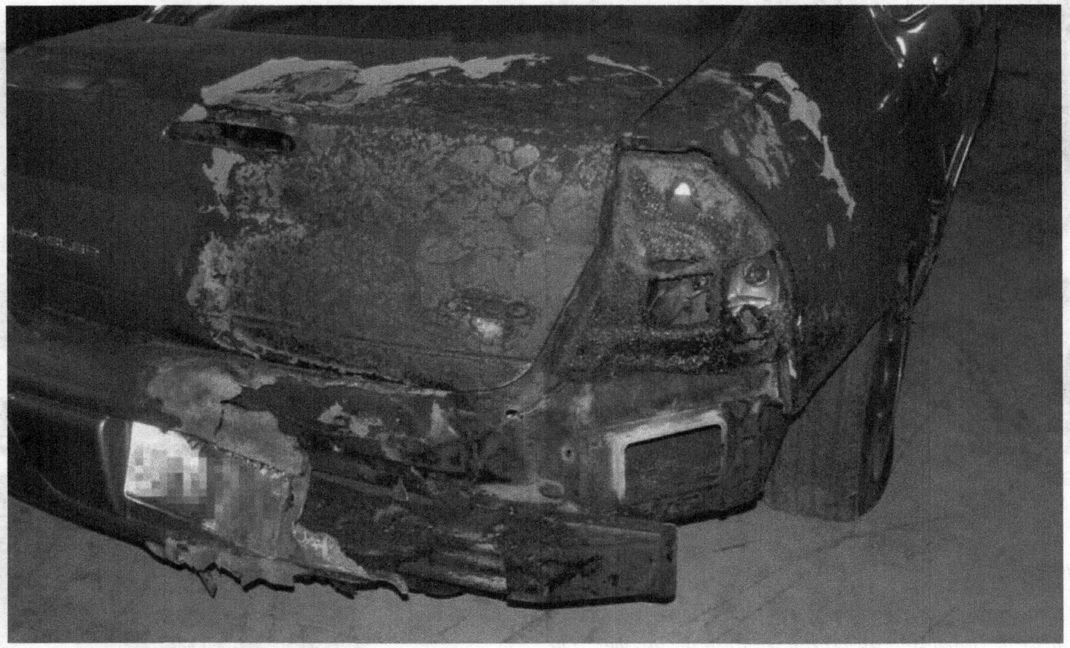

b

Figure 12-19

(b) View of the rear right of the same Chrysler 300 showing a second point of origin, with no link to the first point of origin. Nevertheless, the fact that there are no ignition sources present in these areas of origin it clearly establishes a human intervention. (Photographs courtesy of Mike Carlson, Probe Inc.)

Figure 12-16

In some instances, the heat patterns might be very faint or not at all present. The only heat pattern that can be seen on this Ford F-150 is located at the upper right corner of the door, just under the base of the windshield and consists of a few faint wave patterns, which indicates an interior (no. 2) fire that progressed toward the front.

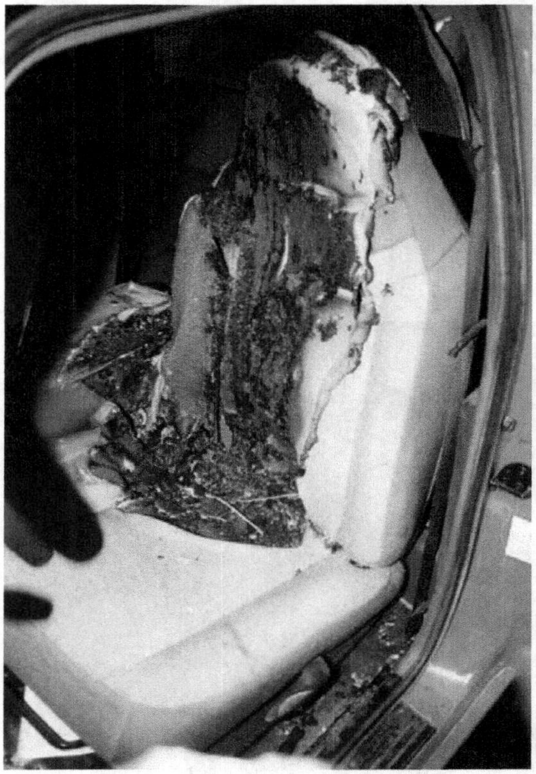

a

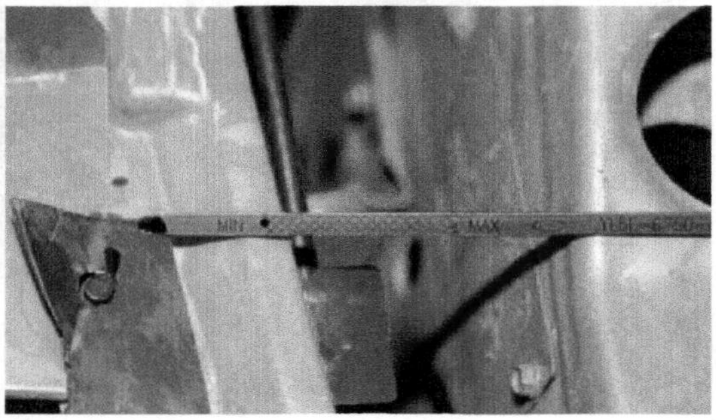

b

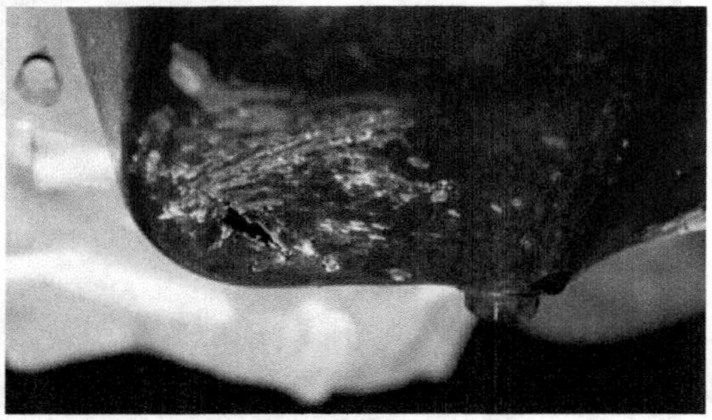

c

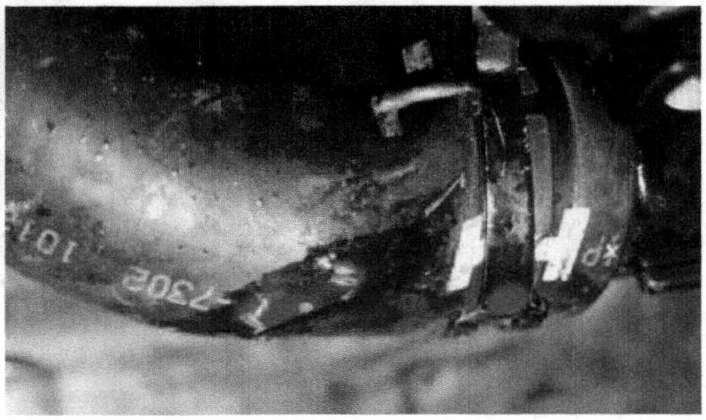

d

Figure 12-21

(a) View of the driver's seat of a Ford Ranger that was set on fire by spreading an ignitable liquid on the seat. The fire died by starvation of oxygen, which preserved all the necessary evidence to establish its incendiary nature.

(b) View of the engine oil dipstick, showing less than minimum level.

(c) View from the undercarriage, showing sabotage action on the oil pan.

(d) View from the undercarriage, showing sabotage action on the lower radiator hose.

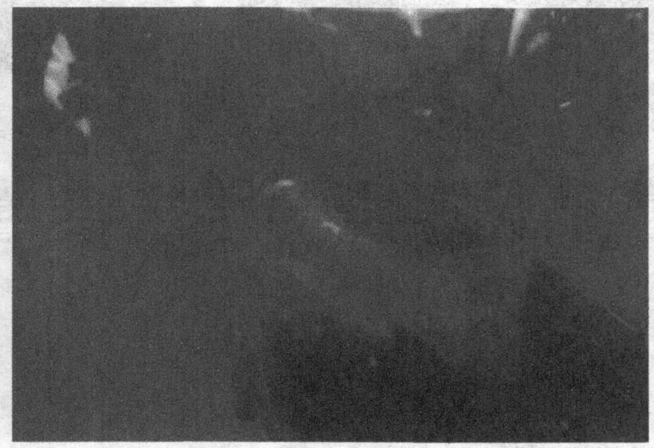

Figure 13-1

Divers of the Gendarmerie Nationale (French military police) ready to dive in "black water" for stolen vehicles search (July 1999, Oise river, France).

Figure 13-2

Example of poor visibility conditions during the examination of a vehicle in which a body was discovered.

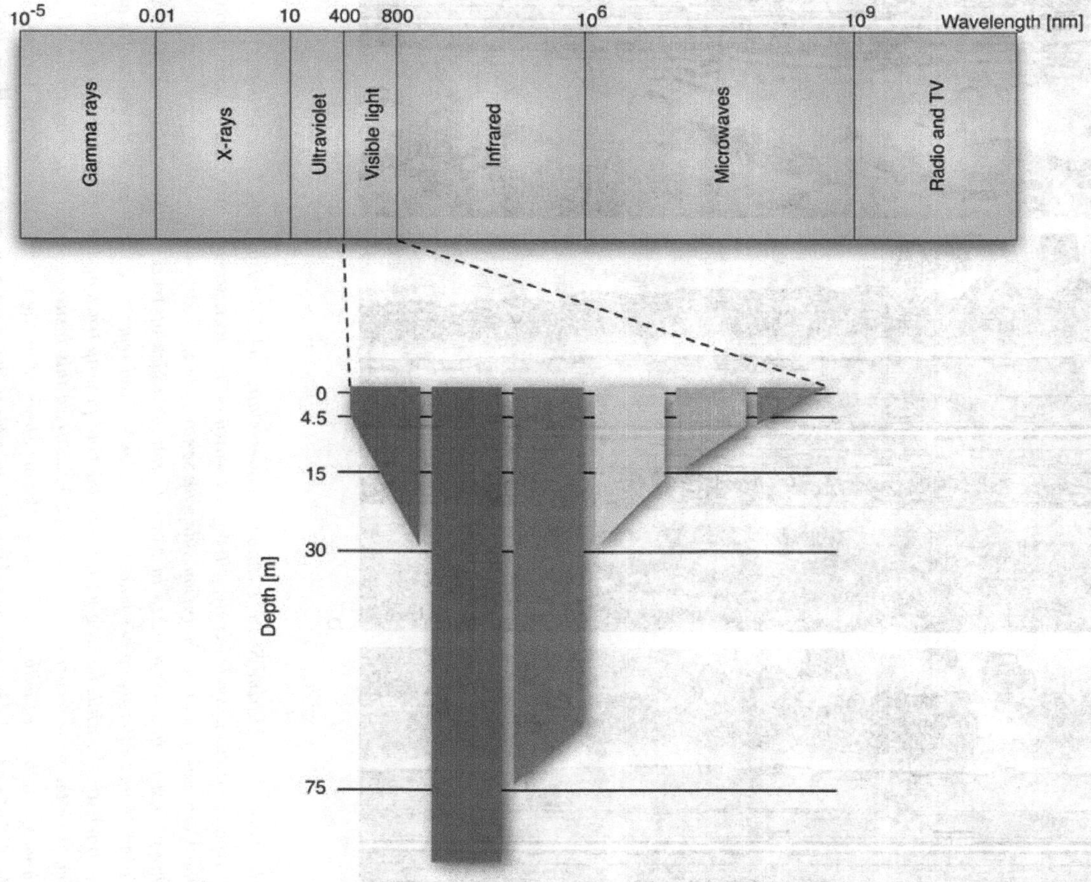

Figure 13-3

Illustration of the electromagnetic spectrum and visible light. The different wavelengths (colors) of the visible spectrum are absorbed by the water at different intensities. Notice how all the colors, except for blue, are completely absorbed after approximately 75 meters. (Diagram courtesy of Eric Stauffer.)

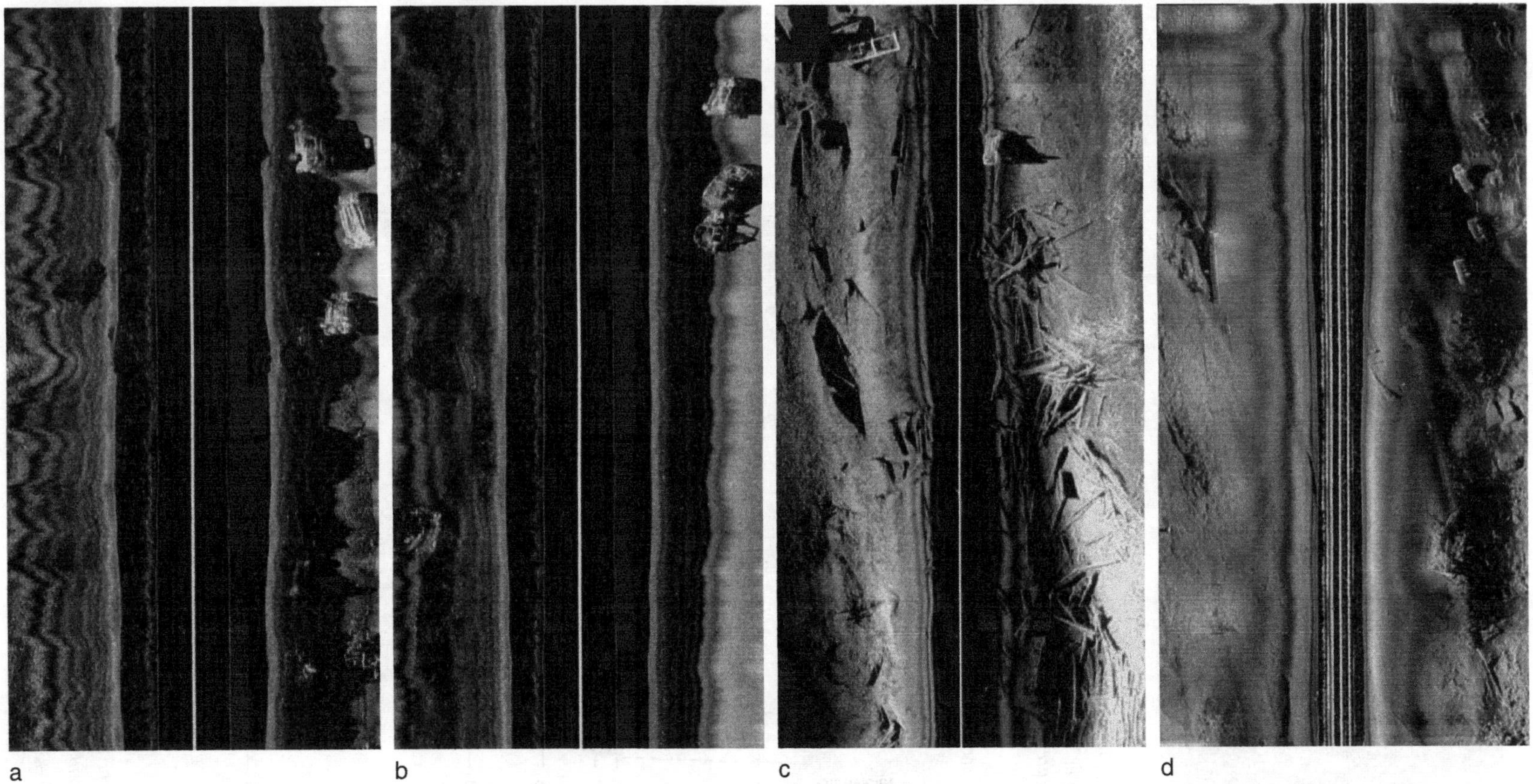

Figure 13-9

(a) and (b) Vehicles found by a Marine Sonic Technology representative while assisting the French police. The police were looking for a missing child that was allegedly placed inside a vehicle dumped in a river. Surprisingly, during the search of the river, several other vehicles were discovered as seen on the images (particularly on the right side of the images). These images were taken with a Centurion side scan sonar system.

(c) Two vehicles are present in this image taken in North Carolina: a pick-up truck on the top left and an up-side-down vehicle slightly below, on the right. The vehicle on the right is a 1996 Nissan, found after its owner placed it in the water and reported it as stolen to the police. He attempted to collect the insurance settlement, however the car was found by the police. Interestingly, while looking for this car, the Ford pick-up truck on the top left was also discovered. It was later determined that the pick-up truck was also stolen several years earlier, from the same owner of the Nissan, who collected this insurance money. The owner was charged with two counts of insurance fraud. (Source (a, b, and c): Marine Sonic Technology. Reprinted with permission of Marine Sonic Technology, Inc., White Marsh, Virginia.)

(d) Several vehicles found in the Arkansas River (US). It is possible to distinguish at least four vehicles on the upper right part of the side scan sonar image. The second vehicle from the top is laying on its back, as it is possible to distinguish the two axles and the four wheels. (Source: Marine Sonic Technology. Reprinted with permission of Marine Sonic Technology, Inc., White Marsh, Virginia.)

Figure 13-11

A firearm discovered by a police diver away from the main crime scene (vehicle) is placed inside an evidence container. This shows the importance of checking the extended area in the direction of the water flow for objects that could have been carried away. (Photograph courtesy of Gendarmerie River Company of the Rhine, March 2004, Strasbourg, France.)

Figure 13-20

Example of collection of a cartridge shell from inside a vehicle. If this evidence was not collected before vehicle removal, it would probably have been lost during the vehicle's extraction. (Photograph courtesy of Gendarmerie River Company of the Rhine, March 2004, Strasbourg, France.)

Figure 13-23

Example of collection of a bottle of an alcoholic beverage found inside the vehicle. This item is susceptible of containing DNA material. (Photograph courtesy of Gendarmerie River Company of the Rhine, March 2004, Strasbourg, France.)

Figure 13-26

Collection of diatoms is performed by sampling the water in a jug at different locations. (Photograph courtesy of Gendarmerie River Company of the Rhine, March 2004, Strasbourg, France.)

a b

Figure 14-1

(a) Front Belgian license plate with the euroband.
(b) Rear Belgian license plate. (Photographs courtesy of Jean-François Chevalley.)

Figure 14-2

Front German license plate. Note the euroband on the left. (Photograph courtesy of Jean-François Chevalley.)

Figure 14-3

Rear Italian license plate issued in Rome, Italy. Note the euroband on the left and the Italian-specific blue strip on the right.

a b c

Figure 14-4

(a) License plate from the State of Florida.
(b) License plate from the State of Pennsylvania.
(c) License plate from the State of North Carolina. (Photographs courtesy of Eric Stauffer.)

Figure 14-5

A personalized (prestige) Georgia license plate. Note the different elements that are found in most tags in the United States. (Photograph courtesy of Eric Stauffer.)

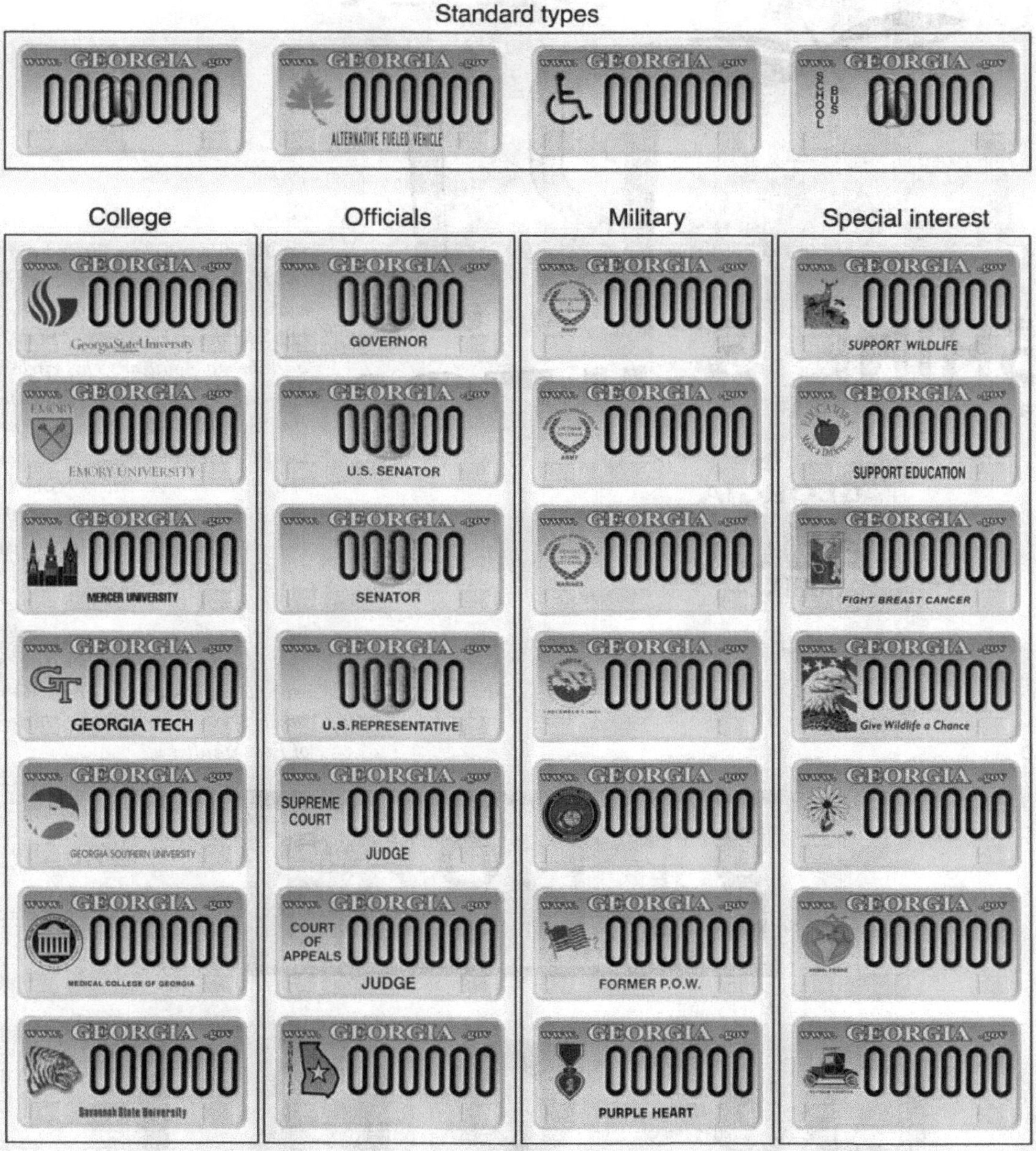

Figure 14-6

Examples of some of the different plates available in Georgia in 2005. (Source: Georgia Department of Revenue, Motor Vehicle Division, available at http://www.dmvs.ga.gov.)

Figure 14-7

(a) Example of the security feature (OVD) located on a Florida tag. It consists of a circle with the letters "FL" followed by a code "BOO2" and the state outline. This circle is repeated vertically across the plate.
(b) Example of the security feature located on a Georgia tag. It consists of a circle with the letters "GA" followed by a code "BOR1" and the state outline. This circle is repeated vertically across the plate.
(c) Example of the security feature located on the yearly renewal sticker of a Georgia tag. It consists of several circles with the letters "GA" inside them. (Photographs courtesy of Eric Stauffer.)

Figure 14-10

(a) Forged Belgian vehicle license plate.
(b) Detailed view of the modification of the characters "0" into the characters "8" by the addition of red and white paints.
(c) Detailed view of the modification of the character "P" into the character "B" by addition of red paint.

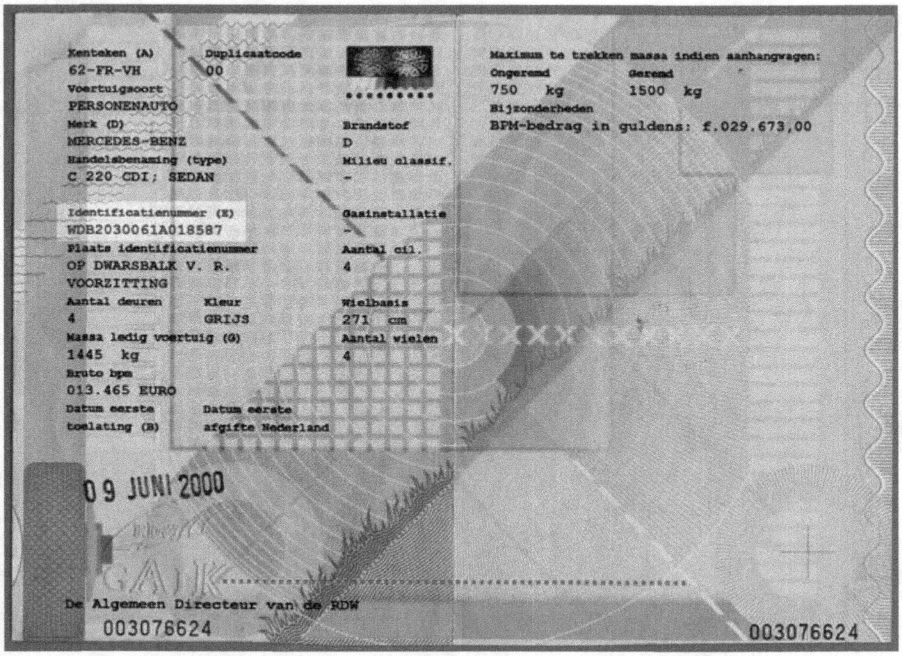

Kenteken (A)
62-FR-VH
Duplicaatcode
00
Voertuigsoort
PERSONENAUTO
Merk (D)
MERCEDES-BENZ
Brandstof
D
Handelsbenaming (type)
C 220 CDI; SEDAN
Milieu classif.
-
Identificatienummer (E)
WDB2030061A018587
Gasinstallatie
-
Plaats identificatienummer
OP DWARSBALK V. R.
VOORZITTING
Aantal cil.
4
Aantal deuren
4
Kleur
GRIJS
Wielbasis
271 cm
Massa ledig voertuig (G)
1445 kg
Aantal wielen
4
Bruto bpm
013.465 EURO
Datum eerste
toelating (B)
Datum eerste
afgifte Nederland
09 JUNI 2000
De Algemeen Directeur van de RDW
003076624

Maximum te trekken massa indien aanhangwagen:
Ongeremd
750 kg
Geremd
1500 kg
Bijzonderheden
BPM-bedrag in guldens: f.029.673,00
003076624

Figure 15-1

The Dutch vehicle registration document part IA (Voertuigbewijs, the vehicle document). The highlighted caption (E), called Identificatienummer, is the VIN of the vehicle.

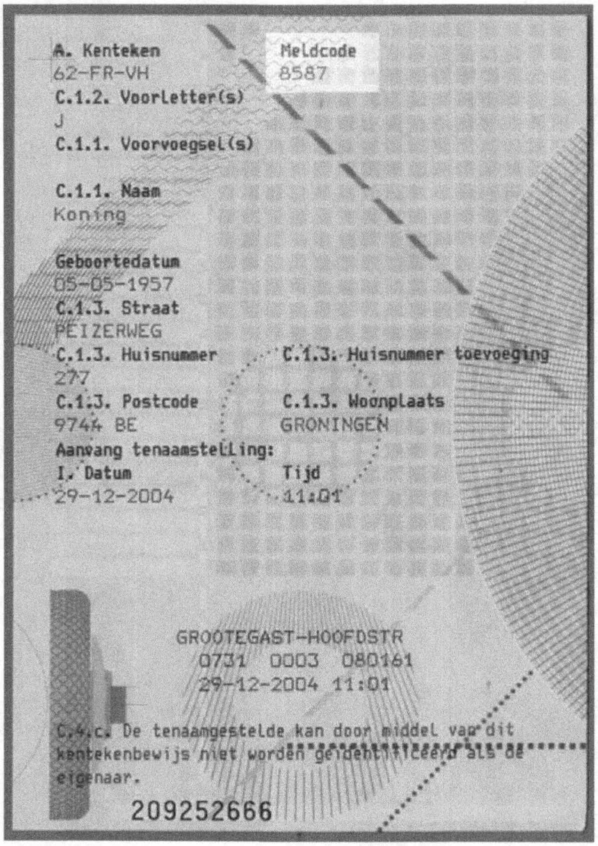

A. Kenteken
62-FR-VH
Meldcode
8587
C.1.2. Voorletter(s)
J
C.1.1. Voorvoegsel(s)
C.1.1. Naam
Koning
Geboortedatum
05-05-1957
C.1.3. Straat
PEIZERWEG
C.1.3. Huisnummer
277
C.1.3. Huisnummer toevoeging
C.1.3. Postcode
9744 BE
C.1.3. Woonplaats
GRONINGEN
Aanvang tenaamstelling:
I. Datum
29-12-2004
Tijd
11:01
GROOTEGAST-HOOFDSTR
0731 0003 080161
29-12-2004 11:01
C.4.c. De tenaamgestelde kan door middel van dit kentekenbewijs niet worden geïdentificeerd als de eigenaar.
209252666

Figure 15-2

The Dutch vehicle registration document, part IB (Tenaamstellingbewijs, the keeper's document). The highlighted caption "Meldcode" shows the last four digits of the VIN (written in full in Figure 15-1).

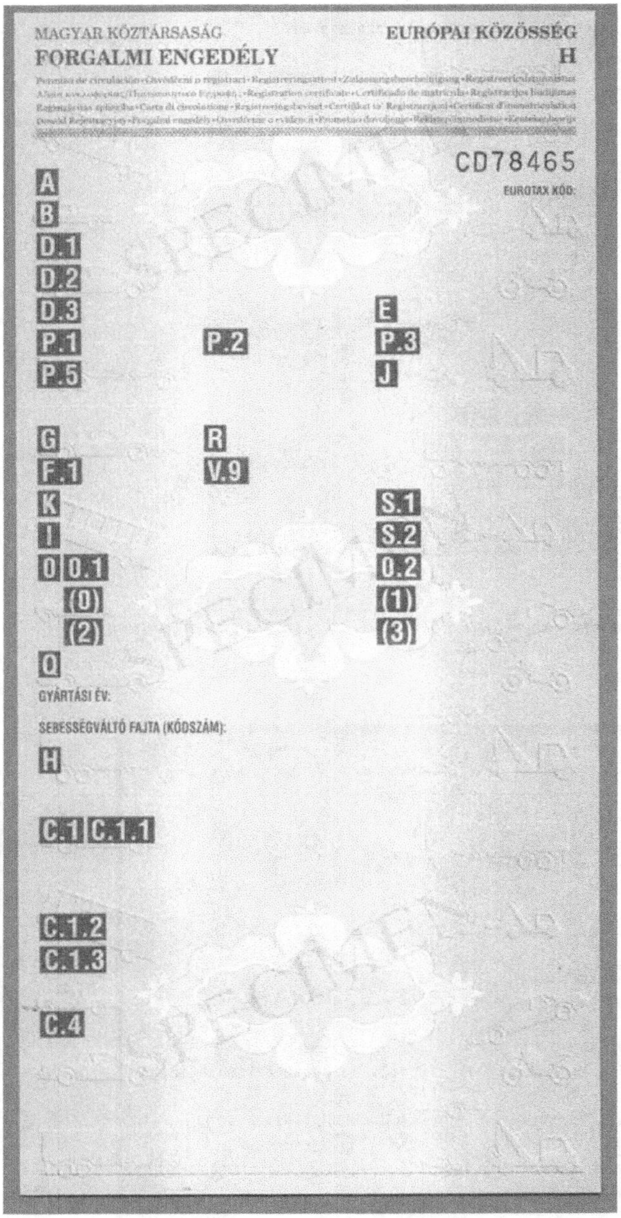

Figure 15-3

The vehicle registration document of Hungary (front).

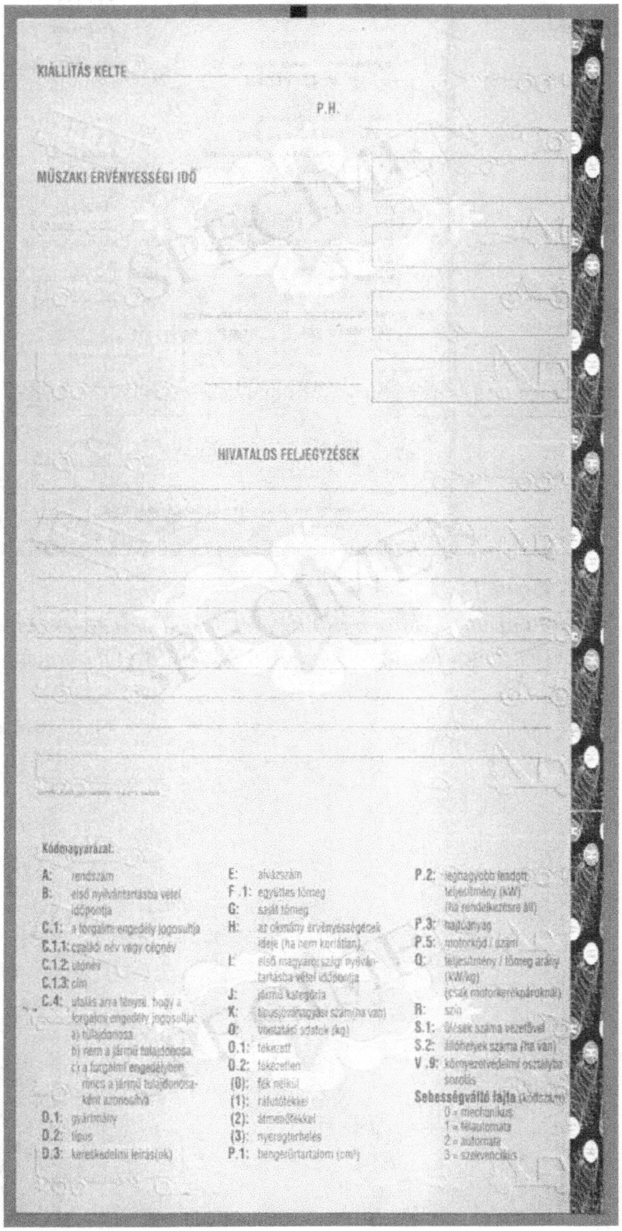

Figure 15-4

The vehicle registration document of Hungary (back).

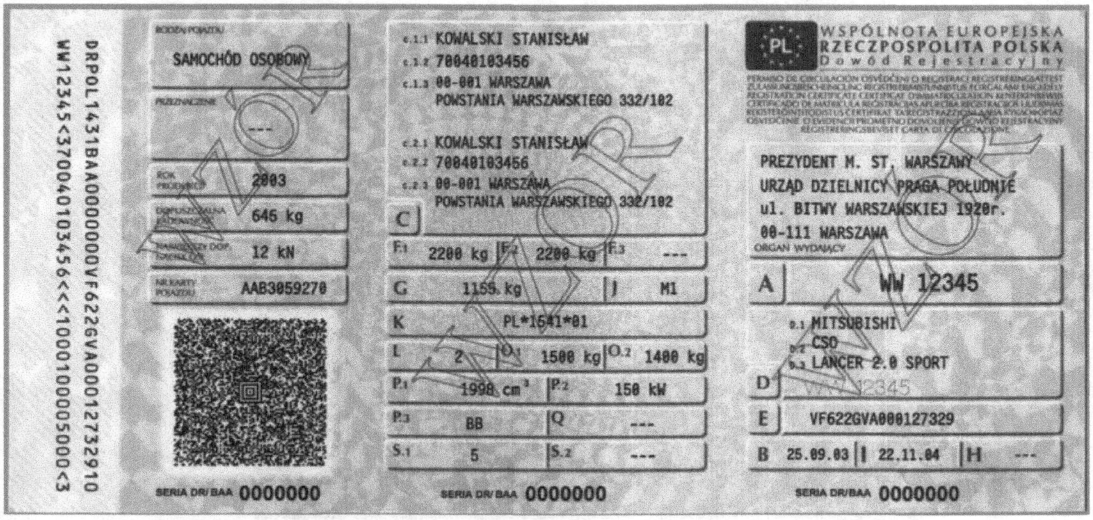

DRPOL1431BAA000000VF622GVA0001273291O
WW12345<3700401O3456<<1000100005000<3

RODZAJ POJAZDU: SAMOCHÓD OSOBOWY
PRZEZNACZENIE: ---
ROK PRODUKCJI: 2003
DOPUSZCZALNA ŁADOWNOŚĆ: 646 kg
NACISK NA OŚ: 12 kN
NR KARTY POJAZDU: AAB3059270
SERIA DR/BAA 0000000

C.1.1 KOWALSKI STANISŁAW
C.1.2 70040103456
C.1.3 00-001 WARSZAWA
POWSTANIA WARSZAWSKIEGO 332/102
C.2.1 KOWALSKI STANISŁAW
C.2.2 70040103456
C.2.3 00-001 WARSZAWA
POWSTANIA WARSZAWSKIEGO 332/102
C
F.1 2200 kg | F.2 2200 kg | F.3 ---
G 1155 kg | J M1
K PL*1541*01
L 2 | O.1 1500 kg | O.2 1400 kg
P.1 1998 cm³ | P.2 150 kW
P.3 BB | Q ---
S.1 5 | S.2 ---
SERIA DR/BAA 0000000

WSPÓLNOTA EUROPEJSKA
RZECZPOSPOLITA POLSKA
Dowód Rejestracyjny
PREZYDENT M. ST. WARSZAWY
URZĄD DZIELNICY PRAGA POŁUDNIE
ul. BITWY WARSZAWSKIEJ 1920r.
00-111 WARSZAWA
ORGAN WYDAJĄCY
A WW 12345
D.1 MITSUBISHI
D.2 CSO
D.3 LANCER 2.0 SPORT
E VF622GVA000127329
B 25.09.03 | I 22.11.04 | H ---
SERIA DR/BAA 0000000

Figure 15-5

The vehicle registration document of Poland (front).

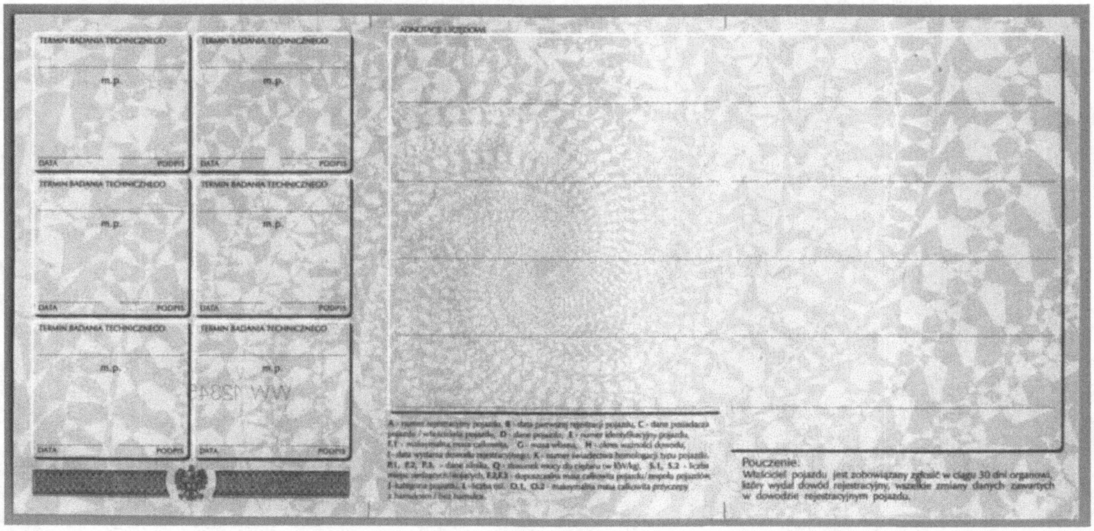

TERMIN BADANIA TECHNICZNEGO
m.p.
DATA PODPIS

Pouczenie:
Właściciel pojazdu jest zobowiązany zgłosić w ciągu 30 dni organowi, który wydał dowód rejestracyjny, wszelkie zmiany danych zawartych w dowodzie rejestracyjnym pojazdu.

Figure 15-6

The vehicle registration document of Poland (back).

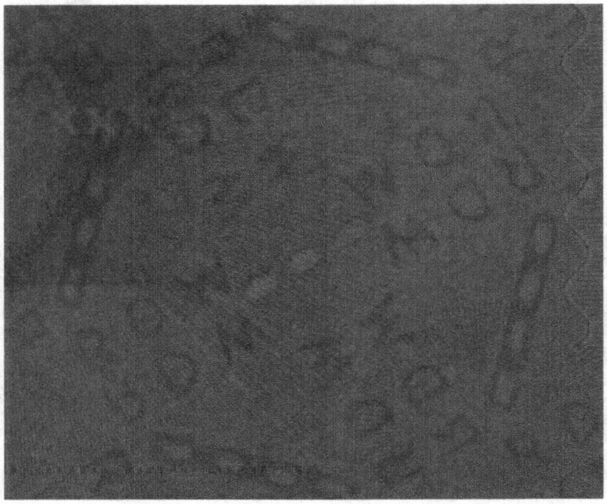

Figure 15-7

The repeated letters RDW and bars with four holes present in the background are from the watermark of the Dutch vehicle registration document at 0.7× magnification. (Courtesy of RDW, The Netherlands.)

Figure 15-8

Blue, yellow, and pink fibers integrated in a paper sample at 2.5× magnification.

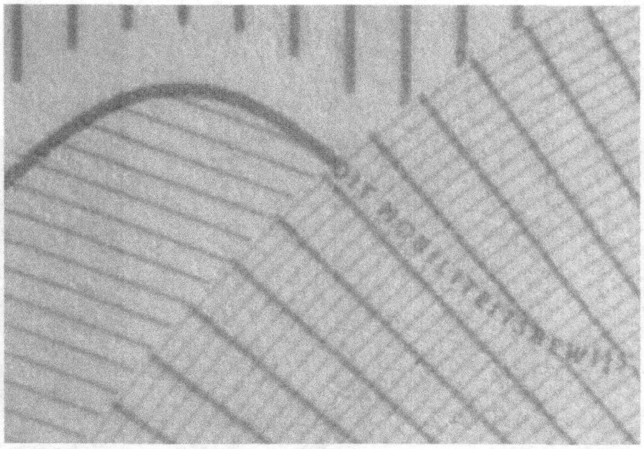

Figure 15-9

Example of microtext found on the Dutch vehicle registration document and seen at 7× magnification. The red line contains some microtext printed using the lithographic technique. (Courtesy of RDW, The Netherlands.)

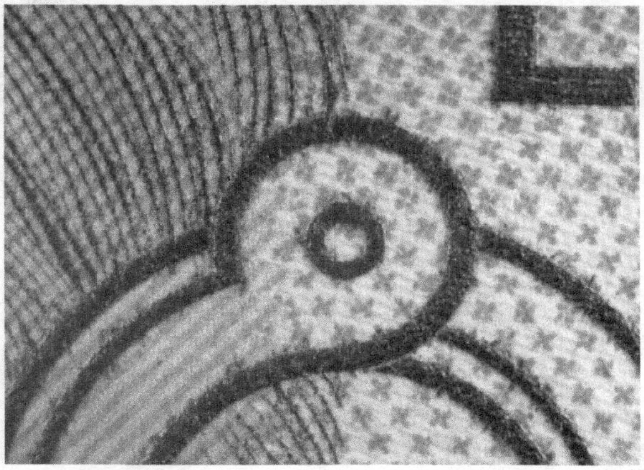

Figure 15-10

Example of a design element printed in intaglio on a Le Mont sample banknote at a 10× magnification. (Courtesy of KBA-GIORI, Switzerland.)

Figure 15-12

Numbering in letterpress of the Dutch vehicle registration document at 9× magnification. Note the border of the characters, showing slight inverted embossment, typical of letterpress technique. (Courtesy of RDW, The Netherlands.)

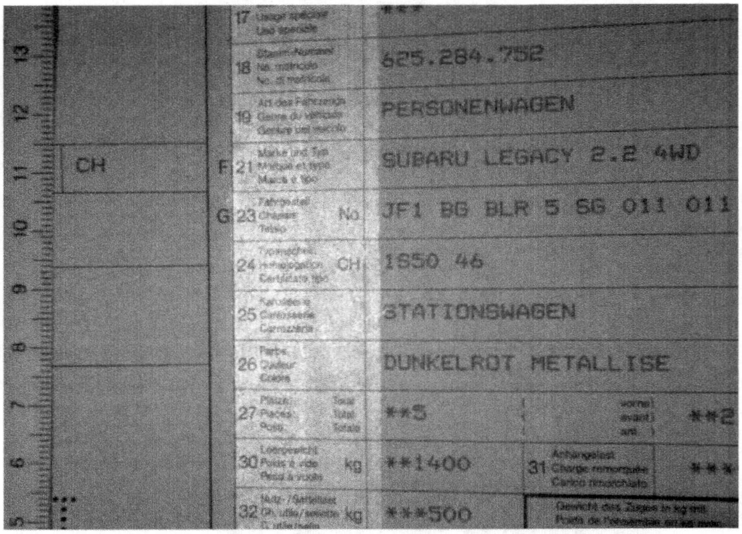

Figure 15-13

View of the iridescent effect present on the Swiss vehicle registration document. The two bands (green and purple) are only visible with the incident light and the observation at specific angles.

Figure 15-14

Detail of the genuine Dutch registration document (Part IA) at 1.5× magnification.

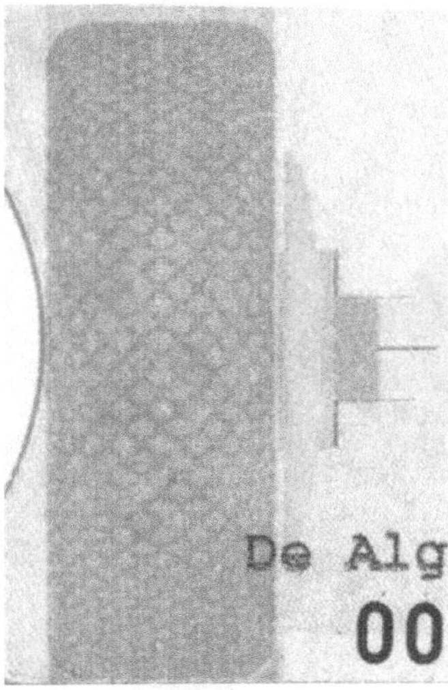

Figure 15-15

Detail of a counterfeit Dutch registration document (Part IA) at 1.5× magnification.

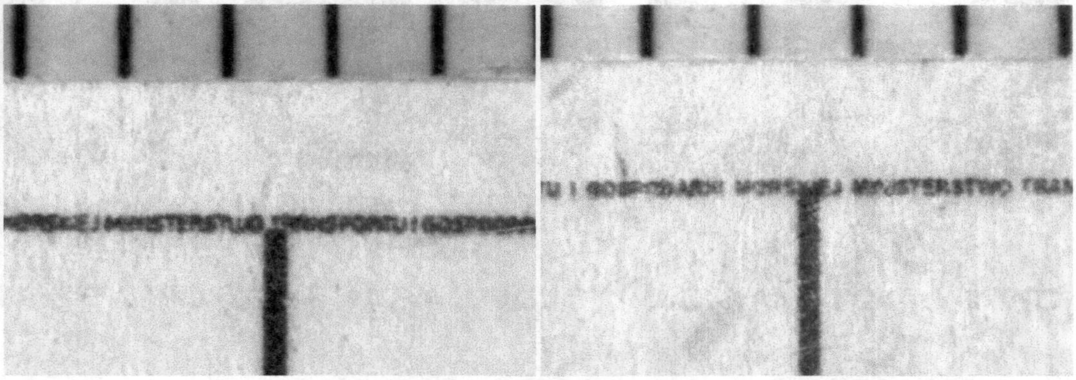

a b

Figure 15-16

(a) View of the microtext on an authentic Polish vehicle registration document. Note that the microimpression is not of a very good quality and could be misinterpreted as not being an original.

(b) View of the microtext on a counterfeit Polish vehicle registration document. Note the quality of the microimpression, which is not too poor compared with the authentic version (a). (Courtesy of The Swiss Border Guards [Corps Suisse des Gardes-Frontières].)

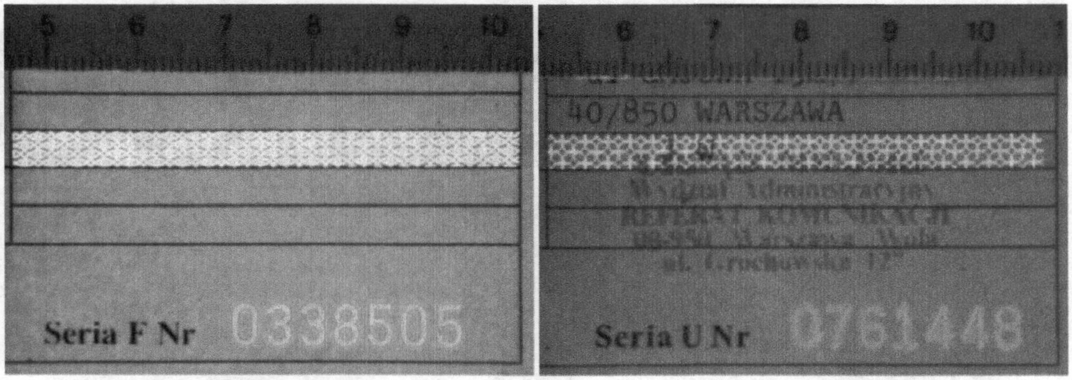

a b

Figure 15-17

(a) View of the UV fluorescent pattern and serial number on an authentic Polish vehicle registration document.

(b) View of the UV fluorescent pattern and serial number on a counterfeit Polish vehicle registration document. Note that the presence of a fluorescent pattern is not enough to guarantee the genuineness of the document; the pattern itself must also match. Also, note the difference in the color of the serial number and the fluorescence of the background of the document. (Courtesy of the Brigade de Police Technique et Scientifique, Geneva police, Switzerland.)

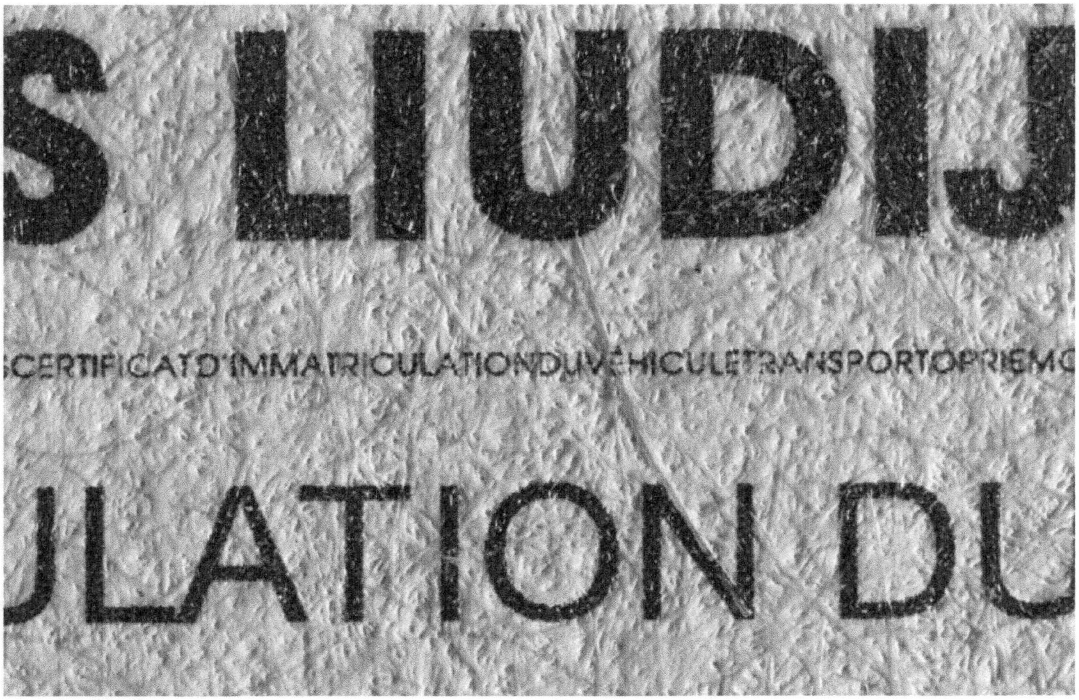

a

b

Figure 15-20

(a) View of a fiber included in the paper on an authentic Lithuanian vehicle registration document.
(b) View of an imitation of a fiber on a counterfeit Lithuanian vehicle registration document. A line was actually drawn on the document to make it look like an included fiber. (Courtesy of The Swiss Border Guards [Corps Suisse des Gardes-Frontières].)

a

b

Figure 15-21

(a) View of an authentic Lithuanian vehicle registration document under UV light.

(b) View of a counterfeit Lithuanian vehicle registration document under UV light. Note the difference in the amount and characteristics of fibers, the serial numbers fluorescence, and the general fluorescence of the document. (Courtesy of The Swiss Border Guards [Corps Suisse des Gardes-Frontières].)

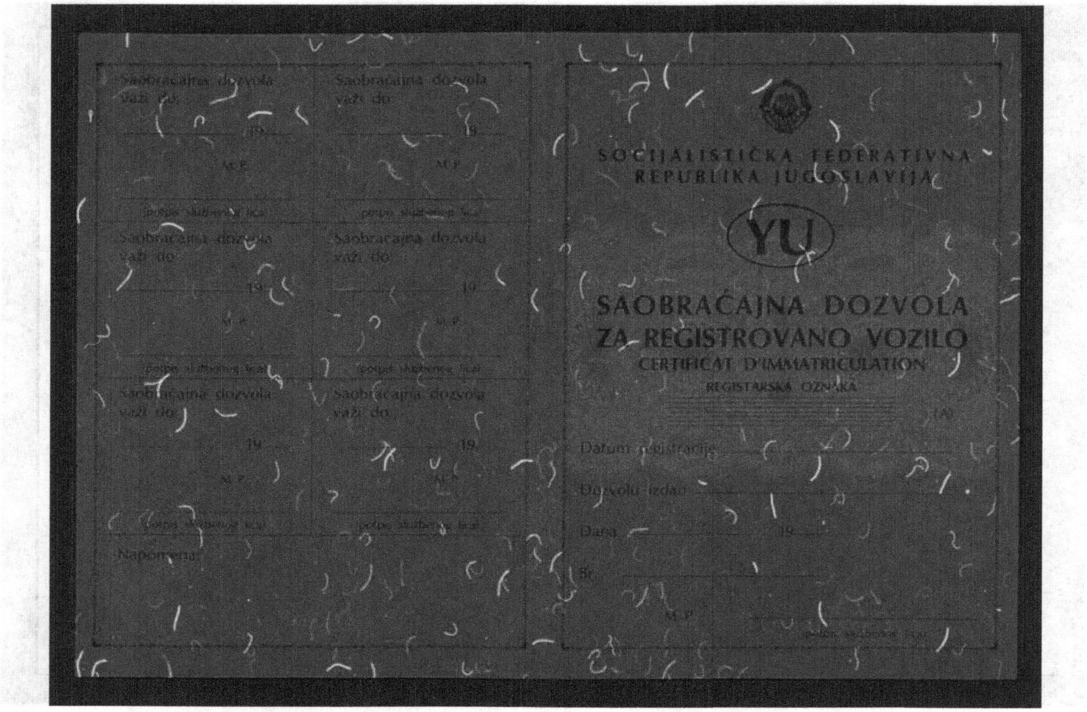

a

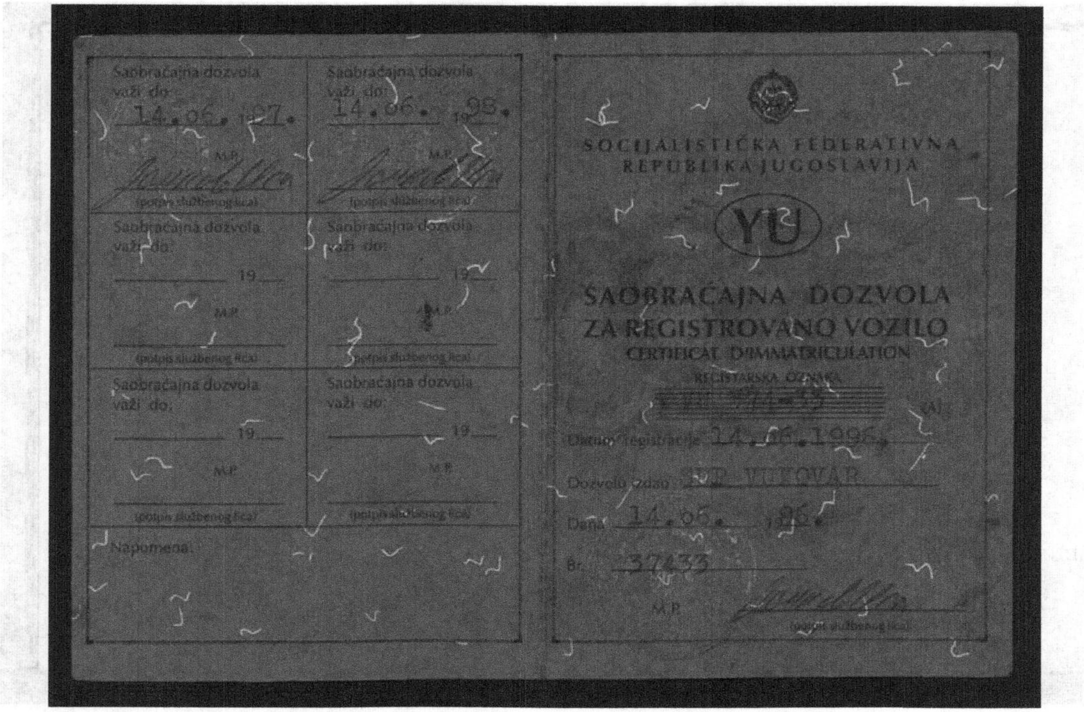

b

Figure 15-22

(a) View of an authentic Yugoslavian vehicle registration document under UV light.

(b) View of a counterfeit Yugoslavian vehicle registration document under UV light. Note the difference in the amount and characteristics of fibers and the general fluorescence of the document. (Courtesy of The Swiss Border Guards [Corps Suisse des Gardes-Frontières].)

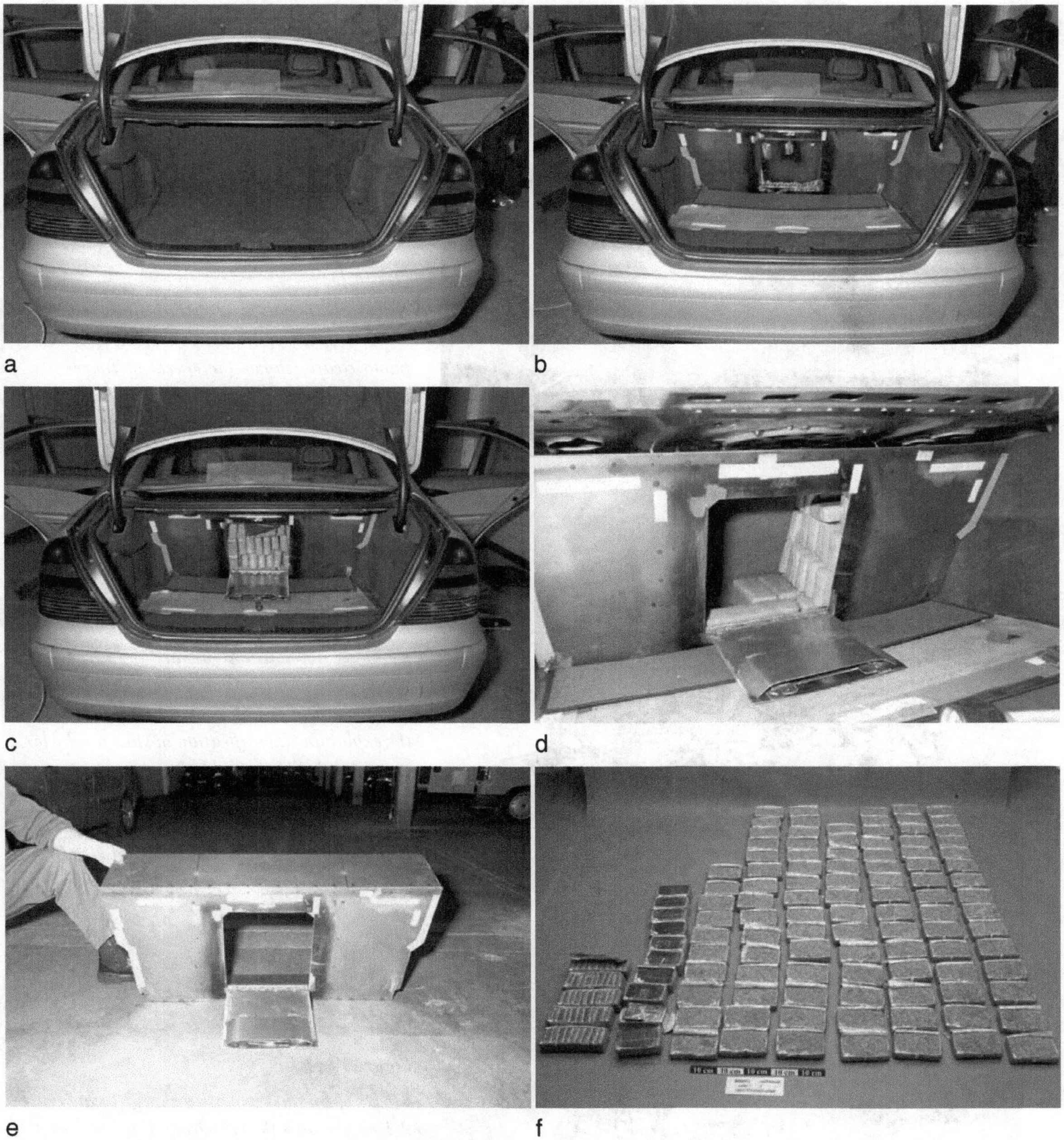

Figure 16-12

Example of a hidden compartment located between the back of the rear seats and the trunk in a Mercedes E220.
(a) When opening the trunk, no particular signs are observable, except for the slightly reduced trunk space.
(b) When the carpet is removed, indication of the presence of a door is directly noticed.
(c) Behind the door, a significant amount of drugs is discovered.
(d) The compartment extends from side to side.
(e) The disassembly of the cache allows for the investigator to understand better how such compartments are built.
(f) The seizure of illegal drugs found in the cache of the Mercedes.

Figure 17-9

A vehicle bomb is a lethal and favorite weapon of terrorist groups worldwide. This photo clearly shows the effects of fragments being propelled at high speed. The passenger door is the largest visible piece, but all can be lethal.

Figure 17-10

Effects of a charge of 2,268 kilograms of calcium ammonium nitrate mixed with sugar. The vehicle is a Toyota Celica and was located about 15 meters from the seat of the explosion. Identification of such a vehicle is rendered difficult, but the main structure of the vehicle is still present. (Copyright British Crown 1999/DERA. Reproduced with permission of the Controller of Her Britannic Majesty's Stationary Office.)

Figure 17-11

Effects of another explosive charge located at 2.1 meters from the vehicle. In this case, most parts disappeared and the vehicle is barely recognizable. In such instance, the identification of the vehicle is a very difficult process, if not impossible. (Copyright British Crown 1999/DERA. Reproduced with permission of the Controller of Her Britannic Majesty's Stationary Office.)

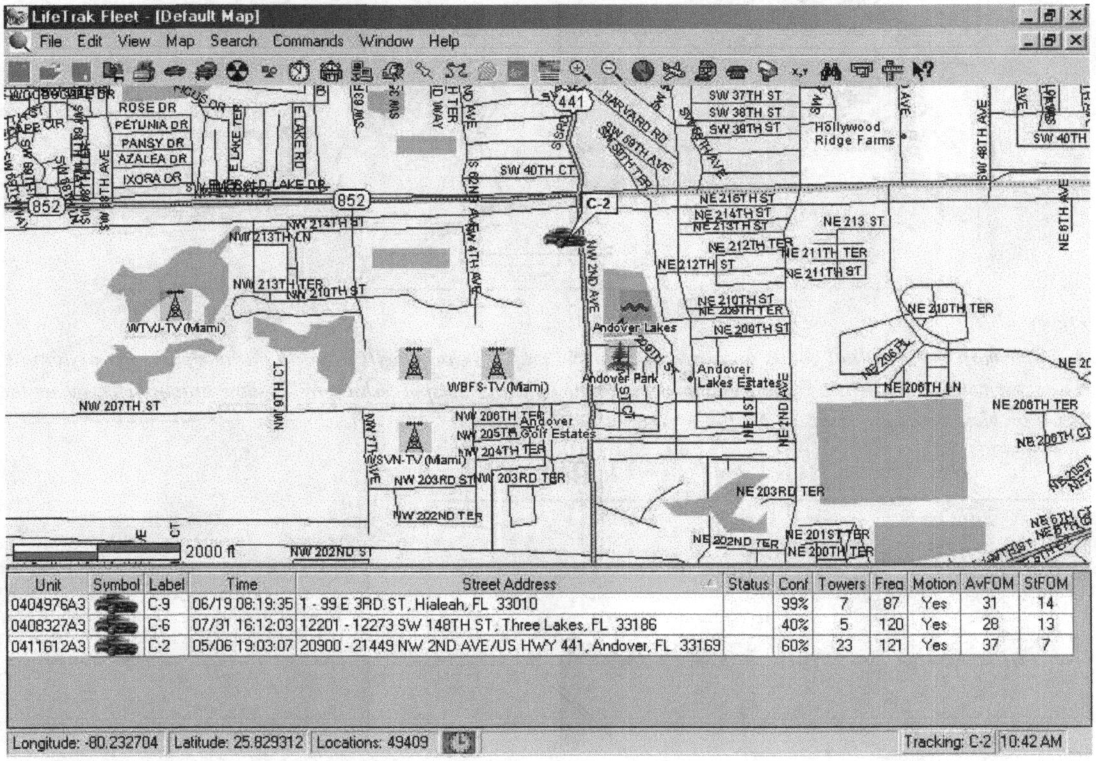

Figure 20-2

Mapping software as seen in this screen shot enables investigators to track stolen vehicles and suspects. Both GPS-based and systems such as Ituran (as shown here) provide the exact location, time, and even speed of a vehicle being tracked. The bottom portion of the screen indicates three active tracks being monitored at one time, C-2, C-6, and C-9. The vehicle C-2 is clearly shown in the center of the map.

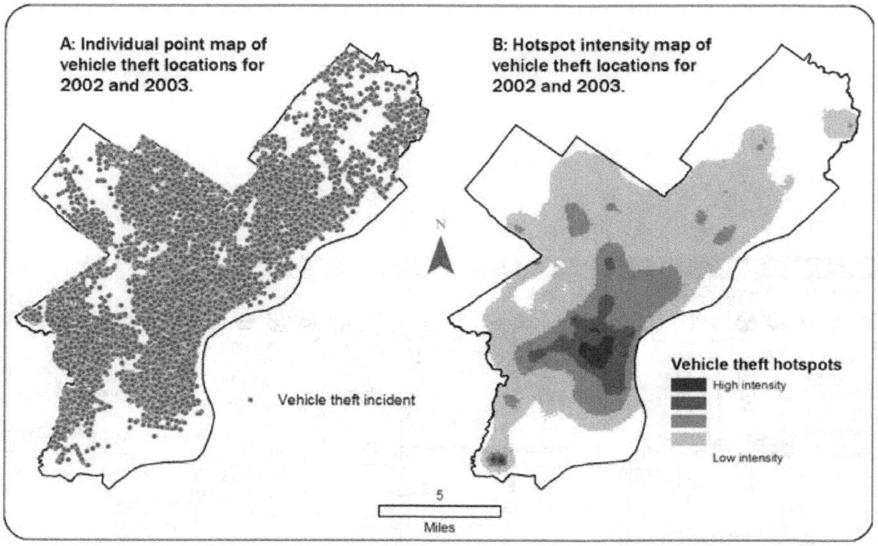

Figure 21-6

A simple map of individual points compared to a hotspot map of vehicle thefts in north Philadelphia, Pennsylvania. The sheer volume of offenses in map A shows the value in using hotspot techniques to show practitioners where the main problem areas are (map B).

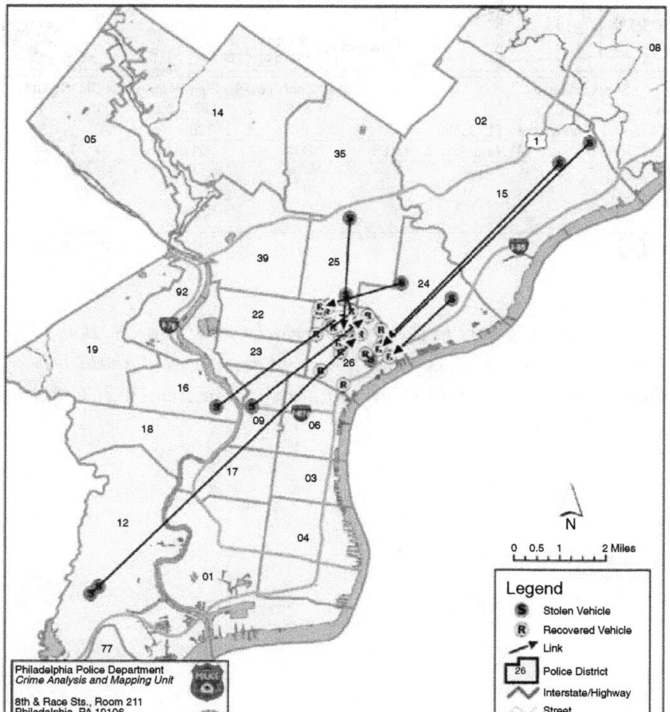

Figure 21-8

The point of theft and the point of recovery for stolen vehicles that were discovered in a police district in Philadelphia, Pennsylvania. Although most of the cars were stolen and recovered within the same police district, it is noticeable that some stolen cars that were recovered in close proximity were stolen from some distance away. In particular, two cars were stolen from many miles away in the southwest part of the city, whereas two other cars were stolen from the northeast. These vehicles may have been stolen to order or were used as transport to get to a particular place by the offender.

Edwards Brothers Malloy
Ann Arbor MI. USA
November 27, 2013